ECE/TRANS/203 (Vol. II)

ECONOMIC COMMISSION FOR EUROPE

Inland Transport Committee

European Agreement concerning the International Carriage of Dangerous Goods by Inland Waterways (ADN)

including the Annexed Regulations, applicable as from 28 February 2009

Volume II

UNITED NATIONS
New York and Geneva, 2008

NOTE

The designations employed and the presentation of the material in this publication do not imply the expression of any opinion whatsoever on the part of the Secretariat of the United Nations concerning the legal status of any country, territory city or area, or of its authorities, or concerning the delimitation of its frontiers or boundaries.

ECE/TRANS/203 (Vol. II)

PUBLICATION DES NATIONS UNIES
Sales No.: E.08.VIII.4
ISBN 978-92-1-139134-3 *(Complete set of 2 volumes)*

Volumes I and II not to be sold separately.

TABLE OF CONTENTS

VOLUME II

Table of contents (cont'd)

ANNEXED REGULATIONS (continued)

PART 2

Classification

CHAPTER 2.1

GENERAL PROVISIONS

2.1.1 Introduction

2.1.1.1 The classes of dangerous goods according to ADN are the following:

Class 1	Explosive substances and articles
Class 2	Gases
Class 3	Flammable liquids
Class 4.1	Flammable solids, self-reactive substances and solid desensitized explosives
Class 4.2	Substances liable to spontaneous combustion
Class 4.3	Substances which, in contact with water, emit flammable gases
Class 5.1	Oxidizing substances
Class 5.2	Organic peroxides
Class 6.1	Toxic substances
Class 6.2	Infectious substances
Class 7	Radioactive material
Class 8	Corrosive substances
Class 9	Miscellaneous dangerous substances and articles

2.1.1.2 Each entry in the different classes has been assigned a UN number. The following types of entries are used:

A. Single entries for well defined substances or articles including entries for substances covering several isomers, e.g.:

UN No. 1090	ACETONE
UN No. 1104	AMYL ACETATES
UN No. 1194	ETHYL NITRITE SOLUTION

B. Generic entries for a well defined group of substances or articles, which are not n.o.s. entries, e.g.:

UN No. 1133	ADHESIVES
UN No. 1266	PERFUMERY PRODUCTS
UN No. 2757	CARBAMATE PESTICIDE, SOLID, TOXIC
UN No. 3101	ORGANIC PEROXIDE TYPE B, LIQUID

C. Specific n.o.s. entries covering a group of substances or articles of a particular chemical or technical nature, not otherwise specified, e.g.:

UN No. 1477	NITRATES, INORGANIC, N.O.S.
UN No. 1987	ALCOHOLS, N.O.S.

D. General n.o.s. entries covering a group of substances or articles having one or more dangerous properties, not otherwise specified, e.g.:

UN No. 1325	FLAMMABLE SOLID, ORGANIC, N.O.S.
UN No. 1993	FLAMMABLE LIQUID, N.O.S.

The entries defined under B., C. and D. are defined as collective entries.

2.1.1.3 For packing purposes, substances other than those of Classes 1, 2, 5.2, 6.2 and 7, and other than self-reactive substances of Class 4.1 are assigned to packing groups in accordance with the degree of danger they present:

Packing group I: Substances presenting high danger;
Packing group II: Substances presenting medium danger;
Packing group III: Substances presenting low danger.

The packing group(s) to which a substance is assigned is (are) indicated in Table A of Chapter 3.2.

2.1.1.4 For the purpose of carriage in tank vessels, some substances may be further subdivided.

2.1.2 Principles of classification

2.1.2.1 The dangerous goods covered by the heading of a class are defined on the basis of their properties according to sub-section 2.2.x.1 of the relevant class. Assignment of dangerous goods to a class and a packing group is made according to the criteria mentioned in the same sub-section 2.2.x.1. Assignment of one or several subsidiary risk(s) to a dangerous substance or article is made according to the criteria of the class or classes corresponding to those risks, as mentioned in the appropriate sub-section(s) 2.2.x.1.

2.1.2.2 All dangerous goods entries are listed in Table A of Chapter 3.2 in the numerical order of their UN Number. This table contains relevant information on the goods listed, such as name, class, packing group(s), label(s) to be affixed, packing and carriage provisions[a].

2.1.2.3 Dangerous goods which are listed or defined in sub-section 2.2.x.2 of each class are not to be accepted for carriage.

2.1.2.4 Goods not mentioned by name, i.e. goods not listed as single entries in Table A of Chapter 3.2 and not listed or defined in one of the above-mentioned sub-sections 2.2.x.2 shall be assigned to the relevant class in accordance with the procedure of section 2.1.3. In addition, the subsidiary risk (if any) and the packing group (if any) shall be determined. Once the class, subsidiary risk (if any) and packing group (if any) have been established the relevant UN number shall be determined. The decision trees in sub-sections 2.2.x.3 (list of collective entries) at the end of each class indicate the relevant parameters for selecting the relevant collective entry (UN number). In all cases the most specific collective entry covering the properties of the substance or article shall be selected, according to the hierarchy indicated in 2.1.1.2 by the letters B, C and D respectively. If the substance or article cannot be classified under entries of type B or C according to 2.1.1.2, then, and only then shall it be classified under an entry of type D.

2.1.2.5 On the basis of the test procedures of Chapter 2.3 and the criteria set out in sub-sections 2.2.x.1 of the various classes when it is so specified, it may be determined that a substance, solution or mixture of a certain class, mentioned by name in Table A of Chapter 3.2, does not meet the criteria of that class. In such a case, the substance, solution or mixture is deemed not to belong to that class.

2.1.2.6 For the purposes of classification, substances with a melting point or initial melting point of 20 °C or lower at a pressure of 101.3 kPa shall be considered to be liquids. A viscous substance for which a specific melting point cannot be determined shall be subjected to the ASTM D 4359-90 test or to the test for determining fluidity (penetrometer test) prescribed in 2.3.4.

---

[a] ***Note by the secretariat:*** *An alphabetic list of these entries has been prepared by the secretariat and is reproduced in Table B of Chapter 3.2. This table is not an official part of the ADN.*

2.1.3 Classification of substances, including solutions and mixtures (such as preparations and wastes), not mentioned by name

2.1.3.1 Substances including solutions and mixtures not mentioned by name shall be classified according to their degree of danger on the basis of the criteria mentioned in sub-section 2.2.x.1 of the various classes. The danger(s) presented by a substance shall be determined on the basis of its physical and chemical characteristics and physiological properties. Such characteristics and properties shall also be taken into account when such experience leads to a more stringent assignment.

2.1.3.2 A substance not mentioned by name in Table A of Chapter 3.2 presenting a single hazard shall be classified in the relevant class under a collective entry listed in sub-section 2.2.x.3 of that class.

2.1.3.3 A solution or mixture containing only one dangerous substance mentioned by name in Table A of Chapter 3.2, together with one or more non-dangerous substance(s), shall be regarded as the dangerous substance listed by name, unless:

(a) the solution or mixture is specifically mentioned by name in Table A of Chapter 3.2; or

(b) it is quite clear from the entry for the dangerous substance that it is applicable only to the pure or technically pure substance; or

(c) the class, physical state or packing group of the solution or mixture is different from that of the dangerous substance.

In the cases referred to under (b) or (c) above, the solution or mixture shall be classified as a substance not mentioned by name in the relevant class under a collective entry listed in sub-section 2.2.x.3 of that class taking account of the subsidiary risks presented by that solution or mixture, if any, unless the solution or mixture do not meet the criteria of any class, in which case they are not subject to ADN.

2.1.3.4 Solutions and mixtures containing a substance belonging to one of the entries mentioned in 2.1.3.4.1 or 2.1.3.4.2 shall be classified in accordance with the provisions of these paragraphs.

2.1.3.4.1 Solutions and mixtures containing one of the following substances mentioned by name shall always be classified under the same entry as the substance they contain, provided they do not have the hazard characteristics as indicated in 2.1.3.5.3:

– Class 3

UN No. 1921 PROPYLENEIMINE, STABILIZED;

UN No. 2481 ETHYL ISOCYANATE;

UN No. 3064 NITROGLYCERIN SOLUTION IN ALCOHOL with more than 1% but not more than 5% nitroglycerin;

– Class 6.1

UN No. 1051 HYDROGEN CYANIDE, STABILIZED, containing less than 3% water;

UN No. 1185 ETHYLENEIMINE, STABILIZED;

UN No. 1259 NICKEL CARBONYL;

UN No. 1613 HYDROCYANIC ACID, AQUEOUS SOLUTION (HYDROGEN CYANIDE, AQUEOUS SOLUTION) with not more than 20% hydrogen cyanide;

UN No. 1614 HYDROGEN CYANIDE, STABILIZED, containing not more than 3% water and absorbed in a porous inert material;

UN No. 1994 IRON PENTACARBONYL;

UN No. 2480 METHYL ISOCYANATE;

UN No. 3294 HYDROGEN CYANIDE, SOLUTION IN ALCOHOL, with not more than 45% hydrogen cyanide;

– Class 8

UN No. 1052 HYDROGEN FLUORIDE, ANHYDROUS;

UN No. 1744 BROMINE or UN No. 1744 BROMINE SOLUTION;

UN No. 1790 HYDROFLUORIC ACID with more than 85% hydrogen fluoride;

UN No. 2576 PHOSPHORUS OXYBROMIDE, MOLTEN.

2.1.3.4.2 Solutions and mixtures containing a substance belonging to one of the following entries of Class 9:

UN No. 2315 POLYCHLORINATED BIPHENYLS, LIQUID;

UN No. 3151 POLYHALOGENATED BIPHENYLS, LIQUID;

UN No. 3151 POLYHALOGENATED TERPHENYLS, LIQUID;

UN No. 3152 POLYHALOGENATED BIPHENYLS, SOLID;

UN No. 3152 POLYHALOGENATED TERPHENYLS, SOLID; or

UN No. 3432 POLYCHLORINATED BIPHENYLS, SOLID

shall always be classified under the same entry of Class 9 provided that:

- they do not contain any additional dangerous component other than components of packing group III of classes 3, 4.1, 4.2, 4.3, 5.1, 6.1 or 8; and

- they do not have the hazard characteristics as indicated in 2.1.3.5.3.

2.1.3.5 Substances not mentioned by name in Table A of Chapter 3.2, having more than one hazard characteristic and solutions or mixtures containing several dangerous substances shall be classified under a collective entry (see 2.1.2.4) and packing group of the appropriate class in accordance with their hazard characteristics. Such classification according to the hazard characteristics shall be carried out as follows:

2.1.3.5.1 The physical and chemical characteristics and physiological properties shall be determined by measurement or calculation and the substance, solution or mixture shall be classified according to the criteria mentioned in sub-section 2.2.x.1 of the various classes.

2.1.3.5.2 If this determination is not possible without disproportionate cost or effort (as for some kinds of wastes), the substance, solution or mixture shall be classified in the class of the component presenting the major hazard.

2.1.3.5.3 If the hazard characteristics of the substance, solution or mixture fall within more than one class or group of substances listed below then the substance, solution or mixture shall be classified in the class or group of substances corresponding to the major hazard on the basis of the following order of precedence:

(a) Material of Class 7 (apart from radioactive material in excepted packages where the other hazardous properties take precedence);

(b) Substances of Class 1;

(c) Substances of Class 2;

(d) Liquid desensitized explosives of Class 3;

(e) Self-reactive substances and solid desensitized explosives of Class 4.1;

(f) Pyrophoric substances of Class 4.2;

(g) Substances of Class 5.2;

(h) Substances of Class 6.1 or Class 3 which, on the basis of their inhalation toxicity, are to be classified under packing group I (Substances meeting the classification criteria of Class 8 and having an inhalation toxicity of dust and mist (LC_{50}) in the range of packing group I and a toxicity through oral ingestion or dermal contact only in the range of packing group III or less, shall be allocated to Class 8);

(i) Infectious substances of Class 6.2.

2.1.3.5.4 If the hazard characteristics of the substance fall within more than one class or group of substances not listed in 2.1.3.5.3 above, the substance shall be classified in accordance with the same procedure but the relevant class shall be selected according to the precedence of hazards table in 2.1.3.10.

2.1.3.5.5 If the substance to be carried is a waste, with a composition that is not precisely known, its assignment to a UN number and packing group in accordance with 2.1.3.5.2 may be based on the consignor's knowledge of the waste, including all available technical and safety data as requested by safety and environmental legislation in force.*

In case of doubt, the highest danger level shall be taken.

If however, on the basis of the knowledge of the composition of the waste and the physical and chemical properties of the identified components, it is possible to demonstrate that the

* *Such legislation is for instance the Commission Decision 2000/532/EC of 3 May 2000 replacing Decision 94/3/EC establishing a list of wastes pursuant to Article 1(a) of Council Directive 75/442/EEC on waste (replaced by the Directive of the European Parliament and of the Council 2006/12/EC (Official Journal of the European Communities No. L 114 of 27 April 2006, page 9) and Council Decision 94/904/EC establishing a list of hazardous wastes pursuant to Article 1(4) of Council Directive 91/689/EEC on hazardous wastes (Official Journal of the European Communities No. L 226 of 6 September 2000, page 3).*

properties of the waste do not correspond to the properties of the packing group I level, the waste may be classified by default in the most appropriate n.o.s. entry of packing group II.

This procedure may not be used for wastes containing substances mentioned in 2.1.3.5.3, substances of Class 4.3, substances of the case mentioned in 2.1.3.7 or substances which are not accepted for carriage in accordance with 2.2.x.2.

2.1.3.6 The most specific applicable collective entry (see 2.1.2.4) shall always be used, i.e. a general n.o.s. entry shall only be used if a generic entry or a specific n.o.s. entry cannot be used.

2.1.3.7 Solutions and mixtures of oxidizing substances or substances with an oxidizing subsidiary risk may have explosive properties. In such a case they are not to be accepted for carriage unless they meet the requirements for Class 1.

2.1.3.8 Substances of Classes 1 to 9, other than UN Nos. 3077 and 3082, meeting the criteria of 2.2.9.1.10 shall, additionally to their hazards of Classes 1 to 9, be considered to be environmentally hazardous substances. Other substances meeting the criteria of 2.2.9.1.10.1 or 2.2.9.1.10.2 shall be assigned to UN Nos. 3077 or 3082 or to identification Nos. 9005 or 9006, as appropriate.

2.1.3.9 Wastes which do not meet the criteria for classification in classes 1 to 9 but are covered by the *Basel Convention on the Control of Transboundary Movements of Hazardous Wastes and their Disposal* may be carried under UN Nos. 3077 or 3082.

2.1.3.10 *Table of precedence of hazards*

Class and packing group	4.1, II	4.1, III	4.2, II	4.2, III	4.3, I	4.3, II	4.3, III	5.1, I	5.1, II	5.1, III	6.1, I DERMAL	6.1, I ORAL	6.1, II	6.1, III	8, I	8, II	8, III	9
3, I	SOL LIQ 4.1 3, I	SOL LIQ 4.1 3, I	SOL LIQ 4.2 3, I	SOL LIQ 4.2 3, I	4.3, I	4.3, I	4.3, I	SOL LIQ 5.1, I 3, I	SOL LIQ 5.1, I 3, I	SOL LIQ 5.1, I 3, I	3, I	3, I	3, I	3, I	3, I	3, I	3, I	3, I
3, II	SOL LIQ 4.1 3, II	SOL LIQ 4.1 3, II	SOL LIQ 4.2 3, II	SOL LIQ 4.2 3, II	4.3, I	4.3, II	4.3, II	SOL LIQ 5.1, I 3, I	SOL LIQ 5.1, II 3, II	SOL LIQ 5.1, II 3, II	3, I	3, I	3, II	3, II	8, I	3, II	3, II	3, II
3, III	SOL LIQ 4.1 3, II	SOL LIQ 4.1 3, III	SOL LIQ 4.2 3, II	SOL LIQ 4.2 3, III	4.3, I	4.3, II	4.3, III	SOL LIQ 5.1, I 3, I	SOL LIQ 5.1, II 3, II	SOL LIQ 5.1, III 3, III	6.1, I	6.1, I	6.1, II	3, III */	8, I	8, II	3, III	3, III
4.1, II			4.2, II	4.2, II	4.3, I	4.3, II	4.3, II	5.1, I	4.1, II	4.1, II	6.1, I	6.1, I	SOL LIQ 4.1, II 6.1, II	SOL LIQ 4.1, II 6.1, II	8, I	SOL LIQ 4.1, II 8, II	SOL LIQ 4.1, II 8, II	4.1, II
4.1, III			4.2, II	4.2, III	4.3, I	4.3, II	4.3, III	5.1, I	4.1, II	4.1, III	6.1, I	6.1, I	6.1, II	SOL LIQ 4.1, III 6.1, III	8, I	8, II	SOL LIQ 4.1, III 8, III	4.1, III
4.2, II					4.3, I	4.3, II	4.3, II	5.1, I	4.2, II	4.2, II	6.1, I	6.1, I	4.2, II	4.2, II	8, I	4.2, II	4.2, II	4.2, II
4.2, III					4.3, I	4.3, II	4.3, III	5.1, I	5.1, II	4.2, III	6.1, I	6.1, I	6.1, II	4.2, III	8, I	8, II	4.2, III	4.2, III
4.3, I								5.1, I	4.3, I	4.3, I	6.1, I	4.3, I	4.3, I	4.3, I	4.3, I	4.3, I	4.3, I	4.3, I
4.3, II								5.1, I	4.3, II	4.3, II	6.1, I	4.3, I	4.3, II	4.3, II	8, I	4.3, II	4.3, II	4.3, II
4.3, III								5.1, I	5.1, II	4.3, III	6.1, I	6.1, I	6.1, II	4.3, III	8, I	8, II	4.3, III	4.3, III
5.1, I											5.1, I	5.1, I	5.1, I	5.1, I	5.1, I	5.1, I	5.1, I	5.1, I
5.1, II											6.1, I	5.1, I	5.1, II	5.1, II	8, I	5.1, II	5.1, II	5.1, II
5.1, III											6.1, I	6.1, I	6.1, II	5.1, III	8, I	8, II	5.1, III	5.1, III
6.1, I DERMA															SOL LIQ 6.1, I 8, I	6.1, I	6.1, I	6.1, I
6.1, I ORAL															SOL LIQ 6.1, I 8, I	6.1, I	6.1, I	6.1, I
6.1, II INHAL															SOL LIQ 6.1, I 8, I	6.1, II	6.1, II	6.1, II
6.1, II DERMA															SOL LIQ 6.1, I 8, I	SOL LIQ 6.1, II 8, II	6.1, II	6.1, II
6.1, II ORAL															8.1	SOL LIQ 6.1, II 8, II	6.1, II	6.1, II
6.1, III															8, I	8, II	8, III	6.1, III
8, I																		8, I
8, II																		8, II
8, III																		8, III

SOL = Solid substances and mixtures
LIQ = Liquid substances, mixtures and solutions
DERMAL = Dermal toxicity
ORAL = Oral toxicity
INHAL = Inhalation toxicity
*/ *Class 6.1 for pesticides*

NOTE 1: Examples to explain the use of the table

Classification of a single substance

Description of the substance to be classified:

An amine not mentioned by name meeting the criteria for Class 3, packing group II as well as those for Class 8, packing group I.

Procedure:

The intersection of line 3 II with column 8 I gives 8 I.
This amine has therefore to be classified in Class 8 under:

UN No. 2734 AMINES LIQUID, CORROSIVE, FLAMMABLE, N.O.S. or UN No. 2734 POLYAMINES, LIQUID, CORROSIVE, FLAMMABLE, N.O.S.
packing group I

Classification of a mixture

Description of the mixture to be classified:

Mixture consisting of a flammable liquid classified in Class 3, packing group III, a toxic substance in Class 6.1, packing group II and a corrosive substance in Class 8, packing group I.

Procedure

The intersection of line 3 III with column 6.1 II gives 6.1 II.
The intersection of line 6.1 II with column 8 I gives 8 I LIQ.
This mixture not further defined has therefore to be classified in Class 8 under:

UN No. 2922 CORROSIVE LIQUID, TOXIC, N.O.S.
packing group I.

NOTE 2: Examples for the classification of mixtures and solutions under a class and a packing group:

A phenol solution of Class 6.1, (II), in benzene of Class 3, (II) is to be classified in Class 3, (II); this solution is to be classified under UN No. 1992 FLAMMABLE LIQUID, TOXIC, N.O.S., Class 3, (II), by virtue of the toxicity of the phenol.

A solid mixture of sodium arsenate of Class 6.1, (II) and sodium hydroxide of Class 8, (II) is to be classified under UN No. 3290 TOXIC SOLID, CORROSIVE, INORGANIC, N.O.S., in Class 6.1 (II).

A solution of crude or refined naphthalene of Class 4.1, (III) in petrol of Class 3, (II), is to be classified under UN No. 3295 HYDROCARBONS, LIQUID, N.O.S. in Class 3, (II).

A mixture of hydrocarbons of Class 3, (III), and of polychlorinated biphenyls (PCB) of Class 9, (II), is to be classified under UN No. 2315 POLYCHLORINATED BIPHENYLS, LIQUID or UN No. 3432 POLYCHLORINATED BIPHENYLS, SOLID in Class 9, (II).

A mixture of propyleneimine of Class 3, and polychlorinated biphenyls (PCB) of Class 9, (II), is to be classified under UN No. 1921 PROPYLENEIMINE, INHIBITED in Class 3.

2.1.4 Classification of samples

2.1.4.1 When the class of a substance is uncertain and it is being carried for further testing, a tentative class, proper shipping name and UN number shall be assigned on the basis of the consignor's knowledge of the substance and application of:

(a) the classification criteria of Chapter 2.2; and

(b) the requirements of this Chapter.

The most severe packing group possible for the proper shipping name chosen shall be used.

Where this provision is used the proper shipping name shall be supplemented with the word "SAMPLE" (e.g., "FLAMMABLE LIQUID, N.O.S., SAMPLE"). In certain instances, where a specific proper shipping name is provided for a sample of a substance considered to meet certain classification criteria (e.g., GAS SAMPLE, NON-PRESSURIZED, FLAMMABLE, UN No. 3167) that proper shipping name shall be used. When an N.O.S. entry is used to carry the sample, the proper shipping name need not be supplemented with the technical name as required by special provision 274 of Chapter 3.3.

2.1.4.2 Samples of the substance shall be carried in accordance with the requirements applicable to the tentative assigned proper shipping name provided:

(a) the substance is not considered to be a substance not accepted for carriage by sub-sections 2.2.x.2 of Chapter 2.2 or by Chapter 3.2;

(b) the substance is not considered to meet the criteria for Class 1 or considered to be an infectious substance or a radioactive material;

(c) the substance is in compliance with 2.2.41.1.15 or 2.2.52.1.9 if it is a self-reactive substance or an organic peroxide, respectively;

(d) the sample is carried in a combination packaging with a net mass per package not exceeding 2.5 kg; and

(e) the sample is not packed together with other goods.

CHAPTER 2.2

CLASS SPECIFIC PROVISIONS

2.2.1 Class 1 Explosive substances and articles

2.2.1.1 ***Criteria***

2.2.1.1.1 The heading of Class 1 covers:

(a) Explosive substances: solid or liquid substances (or mixtures of substances) capable by chemical reaction of producing gases at such a temperature and pressure and at such a speed as to cause damage to the surroundings.

Pyrotechnic substances: substances or mixtures of substances designed to produce an effect by heat, light, sound, gas or smoke or a combination of these as the result of non-detonating self-sustaining exothermic chemical reactions.

***NOTE 1**: Substances which are not themselves explosive but which may form an explosive mixture of gas, vapour or dust are not substances of Class 1.*

***NOTE 2**: Also excluded from Class 1 are: water- or alcohol-wetted explosives of which the water or alcohol content exceeds the limits specified and those containing plasticizers - these explosives are assigned to Class 3 or Class 4.1 - and those explosives which, on the basis of their predominant hazard, are assigned to Class 5.2.*

(b) Explosive articles: articles containing one or more explosive or pyrotechnic substances.

***NOTE**: Devices containing explosive or pyrotechnic substances in such small quantity or of such a character that their inadvertent or accidental ignition or initiation during carriage would not cause any manifestation external to the device by projection, fire, smoke, heat or loud noise are not subject to the requirements of Class 1.*

(c) Substances and articles not mentioned above which are manufactured with a view to producing a practical effect by explosion or a pyrotechnic effect.

2.2.1.1.2 Any substance or article having or suspected of having explosive properties shall be considered for assignment to Class 1 in accordance with the tests, procedures and criteria prescribed in Part I, Manual of Tests and Criteria.

A substance or article assigned to Class 1 can only be accepted for carriage when it has been assigned to a name or n.o.s. entry listed in Table A of Chapter 3.2 and meets the criteria of the Manual of Tests and Criteria.

2.2.1.1.3 The substances and articles of Class 1 shall be assigned to a UN Number and a name or n.o.s. entry listed in Table A of Chapter 3.2. Interpretation of the names of substances and articles in Table A of Chapter 3.2 shall be based upon the glossary in 2.2.1.1.8.

Samples of new or existing explosive substances or articles carried for purposes including: testing, classification, research and development quality control, or as a commercial sample, other than initiating explosive, may be assigned to UN No. 0190 SAMPLES, EXPLOSIVE.

The assignment of explosive substances and articles not mentioned by name as such in Table A of Chapter 3.2 to an n.o.s entry of Class 1 or UN No. 0190 SAMPLES, EXPLOSIVE as well as the assignment of certain substances the carriage of which is subject to a specific authorization by the competent authority according to the special provisions referred to in Column (6) of Table A of Chapter 3.2 shall be made by the competent authority of the country of origin. This competent authority shall also approve in writing the conditions of carriage of these substances and articles. If the country of origin is not a Contracting Party to ADN, the classification and the conditions of carriage shall be recognized by the competent authority of the first country Contracting Party to ADN reached by the consignment.

2.2.1.1.4 Substances and articles of Class 1 shall have been assigned to a division in accordance with 2.2.1.1.5 and to a compatibility group in accordance with 2.2.1.1.6. The division shall be based on the results of the tests described in section 2.3.1 applying the definitions in 2.2.1.1.5. The compatibility group shall be determined in accordance with the definitions in 2.2.1.1.6. The classification code shall consist of the division number and the compatibility group letter.

2.2.1.1.5 *Definition of divisions*

Division 1.1 Substances and articles which have a mass explosion hazard (a mass explosion is an explosion which affects almost the entire load virtually instantaneously).

Division 1.2 Substances and articles which have a projection hazard but not a mass explosion hazard.

Division 1.3 Substances and articles which have a fire hazard and either a minor blast hazard or a minor projection hazard or both, but not a mass explosion hazard:

(a) combustion of which gives rise to considerable radiant heat; or

(b) which burn one after another, producing minor blast or projection effects or both.

Division 1.4 Substances and articles which present only a slight risk of explosion in the event of ignition or initiation during carriage. The effects are largely confined to the package and no projection of fragments of appreciable size or range is to be expected. An external fire shall not cause virtually instantaneous explosion of almost the entire contents of the package.

Division 1.5 Very insensitive substances having a mass explosion hazard which are so insensitive that there is very little probability of initiation or of transition from burning to detonation under normal conditions of carriage. As a minimum requirement they must not explode in the external fire test.

Division 1.6 Extremely insensitive articles which do not have a mass explosion hazard. The articles contain only extremely insensitive detonating substances and demonstrate a negligible probability of accidental initiation or propagation.

***NOTE**: The risk from articles of Division 1.6 is limited to the explosion of a single article.*

2.2.1.1.6 *Definition of compatibility groups of substances and articles*

A Primary explosive substance.

B Article containing a primary explosive substance and not having two or more effective protective features. Some articles, such as detonators for blasting, detonator assemblies for blasting and primers, cap-type, are included, even though they do not contain primary explosives.

C Propellant explosive substance or other deflagrating explosive substance or article containing such explosive substance.

D Secondary detonating explosive substance or black powder or article containing a secondary detonating explosive substance, in each case without means of initiation and without a propelling charge, or article containing a primary explosive substance and having two or more effective protective features.

E Article containing a secondary detonating explosive substance, without means of initiation, with a propelling charge (other than one containing a flammable liquid or gel or hypergolic liquids).

F Article containing a secondary detonating explosive substance with its own means of initiation, with a propelling charge (other than one containing a flammable liquid or gel or hypergolic liquids) or without a propelling charge.

G Pyrotechnic substance, or article containing a pyrotechnic substance, or article containing both an explosive substance and an illuminating, incendiary, tear- or smoke-producing substance (other than a water-activated article or one which contains white phosphorus, phosphides, a pyrophoric substance, a flammable liquid or gel or hypergolic liquids).

H Article containing both an explosive substance and white phosphorus.

J Article containing both an explosive substance and a flammable liquid or gel.

K Article containing both an explosive substance and a toxic chemical agent.

L Explosive substance or article containing an explosive substance and presenting a special risk (e.g. due to water activation or the presence of hypergolic liquids, phosphides or a pyrophoric substance) necessitating isolation of each type.

N Articles containing only extremely insensitive detonating substances.

S Substance or article so packed or designed that any hazardous effects arising from accidental functioning are confined within the package unless the package has been degraded by fire, in which case all blast or projection effects are limited to the extent that they do not significantly hinder or prevent fire-fighting or other emergency response efforts in the immediate vicinity of the package.

NOTE 1*: Each substance or article, packed in a specified packaging, may be assigned to one compatibility group only. Since the criterion of compatibility group S is empirical, assignment to this group is necessarily linked to the tests for assignment of a classification code.*

NOTE 2*: Articles of compatibility groups D and E may be fitted or packed together with their own means of initiation provided that such means have at least two effective protective* *features designed to prevent an explosion in the event of accidental functioning of the means of initiation. Such packages shall be assigned to compatibility groups D or E.*

***NOTE 3**: Articles of compatibility groups D and E may be packed together with their own means of initiation, which do not have two effective protective features (i.e. means of initiation assigned to compatibility group B), provided that they comply with mixed packing provision MP 21 of Section 4.1.10 of ADR.. Such packages shall be assigned to compatibility groups D or E.*

***NOTE 4**: Articles may be fitted or packed together with their own means of ignition provided that the means of ignition cannot function during normal conditions of carriage.*

***NOTE 5**: Articles of compatibility groups C, D and E may be packed together. Such packages shall be assigned to compatibility group E.*

2.2.1.1.7 *Assignment of fireworks to divisions*

2.2.1.1.7.1 Fireworks shall normally be assigned to divisions 1.1, 1.2, 1.3, and 1.4 on the basis of test data derived from Test Series 6 of the Manual of Tests and Criteria. However, since the range of such articles is very extensive and the availability of test facilities may be limited, assignment to divisions may also be made in accordance with the procedure in 2.2.1.1.7.2.

2.2.1.1.7.2 Assignment of fireworks to UN No. 0333, 0334, 0335 or 0336 may be made on the basis of analogy, without the need for Test Series 6 testing, in accordance with the default fireworks classification table in 2.2.1.1.7.5. Such assignment shall be made with the agreement of the competent authority. Items not specified in the table shall be classified on the basis of test data derived from Test Series 6.

***NOTE 1**: The addition of other types of fireworks to column 1 of the table in 2.2.1.1.7.5 shall only be made on the basis of full test data submitted to the UN Sub-Committee of Experts on the Transport of Dangerous Goods for consideration.*

***NOTE 2**: Test data derived by competent authorities which validates, or contradicts the assignment of fireworks specified in column 4 of the table in 2.2.1.1.7.5 to divisions in column 5 should be submitted to the UN Sub-Committee of Experts on the Transport of Dangerous Goods for information.*

2.2.1.1.7.3 Where fireworks of more than one division are packed in the same package they shall be classified on the basis of the highest division unless test data derived from Test Series 6 indicate otherwise.

2.2.1.1.7.4 The classification shown in the table in 2.2.1.1.7.5 applies only for articles packed in fibreboard boxes (4G).

2.2.1.1.7.5 *Default fireworks classification table* [1]

***NOTE 1**: References to percentages in the table, unless otherwise stated, are to the mass of all pyrotechnic composition (e.g. rocket motors, lifting charge, bursting charge and effect charge).*

***NOTE 2**: "Flash composition" in this table refers to pyrotechnic compositions in powder form or as pyrotechnic units as presented in the fireworks, that are used to produce an aural effect, or used as a bursting charge or lifting charge, unless the time taken for the pressure*

---

[1] *This table contains a list of firework classifications which may be used in the absence of Test Series 6 data (see 2.2.1.1.7.2).*

rise is demonstrated to be more than 8 ms for 0.5 g of pyrotechnic composition in Test Series 2(c)(i) "Time/pressure test" of the Manual of Tests and Criteria.

NOTE 3: *Dimensions in mm refer to:*

- *for spherical and peanut shells the diameter of the sphere of the shell;*
- *for cylinder shells the length of the shell;*
- *for a shell in mortar, Roman candle, shot tube firework or mine the inside diameter of the tube comprising or containing the firework;*
- *for a bag mine or cylinder mine, the inside diameter of the mortar intended to contain the mine.*

Type	Includes: / Synonym:	Definition	Specification	Classification
Shell, spherical or cylindrical	Spherical display shell: aerial shell, colour shell, dye shell, multi-break shell, multi-effect shell, nautical shell, parachute shell, smoke shell, star shell; report shell: maroon, salute, sound shell, thunderclap, aerial shell kit	Device with or without propellant charge, with delay fuse and bursting charge, pyrotechnic unit(s) or loose pyrotechnic composition and designed to be projected from a mortar	All report shells	1.1G
			Colour shell: ≥ 180 mm	1.1G
			Colour shell: < 180 mm with > 25% flash composition, as loose powder and/or report effects	1.1G
			Colour shell: < 180 mm with ≤ 25% flash composition, as loose powder and/or report effects	1.3G
			Colour shell: ≤ 50 mm, or ≤ 60 g pyrotechnic composition, with ≤ 2% flash composition as loose powder and/or report effects	1.4G
	Peanut shell	Device with two or more spherical aerial shells in a common wrapper propelled by the same propellant charge with separate external delay fuses	The most hazardous spherical aerial shell determines the classification	
	Preloaded mortar, shell in mortar	Assembly comprising a spherical or cylindrical shell inside a mortar from which the shell is designed to be projected	All report shells	1.1G
			Colour shell: ≥ 180 mm	1.1G
			Colour shell: > 25% flash composition as loose powder and/or report effects	1.1G
			Colour shell: > 50 mm and < 180 mm	1.2G
			Colour shell: ≤ 50 mm, or ≤ 60 g pyrotechnic composition, with ≤ 25% flash composition as loose powder and/or report effects	1.3G

Type	Includes: / Synonym:	Definition	Specification	Classification
Shell, spherical or cylindrical *(cont'd)*	Shell of shells (spherical) (*Reference to percentages for shell of shells are to the gross mass of the fireworks article*)	Device without propellant charge, with delay fuse and bursting charge, containing report shells and inert materials and designed to be projected from a mortar	> 120 mm	1.1G
		Device without propellant charge, with delay fuse and bursting charge, containing report shells ≤ 25g flash composition per report unit, with ≤ 33% flash composition and ≥ 60% inert materials and designed to be projected from a mortar	≤ 120 mm	1.3G
		Device without propellant charge, with delay fuse and bursting charge, containing colour shells and/or pyrotechnic units and designed to be projected from a mortar	> 300 mm	1.1G
		Device without propellant charge, with delay fuse and bursting charge, containing colour shells ≤ 70mm and/or pyrotechnic units, with ≤ 25% flash composition and ≤ 60% pyrotechnic composition and designed to be projected from a mortar	> 200 mm and ≤ 300 mm	1.3G
		Device with propellant charge, with delay fuse and bursting charge, containing colour shells ≤ 70 mm and/or pyrotechnic units, with ≤ 25% flash composition and ≤ 60% pyrotechnic composition and designed to be projected from a mortar	≤ 200 mm	1.3G
Battery/ combination	Barrage, bombardos, cakes, finale box, flowerbed, hybrid, multiple tubes, shell cakes, banger batteries, flash banger batteries	Assembly including several elements either containing the same type or several types each corresponding to one of the types of fireworks listed in this table, with one or two points of ignition	The most hazardous firework type determines the classification	

Type	Includes: / Synonym:	Definition	Specification	Classification
Roman candle	Exhibition candle, candle, bombettes	Tube containing a series of pyrotechnic units consisting of alternate pyrotechnic composition, propellant charge, and transmitting fuse	≥ 50 mm inner diameter, containing flash composition, or < 50 mm with > 25% flash composition	1.1G
			≥ 50 mm inner diameter, containing no flash composition	1.2G
			< 50 mm inner diameter and ≤ 25% flash composition	1.3G
			≤ 30 mm inner diameter, each pyrotechnic unit ≤ 25 g and ≤ 5% flash composition	1.4G
Shot tube	Single shot Roman candle, small preloaded mortar	Tube containing a pyrotechnic unit consisting of pyrotechnic composition, propellant charge with or without transmitting fuse	≤ 30 mm inner diameter and pyrotechnic unit > 25 g, or > 5% and ≤ 25% flash composition	1.3G
			≤ 30 mm inner diameter, pyrotechnic unit ≤ 25 g and ≤ 5% flash composition	1.4G
Rocket	Avalanche rocket, signal rocket, whistling rocket, bottle rocket, sky rocket, missile type rocket, table rocket	Tube containing pyrotechnic composition and/or pyrotechnic units, equipped with stick(s) or other means for stabilization of flight, and designed to be propelled into the air	Flash composition effects only	1.1G
			Flash composition > 25% of the pyrotechnic composition	1.1G
			> 20 g pyrotechnic composition and flash composition ≤ 25%	1.3G
			≤ 20 g pyrotechnic composition, black powder bursting charge and ≤ 0.13 g flash composition per report and ≤ 1 g in total	1.4G

Type	Includes: / Synonym:	Definition	Specification	Classification
Mine	Pot-a-feu, ground mine, bag mine, cylinder mine	Tube containing propellant charge and pyrotechnic units and designed to be placed on the ground or to be fixed in the ground. The principal effect is ejection of all the pyrotechnic units in a single burst producing a widely dispersed visual and/or aural effect in the air or: Cloth or paper bag or cloth or paper cylinder containing propellant charge and pyrotechnic units, designed to be placed in a mortar and to function as a mine	> 25% flash composition, as loose powder and/ or report effects	1.1G
			≥ 180 mm and ≤ 25% flash composition, as loose powder and/ or report effects	1.1G
			< 180 mm and ≤ 25% flash composition, as loose powder and/ or report effects	1.3G
			≤ 150 g pyrotechnic composition, containing ≤ 5% flash composition as loose powder and/ or report effects. Each pyrotechnic unit ≤ 25 g, each report effect < 2g; each whistle, if any, ≤ 3 g	1.4G
Fountain	Volcanos, gerbs, showers, lances, Bengal fire, flitter sparkle, cylindrical fountains, cone fountains, illuminating torch	Non-metallic case containing pressed or consolidated pyrotechnic composition producing sparks and flame	≥ 1 kg pyrotechnic composition	1.3G
			< 1 kg pyrotechnic composition	1.4G
Sparkler	Handheld sparklers, non-handheld sparklers, wire sparklers	Rigid wire partially coated (along one end) with slow burning pyrotechnic composition with or without an ignition tip	Perchlorate based sparklers: > 5 g per item or > 10 items per pack	1.3G
			Perchlorate based sparklers: ≤ 5 g per item and ≤ 10 items per pack; Nitrate based sparklers: ≤ 30 g per item	1.4G
Bengal stick	Dipped stick	Non-metallic stick partially coated (along one end) with slow-burning pyrotechnic composition and designed to be held in the hand	Perchlorate based items: > 5 g per item or > 10 items per pack	1.3 G
			Perchlorate based items: ≤ 5 g per item and ≤ 10 items per pack; nitrate based items: ≤ 30 g per item	1.4G

Type	Includes: / Synonym:	Definition	Specification	Classification
Low hazard fireworks and novelties	Table bombs, throwdowns, crackling granules, smokes, fog, snakes, glow worm, serpents, snaps, party poppers	Device designed to produce very limited visible and/ or audible effect which contains small amounts of pyrotechnic and/or explosive composition.	Throwdowns and snaps may contain up to 1.6 mg of silver fulminate; snaps and party poppers may contain up to 16 mg of potassium chlorate/red phosphorous mixture; other articles may contain up to 5 g of pyrotechnic composition, but no flash composition	1.4G
Spinner	Aerial spinner, helicopter, chaser, ground spinner	Non-metallic tube or tubes containing gas- or spark-producing pyrotechnic composition, with or without noise producing composition, with or without aerofoils attached	Pyrotechnic composition per item > 20 g, containing ≤ 3% flash composition as report effects, or whistle composition ≤ 5 g	1.3G
			Pyrotechnic composition per item ≤ 20 g, containing ≤ 3% flash composition as report effects, or whistle composition ≤ 5 g	1.4G
Wheels	Catherine wheels, Saxon	Assembly including drivers containing pyrotechnic composition and provided with a means of attaching it to a support so that it can rotate	≥ 1 kg total pyrotechnic composition, no report effect, each whistle (if any) ≤ 25 g and ≤ 50 g whistle composition per wheel	1.3G
			< 1 kg total pyrotechnic composition, no report effect, each whistle (if any) ≤ 5 g and ≤ 10 g whistle composition per wheel	1.4G
Aerial wheel	Flying Saxon, UFO's, rising crown	Tubes containing propellant charges and sparks-flame- and/or noise producing pyrotechnic compositions, the tubes being fixed to a supporting ring	> 200 g total pyrotechnic composition or > 60 g pyrotechnic composition per driver, ≤ 3% flash composition as report effects, each whistle (if any) ≤ 25 g and ≤ 50 g whistle composition per wheel	1.3G

Type	Includes: / Synonym:	Definition	Specification	Classification
			≤ 200 g total pyrotechnic composition and ≤ 60 g pyrotechnic composition per driver, ≤ 3% flash composition as report effects, each whistle (if any) ≤ 5 g and ≤ 10 g whistle composition per wheel	1.4G
Selection pack	Display selection box, display selection pack, garden selection box, indoor selection box; assortment	A pack of more than one type each corresponding to one of the types of fireworks listed in this table	The most hazardous firework type determines the classification	
Firecracker	Celebration cracker, celebration roll, string cracker	Assembly of tubes (paper or cardboard) linked by a pyrotechnic fuse, each tube intended to produce an aural effect	Each tube ≤ 140 mg of flash composition or ≤ 1 g black powder	1.4G
Banger	Salute, flash banger, lady cracker	Non-metallic tube containing report composition intended to produce an aural effect	> 2 g flash composition per item	1.1G
			≤ 2 g flash composition per item and ≤ 10 g per inner packaging	1.3G
			≤ 1 g flash composition per item and ≤ 10 g per inner packaging or ≤ 10 g black powder per item	1.4G

2.2.1.1.8 *Glossary of names*

NOTE 1: *The descriptions in the glossary are not intended to replace the test procedures, nor to determine the hazard classification of a substance or article of Class 1. Assignment to the correct division and a decision on whether Compatibility Group S is appropriate shall be based on testing of the product in accordance with the Manual of Tests and Criteria, Part I or by analogy with similar products which have already been tested and assigned in accordance with the procedures of the Manual of Tests and Criteria.*

NOTE 2*: The figures given after the names refer to the relevant UN numbers (Column (1) of Table A of Chapter 3.2). For the classification code, see 2.2.1.1.4.*

AIR BAG INFLATORS or AIR BAG MODULES or SEAT-BELT PRETENSIONERS: UN No. 0503

Articles which contain pyrotechnic substances and are used as life-saving vehicle airbags or seat-belts.

AMMUNITION, ILLUMINATING, with or without burster, expelling charge or propelling charge: UN Nos. 0171, 0254, 0297

Ammunition designed to produce a single source of intense light for lighting up an area. The term includes illuminating cartridges, grenades and projectiles; and illuminating and target identification bombs.

NOTE*: The following articles: CARTRIDGES, SIGNAL; SIGNAL DEVICES HAND; SIGNALS, DISTRESS; FLARES, AERIAL; FLARES, SURFACE are not included in this definition. They are listed separately.*

AMMUNITION, INCENDIARY, liquid or gel, with burster, expelling charge or propelling charge: UN No. 0247

Ammunition containing liquid or gelatinous incendiary substance. Except when the incendiary substance is an explosive per se, it also contains one or more of the following: a propelling charge with primer and igniter charge; a fuze with burster or expelling charge.

AMMUNITION, INCENDIARY, WHITE PHOSPHORUS with burster, expelling charge or propelling charge: UN Nos. 0243, 0244

Ammunition containing white phosphorus as incendiary substance. It also contains one or more of the following: a propelling charge with primer and igniter charge; a fuze with burster or expelling charge.

AMMUNITION, INCENDIARY with or without burster, expelling charge or propelling charge: UN Nos. 0009, 0010, 0300

Ammunition containing incendiary composition. Except when the composition is an explosive per se, it also contains one or more of the following: a propelling charge with primer and igniter charge; a fuze with burster or expelling charge.

AMMUNITION, PRACTICE: UN Nos. 0362, 0488

Ammunition without a main bursting charge, containing a burster or expelling charge. Normally it also contains a fuze and a propelling charge.

***NOTE**: GRENADES, PRACTICE are not included in this definition. They are listed separately.*

AMMUNITION, PROOF: UN No. 0363

Ammunition containing pyrotechnic substances, used to test the performance or strength of new ammunition, weapon components or assemblies.

AMMUNITION, SMOKE, WHITE PHOSPHORUS, with burster, expelling charge or propelling charge: UN Nos. 0245, 0246

Ammunition containing white phosphorus as a smoke-producing substance. It also contains one or more of the following: a propelling charge with primer and igniter charge; a fuze with burster or expelling charge. The term includes grenades, smoke.

AMMUNITION, SMOKE with or without burster, expelling charge or propelling charge: UN Nos. 0015, 0016, 0303

Ammunition containing a smoke-producing substance such as chlorosulphonic acid mixture or titanium tetrachloride; or a smoke-producing pyrotechnic composition based on hexachloroethane or red phosphorus. Except when the substance is an explosive per se, the ammunition also contains one or more of the following: a propelling charge with primer and igniter charge; a fuze with burster or expelling charge. The term includes grenades, smoke.

***NOTE:** SIGNALS, SMOKE are not included in this definition. They are listed separately.*

AMMUNITION, TEAR-PRODUCING, with burster, expelling charge or propelling charge: UN Nos. 0018, 0019, 0301

Ammunition containing a tear-producing substance. It also contains one or more of the following: a pyrotechnic substance; a propelling charge with primer and igniter charge; a fuze with burster or expelling charge.

ARTICLES, EXPLOSIVE, EXTREMELY INSENSITIVE (ARTICLES EEI): UN No. 0486

Articles containing only extremely insensitive detonating substances (EIDS) which demonstrate a negligible probability of accidental initiation or propagation under normal conditions of transport, and which have passed Test Series 7.

ARTICLES, PYROPHORIC: UN No. 0380

Articles which contain a pyrophoric substance (capable of spontaneous ignition when exposed to air) and an explosive substance or component. The term excludes articles containing white phosphorus.

ARTICLES, PYROTECHNIC, for technical purposes: UN Nos. 0428, 0429, 0430, 0431, 0432

Articles which contain pyrotechnic substances and are used for technical purposes such as heat generation, gas generation, theatrical effects, etc.

***NOTE**: The following articles: all ammunition; CARTRIDGES, SIGNAL; CUTTERS, CABLE, EXPLOSIVE; FIREWORKS; FLARES, AERIAL; FLARES, SURFACE; RELEASE DEVICES, EXPLOSIVE; RIVETS, EXPLOSIVE; SIGNAL DEVICES, HAND; SIGNALS,*

DISTRESS; SIGNALS, RAILWAY TRACK, EXPLOSIVES; SIGNALS, SMOKE are not included in this definition. They are listed separately.

BLACK POWDER (GUNPOWDER), COMPRESSED or BLACK POWDER (GUNPOWDER), IN PELLETS: UN No. 0028

Substance consisting of a pelletized form of black powder.

BLACK POWDER (GUNPOWDER), granular or as meal: UN No. 0027

Substance consisting of an intimate mixture of charcoal or other carbon and either potassium nitrate or sodium nitrate, with or without sulphur.

BOMBS, WITH FLAMMABLE LIQUID, with bursting charge: UN Nos. 0399, 0400

Articles which are dropped from aircraft, consisting of a tank filled with inflammable liquid and bursting charge.

BOMBS, PHOTO-FLASH: UN No. 0038

Explosive articles which are dropped from aircraft to provide brief, intense illumination for photography. They contain a charge of detonating explosive without means of initiation or with means of initiation containing two or more effective protective features.

BOMBS, PHOTO-FLASH: UN No. 0037

Explosive articles which are dropped from aircraft to provide brief, intense illumination for photography. They contain a charge of detonating explosive with means of initiation not containing two or more effective protective features.

BOMBS, PHOTO-FLASH: UN Nos. 0039, 0299

Explosive articles which are dropped from aircraft to provide brief, intense illumination for photography. They contain a photo-flash composition.

BOMBS with bursting charge: UN Nos. 0034; 0035

Explosive articles which are dropped from aircraft, without means of initiation or with means of initiation containing two or more effective protective features.

BOMBS with bursting charge: UN Nos. 0033, 0291

Explosive articles which are dropped from aircraft, with means of initiation not containing two or more effective protective features.

BOOSTERS WITH DETONATOR: UN Nos. 0225, 0268

Articles consisting of a charge of detonating explosive with means of initiation. They are used to increase the initiating power of detonators or detonating cord.

BOOSTERS without detonator: UN Nos. 0042, 0283

Articles consisting of a charge of detonating explosive without means of initiation. They are used to increase the initiating power of detonators or detonating cord.

BURSTERS, explosive: UN No. 0043

Articles consisting of a small charge of explosive used to open projectiles or other ammunition in order to disperse their contents.

CARTRIDGES, FLASH: UN Nos. 0049, 0050

Articles consisting of a casing, a primer and flash powder, all assembled in one piece ready for firing.

CARTRIDGES FOR WEAPONS, BLANK: UN Nos. 0326, 0413, 0327, 0338, 0014

Ammunition consisting of a closed cartridge case with a centre or rim fire primer and a charge of smokeless or black powder but no projectile. It produces a loud noise and is used for training, saluting, propelling charge, starter pistols, etc. The term includes ammunition, blank.

CARTRIDGES FOR WEAPONS, INERT PROJECTILE: UN Nos. 0328, 0417, 0339, 0012

Ammunition consisting of a projectile without bursting charge but with a propelling charge with or without a primer. The articles may include a tracer, provided that the predominant hazard is that of the propelling charge.

CARTRIDGES FOR WEAPONS with bursting charge: UN Nos. 0006, 0321, 0412

Ammunition consisting of a projectile with a bursting charge without means of initiation or with means of initiation containing two or more effective protective features; and a propelling charge with or without a primer. The term includes fixed (assembled) ammunition, semi-fixed (partially assembled) ammunition and separate loading ammunition when the components are packed together.

CARTRIDGES FOR WEAPONS with bursting charge: UN Nos. 0005, 0007, 0348

Ammunition consisting of a projectile with a bursting charge with means of initiation not containing two or more effective protective features; and a propelling charge with or without a primer. The term includes fixed (assembled) ammunition, semi-fixed (partially assembled) ammunition and separate loading ammunition when the components are packed together.

CARTRIDGES, OIL WELL: UN Nos. 0277, 0278

Articles consisting of a thin casing of fibreboard, metal or other material containing only propellant powder which projects a hardened projectile to perforate an oil well casing.

***NOTE**: CHARGES, SHAPED are not included in this definition. They are listed separately.*

CARTRIDGES, POWER DEVICE: UN Nos. 0275, 0276, 0323, 0381

Articles designed to accomplish mechanical actions. They consist of a casing with a charge of deflagrating explosive and a means of ignition. The gaseous products of the deflagration produce inflation, linear or rotary motion or activate diaphragms, valves or switches or project fastening devices or extinguishing agents.

CARTRIDGES, SIGNAL: UN Nos. 0054, 0312, 0405

Articles designed to fire coloured flares or other signals from signal pistols, etc.

CARTRIDGES, SMALL ARMS: UN Nos. 0417, 0339, 0012

Ammunition consisting of a cartridge case fitted with a centre or rim fire primer and containing both a propelling charge and solid projectile. They are designed to be fired in weapons of calibre not larger than 19.1 mm. Shot-gun cartridges of any calibre are included in this description.

***NOTE**: CARTRIDGES, SMALL ARMS, BLANK, are not included in this definition. They are listed separately. Some military small arms cartridges are not included in this definition. They are listed under CARTRIDGES FOR WEAPONS, INERT PROJECTILE.*

CARTRIDGES, SMALL ARMS, BLANK: UN Nos. 0014, 0327, 0338

Ammunition consisting of a closed cartridge case with a centre or rim fire primer and a charge of smokeless or black powder. The cartridge cases contain no projectiles. The cartridges are designed to be fired from weapons with a calibre of at most 19.1 mm and serve to produce a loud noise and are used for training, saluting, propelling charge, starter pistols, etc.

CASES, CARTRIDGE, EMPTY, WITH PRIMER: UN Nos. 0379; 0055

Articles consisting of a cartridge case made from metal, plastics or other non-inflammable material, in which the only explosive component is the primer.

CASES, COMBUSTIBLE, EMPTY, WITHOUT PRIMER: UN Nos. 0447, 0446

Articles consisting of a cartridge case made partly or entirely from nitrocellulose.

CHARGES, BURSTING, PLASTICS BONDED: UN Nos. 0457, 0458, 0459, 0460

Articles consisting of a charge of detonating explosive, plastics bonded, manufactured in a specific form without a casing and without means of initiation. They are designed as components of ammunition such as warheads.

CHARGES, DEMOLITION: UN No. 0048

Articles containing a charge of a detonating explosive in a casing of fibreboard, plastics, metal or other material. The articles are without means of initiation or with means of initiation containing two or more effective protective features.

***NOTE**: The following articles: BOMBS; MINES; PROJECTILES are not included in this definition. They are listed separately.*

CHARGES, DEPTH: UN No. 0056

Articles consisting of a charge of detonating explosive contained in a drum or projectile without means of initiation or with means of initiation containing two or more effective protective features. They are designed to detonate under water.

CHARGES, EXPLOSIVE, COMMERCIAL without detonator: UN Nos. 0442, 0443, 0444, 0445

Articles consisting of a charge of detonating explosive without means of initiation, used for explosive welding, jointing, forming and other metallurgical processes.

CHARGES, PROPELLING, FOR CANNON: UN Nos. 0242, 0279, 0414

Charges of propellant in any physical form for separate-loading ammunition for cannon.

CHARGES, PROPELLING: UN Nos. 0271, 0272, 0415, 0491

Articles consisting of a charge of a propellant charge in any physical form, with or without a casing, as a component of rocket motors or for reducing the drag of projectiles.

CHARGES, SHAPED, without detonator: UN Nos. 0059, 0439, 0440, UN 0441

Articles consisting of a casing containing a charge of detonating explosive with a cavity lined with rigid material, without means of initiation. They are designed to produce a powerful, penetrating jet effect.

CHARGES, SHAPED, FLEXIBLE, LINEAR: UN Nos. 0237, 0288

Articles consisting of a V-shaped core of a detonating explosive clad by a flexible sheath.

CHARGES, SUPPLEMENTARY, EXPLOSIVE: UN No. 0060

Articles consisting of a small removable booster placed in the cavity of a projectile between the fuze and the bursting charge.

COMPONENTS, EXPLOSIVE TRAIN, N.O.S.: UN Nos. 0382, 0383, 0384, 0461

Articles containing an explosive designed to transmit detonation or deflagration within an explosive train.

CONTRIVANCES, WATER-ACTIVATED with burster, expelling charge or propelling charge: UN Nos. 0248, 0249

Articles whose functioning depends upon physico-chemical reaction of their contents with water.

CORD, DETONATING, flexible: UN Nos. 0065, 0289

Article consisting of a core of detonating explosive enclosed in spun fabric and a plastics or other covering. The covering is not necessary if the spun fabric is sift-proof.

CORD (FUSE) DETONATING, metal clad: UN Nos. 0102, 0290

Article consisting of a core of detonating explosive clad by a soft metal tube with or without protective covering.

CORD (FUSE) DETONATING, MILD EFFECT, metal clad: UN No. 0104

Article consisting of a core of detonating explosive clad by a soft metal tube with or without a protective covering. The quantity of explosive substance is so small that only a mild effect is manifested outside the cord.

CORD, IGNITER: UN No. 0066

Article consisting of textile yarns covered with black powder or another fast burning pyrotechnic composition and of a flexible protective covering; or it consists of a core of black powder surrounded by a flexible woven fabric. It burns progressively along its length with an external flame and is used to transmit ignition from a device to a charge or primer.

CUTTERS, CABLE, EXPLOSIVE: UN No. 0070

Articles consisting of a knife-edged device which is driven by a small charge of deflagrating explosive into an anvil.

DETONATOR ASSEMBLIES, NON-ELECTRIC for blasting: UN Nos. 0360, 0361, 0500

Non-electric detonators assembled with and activated by such means as safety fuse, shock tube, flash tube or detonating cord. They may be of instantaneous design or incorporate delay elements. Detonating relays incorporating detonating cord are included.

DETONATORS, ELECTRIC for blasting: UN Nos. 0030, 0255, 0456

Articles specially designed for the initiation of blasting explosives. These detonators may be constructed to detonate instantaneously or may contain a delay element. Electric detonators are activated by an electric current.

DETONATORS FOR AMMUNITION: UN Nos. 0073, 0364, 0365, 0366

Articles consisting of a small metal or plastics tube containing explosives such as lead azide, PETN or combinations of explosives. They are designed to start a detonation train.

DETONATORS, NON-ELECTRIC for blasting: UN Nos. 0029, 0267, 0455

Articles specially designed for the initiation of blasting explosives. These detonators may be constructed to detonate instantaneously or may contain a delay element. Non-electric detonators are activated by such means as shock tube, flash tube, safety fuse, other igniferous device or flexible detonating cord. Detonating relays without detonating cord are included.

EXPLOSIVE, BLASTING, TYPE A: UN No. 0081

Substances consisting of liquid organic nitrates such as nitroglycerine or a mixture of such ingredients with one or more of the following: nitrocellulose; ammonium nitrate or other inorganic nitrates; aromatic nitro-derivatives, or combustible materials, such as wood-meal and aluminium powder. They may contain inert components such as kieselguhr, and additives such as colouring agents and stabilizers. Such explosives shall be in powdery, gelatinous or elastic form. The term includes dynamite; gelatine, blasting and gelatine dynamites.

EXPLOSIVE, BLASTING, TYPE B: UN Nos. 0082, 0331

Substances consisting of

(a) a mixture of ammonium nitrate or other inorganic nitrates with an explosive such as trinitrotoluene, with or without other substances such as wood-meal and aluminium powder; or

(b) a mixture of ammonium nitrate or other inorganic nitrates with other combustible substances which are not explosive ingredients. In both cases they may contain inert components such as kieselguhr, and additives such as colouring agents and stabilizers. Such explosives must not contain nitroglycerine, similar liquid organic nitrates or chlorates.

EXPLOSIVE, BLASTING, TYPE C: UN No. 0083

Substances consisting of a mixture of either potassium or sodium chlorate or potassium, sodium or ammonium perchlorate with organic nitro-derivatives or combustible materials such as wood-meal or aluminium powder or a hydrocarbon. They may contain inert components such as kieselguhr and additives such as colouring agents and stabilizers. Such explosives must not contain nitroglycerine or similar liquid organic nitrates.

EXPLOSIVE, BLASTING, TYPE D: UN No. 0084

Substances consisting of a mixture of organic nitrated compounds and combustible materials such as hydrocarbons and aluminium powder. They may contain inert components such as kieselguhr and additives such as colouring agents and stabilizers. Such explosives must not contain nitroglycerine, similar liquid organic nitrates, chlorates and ammonium nitrate. The term generally includes plastic explosives.

EXPLOSIVES, BLASTING, TYPE E: UN Nos. 0241, 0332

Substances consisting of water as an essential ingredient and high proportions of ammonium nitrate or other oxidizers, some or all of which are in solution. The other constituents may include nitro-derivatives such as trinitrotoluene, hydrocarbons or aluminium powder. They may contain inert components such as kieselguhr and additives such as colouring agents and stabilizers. The term includes explosives, emulsion, explosives, slurry and explosives, watergel.

FIREWORKS: UN Nos. 0333, 0334, 0335, 0336, 0337

Pyrotechnic articles designed for entertainment.

FLARES, AERIAL: UN Nos. 0093, 0403, 0404, 0420, 0421;

Articles containing pyrotechnic substances which are designed to be dropped from an aircraft to illuminate, identify, signal or warn.

FLARES, SURFACE: UN Nos. 0092, 0418, 0419

Articles containing pyrotechnic substances which are designed for use on the surface to illuminate, identify, signal or warn.

FLASH POWDER: UN Nos. 0094, 0305

Pyrotechnic substance which, when ignited, produces an intense light.

FRACTURING DEVICES, EXPLOSIVE without detonator, for oil wells: UN No. 0099

Articles consisting of a charge of detonating explosive contained in a casing without means of initiation. They are used to fracture the rock around a drill shaft to assist the flow of crude oil from the rock.

FUSE, IGNITER, tubular, metal clad: UN No. 0103

Article consisting of a metal tube with a core of deflagrating explosive.

FUSE, NON-DETONATING: UN No. 0101

Article consisting of cotton yarns impregnated with fine black powder (quickmatch). It burns with an external flame and is used in ignition trains for fireworks, etc.

FUSE, SAFETY: UN No. 0105

Article consisting of a core of fine grained black powder surrounded by a flexible woven fabric with one or more protective outer coverings. When ignited, it burns at a predetermined rate without any external explosive effect.

FUZES, DETONATING: UN Nos. 0106, 0107, 0257, 0367

Articles with explosive components designed to produce a detonation in ammunition. They incorporate mechanical, electrical, chemical or hydrostatic components to initiate the detonation. They generally incorporate protective features.

FUZES, DETONATING with protective features: UN Nos. 0408, 0409, 0410

Articles with explosive components designed to produce a detonation in ammunition. They incorporate mechanical, electrical, chemical or hydrostatic components to initiate the detonation. The detonating fuze must incorporate two or more effective protective features.

FUZES, IGNITING: UN Nos. 0316, 0317, 0368

Articles with primary explosive components designed to produce a deflagration in ammunition. They incorporate mechanical, electrical, chemical or hydrostatic components to start the deflagration. They generally incorporate protective features.

GRENADES, hand or rifle, with bursting charge: UN Nos. 0284, 0285

Articles which are designed to be thrown by hand or to be projected by a rifle. They are without means of initiation or with means of initiation containing two or more effective protective features.

GRENADES, hand or rifle, with bursting charge: UN Nos. 0292, 0293

Articles which are designed to be thrown by hand or to be projected by a rifle. They are with means of initiation not containing two or more effective protective features.

GRENADES, PRACTICE, hand or rifle: UN Nos. 0110, 0372, 0318, 0452

Articles without a main bursting charge which are designed to be thrown by hand or to be projected by a rifle. They contain the priming device and may contain a spotting charge.

HEXOTONAL: UN No. 0393

Substance consisting of an intimate mixture of cyclotrimethylenetrinitramine (RDX), trinitrotoluene (TNT) and aluminium.

HEXOLITE (HEXOTOL), dry or wetted with less than 15 % water, by mass: UN No. 0118

Substance consisting of an intimate mixture of cyclotrimethylenetrinitramine (RDX) and trinitrotoluene (TNT). The term includes "Composition B".

IGNITERS: UN Nos. 0121, 0314, 0315, 0325, 0454

Articles containing one or more explosive substances designed to produce a deflagration in an explosive train. They may be actuated chemically, electrically or mechanically.

NOTE: *The following articles: CORD, IGNITER; FUSE, IGNITER; FUSE, NON-DETONATING; FUZES, IGNITING; LIGHTERS, FUSE; PRIMERS, CAP TYPE; PRIMERS, TUBULAR are not included in this definition. They are listed separately.*

JET PERFORATING GUNS, CHARGED, oil well, without detonator: UN Nos. 0124, 0494

Articles consisting of a steel tube or metallic strip, into which are inserted shaped charges connected by detonating cord, without means of initiation.
LIGHTERS, FUSE: UN No. 0131

Articles of various design actuated by friction, percussion or electricity and used to ignite safety fuse.

MINES with bursting charge: UN Nos. 0137, 0138

Articles consisting normally of metal or composition receptacles filled with a detonating explosive, without means of initiation or with means of initiation containing two or more effective protective features. They are designed to be operated by the passage of ships, vehicles or personnel. The term includes "Bangalore torpedoes".

MINES with bursting charge: UN Nos. 0136, 0294

Articles consisting normally of metal or composition receptacles filled with a detonating explosive, with means of initiation not containing two or more effective protective features. They are designed to be operated by the passage of ships, vehicles or personnel. The term includes "Bangalore torpedoes".

OCTOLITE (OCTOL), dry or wetted with less than 15 % water, by mass: UN No. 0266

Substance consisting of an intimate mixture of cyclotetramethylenetetranitramine (HMX) and trinitrotoluene (TNT).

OCTONAL: UN No. 0496

Substance consisting of an intimate mixture of cyclotetramethylenetetranitramine (HMX), trinitrotoluene (TNT) and aluminium.

PENTOLITE, dry or wetted with less than 15 % water, by mass: UN No. 0151

Substance consisting of an intimate mixture of pentaerythrite tetranitrate (PETN) and trinitrotoluene (TNT).

POWDER CAKE (POWDER PASTE), WETTED with not less than 17 % alcohol, by mass; POWDER CAKE (POWDER PASTE), WETTED with not less than 25 % water, by mass: UN Nos. 0433, 0159

Substance consisting of nitrocellulose impregnated with not more than 60 % of nitroglycerine or other liquid organic nitrates or a mixture of these.

POWDER, SMOKELESS: UN Nos. 0160, 0161

Substance based on nitrocellulose used as propellant. The term includes propellants with a single base (nitrocellulose (NC) alone), those with a double base (such as NC and nitroglycerine (NG)) and those with a triple base (such as NC/NG/nitroguanidine).

NOTE: *Cast, pressed or bag-charges of smokeless powder are listed under CHARGES, PROPELLING or CHARGES, PROPELLING, FOR CANON.*

PRIMERS, CAP TYPE: UN Nos. 0044, 0377, 0378

Articles consisting of a metal or plastics cap containing a small amount of primary explosive mixture that is readily ignited by impact. They serve as igniting elements in small arms cartridges and in percussion primers for propelling charges.

PRIMERS, TUBULAR: UN Nos. 0319, 0320, 0376

Articles consisting of a primer for ignition and an auxiliary charge of deflagrating explosive such as black powder used to ignite the propelling charge in a cartridge case for cannon, etc.

PROJECTILES, inert with tracer: UN Nos. 0345, 0424, 0425

Articles such as a shell or bullet, which are projected from a cannon or other gun, rifle or other small arm.

PROJECTILES with burster or expelling charge: UN Nos. 0346, 0347

Articles such as a shell or bullet, which are projected from a cannon or other gun. They are without means of initiation or with means of initiation containing two or more effective protective features. They are used to scatter dyes for spotting or other inert materials.

PROJECTILES with burster or expelling charge: UN Nos. 0426, 0427

Articles such as a shell or bullet, which are projected from a cannon or other gun. They are with means of initiation not containing two or more effective protective features. They are used to scatter dyes for spotting or other inert materials.

PROJECTILES with burster or expelling charge: UN Nos. 0434, 0435

Articles such as a shell or bullet, which are projected from a cannon or other gun, rifle or other small arm. They are used to scatter dyes for spotting or other inert materials.

PROJECTILES with bursting charge: UN Nos. 0168, 0169, 0344

Articles such as a shell or bullet, which are projected from a cannon or other gun. They are without means of initiation or with means of initiation containing two or more effective protective features.

PROJECTILES with bursting charge: UN Nos. 0167, 0324

Articles such as a shell or bullet, which are projected from a cannon or other gun. They are with means of initiation not containing two or more effective protective features.

PROPELLANT, LIQUID: UN Nos. 0495, 0497

Substance consisting of a deflagrating liquid explosive, used for propulsion.

PROPELLANT, SOLID: UN Nos. 0498, 0499, 0501

Substance consisting of a deflagrating solid explosive, used for propulsion.

RELEASE DEVICES, EXPLOSIVE: UN No. 0173

Articles consisting of a small charge of explosive with means of initiation and rods or links. They sever the rods or links to release equipment quickly.

RIVETS, EXPLOSIVE: UN No. 0174

Articles consisting of a small charge of explosive inside a metallic rivet.

ROCKET MOTORS: UN Nos. 0186, 0280, 0281

Articles consisting of a charge of explosive, generally a solid propellant, contained in a cylinder fitted with one or more nozzles. They are designed to propel a rocket or a guided missile.

ROCKET MOTORS, LIQUID FUELLED: UN Nos. 0395, 0396

Articles consisting of a liquid fuel within a cylinder fitted with one or more nozzles. They are designed to propel a rocket or a guided missile.

ROCKET MOTORS WITH HYPERGOLIC LIQUIDS with or without expelling charge: UN Nos. 0322, 0250

Articles consisting of a hypergolic fuel contained in a cylinder fitted with one or more nozzles. They are designed to propel a rocket or a guided missile.

ROCKETS, LINE THROWING: UN Nos. 0238, 0240, 0453

Articles consisting of a rocket motor which is designed to extend a line.

ROCKETS, LIQUID FUELLED with bursting charge: UN Nos. 0397, 0398

Articles consisting of a liquid fuel within a cylinder fitted with one or more nozzles and fitted with a warhead. The term includes guided missiles.

ROCKETS with bursting charge: UN Nos. 0181, 0182

Articles consisting of a rocket motor and a warhead without means of initiation or with means of initiation containing two or more effective protective features. The term includes guided missiles.

ROCKETS with bursting charge: UN Nos. 0180, 0295

Articles consisting of a rocket motor and a warhead with means of initiation not containing two or more effective protective features. The term includes guided missiles.

ROCKETS with expelling charge: UN Nos. 0436, 0437, 0438

Articles consisting of a rocket motor and a charge to expel the payload from a rocket head. The term includes guided missiles.

ROCKETS with inert head: UN Nos. 0183, 0502

Articles consisting of a rocket motor and an inert head. The term includes guided missiles.

SAMPLES, EXPLOSIVE, other than initiating explosive UN No. 0190

New or existing explosive substances or articles, not yet assigned to a name in Table A of Chapter 3.2 and carried in conformity with the instructions of the competent authority and generally in small quantities, inter alia, for the purposes of testing, classification, research and development, or quality control, or as commercial samples.

NOTE: *Explosive substances or articles already assigned to another name in Table A of Chapter 3.2 are not included in this definition.*

SIGNAL DEVICES, HAND: UN Nos. 0191, 0373

Portable articles containing pyrotechnic substances which produce visual signals or warnings. The term includes small surface flares such as highway or railway flares and small distress flares.

SIGNALS, DISTRESS, ship: UN Nos. 0194, 0195, 0505, 0506

Articles containing pyrotechnic substances designed to produce signals by means of sound, flame or smoke or any combination thereof.

SIGNALS, RAILWAY TRACK, EXPLOSIVE: UN Nos. 0192, 0193, 0492, 0493

Articles containing a pyrotechnic substance which explodes with a loud report when the article is crushed. They are designed to be placed on a rail.

SIGNALS, SMOKE: UN Nos. 0196, 0197, 0313, 0487, 0507

Articles containing pyrotechnic substances which emit smoke. In addition they may contain devices for emitting audible signals.

SOUNDING DEVICES, EXPLOSIVE: UN Nos. 0374, 0375

Articles consisting of a charge of detonating explosive, without means of initiation or with means of initiation containing two or more effective protective features. They are dropped from ships and function when they reach a predetermined depth or the sea bed.

SOUNDING DEVICES, EXPLOSIVE: UN Nos. 0204, 0296

Articles consisting of a charge of detonating explosive with means of initiation not containing two or more effective protective features. They are dropped from ships and function when they reach a predetermined depth or the sea bed.

SUBSTANCES, EXPLOSIVE, VERY INSENSITIVE (Substances, EVI), N.O.S.: UN No. 0482

Substances presenting a mass explosion hazard but which are so insensitive that there is very little probability of initiation or of transition from burning to detonation under normal conditions of transport, and which have passed Test Series 5.

TORPEDOES, LIQUID FUELLED with inert head: UN No. 0450

Articles consisting of a liquid explosive system to propel the torpedo through the water, with an inert head.

TORPEDOES, LIQUID FUELLED with or without bursting charge: UN No. 0449

Articles consisting of either a liquid explosive system to propel the torpedo through the water, with or without a warhead; or a liquid non-explosive system to propel the torpedo through the water, with a warhead.

TORPEDOES with bursting charge: UN No. 0451

Articles consisting of a non-explosive system to propel the torpedo through the water, and a warhead without means of initiation or with means of initiation containing two or more effective protective features.

TORPEDOES with bursting charge: UN No. 0329

Articles consisting of an explosive system to propel the torpedo through the water, and a warhead without means of initiation or with means of initiation containing two or more effective protective features.

TORPEDOES with bursting charge: UN No. 0330

Articles consisting of an explosive or non-explosive system to propel the torpedo through the water, and a warhead with means of initiation not containing two or more effective protective features.

TRACERS FOR AMMUNITION: UN Nos. 0212, 0306

Sealed articles containing pyrotechnic substances, designed to reveal the trajectory of a projectile.

TRITONAL: UN No. 0390

Substance consisting of trinitrotoluene (TNT) mixed with aluminium.

WARHEADS, ROCKET with burster or expelling charge: UN No. 0370

Articles consisting of an inert payload and a small charge of detonating or deflagrating explosive, without means of initiation or with means of initiation containing two or more effective protective features. They are designed to be fitted to a rocket motor to scatter inert material. The term includes warheads for guided missiles.

WARHEADS, ROCKET with burster or expelling charge: UN No. 0371

Articles consisting of an inert payload and a small charge of detonating or deflagrating explosive, with means of initiation not containing two or more effective protective features. They are designed to be fitted to a rocket motor to scatter inert material. The term includes warheads for guided missiles.

WARHEADS, ROCKET with bursting charge: UN Nos. 0286, 0287

Articles consisting of a detonating explosive, without means of initiation or with means of initiation containing two or more effective protective features. They are designed to be fitted to a rocket. The term includes warheads for guided missiles.

WARHEADS, ROCKET with bursting charge: UN No. 0369

Articles consisting of a detonating explosive, with means of initiation not containing two or more effective protective features. They are designed to be fitted to a rocket. The term includes warheads for guided missiles.

WARHEADS, TORPEDO with bursting charge: UN No. 0221

Articles consisting of a detonating explosive, without means of initiation or with means of initiation containing two or more effective protective features. They are designed to be fitted to a torpedo.

2.2.1.2 ***Substances and articles not accepted for carriage***

2.2.1.2.1 Explosive substances which are unduly sensitive according to the criteria of the Manual of Tests and Criteria, Part I, or are liable to spontaneous reaction, as well as explosive substances and articles which cannot be assigned to a name or n.o.s. entry listed in Table A of Chapter 3.2, shall not be accepted for carriage.

2.2.1.2.2 Articles of compatibility group K shall not be accepted for carriage (1.2K, UN No. 0020 and 1.3K, UN No. 0021).

2.2.1.3 *List of collective entries*

Classification code (see 2.2.1.1.4)	UN No	Name of the substance or article
1.1A	0473	SUBSTANCES, EXPLOSIVE, N.O.S.
1.1B	0461	COMPONENTS, EXPLOSIVE TRAIN, N.O.S.
1.1C	0474	SUBSTANCES, EXPLOSIVE, N.O.S.
	0497	PROPELLANT, LIQUID
	0498	PROPELLANT, SOLID
	0462	ARTICLES, EXPLOSIVE, N.O.S.
1.1D	0475	SUBSTANCES, EXPLOSIVE, N.O.S.
	0463	ARTICLES, EXPLOSIVE, N.O.S.
1.1E	0464	ARTICLES, EXPLOSIVE, N.O.S.
1.1F	0465	ARTICLES, EXPLOSIVE, N.O.S.
1.1G	0476	SUBSTANCES, EXPLOSIVE, N.O.S.
1.1L	0357	SUBSTANCES, EXPLOSIVE, N.O.S.
	0354	ARTICLES, EXPLOSIVE, N.O.S.
1.2B	0382	COMPONENTS, EXPLOSIVE TRAIN, N.O.S.
1.2C	0466	ARTICLES, EXPLOSIVE, N.O.S.
1.2D	0467	ARTICLES, EXPLOSIVE, N.O.S.
1.2E	0468	ARTICLES, EXPLOSIVE, N.O.S.
1.2F	0469	ARTICLES, EXPLOSIVE, N.O.S.
1.2L	0358	SUBSTANCES, EXPLOSIVE, N.O.S.
	0248	CONTRIVANCES, WATER-ACTIVATED with burster, expelling charge or propelling charge
	0355	ARTICLES, EXPLOSIVE, N.O.S.
1.3C	0132	DEFLAGRATING METAL SALTS OF AROMATIC NITRO-DERIVATIVES, N.O.S.
	0477	SUBSTANCES, EXPLOSIVE, N.O.S.
	0495	PROPELLANT, LIQUID
	0499	PROPELLANT, SOLID
	0470	ARTICLES, EXPLOSIVE, N.O.S.
1.3G	0478	SUBSTANCES, EXPLOSIVE, N.O.S.
1.3L	0359	SUBSTANCES, EXPLOSIVE, N.O.S.
	0249	CONTRIVANCES, WATER-ACTIVATED with burster, expelling charge or propelling charge
	0356	ARTICLES, EXPLOSIVE, N.O.S.
1.4B	0350	ARTICLES, EXPLOSIVE, N.O.S.
	0383	COMPONENTS, EXPLOSIVE TRAIN, N.O.S.
1.4C	0479	SUBSTANCES, EXPLOSIVE, N.O.S.
	0351	ARTICLES, EXPLOSIVE, N.O.S.
	0501	PROPELLANT, SOLID
1.4D	0480	SUBSTANCES, EXPLOSIVE, N.O.S.
	0352	ARTICLES, EXPLOSIVE, N.O.S.
1.4E	0471	ARTICLES, EXPLOSIVE, N.O.S.
1.4F	0472	ARTICLES, EXPLOSIVE, N.O.S.
1.4G	0485	SUBSTANCES, EXPLOSIVE, N.O.S.
	0353	ARTICLES, EXPLOSIVE, N.O.S.
1.4S	0481	SUBSTANCES, EXPLOSIVE, N.O.S.
	0349	ARTICLES, EXPLOSIVE, N.O.S.
	0384	COMPONENTS, EXPLOSIVE TRAIN, N.O.S.
1.5D	0482	SUBSTANCES, EXPLOSIVE, VERY INSENSITIVE (SUBSTANCES, EVI) N.O.S.

Classification code (see 2.2.1.1.4)	UN No	Name of the substance or article
1.6N	0486	ARTICLES, EXPLOSIVE, EXTREMELY INSENSITIVE (ARTICLES, EEI)
	0190	SAMPLES, EXPLOSIVE other than initiating explosive ***NOTE:*** *Division and Compatibility Group shall be defined as directed by the competent authority and according to the principles in 2.2.1.1.4.*

2.2.2 Class 2 Gases

2.2.2.1 *Criteria*

2.2.2.1.1 The heading of Class 2 covers pure gases, mixtures of gases, mixtures of one or more gases with one or more other substances and articles containing such substances.

A gas is a substance which:

(a) at 50 °C has a vapour pressure greater than 300 kPa (3 bar); or

(b) is completely gaseous at 20° C at the standard pressure of 101.3 kPa .

NOTE 1: *UN No. 1052 HYDROGEN FLUORIDE, ANHYDROUS is nevertheless classified in Class 8.*

NOTE 2: *A pure gas may contain other components deriving from its production process or added to preserve the stability of the product, provided that the level of these components does not change its classification or its conditions of carriage, such as filling ratio, filling pressure, test pressure.*

NOTE 3: *N.O.S. entries in 2.2.2.3 may cover pure gases as well as mixtures.*

NOTE 4: *Carbonated beverages are not subject to the provisions of ADN.*

2.2.2.1.2 The substances and articles of Class 2 are subdivided as follows:

1. *Compressed gas:* a gas which when packaged under pressure for carriage is entirely gaseous at -50 °C; this category includes all gases with a critical temperature less than or equal to -50 °C;

2. *Liquefied gas:* a gas which when packaged under pressure for carriage is partially liquid at temperatures above -50 °C. A distinction is made between:

   High pressure liquefied gas: a gas with a critical temperature above -50 °C and equal to or below +65 °C; and

   Low pressure liquefied gas: a gas with a critical temperature above +65 °C;

3. *Refrigerated liquefied gas*: a gas which when packaged for carriage is made partially liquid because of its low temperature;

4. *Dissolved gas:* a gas which when packaged under pressure for carriage is dissolved in a liquid phase solvent;

5. Aerosol dispensers and receptacles, small, containing gas (gas cartridges);

6. Other articles containing gas under pressure;

7. Non-pressurized gases subject to special requirements (gas samples).

2.2.2.1.3 Substances and articles (except aerosols) of Class 2 are assigned to one of the following groups according to their hazardous properties, as follows:

A asphyxiant;

O oxidizing;

F flammable;

T toxic;

TF toxic, flammable;

TC toxic, corrosive;

TO toxic, oxidizing;

TFC toxic, flammable, corrosive;

TOC toxic, oxidizing, corrosive.

For gases and gas mixtures presenting hazardous properties associated with more than one group according to the criteria, the groups designated by letter T take precedence over all other groups. The groups designated by letter F take precedence over the groups designated by letters A or O.

NOTE 1: *In the UN Model Regulations, the IMDG Code and the ICAO Technical Instructions, gases are assigned to one of the following three divisions, based on the primary hazard:*

Division 2.1: *flammable gases (corresponding to the groups designated by the capital letter F);*

Division 2.2: *non-flammable, non-toxic gases (corresponding to the groups designated by the capital letters A or O);*

Division 2.3: *toxic gases (corresponding to the groups designated by the capital letter T (i.e. T, TF, TC, TO, TFC and TOC).*

NOTE 2: *Receptacles, small containing gas (UN No. 2037) shall be assigned to the groups A to TOC according to the hazard of the contents. For aerosols (UN No. 1950), see 2.2.2.1.6.*

NOTE 3: *Corrosive gases are considered to be toxic, and are therefore assigned to the group TC, TFC or TOC.*

NOTE 4: *Mixtures containing more than 21% oxygen by volume shall be classified as oxidizing.*

2.2.2.1.4 If a mixture of Class 2 mentioned by name in Table A of Chapter 3.2 meets different criteria as mentioned in 2.2.2.1.2 and 2.2.2.1.5, this mixture shall be classified according to the criteria and assigned to an appropriate N.O.S. entry.

2.2.2.1.5 Substances and articles (except aerosols) of Class 2 which are not mentioned by name in Table A of Chapter 3.2 shall be classified under a collective entry listed in 2.2.2.3 in accordance with 2.2.2.1.2 and 2.2.2.1.3. The following criteria shall apply:

Asphyxiant gases

Gases which are non-oxidizing, non-flammable and non-toxic and which dilute or replace oxygen normally in the atmosphere.

Flammable gases

Gases which at 20 °C and a standard pressure of 101.3 kPa:

(a) are ignitable when in a mixture of 13% or less by volume with air; or

(b) have a flammable range with air of at least 12 percentage points regardless of the lower flammable limit.

Flammability shall be determined by tests or by calculation, in accordance with methods adopted by ISO (see ISO 10156:1996).

Where insufficient data are available to use these methods, tests by a comparable method recognized by the competent authority of the country of origin may be used.

If the country of origin is not a Contracting Party to ADN these methods shall be recognized by the competent authority of the first country Contracting Party to ADN reached by the consignment.

Oxidizing gases

Gases, which may, generally by providing oxygen, cause or contribute to the combustion of other material more than air does. Oxidizing ability is determined either by tests or by calculation methods adopted by ISO (see ISO 10156:1996 and ISO 10156-2:2005).

Toxic gases

NOTE: *Gases meeting the criteria for toxicity in part or completely owing to their corrosivity are to be classified as toxic. See also the criteria under the heading "Corrosive gases" for a possible subsidiary corrosivity risk.*

Gases which:

(a) are known to be so toxic or corrosive to humans as to pose a hazard to health; or

(b) are presumed to be toxic or corrosive to humans because they have a LC_{50} value for acute toxicity equal to or less than 5 000 ml/m³ (ppm) when tested in accordance with 2.2.61.1.

In the case of gas mixtures (including vapours of substances from other classes) the following formula may be used:

$$LC_{50}\ \text{Toxic (mixture)} = \frac{1}{\sum_{i=1}^{n} \frac{f_i}{T_i}}$$

where f_i = mole fraction of the ith component substance of the mixture;

T_i = toxicity index of the ith component substance of the mixture. The T_i equals the LC_{50} value as found in packing instruction P200 of 4.1.4.1 of ADR. When no LC_{50} value is listed in packing instruction P200 of 4.1.4.1 of ADR, a LC_{50} value available in scientific literature shall be used.

When the LC_{50} value is unknown, the toxicity index is determined by using the lowest LC_{50} value of substances of similar physiological and chemical effects, or through testing if this is the only practical possibility.

Corrosive gases

Gases or gas mixtures meeting the criteria for toxicity completely owing to their corrosivity are to be classified as toxic with a subsidiary corrosivity risk.

A gas mixture that is considered to be toxic due to the combined effects of corrosivity and toxicity has a subsidiary risk of corrosivity when the mixture is known by human experience to be destructive to the skin, eyes or mucous membranes or when the LC_{50} value of the corrosive components of the mixture is equal to or less than 5 000 ml/m^3 (ppm) when the LC_{50} is calculated by the formula:

$$LC_{50}\text{ Corrosive (mixture)} = \frac{1}{\sum_{i=1}^{n} \frac{f_{ci}}{T_{ci}}}$$

where fc_i = mole fraction of the ith corrosive component substance of the mixture;

Tc_i = toxicity index of the ith corrosive component substance of the mixture.
The Tci equals the LC50 value as found in packing instruction P200 of 4.1.4.1 of ADR.
When no LC_{50} value is listed in packing instruction P200 of 4.1.4.1 of ADR, a LC_{50} value available in scientific literature shall be used.
When the LC_{50} value is unknown the toxicity index is determined by using the lowest LC_{50} value of substances of similar physiological and chemical effects, or through testing if this is the only practical possibility.

2.2.2.1.6 *Aerosols*

Aerosols (UN No. 1950) are assigned to one of the following groups according to their hazardous properties, as follows:

A asphyxiant;

O oxidizing;

F flammable;

T toxic;

C corrosive;

CO corrosive, oxidizing;

FC flammable, corrosive;

TF toxic, flammable;

TC toxic, corrosive;

TO toxic, oxidizing;

TFC toxic, flammable, corrosive

TOC toxic, oxidizing, corrosive.

The classification depends on the nature of the contents of the aerosol dispenser.

NOTE: *Gases, which meet the definition of toxic gases according to 2.2.2.1.5 or of pyrophoric gases according to packing instruction P200 in 4.1.4.1 of ADR, shall not be used as a propellant in an aerosol dispenser. Aerosols with contents meeting the criteria for packing group I for toxicity or corrosivity shall not be accepted for carriage (see also 2.2.2.2.2).*

The following criteria shall apply:

(a) Assignment to group A shall apply when the contents do not meet the criteria for any other group according to sub-paragraphs (b) to (f) below;

(b) Assignment to group O shall apply when the aerosol contains an oxidizing gas according to 2.2.2.1.5;

(c) Assignment to group F shall apply if the contents include 85% by mass or more flammable components and the chemical heat of combustion is 30 kJ/g or more.

It shall not apply if the contents contain 1% by mass or less flammable components and the heat of combustion is less than 20 kJ/g.

Otherwise the aerosol shall be tested for flammability in accordance with the tests described in the *Manual of Tests and Criteria*, Part III, section 31. Extremely flammable and flammable aerosols shall be assigned to group F;

NOTE*:* *Flammable components are flammable liquids, flammable solids or flammable gases and gas mixtures as defined in Notes 1 to 3 of sub-section 31.1.3 of Part III of the Manual of Tests and Criteria. This designation does not cover pyrophoric, self-heating or water-reactive substances. The chemical heat of combustion shall be determined by one of the following methods ASTM D 240, ISO/FDIS 13943: 1999 (E/F) 86.1 to 86.3 or NFPA 30B.*

(d) Assignment to group T shall apply when the contents, other than the propellant of aerosol dispensers to be ejected, are classified as Class 6.1, packing groups II or III;

(e) Assignment to group C shall apply when the contents, other than the propellant of aerosol dispensers to be ejected, meet the criteria for Class 8, packing groups II or III;

(f) When the criteria for more than one group amongst groups O, F, T, and C are met, assignment to groups CO, FC, TF, TC TO, TFC or TOC shall apply, as relevant.

2.2.2.2 ***Gases not accepted for carriage***

2.2.2.2.1 Chemically unstable substances of Class 2 shall not be accepted for carriage, unless the necessary steps have been taken to prevent all possibility of a dangerous reaction e.g. decomposition, dismutation or polymerisation under normal conditions during transport.

To this end particular care shall be taken to ensure that receptacles and tanks do not contain any substances liable to promote these reactions.

2.2.2.2.2 The following substances and mixtures shall not be accepted for carriage:

- UN No. 2186 HYDROGEN CHLORIDE, REFRIGERATED LIQUID;
- UN No. 2421 NITROGEN TRIOXIDE;
- UN No. 2455 METHYL NITRITE;
- Refrigerated liquefied gases which cannot be assigned to classification codes 3A, 3O or 3F;
- Dissolved gases which cannot be classified under UN Nos. 1001, 2073 or 3318;
- Aerosols where gases which are toxic according to 2.2.2.1.5 or pyrophoric according to packing instruction P200 in 4.1.4.1 of ADR are used as propellants;
- Aerosols with contents meeting the criteria for packing group I for toxicity or corrosivity (see 2.2.61 and 2.2.8);
- Receptacles, small, containing gases which are very toxic (LC_{50} lower than 200 ppm) or pyrophoric according to packing instruction P200 in 4.1.4.1 of ADR.

2.2.2.3 *List of collective entries*

Compressed gases		
Classification code	**UN No**	**Name and description**
1 A	1956	COMPRESSED GAS, N.O.S.
1 O	3156	COMPRESSED GAS, OXIDIZING, N.O.S.
1 F	1964	HYDROCARBON GAS MIXTURE, COMPRESSED, N.O.S.
	1954	COMPRESSED GAS, FLAMMABLE, N.O.S.
1T	1955	COMPRESSED GAS, TOXIC, N.O.S.
1 TF	1953	COMPRESSED GAS, TOXIC, FLAMMABLE, N.O.S.
1 TC	3304	COMPRESSED GAS, TOXIC, CORROSIVE, N.O.S.
1 TO	3303	COMPRESSED GAS, TOXIC, OXIDIZING, N.O.S.
1 TFC	3305	COMPRESSED GAS, TOXIC, FLAMMABLE, CORROSIVE, N.O.S.
1 TOC	3306	COMPRESSED GAS, TOXIC, OXIDIZING, CORROSIVE, N.O.S.

Liquefied gases		
Classification code	**UN No**	**Name and description**
2 A	1058	LIQUEFIED GASES, non-flammable, charged with nitrogen, carbon dioxide or air
	1078	REFRIGERANT GAS, N.O.S. such as mixtures of gases, indicated by the letter R, which as: Mixture F1, have a vapour pressure at 70 °C not exceeding 1.3 MPa (13 bar) and a density at 50 °C not lower than that of dichlorofluoromethane (1.30 kg/l); Mixture F2, have a vapour pressure at 70 °C not exceeding 1.9 MPa (19 bar) and a density at 50 °C not lower than that of dichlorodifluoromethane (1.21 kg/l); Mixture F3, have a vapour pressure at 70 °C not exceeding 3 MPa (30 bar) and a density at 50 °C not lower than that of chlorodifluoromethane (1.09 kg/l). ***NOTE:*** *Trichlorofluoromethane (Refrigerant R 11), 1,1,2-trichloro-1,2,2-trifluoroethane (Refrigerant R 113), 1,1,1-trichloro-2,2,2-trifluoroethane (Refrigerant R 113a), 1-chloro-1,2,2-trifluoroethane (Refrigerant R 133) and 1-chloro-1,1,2-trifluoroethane (Refrigerant R 133b) are not substances of Class 2. They may, however, enter into the composition of mixtures F1 to F3.*
	1968	INSECTICIDE GAS, N.O.S.
	3163	LIQUEFIED GAS, N.O.S.
2 O	3157	LIQUEFIED GAS, OXIDIZING, N.O.S.

Liquefied gases (cont'd)		
Classification code	**UN No**	**Name and description**
2 F	1010	BUTADIENES AND HYDROCARBON MIXTURE, STABILIZED, having a vapour pressure at 70 °C not exceeding 1.1 MPa (11 bar) and a density at 50 °C not lower than 0.525 kg/l. ***NOTE**: Butadienes, stabilized are also classified under UN No. 1010, see Table A of Chapter 3.2.*
	1060	METHYLACETYLENE AND PROPADIENE MIXTURE, STABILIZED such as mixtures of methylacetylene and propadiene with hydrocarbons, which as: Mixture P1, contain not more than 63% methylacetylene and propadiene by volume and not more than 24% propane and propylene by volume, the percentage of C_4- saturated hydrocarbons being not less than 14% by volume; and as Mixture P2, contain not more than 48% methylacetylene and propadiene by volume and not more than 50% propane and propylene by volume, the percentage of C_4- saturated hydrocarbons being not less than 5% by volume, as well as mixtures of propadiene with 1 to 4% methylacetylene.
	1965	HYDROCARBON GAS MIXTURE, LIQUEFIED, N.O.S such as mixtures, which as: Mixture A, have a vapour pressure at 70 °C not exceeding 1.1 MPa (11 bar) and a density at 50 °C not lower than 0.525 kg/l; Mixture A01, have a vapour pressure at 70 °C not exceeding 1.6 MPa (16 bar) and a relative density at 50 °C not lower than 0.516 kg/l; Mixture A02, have a vapour pressure at 70 °C not exceeding 1.6 MPa (16 bar) and a relative density at 50 °C not lower than 0.505 kg/l; Mixture A0, have a vapour pressure at 70 °C not exceeding 1.6 MPa (16 bar) and a density at 50 °C not lower than 0.495 kg/l; Mixture A1, have a vapour pressure at 70 °C not exceeding 2.1 MPa (21 bar) and a density at 50 °C not lower than 0.485 kg/l; Mixture B1 have a vapour pressure at 70 °C not exceeding 2.6 MPa (26 bar) and a relative density at 50 °C not lower than 0.474 kg/l; Mixture B2 have a vapour pressure at 70 °C not exceeding 2.6 MPa (26 bar) and a relative density at 50 °C not lower than 0.463 kg/l; Mixture B, have a vapour pressure at 70 °C not exceeding 2.6 MPa (26 bar) and a density at 50 °C not lower than 0.450 kg/l; Mixture C, have a vapour pressure at 70 °C not exceeding 3.1 MPa (31 bar) and a relative density at 50 °C not lower than 0.440 kg/l; ***NOTE 1:** In the case of the foregoing mixtures, the use of the following names customary in the trade is permitted for describing these substances: for mixture A01, A02 and A0: BUTANE; for mixture C: PROPANE.* ***NOTE 2:** UN No. 1075 PETROLEUM GASES, LIQUEFIED may be used as an alternative entry for UN No. 1965 HYDROCARBON GAS MIXTURE LIQUEFIED, N.O.S. for carriage prior to or following maritime or air carriage.*
	3354	INSECTICIDE GAS, FLAMMABLE, N.O.S.
	3161	LIQUEFIED GAS, FLAMMABLE, N.O.S.

Liquefied gases (cont'd)		
Classification code	**UN No**	**Name and description**
2 T	1967	INSECTICIDE GAS, TOXIC, N.O.S.
	3162	LIQUEFIED GAS, TOXIC, N.O.S.
2 TF	3355	INSECTICIDE GAS, TOXIC, FLAMMABLE, N.O.S.
	3160	LIQUEFIED GAS, TOXIC, FLAMMABLE, N.O.S.
2 TC	3308	LIQUEFIED GAS, TOXIC, CORROSIVE, N.O.S.
2 TO	3307	LIQUEFIED GAS, TOXIC, OXIDIZING, N.O.S.
2 TFC	3309	LIQUEFIED GAS, TOXIC, FLAMMABLE, CORROSIVE, N.O.S.
2 TOC	3310	LIQUEFIED GAS, TOXIC, OXIDIZING, CORROSIVE, N.O.S.

Refrigerated liquefied gases		
Classification code	**UN No**	**Name and description**
3 A	3158	GAS, REFRIGERATED LIQUID, N.O.S.
3 O	3311	GAS, REFRIGERATED LIQUID, OXIDIZING, N.O.S.
3 F	3312	GAS, REFRIGERATED LIQUID, FLAMMABLE, N.O.S.

Dissolved gases		
Classification code	**UN No**	**Name and description**
4		Only substances listed in Table A of Chapter 3.2 are to be accepted for carriage.

Aerosols and receptacles, small, containing gas		
Classification code	**UN No**	**Name and description**
5	1950	AEROSOLS
	2037	RECEPTACLES, SMALL CONTAINING GAS (GAS CARTRIDGES) without a release device, non-refillable

Other articles containing gas under pressure		
Classification code	**UN No**	**Name and description**
6A	2857	REFRIGERATING MACHINES containing non-flammable, non-toxic gases or ammonia solutions (UN 2672)
	3164	ARTICLES, PRESSURIZED, PNEUMATIC (containing non-flammable gas) or
	3164	ARTICLES, PRESSURIZED, HYDRAULIC (containing non-flammable gas)
6F	3150	DEVICES, SMALL, HYDROCARBON GAS POWERED or
	3150	HYDROCARBON GAS REFILLS FOR SMALL DEVICES, with release device
	3478	FUEL CELL CARTRIDGES, containing liquefied flammable gas or
	3478	FUEL CELL CARTRIDGES CONTAINED IN EQUIPMENT, containing liquefied flammable gas or
	3478	FUEL CELL CARTRIDGES PACKED WITH EQUIPMENT, containing liquefied flammable gas
	3479	FUEL CELL CARTRIDGES, containing hydrogen in metal hydride or
	3479	FUEL CELL CARTRIDGES CONTAINED IN EQUIPMENT, containing hydrogen in metal hydride or
	3479	FUEL CELL CARTRIDGES PACKED WITH EQUIPMENT, containing hydrogen in metal hydride

Gas samples		
Classification code	**UN No**	**Name and description**
7 F	3167	GAS SAMPLE, NON-PRESSURIZED, FLAMMABLE, N.O.S., not refrigerated liquid
7 T	3169	GAS SAMPLE, NON-PRESSURIZED, TOXIC, N.O.S., not refrigerated liquid
7 TF	3168	GAS SAMPLE, NON-PRESSURIZED, TOXIC, FLAMMABLE, N.O.S., not refrigerated liquid

2.2.3 **Class 3 Flammable liquids**

2.2.3.1 ***Criteria***

2.2.3.1.1 The heading of Class 3 covers substances and articles containing substances of this Class which:

– are liquids according to subparagraph (a) of the definition for "liquid" in 1.2.1;

– have at 50 °C a vapour pressure of not more than 300 kPa (3 bar) and are not completely gaseous at 20 °C and at standard pressure of 101.3 kPa; and

– have a flash-point of not more than 60 °C (see 2.3.3.1 for the relevant test).

The heading of Class 3 also covers liquid substances and molten solid substances with a flash-point of more than 60°C and which are carried or handed over for carriage whilst heated at temperatures equal to or higher than their flash-point. These substances are assigned to UN No. 3256.

The heading of Class 3 also covers liquid desensitized explosives. Liquid desensitized explosives are explosive substances which are dissolved or suspended in water or other liquid substances, to form an homogeneous liquid mixture to suppress their explosive properties. Such entries in Table A of Chapter 3.2 are UN Nos. 1204, 2059, 3064, 3343, 3357 and 3379.

For the purpose of carriage in tank vessels, the heading of Class 3 also covers the following substances which:

– have a flash-point above 60° C and which are carried or handed over for carriage at a temperature within a range of 15 K below the flash-point;

– have an auto-ignition temperature of 200° C or below and which are not mentioned elsewhere.

***NOTE 1**: Substances having a flash-point above 35 °C, non-toxic and non-corrosive, which, do not sustain combustion according to the criteria of 32.2.5 of Part III of the Manual of Tests and Criteria are not substances of Class 3; if, however, these substances are handed over for carriage and carried whilst heated at temperatures equal to or higher than their flash-point, they are substances of Class 3.*

***NOTE 2**: By derogation from paragraph 2.2.3.1.1 above, diesel fuel, gasoil, heating oil (light) having a flash-point above 60 °C and not more than 100 °C shall be deemed substances of Class 3, UN No. 1202.*

***NOTE 3:** Liquids which are highly toxic on inhalation, having a flash-point below 23 °C and toxic substances, having a flash-point of 23 °C or above are substances of Class 6.1 (see 2.2.61.1).*

***NOTE 4:** Flammable liquid substances and preparations used as pesticides, which are highly toxic, toxic or slightly toxic and have a flash-point of 23 °C or above are substances of Class 6.1 (see 2.2.61.1).*

***NOTE 5:** For the purpose of carriage in tank vessels, substances having a flash-point above 60° C and not more than 100° C are substances of Class 9 (identification number 9003).*

2.2.3.1.2 The substances and articles of Class 3 are subdivided as follows:

F Flammable liquids, without subsidiary risk:

F1 Flammable liquids having a flash-point of or below 60 °C;
F2 Flammable liquids having a flash-point above 60 °C which are carried or handed over for carriage at or above their flash-point (elevated temperature substances);
F3 Substances having a flash-point above 60° C which are carried or handed over for carriage at a temperature within a range of 15 K below the flash-point;
F4 Substances having an auto-ignition temperature of 200° C or below and which are not mentioned elsewhere.

FT Flammable liquids, toxic:

FT1 Flammable liquids, toxic;
FT2 Pesticides;

FC Flammable liquids, corrosive;

FTC Flammable liquids, toxic, corrosive;

D Liquid desensitized explosives.

2.2.3.1.3 Substances and articles classified in Class 3 are listed in Table A of Chapter 3.2. Substances not mentioned by name in Table A of Chapter 3.2 shall be assigned to the relevant entry of 2.2.3.3 and the relevant packing group in accordance with the provisions of this section. Flammable liquids shall be assigned to one of the following packing groups according to the degree of danger they present for carriage:

Packing Group	Flash-point (closed cup)	Initial boiling point
I	--	≤ 35°C
II [a]	< 23°C	> 35°C
III [a]	≥ 23°C and ≤ 60°C	> 35°C

a See also 2.2.3.1.4

For a liquid with (a) subsidiary risk(s), the packing group determined in accordance with the table above and the packing group based on the severity of the subsidiary risk(s) shall be considered; the classification and packing group shall then be determined in accordance with the table of precedence of hazards in 2.1.3.10.

2.2.3.1.4 Liquid or viscous mixtures and preparations, including those containing no more than 20% nitrocellulose with a nitrogen content not exceeding 12.6% (by dry mass), shall be assigned to packing group III only if the following requirements are met:

(a) the height of the separated layer of solvent is less than 3 % of the total height of the sample in the solvent-separation test (see Manual of Tests and Criteria, Part III, sub-section 32.5.1); and

(b) the viscosity [2] and flash-point are in accordance with the following table:

---

2 *Viscosity determination: Where the substance concerned is non-Newtonian, or where a flow cup method of viscosity determination is otherwise unsuitable, a variable shear-rate viscometer shall be used to determine the dynamic viscosity coefficient of the substance, at 23 °C, at a number of shear rates. The values obtained are plotted against shear rate and then extrapolated to zero shear rate. The dynamic viscosity thus obtained, divided by the density, gives the apparent kinematic viscosity at near-zero shear rate.*

Kinematic viscosity (extrapolated) ν (at near-zero shear rate) mm^2/s at 23 °C	Flow time t in accordance with ISO 2431:1993		Flash-point in °C
	in s	Jet diameter in mm	
$20 < \nu \leq 80$	$20 < t \leq 60$	4	above 17
$80 < \nu \leq 135$	$60 < t \leq 100$	4	above 10
$135 < \nu \leq 220$	$20 < t \leq 32$	6	above 5
$220 < \nu \leq 300$	$32 < t \leq 44$	6	above -1
$300 < \nu \leq 700$	$44 < t \leq 100$	6	above -5
700 < <	$100 < t$	6	-5 and below

***NOTE**: Mixtures containing more than 20% but not more than 55% nitrocellulose with a nitrogen content not exceeding 12.6% by dry mass are substances assigned to UN No. 2059.*

Mixtures having a flash-point below 23 °C and containing:

– *more than 55 % nitrocellulose, whatever their nitrogen content; or*

– *not more than 55 % nitrocellulose with a nitrogen content above 12.6 % by dry mass,*

are substances of Class 1 (UN Nos. 0340 or 0342) or of Class 4.1 (UN Nos. 2555, 2556 or 2557).

2.2.3.1.5 Non-toxic, non-corrosive and non-environmentally hazardous solutions and homogeneous mixtures having a flash-point of 23 °C or above (viscous substances, such as paints or varnishes, excluding substances containing more than 20 % nitrocellulose) packed in receptacles of less than 450 litres capacity, are not subject to ADN if, in the solvent-separation test (see Manual of Tests and Criteria, Part III, sub-section 32.5.1), the height of the separated layer of solvent is less than 3 % of the total height, and if the substances at 23 °C have, in the flow cup conforming to ISO 2431:1993 having a jet 6 mm in diameter, a flow time of:

(a) not less than 60 seconds, or

(b) not less than 40 seconds and contain not more than 60 % of substances of Class 3.

2.2.3.1.6 If substances of Class 3, as a result of admixtures, come into categories of risk different from those to which the substances mentioned by name in Table A of Chapter 3.2 belong, these mixtures or solutions shall be assigned to the entries to which they belong on the basis of their actual degree of danger.

***NOTE**: For the classification of solutions and mixtures (such as preparations and wastes) see also 2.1.3.*

2.2.3.1.7 On the basis of the test procedures in accordance with 2.3.3.1 and 2.3.4, and the criteria set out in 2.2.3.1.1, it may also be determined whether the nature of a solution or a mixture mentioned by name or containing a substance mentioned by name is such that the solution or mixture is not subject to the provisions for this Class (see also 2.1.3).

2.2.3.2 ***Substances not accepted for carriage***

2.2.3.2.1 Substances of Class 3 which are liable to form peroxides easily (as happens with ethers or with certain heterocyclic oxygenated substances) shall not be accepted for carriage if their peroxide content, calculated as hydrogen peroxide (H_2O_2), exceeds 0.3%. The peroxide content shall be determined as indicated in 2.3.3.2.

2.2.3.2.2 The chemically unstable substances of Class 3 shall not be accepted for carriage unless the necessary steps have been taken to prevent their dangerous decomposition or polymerization during carriage. To this end, it shall be ensured in particular that receptacles and tanks do not contain any substance liable to promote these reactions.

2.2.3.2.3 Liquid desensitized explosives other than those listed in Table A of Chapter 3.2 shall not be accepted for carriage as substances of Class 3.

2.2.3.3 *List of collective entries*

Group	Subgroup	Classification code	UN No.	Name and description
Flammable liquids				
Without subsidiary risk		**F1**	1133	ADHESIVES containing flammable liquid
			1136	COAL TAR DISTILLATES, FLAMMABLE
			1139	COATING SOLUTION (includes surface treatments or coatings used for industrial or other purposes such as vehicle undercoating, drum or barrel lining)
			1169	EXTRACTS, AROMATIC, LIQUID
			1197	EXTRACTS, FLAVOURING, LIQUID
			1210	PRINTING INK, flammable or
			1210	PRINTING INK RELATED MATERIAL (including printing ink thinning or reducing compound), flammable
			1263	PAINT (including paint, lacquer, enamel, stain, shellac, varnish, polish, liquid filler and liquid lacquer base) or
			1263	PAINT RELATED MATERIAL (including paint thinning or reducing compound)
			1266	PERFUMERY PRODUCTS with flammable solvents
			1293	TINCTURES, MEDICINAL
			1306	WOOD PRESERVATIVES, LIQUID
			1866	RESIN SOLUTION, flammable
			1999	TARS, LIQUID, including road asphalt and oils, bitumen and cut backs
			3065	ALCOHOLIC BEVERAGES
			3269	POLYESTER RESIN KITS
			1224	KETONES, LIQUID, N.O.S.
			1268	PETROLEUM DISTILLATES, N.O.S. or
			1268	PETROLEUM PRODUCTS, N.O.S.
			1987	ALCOHOLS, N.O.S.
			1989	ALDEHYDES, N.O.S.
			2319	TERPENE HYDROCARBONS, N.O.S.
			3271	ETHERS, N.O.S.
			3272	ESTERS, N.O.S.
			3295	HYDROCARBONS, LIQUID, N.O.S.
			3336	MERCAPTANS, LIQUID, FLAMMABLE, N.O.S. or
			3336	MERCAPTANS MIXTURE, LIQUID, FLAMMABLE, N.O.S.
			1993	FLAMMABLE LIQUID, N.O.S.
	elevated temperature	**F2**	3256	ELEVATED TEMPERATURE LIQUID, FLAMMABLE, N.O.S., with flash-point above 60 °C, at or above its flash-point
		F3	9001	SUBSTANCES HAVING A FLASH-POINT ABOVE 60 °C carried or handed over for carriage at a TEMPERATURE WITHIN A RANGE OF 15 K BELOW THE FLASH-POINT
		F4	9002	SUBSTANCES WITH A SELF-IGNITION TEMPERATURE OF 200 °C AND BELOW, n.o.s.
Toxic FT		**FT1**	1228	MERCAPTANS, LIQUID, FLAMMABLE, TOXIC, N.O.S. or
			1228	MERCAPTAN MIXTURE, LIQUID, FLAMMABLE, TOXIC, N.O.S.
			1986	ALCOHOLS, FLAMMABLE, TOXIC, N.O.S.
			1988	ALDEHYDES, FLAMMABLE, TOXIC, N.O.S.
			2478	ISOCYANATES, FLAMMABLE, TOXIC, N.O.S. or
			2478	ISOCYANATE SOLUTION, FLAMMABLE, TOXIC, N.O.S.
			3248	MEDICINE, LIQUID, FLAMMABLE, TOXIC, N.O.S.
			3273	NITRILES, FLAMMABLE, TOXIC, N.O.S.
			1992	FLAMMABLE LIQUID, TOXIC, N.O.S.

(cont'd)

2.2.3.3 ***List of collective entries (cont'd)***

Subsidiary risk		Classification code	UN No.	Name of the substance or article
Toxic FT	**Pesticide (f.p<23 °C)**	**FT2**	2758	CARBAMATE PESTICIDE, LIQUID, FLAMMABLE, TOXIC
			2760	ARSENICAL PESTICIDE, LIQUID, FLAMMABLE, TOXIC
			2762	ORGANOCHLORINE PESTICIDE, LIQUID, FLAMMABLE, TOXIC
			2764	TRIAZINE PESTICIDE, LIQUID, FLAMMABLE, TOXIC
			2772	THIOCARBAMATE PESTICIDE, LIQUID, FLAMMABLE, TOXIC
			2776	COPPER BASED PESTICIDE, LIQUID, FLAMMABLE, TOXIC
			2778	MERCURY BASED PESTICIDE, LIQUID, FLAMMABLE, TOXIC
			2780	SUBSTITUTED NITROPHENOL PESTICIDE, LIQUID, FLAMMABLE, TOXIC
			2782	BIPYRIDILIUM PESTICIDE, LIQUID, FLAMMABLE, TOXIC
			2784	ORGANOPHOSPHORUS PESTICIDE, LIQUID, FLAMMABLE, TOXIC
			2787	ORGANOTIN PESTICIDE, LIQUID, FLAMMABLE, TOXIC
			3024	COUMARIN DERIVATIVE PESTICIDE, LIQUID, FLAMMABLE, TOXIC
			3346	PHENOXYACETIC ACID DERIVATIVE PESTICIDE, LIQUID, FLAMMABLE, TOXIC
			3350	PYRETHROID PESTICIDE, LIQUID, FLAMMABLE TOXIC
			3021	PESTICIDE, LIQUID, FLAMMABLE, TOXIC, N.O.S.
				***NOTE** : The classification of a pesticide under an entry shall be effected on the basis of the active ingredient, of the physical state of the pesticide and any subsidiary risks it may exhibit.*
Corrosive		**FC**	3469	PAINT, FLAMMABLE, CORROSIVE (including paint, lacquer, enamel, stain, shellac, varnish, polish, liquid filler and liquid lacquer base) or
			3469	PAINT RELATED MATERIAL, FLAMMABLE, CORROSIVE (including paint thinning or reducing compound)
			2733	AMINES, FLAMMABLE, CORROSIVE, N.O.S. or
			2733	POLYAMINES, FLAMMABLE, CORROSIVE, N.O.S.
			2985	CHLOROSILANES, FLAMMABLE, CORROSIVE, N.O.S.
			3274	ALCOHOLATES SOLUTION, N.O.S., in alcohol
			2924	FLAMMABLE LIQUID, CORROSIVE, N.O.S.
Toxic, corrosive		**FTC**	3286	FLAMMABLE LIQUID, TOXIC, CORROSIVE, N.O.S.
Liquid desensitised Explosive		**D**	3343	NITROGLYCERIN MIXTURE, DESENSITIZED, LIQUID, FLAMMABLE, N.O.S. with not more than 30% nitroglycerin by mass
			3357	NITROGLYCERIN MIXTURE, DESENSITIZED, LIQUID, N.O.S. with not more than 30% nitroglycerin by mass
			3379	DESENSITIZED EXPLOSIVE, LIQUID, N.O.S.

2.2.41 Class 4.1 Flammable solids, self-reactive substances and solid desensitized explosives

2.2.41.1 *Criteria*

2.2.41.1.1 The heading of Class 4.1 covers flammable substances and articles, desensitized explosives which are solids according to subparagraph (a) of the definition "solid" in 1.2.1 and self-reactive liquids or solids.

The following are assigned to Class 4.1:

- readily flammable solid substances and articles (see paragraphs 2.2.41.1.3 to 2.2.41.1.8);
- self-reactive solids or liquids (see paragraphs 2.2.41.1.9 to 2.2.41.1.17);
- solid desensitized explosives (see 2.2.41.1.18);
- substances related to self-reactive substances (see 2.2.41.1.19).

2.2.41.1.2 The substances and articles of Class 4.1 are subdivided as follows:

F Flammable solids, without subsidiary risk:

F1 Organic;
F2 Organic, molten;
F3 Inorganic;

FO Flammable solids, oxidizing;

FT Flammable solids, toxic:

FT1 Organic, toxic;
FT2 Inorganic, toxic;

FC Flammable solids, corrosive:

FC1 Organic, corrosive;
FC2 Inorganic, corrosive;

D Solid desensitized explosives without subsidiary risk;

DT Solid desensitized explosives, toxic;

SR Self-reactive substances:

SR1 Not requiring temperature control;
SR2 Requiring temperature control.

Flammable solids

Definition and properties

2.2.41.1.3 *Flammable solids* are readily combustible solids and solids which may cause fire through friction.

Readily combustible solids are powdered, granular, or pasty substances which are dangerous if they can be easily ignited by brief contact with an ignition source, such as a burning match, and if the flame spreads rapidly. The danger may come not only from the fire but also from toxic combustion products. Metal powders are especially dangerous because of the difficulty of extinguishing a fire since normal extinguishing agents such as carbon dioxide or water can increase the hazard.

Classification

2.2.41.1.4 Substances and articles classified as flammable solids of Class 4.1 are listed in Table A of Chapter 3.2. The assignment of organic substances and articles not mentioned by name in Table A of Chapter 3.2 to the relevant entry of sub-section 2.2.41.3 in accordance with the provisions of Chapter 2.1 can be based on experience or on the results of the test procedures in accordance with Part III, sub-section 33.2.1 of the Manual of Tests and Criteria. The assignment of inorganic substances not mentioned by name shall be based on the results of the test procedures in accordance with Part III, sub-section 33.2.1 of the Manual of Tests and Criteria; experience shall also be taken into account when it leads to a more stringent assignment.

2.2.41.1.5 When substances not mentioned by name are assigned to one of the entries listed in 2.2.41.3 on the basis of the test procedures in accordance with the Manual of Tests and Criteria, Part III, sub-section 33.2.1, the following criteria apply:

(a) With the exception of metal powders or powders of metal alloys, powdery, granular or pasty substances shall be classified as readily flammable substances of Class 4.1 if they can be easily ignited by brief contact with an ignition source (e.g. a burning match), or if, in the event of ignition, the flame spreads rapidly, the burning time is less than 45 seconds for a measured distance of 100 mm or the rate of burning is greater than 2.2 mm/s.

(b) Metal powders or powders of metal alloys shall be assigned to Class 4.1 if they can be ignited by a flame and the reaction spreads over the whole length of the sample in 10 minutes or less.

Solids which may cause fire through friction shall be classified in Class 4.1 by analogy with existing entries (e.g. matches) or in accordance with any appropriate special provision.

2.2.41.1.6 On the basis of the test procedure in accordance with the Manual of Tests and Criteria, Part III, Section 33.2.1 and the criteria set out in 2.2.41.1.4 and 2.2.41.1.5, it may also be determined whether the nature of a substance mentioned by name is such that the substance is not subject to the provisions for this Class.

2.2.41.1.7 If substances of Class 4.1, as a result of admixtures, come into different categories of risk from those to which the substances mentioned by name in Table A of Chapter 3.2 belong, these mixtures shall be assigned to the entries to which they belong on the basis of their actual degree of danger.

NOTE: *For the classification of solutions and mixtures (such as preparations and wastes), see also 2.1.3.*

Assignment of packing groups

2.2.41.1.8 Flammable solids classified under the various entries in Table A of Chapter 3.2 shall be assigned to packing groups II or III on the basis of test procedures of the Manual of Tests and Criteria, Part III, sub-section 33.2.1, in accordance with the following criteria:

(a) Readily flammable solids which, when tested, have a burning time of less than 45 seconds over a measured distance of 100 mm shall be assigned to:

Packing group II: if the flame passes the wetted zone;

Packing group III: if the wetted zone stops the flame for at least four minutes;

(b) Metal powders or powders of metal alloys shall be assigned to:

Packing group II: if, when tested, the reaction spreads over the whole length of the sample in five minutes or less;

Packing group III: if, when tested, the reaction spreads over the whole length of the sample in more than five minutes.

For solids which may cause fire through friction, the packing group shall be assigned by analogy with existing entries or in accordance with any special provision.

Self-reactive substances

Definitions

2.2.41.1.9 For the purposes of ADN, *self-reactive substances* are thermally unstable substances liable to undergo a strongly exothermic decomposition even without participation of oxygen (air). Substances are not considered to be self-reactive substances of Class 4.1, if:

(a) they are explosives according to the criteria of Class 1;

(b) they are oxidizing substances according to the classification procedure for Class 5.1 (see 2.2.51.1) except that mixtures of oxidizing substances which contain 5.0% or more of combustible organic substances shall be subjected to the classification procedure defined in Note 2;

(c) they are organic peroxides according to the criteria of Class 5.2 (see 2.2.52.1);

(d) their heat of decomposition is less than 300 J/g; or

(e) their self-accelerating decomposition temperature (SADT) (see NOTE 3 below) is greater than 75 °C for a 50 kg package.

NOTE 1: *The heat of decomposition can be determined using any internationally recognised method e.g. differential scanning calorimetry and adiabatic calorimetry.*

***NOTE 2**:* *Mixtures of oxidizing substances meeting the criteria of Class 5.1 which contain 5.0% or more of combustible organic substances, which do not meet the criteria mentioned in (a), (c), (d) or (e) above, shall be subjected to the self-reactive substance classification procedure.*

A mixture showing the properties of a self-reactive substance, type B to F, shall be classified as a self-reactive substance of Class 4.1.

A mixture showing the properties of a self-reactive substance, type G, according to the principle given in 20.4.3 (g) of Part II of the Manual of Tests and Criteria shall be considered for classification as a substance of Class 5.1 (see 2.2.51.1).

***NOTE 3**: The self-accelerating decomposition temperature (SADT) is the lowest temperature at which self-accelerating decomposition may occur with a substance in the packaging as used during carriage. Requirements for the determination of the SADT are given in the Manual of Tests and Criteria, Part II, Chapter 20 and section 28.4.*

***NOTE 4**: Any substance which shows the properties of a self-reactive substance shall be classified as such, even if this substance gives a positive test result according to 2.2.42.1.5 for inclusion in Class 4.2.*

Properties

2.2.41.1.10 The decomposition of self-reactive substances can be initiated by heat, contact with catalytic impurities (e.g. acids, heavy-metal compounds, bases), friction or impact. The rate of decomposition increases with temperature and varies with the substance. Decomposition, particularly if no ignition occurs, may result in the evolution of toxic gases or vapours. For certain self-reactive substances, the temperature shall be controlled. Some self-reactive substances may decompose explosively, particularly if confined. This characteristic may be modified by the addition of diluents or by the use of appropriate packagings. Certain self-reactive substances burn vigorously. Self-reactive substances are, for example, some compounds of the types listed below:

aliphatic azo compounds (-C-N=N-C-);
organic azides ($-C-N_3$);
diazonium salts ($-CN_2^+ Z^-$);
N-nitroso compounds (-N-N=O); and
aromatic sulphohydrazides ($-SO_2-NH-NH_2$).

This list is not exhaustive and substances with other reactive groups and some mixtures of substances may have similar properties.

Classification

2.2.41.1.11 Self-reactive substances are classified into seven types according to the degree of danger they present. The types of self-reactive substances range from type A, which is not accepted for carriage in the packaging in which it is tested, to type G, which is not subject to the provisions for self-reactive substances of Class 4.1. The classification of types B to F is directly related to the maximum quantity allowed in one packaging. The principles to be applied for classification as well as the applicable classification procedures, test methods and criteria and an example of a suitable test report are given in Part II of the Manual of Tests and Criteria.

2.2.41.1.12 Self-reactive substances which have already been classified and are already permitted for carriage in packagings are listed in 2.2.41.4, those already permitted for carriage in IBCs are listed in 4.1.4.2 of ADR, packing instruction IBC520 and those already permitted for carriage in tanks according to Chapter 4.2 of ADR are listed in 4.2.5.2 of ADR, portable tank instruction T23. Each permitted substance listed is assigned to a generic entry of Table A of Chapter 3.2 (UN Nos. 3221 to 3240), and appropriate subsidiary risks and remarks providing relevant transport information are given.

The collective entries specify:

– self-reactive substances types B to F, see 2.2.41.1.11 above;

– physical state (liquid/solid); and

– temperature control (when required), see 2.2.41.1.17 below.

The classification of the self-reactive substances listed in 2.2.41.4 is based on the technically pure substance (except where a concentration of less than 100% is specified).

2.2.41.1.13 Classification of self-reactive substances not listed in 2.2.41.4, 4.1.4.2 of ADR, packing instruction IBC520 or 4.2.5.2 of ADR, portable tank instruction T23 and assignment to a collective entry shall be made by the competent authority of the country of origin on the basis of a test report. The statement of approval shall contain the classification and the relevant conditions of carriage. If the country of origin is not a Contracting Party to ADN, the classification and the conditions of carriage shall be recognized by the competent authority of the first country Contracting Party to ADN reached by the consignment.

2.2.41.1.14 Activators, such as zinc compounds, may be added to some self-reactive substances to change their reactivity. Depending on both the type and the concentration of the activator, this may result in a decrease in thermal stability and a change in explosive properties. If either of these properties is altered, the new formulation shall be assessed in accordance with the classification procedure.

2.2.41.1.15 Samples of self-reactive substances or formulations of self-reactive substances not listed in 2.2.41.4, for which a complete set of test results is not available and which are to be carried for further testing or evaluation, shall be assigned to one of the appropriate entries for self-reactive substances type C provided the following conditions are met:

– the available data indicate that the sample would be no more dangerous than self-reactive substances type B;

– the sample is packaged in accordance with packing method OP2 of 4.1.4.1 of ADR and the quantity per cargo transport unit and per transport unit is limited to 10 kg;

– the available data indicate that the control temperature, if any, is sufficiently low to prevent any dangerous decomposition and sufficiently high to prevent any dangerous phase separation.

Desensitization

2.2.41.1.16 In order to ensure safety during carriage, self-reactive substances are in many cases desensitized by use of a diluent. Where a percentage of a substance is stipulated, this refers to the percentage by mass, rounded to the nearest whole number. If a diluent is used, the self-reactive substance shall be tested with the diluent present in the concentration and form used in carriage. Diluents which may allow a self-reactive substance to concentrate to a dangerous extent in the event of leakage from a packaging shall not be used. Any diluent shall be compatible with the self-reactive substance. In this regard, compatible diluents are those solids or liquids which have no detrimental influence on the thermal stability and hazard type of the self-reactive substance. Liquid diluents in formulations requiring temperature control (see 2.2.41.1.14) shall have a boiling point of at least 60 °C and a flash-point not less than 5 °C. The boiling point of the liquid shall be at least 50 °C higher than the control temperature of the self-reactive substance.

Temperature control requirements

2.2.41.1.17 Certain self-reactive substances may only be carried under temperature controlled conditions. The control temperature is the maximum temperature at which the self-reactive substance can be safely carried. It is assumed that the temperature of the immediate surroundings of a package only exceeds 55 °C during carriage for a relatively short time in a 24 hour period. In the event of loss of temperature control, it may be necessary to implement emergency procedures. The emergency temperature is the temperature at which such procedures shall be implemented.

The control and emergency temperatures are derived from the SADT (see table 1). The SADT shall be determined in order to decide whether a substance shall be subjected to temperature control during carriage. Provisions for the determination of the SADT are given in the Manual of Tests and Criteria, Part II, Chapter 20 and Section 28.4.

Table 1 Derivation of control and emergency temperatures

Type of receptacle	SADT [a]	Control temperature	Emergency temperature
Single packagings and IBCs	20 °C or less	20 °C below SADT	10 °C below SADT
	over 20 °C to 35 °C	15 °C below SADT	10 °C below SADT
	over 35 °C	10 °C below SADT	5 °C below SADT
Tanks	not greater than 50 °C	10 °C below SADT	5 °C below SADT

[a] *SADT of the substance as packaged for carriage.*

Self-reactive substances with an SADT not greater than 55 °C shall be subject to temperature control during carriage. Where applicable, control and emergency temperatures are listed in 2.2.41.4. The actual temperature during carriage may be lower than the control temperature but shall be selected so as to avoid dangerous separation of phases.

Solid desensitized explosives

2.2.41.1.18 Solid desensitized explosives are substances which are wetted with water or alcohols or are diluted with other substances to suppress their explosive properties. Such entries in Table A of Chapter 3.2 are: UN Nos. 1310, 1320, 1321, 1322, 1336, 1337, 1344, 1347, 1348, 1349, 1354, 1355, 1356, 1357, 1517, 1571, 2555, 2556, 2557, 2852, 2907, 3317, 3319, 3344, 3364, 3365, 3366, 3367, 3368, 3369, 3370, 3376, 3380 and 3474.

Substances related to self-reactive substances

2.2.41.1.19 Substances that:

(a) have been provisionally accepted into Class 1 according to Test Series 1 and 2 but exempted from Class 1 by Test Series 6;

(b) are not self-reactive substances of Class 4.1; and

(c) are not substances of Classes 5.1 or 5.2

are also assigned to Class 4.1. UN Nos. 2956, 3241, 3242 and 3251 are such entries.

2.2.41.2 ***Substances not accepted for carriage***

2.2.41.2.1 The chemically unstable substances of Class 4.1 shall not be accepted for carriage unless the necessary steps have been taken to prevent their dangerous decomposition or polymerization during carriage. To this end, it shall in particular be ensured that receptacles and tanks do not contain any substance liable to promote these reactions.

2.2.41.2.2 Flammable solids, oxidizing, assigned to UN No. 3097 shall not be accepted for carriage unless they meet the requirements for Class 1 (see also 2.1.3.7).

2.2.41.2.3 The following substances shall not be accepted for carriage:

- Self-reactive substances of type A (see Manual of Tests and Criteria, Part II, paragraph 20.4.2 (a));
- Phosphorus sulphides which are not free from yellow and white phosphorus;
- Solid densitized explosives other than those listed in Table A of Chapter 3.2;
- Inorganic flammable substances in the molten form other than UN No. 2448 SULPHUR, MOLTEN;
- Barium azide with a water content less than 50% (mass).

2.2.41.3 *List of collective entries*

Classification	Subdivision	Further subdivision	Code	UN No.	Name and description
Flammable solids	**without subsidiary risk**	**organic**	**F1**	3175	SOLIDS CONTAINING FLAMMABLE LIQUID, N.O.S.
				1353	FIBRES IMPREGNATED WITH WEAKLY NITRATED NITROCELLULOSE, N.O.S. or
				1353	FABRICS IMPREGNATED WITH WEAKLY NITRATED NITROCELLULOSE, N.O.S.
				1325	FLAMMABLE SOLID, ORGANIC, N.O.S.
		organic molten	**F2**	3176	FLAMMABLE SOLID, ORGANIC, MOLTEN, N.O.S.
		inorganic	**F3**	3089	METAL POWDER, FLAMMABLE, N.O.S. [a] [b]
				3181	METAL SALTS OF ORGANIC COMPOUNDS, FLAMMABLE, N.O.S.
				3182	METAL HYDRIDES, FLAMMABLE,N.O.S. [c]
				3178	FLAMMABLE SOLID, INORGANIC, N.O.S.
	oxidizing		**FO**	3097	FLAMMABLE SOLID, OXIDIZING, N.O.S. (not allowed, see Para. 2.2.41.2.2)
	toxic FT	**organic**	**FT1**	2926	FLAMMABLE SOLID, TOXIC, ORGANIC, N.O.S.
		inorganic	**FT2**	3179	FLAMMABLE SOLID, TOXIC, INORGANIC, N.O.S.
	corrosive FC	**organic**	**FC1**	2925	FLAMMABLE SOLID, CORROSIVE, ORGANIC, N.O.S.
		inorganic	**FC2**	3180	FLAMMABLE SOLID, CORROSIVE, INORGANIC, N.O.S.
Solid desen-sitized explosives	**without subsidiary risk**		**D**	3319	NITROGLYCERIN MIXTURE, DESENSITIZED, SOLID, N.O.S. with more than 2% but not more than 10% nitroglycerin by mass
				3344	PENTAERYTHRITE TETRANITRATE (PENTAERYTHRITOL TETRANITRATE, PETN) MIXTURE, DESENSITIZED, SOLID, N.O.S. with more than 10% but not more than 20% PETN by mass
				3380	DESENSITIZED EXPLOSIVE, SOLID, N.O.S.
	toxic		**DT**		Only substances listed in Table A of Chapter 3.2 are to be accepted for carriage as substances of Class 4.1
Self-reactive substances SR	**not requiring temperature control**		**SR1**		SELF-REACTIVE LIQUID TYPE A } Not accepted for carriage, see 2.2.41.2.3
					SELF-REACTIVE SOLID TYPE A } Not accepted for carriage, see 2.2.41.2.3
				3221	SELF-REACTIVE LIQUID TYPE B
				3222	SELF-REACTIVE SOLID TYPE B
				3223	SELF-REACTIVE LIQUID TYPE C
				3224	SELF-REACTIVE SOLID TYPE C
				3225	SELF-REACTIVE LIQUID TYPE D
				3226	SELF-REACTIVE SOLID TYPE D
				3227	SELF-REACTIVE LIQUID TYPE E
				3228	SELF-REACTIVE SOLID TYPE E
				3229	SELF-REACTIVE LIQUID TYPE F
				3230	SELF-REACTIVE SOLID TYPE F
					SELF-REACTIVE LIQUID TYPE G } Not subject to the provisions applicable to Class 4.1, see 2.2.41.1.11
					SELF-REACTIVE SOLID TYPE G } Not subject to the provisions applicable to Class 4.1, see 2.2.41.1.11
	requiring temperature control		**SR2**	3231	SELF-REACTIVE LIQUID TYPE B, TEMPERATURE CONTROLLED
				3232	SELF-REACTIVE SOLID TYPE B, TEMPERATURE CONTROLLED
				3233	SELF-REACTIVE LIQUID TYPE C, TEMPERATURE CONTROLLED
				3234	SELF-REACTIVE SOLID TYPE C, TEMPERATURE CONTROLLED
				3235	SELF-REACTIVE LIQUID TYPE D, TEMPERATURE CONTROLLED
				3236	SELF-REACTIVE SOLID TYPE D, TEMPERATURE CONTROLLED
				3237	SELF-REACTIVE LIQUID TYPE E, TEMPERATURE CONTROLLED
				3238	SELF-REACTIVE SOLID TYPE E, TEMPERATURE CONTROLLED
				3239	SELF-REACTIVE LIQUID TYPE F, TEMPERATURE CONTROLLED
				3240	SELF-REACTIVE SOLID TYPE F, TEMPERATURE CONTROLLED

---

[a] *Metals and metal alloys in powdered or other flammable form, liable to spontaneous combustion, are substances of Class 4.2.*

[b] *Metals and metal alloys in powdered or other flammable form, which in contact with water, emit flammable gases, are substances of Class 4.3.*

[c] *Metals hydrides which, in contact with water, emit flammable gases, are substances of Class 4.3. Aluminium borohydride or aluminium borohydride in devices are substances of Class 4.2, UN No. 2870.*

2.2.41.4 List of currently assigned self-reactive substances in packagings

In the column "Packing Method" codes "OP1" to "OP8" refer to packing methods in 4.1.4.1 of ADR, packing instruction P520 (see also 4.1.7.1 of ADR). Self-reactive substances to be carried shall fulfil the classification and the control and emergency temperatures (derived from the SADT) as listed. For substances permitted in IBCs, see 4.1.4.2 of ADR, packing instruction IBC520 and, for those permitted in tanks according Chapter 4.2 of ADR, see 4.2.5.2 of ADR, portable tank instruction T23.

***NOTE**: The classification given in this table is based on the technically pure substance (except where a concentration of less than 100 % is specified). For other concentrations, the substance may be classified differently following the procedures given in Part II of the Manual of Tests and Criteria and in 2.2.41.1.17.*

SELF-REACTIVE SUBSTANCE	Concentration (%)	Packing method	Control temperature (°C)	Emergency temperature (°C)	UN generic entry	Remarks
ACETONE-PYROGALLOL COPOLYMER 2-DIAZO-1-NAPHTHOL-5-SULPHONATE	100	OP8			3228	
AZODICARBONAMIDE FORMULATION TYPE B, TEMPERATURE CONTROLLED	< 100	OP5			3232	(1) (2)
AZODICARBONAMIDE FORMULATION TYPE C	< 100	OP6			3224	(3)
AZODICARBONAMIDE FORMULATION TYPE C, TEMPERATURE CONTROLLED	< 100	OP6			3234	(4)
AZODICARBONAMIDE FORMULATION TYPE D	< 100	OP7			3226	(5)
AZODICARBONAMIDE FORMULATION TYPE D, TEMPERATURE CONTROLLED	< 100	OP7			3236	(6)
2,2' -AZODI(2,4-DIMETHYL-4-METHOXYVALERONITRILE)	100	OP7	-5	+5	3236	
2,2' -AZODI(2,4-DIMETHYL-VALERONITRILE)	100	OP7	+10	+15	3236	
2,2' -AZODI(ETHYL-2-METHYLPROPIONATE)	100	OP7	+20	+25	3235	
1,1-AZODI(HEXAHYDROBENZONITRILE)	100	OP7			3226	
2,2' -AZODI(ISOBUTYRONITRILE	100	OP6	+40	+45	3234	
2,2' -AZODI(ISOBUTYRONITRILE) as a water based paste	≤ 50	OP6			3224	
2,2' -AZODI(2-METHYLBUTYRO-NITRILE)	100	OP7	+35	+40	3236	
BENZENE-1,3-DISULPHONYL HYDRAZIDE, as a paste	52	OP7			3226	
BENZENE SULPHONYL HYDRAZIDE	100	OP7			3226	
4-(BENZYL(ETHYL)AMINO)-3-ETHOXY-BENZENEDIAZONIUM ZINC CHLORIDE	100	OP7			3226	
4-(BENZYL(METHYL)AMINO)-3-ETHOXYBENZENEDIAZONIUM ZINC CHLORIDE	100	OP7	+40	+45	3236	
3-CHLORO-4-DIETHYLAMINOBENZENE-DIAZONIUM ZINC CHLORIDE	100	OP7			3226	

2.2.41.4 *List of currently assigned self-reactive substances in packagings (cont'd)*

SELF-REACTIVE SUBSTANCE	Concentration (%)	Packing method	Control temperature (°C)	Emergency temperature (°C)	UN generic entry	Remarks
2-DIAZO-1-NAPHTHOL-4-SULPHONYL CHLORIDE	100	OP5			3222	(2)
2-DIAZO-1-NAPHTHOL-5-SULPHONYL CHLORIDE	100	OP5			3222	(2)
2-DIAZO-1-NAPHTHOL SULPHONIC ACID ESTER MIXTURE, TYPE D	< 100	OP7			3226	(9)
2,5-DIBUTOXY-4-(4-MORPHOLINYL)-BENZENEDIAZONIUM, TETRACHLOROZINCATE (2:1)	100	OP8			3228	
2,5-DIETHOXY-4-MORPHOLINO-BENZENEDIAZONIUM ZINC CHLORIDE	67-100	OP7	+35	+40	3236	
2,5-DIETHOXY-4-MORPHOLINO-BENZENEDIAZONIUM ZINC CHLORIDE	66	OP7	+40	+45	3236	
2,5-DIETHOXY-4-MORPHOLINO-BENZENEDIAZONIUM TETRAFLUOROBORATE	100	OP7	+30	+35	3236	
2,5-DIETHOXY-4-(4-MORPHOLINYL)-BENZENEDIAZONIUM SULPHATE	100	OP7			3226	
2,5-DIETHOXY-4-(PHENYLSULPHONYL)-BENZENEDIAZONIUM ZINC CHLORIDE	67	OP7	+40	+45	3236	
DIETHYLENEGLYCOL BIS (ALLYL CARBONATE) + DI-ISOPROPYLPEROXYDICARBONATE	≥ 88 + ≤ 12	OP8	-10	0	3237	
2,5-DIMETHOXY-4-(4-METHYL-PHENYLSULPHONYL)BENZENE-DIAZONIUM ZINC CHLORIDE	79	OP7	+40	+45	3236	
4-(DIMETHYLAMINO)-BENZENEDIAZONIUM TRICHLOROZINCATE (-1)	100	OP8			3228	
4-DIMETHYLAMINO-6-(2-DIMETHYL-AMINOETHOXY) TOLUENE-2-DIAZONIUM ZINC CHLORIDE	100	OP7	+40	+45	3236	
N,N'-DINITROSO-N,N'- DIMETHYL TEREPHTHALAMIDE, as a paste	72	OP6			3224	
N,N'-DINITROSOPENTAMETHYLENE-TETRAMINE	82	OP6			3224	(7)
DIPHENYLOXIDE-4,4'-DISULPHONYL HYDRAZIDE	100	OP7			3226	
4-DIPROPYLAMINOBENZENE-DIAZONIUM ZINC CHLORIDE	100	OP7			3226	
2-(N,N-ETHOXYCARBONYL-PHENYLAMINO)-3-METHOXY-4-(N-METHYL-N-CYCLOHEXYLAMINO) BENZENEDIAZONIUM ZINC CHLORIDE	63-92	OP7	+ 40	+ 45	3236	
2-(N,N-ETHOXYCARBONYL-PHENYLAMINO)-3-METHOXY-4-(N-METHYL-N-CYCLOHEXYLAMINO) BENZENEDIAZONIUM ZINC CHLORIDE	62	OP7	+ 35	+ 40	3236	

2.2.41.4 ***List of currently assigned self-reactive substances in packagings (cont'd)***

SELF-REACTIVE SUBSTANCE	Concentration (%)	Packing method	Control temperature (°C)	Emergency temperature (°C)	UN generic entry	Remarks
N-FORMYL-2-(NITROMETHYLENE) -1,3-PERHYDROTHIAZINE	100	OP7	+45	+50	3236	
2-(2-HYDROXYETHOXY)-1-(PYRROLIDIN-1-YL)BENZENE-4-DIAZONIUM ZINC CHLORIDE	100	OP7	+ 45	+ 50	3236	
3-(2-HYDROXYETHOXY)-4-(PYRROLIDIN-1-YL)BENZENE DIAZONIUM ZINC CHLORIDE	100	OP7	+40	+45	3236	
2-(N,N-METHYLAMINOETHYL-CARBONYL)-4-(3,4-DIMETHYL-PHENYLSULPHONYL)BENZENE-DIAZONIUM HYDROGEN SULPHATE	96	OP7	+45	+50	3236	
4-METHYLBENZENESULPHONYL-HYDRAZIDE	100	OP7			3226	
3-METHYL-4-(PYRROLIDIN-1-YL) BENZENEDIAZONIUM TETRAFLUOROBORATE	95	OP6	+45	+50	3234	
4-NITROSOPHENOL	100	OP7	+35	+40	3236	
SELF-REACTIVE LIQUID, SAMPLE		OP2			3223	(8)
SELF-REACTIVE LIQUID, SAMPLE, TEMPERATURE CONTROLLED		OP2			3233	(8)
SELF-REACTIVE SOLID, SAMPLE		OP2			3224	(8)
SELF-REACTIVE SOLID, SAMPLE, TEMPERATURE CONTROLLED		OP2			3234	(8)
SODIUM 2-DIAZO-1-NAPHTHOL-4-SULPHONATE	100	OP7			3226	
SODIUM 2-DIAZO-1-NAPHTHOL-5-SULPHONATE	100	OP7			3226	
TETRAMINE PALLADIUM (II) NITRATE	100	OP6	+30	+35	3234	

Remarks

(1) Azodicarbonamide formulations which fulfil the criteria of paragraph 20.4.2 (b) of the Manual of Tests and Criteria. The control and emergency temperatures shall be determined by the procedure given in 2.2.41.1.17.

(2) "EXPLOSIVE" subsidiary risk label required (Model No. 1, see 5.2.2.2.2).

(3) Azodicarbonamide formulations which fulfil the criteria of paragraph 20.4.2 (c) of the Manual of Tests and Criteria.

(4) Azodicarbonamide formulations which fulfil the criteria of paragraph 20.4.2 (c) of the Manual of Tests and Criteria. The control and emergency temperatures shall be determined by the procedure given in 2.2.41.1.17.

(5) Azodicarbonamide formulations which fulfil the criteria of paragraph 20.4.2 (d) of the Manual of Tests and Criteria.

(6) Azodicarbonamide formulations which fulfil the criteria of paragraph 20.4.2 (d) of the Manual of Tests and Criteria. The control and emergency temperatures shall be determined by the procedure given in 2.2.41.1.17.

(7) With a compatible diluent having a boiling point of not less than 150 °C.

(8) See 2.2.41.1.15.

(9) This entry applies to mixtures of esters of 2-diazo-1-naphthol-4-sulphonic acid and 2-diazo-1-naphthol-5-sulphonic acid which fulfil the criteria of paragraph 20.4.2 (d) of the *Manual of Test and Criteria*.

2.2.42 Class 4.2 Substances liable to spontaneous combustion

2.2.42.1 *Criteria*

2.2.42.1.1 The heading of Class 4.2 covers:

- *Pyrophoric substances* which are substances, including mixtures and solutions (liquid or solid), which even in small quantities ignite on contact with air within five minutes. These are the Class 4.2 substances the most liable to spontaneous combustion; and

- *Self-heating substances and articles* which are substances and articles, including mixtures and solutions, which, on contact with air, without energy supply, are liable to self-heating. These substances will ignite only in large amounts (kilogrammes) and after long periods of time (hours or days).

2.2.42.1.2 The substances and articles of Class 4.2 are subdivided as follows:

S Substances liable to spontaneous combustion, without subsidiary risk:

S1 Organic, liquid;
S2 Organic, solid;
S3 Inorganic, liquid;
S4 Inorganic, solid;
S5 Organometallic;

SW Substances liable to spontaneous combustion, which, in contact with water, emit flammable gases;

SO Substances liable to spontaneous combustion, oxidizing;

ST Substances liable to spontaneous combustion, toxic:

ST1 Organic, toxic, liquid;
ST2 Organic, toxic, solid;
ST3 Inorganic, toxic, liquid;
ST4 Inorganic, toxic, solid;

SC Substances liable to spontaneous combustion, corrosive:

SC1 Organic, corrosive, liquid;
SC2 Organic, corrosive, solid;
SC3 Inorganic, corrosive, liquid;
SC4 Inorganic, corrosive, solid.

Properties

2.2.42.1.3 Self-heating of these substances, leading to spontaneous combustion, is caused by reaction of the substance with oxygen (in the air) and the heat developed not being conducted away rapidly enough to the surroundings. Spontaneous combustion occurs when the rate of heat production exceeds the rate of heat loss and the auto-ignition temperature is reached.

Classification

2.2.42.1.4 Substances and articles classified in Class 4.2 are listed in Table A of Chapter 3.2. The assignment of substances and articles not mentioned by name in Table A of Chapter 3.2 to the relevant specific N.O.S. entry of 2.2.42.3 in accordance with the provisions of Chapter 2.1 can be based on experience or the results of the test procedures in accordance with the Manual of Tests and Criteria, Part III, Section 33.3. Assignment to general N.O.S. entries of Class 4.2 shall be based on the results of the test procedures in accordance with the Manual of Tests and Criteria, Part III, section 33.3; experience shall also be taken into account when it leads to a more stringent assignment.

2.2.42.1.5 When substances or articles not mentioned by name are assigned to one of the entries listed in 2.2.42.3 on the basis of the test procedures in accordance with the Manual of Tests and Criteria, Part III, section 33.3, the following criteria shall apply:

(a) Solids liable to spontaneous combustion (pyrophoric) shall be assigned to Class 4.2 when they ignite on falling from a height of 1 m or within five minutes;

(b) Liquids liable to spontaneous combustion (pyrophoric) shall be assigned to Class 4.2 when:

(i) on being poured on an inert carrier, they ignite within five minutes, or

(ii) in the event of a negative result of the test according to (i), when poured on a dry, indented filter paper (Whatman No. 3 filter), they ignite or carbonize it within five minutes;

(c) Substances in which, in a 10 cm sample cube, at 140 °C test temperature, spontaneous combustion or a rise in temperature to over 200 °C is observed within 24 hours shall be assigned to Class 4.2. This criterion is based on the temperature of the spontaneous combustion of charcoal, which is at 50 °C for a sample cube of 27 m^3. Substances with a temperature of spontaneous combustion higher than 50 °C for a volume of 27 m^3 are not to be assigned to Class 4.2.

***NOTE 1**: Substances carried in packages with a volume of not more than 3 m^3 are exempted from Class 4.2 if, tested with a 10 cm sample cube at 120 °C, no spontaneous combustion nor a rise in temperature to over 180 °C is observed within 24 hours.*

***NOTE 2:** Substances carried in packages with a volume of not more than 450 litres are exempted from Class 4.2 if, tested with a 10 cm sample cube at 100 °C, no spontaneous combustion nor a rise in temperature to over 160 °C is observed within 24 hours.*

***NOTE 3:** Since organometallic substances can be classified in Class 4.2 or 4.3 with additional subsidiary risks, depending on their properties, a specific classification flow chart for these substances is given in 2.3.5.*

2.2.42.1.6 If substances of Class 4.2, as a result of admixtures, come into different categories of risk from those to which the substances mentioned by name in Table A of Chapter 3.2 belong, these mixtures shall be assigned to the entries to which they belong on the basis of their actual degree of danger.

***NOTE**: For the classification of solutions and mixtures (such as preparations and wastes), see also 2.1.3.*

2.2.42.1.7 On the basis of the test procedure in the Manual of Tests and Criteria, Part III, section 33.3 and the criteria set out in 2.2.42.1.5, it may also be determined whether the nature of a substance mentioned by name is such that the substance is not subject to the provisions for this Class.

Assignment of packing groups

2.2.42.1.8 Substances and articles classified under the various entries in Table A of Chapter 3.2 shall be assigned to packing groups I, II or III on the basis of test procedures of the Manual of Tests and Criteria, Part III, section 33.3, in accordance with the following criteria:

(a) Substances liable to spontaneous combustion (pyrophoric) shall be assigned to packing group I;

(b) Self-heating substances and articles in which, in a 2.5 cm sample cube, at 140 °C test temperature, spontaneous combustion or a rise in temperature to over 200 °C is observed within 24 hours, shall be assigned to packing group II;
Substances with a temperature of spontaneous combustion higher than 50 °C for a volume of 450 litres are not to be assigned to packing group II;

(c) Slightly self-heating substances in which, in a 2.5 cm sample cube, the phenomena referred to under (b) are not observed, in the given conditions, but in which in a 10 cm sample cube at 140 °C test temperature spontaneous combustion or a rise in temperature to over 200 °C is observed within 24 hours, shall be assigned to packing group III.

2.2.42.2 ***Substances not accepted for carriage***

The following substances shall not be accepted for carriage:

– UN No. 3255 tert-BUTYL HYPOCHLORITE; and

– Self-heating solids, oxidizing, assigned to UN No. 3127 unless they meet the requirements for Class 1 (see 2.1.3.7).

2.2.42.3 *List of collective entries*

Substances liable to spontaneous combustion			Classification code	UN No.	Name of the substance or article
Without subsidiary risk S	organic	liquid	S1	2845	PYROPHORIC LIQUID, ORGANIC, N.O.S.
				3183	SELF-HEATING LIQUID, ORGANIC, N.O.S.
		solid	S2	1373	FIBRES or FABRICS, ANIMAL or VEGETABLE or SYNTHETIC, N.O.S. with oil
				2006	PLASTICS, NITROCELLULOSE-BASED, SELF-HEATING, N.O.S.
				3313	ORGANIC PIGMENTS, SELF HEATING
				2846	PYROPHORIC SOLID, ORGANIC, N.O.S.
				3088	SELF-HEATING SOLID, ORGANIC, N.O.S.
	inorganic	liquid	S3	3194	PYROPHORIC LIQUID, INORGANIC, N.O.S.
				3186	SELF-HEATING LIQUID, INORGANIC, N.O.S.
		solid	S4	1383	PYROPHORIC METAL, N.O.S. or
				1383	PYROPHORIC ALLOY, N.O.S.
				1378	METAL CATALYST, WETTED with a visible excess of liquid
				2881	METAL CATALYST, DRY
				3189[a]	METAL POWDER, SELF-HEATING, N.O.S.
				3205	ALKALINE EARTH METAL ALCOHOLATES, N.O.S.
				3200	PYROPHORIC SOLID, INORGANIC, N.O.S.
				3190	SELF-HEATING SOLID, INORGANIC, N.O.S.
	organometallic		S5	3392	ORGANOMETALLIC SUBSTANCE, LIQUID, PYROPHORIC
				3391	ORGANOMETALLIC SUBSTANCE, SOLID, PYROPHORIC
				3400	ORGANOMETALLIC SUBSTANCE, SOLID, SELF-HEATING
Water-reactive			SW	3394	ORGANOMETALLIC SUBSTANCE, LIQUID, PYROPHORIC, WATER-REACTIVE
				3393	ORGANOMETALLIC SUBSTANCE, SOLID, PYROPHORIC, WATER-REACTIVE
Oxidizing			SO	3127	SELF-HEATING SOLID, OXIDIZING, N.O.S. (not allowed, see 2.2.42.2)
Toxic ST	organic	liquid	ST1	3184	SELF-HEATING LIQUID, TOXIC, ORGANIC, N.O.S.
		solid	ST2	3128	SELF-HEATING SOLID, TOXIC, ORGANIC, N.O.S.
	inorganic	liquid	ST3	3187	SELF-HEATING LIQUID, TOXIC, INORGANIC, N.O.S.
		solid	ST4	3191	SELF-HEATING SOLID, TOXIC, INORGANIC, N.O.S.
Corrosive SC	organic	liquid	SC1	3185	SELF-HEATING LIQUID, CORROSIVE, ORGANIC, N.O.S.
		solid	SC2	3126	SELF-HEATING SOLID, CORROSIVE, ORGANIC, N.O.S.
	inorganic	liquid	SC3	3188	SELF-HEATING LIQUID, CORROSIVE, INORGANIC, N.O.S.
		solid	SC4	3206	ALKALI METAL ALCOHOLATES, SELF-HEATING, CORROSIVE, N.O.S.
				3192	SELF-HEATING SOLID, CORROSIVE, INORGANIC, N.O.S.

[a] *Dust and powder of metals, non toxic in a non-spontaneous combustible form which nevertheless, in contact with water, emit flammable gases, are substances of Class 4.3.*

2.2.43 Class 4.3 Substances which, in contact with water, emit flammable gases

2.2.43.1 *Criteria*

2.2.43.1.1 The heading of Class 4.3 covers substances which react with water to emit flammable gases liable to form explosive mixtures with air, and articles containing such substances.

2.2.43.1.2 Substances and articles of Class 4.3 are subdivided as follows:

W Substances which, in contact with water, emit flammable gases, without subsidiary risk, and articles containing such substances:

W1 Liquid;
W2 Solid;
W3 Articles;

WF1 Substances which, in contact with water, emit flammable gases, liquid, flammable;

WF2 Substances which, in contact with water, emit flammable gases, solid, flammable;

WS Substances which, in contact with water, emit flammable gases, solid, self-heating;

WO Substances which, in contact with water, emit flammable gases, oxidizing, solid;

WT Substances which, in contact with water, emit flammable gases, toxic:

WT1 Liquid;
WT2 Solid;

WC Substances which, in contact with water, emit flammable gases, corrosive:

WC1 Liquid;
WC2 Solid;

WFC Substances which, in contact with water, emit flammable gases, flammable, corrosive.

Properties

2.2.43.1.3 Certain substances in contact with water may emit flammable gases that can form explosive mixtures with air. Such mixtures are easily ignited by all ordinary sources of ignition, for example naked lights, sparking handtools or unprotected light bulbs. The resulting blast wave and flames may endanger people and the environment. The test method referred to in 2.2.43.1.4 below is used to determine whether the reaction of a substance with water leads to the development of a dangerous amount of gases which may be flammable. This test method shall not be applied to pyrophoric substances.

Classification

2.2.43.1.4 Substances and articles classified in Class 4.3 are listed in Table A of Chapter 3.2. The assignment of substances and articles not mentioned by name in Table A of Chapter 3.2 to the relevant entry of 2.2.43.3 in accordance with the provisions of Chapter 2.1 shall be based on the results of the test procedure in accordance with the Manual of Tests and Criteria, Part III, Section 33.4; experience shall also be taken into account when it leads to a more stringent assignment.

2.2.43.1.5 When substances not mentioned by name are assigned to one of the entries listed in 2.2.43.3 on the basis of the test procedure in accordance with the Manual of Tests and Criteria, Part III, Section 33.4, the following criteria shall apply:

A substance shall be assigned to Class 4.3 if:

(a) spontaneous ignition of the gas emitted takes place in any step of the test procedure; or

(b) there is an evolution of flammable gas at a rate greater than 1 litre per kilogram of the substance to be tested per hour.

NOTE: *Since organometallic substances can be classified in Classes 4.2 or 4.3 with additional subsidiary risks, depending on their properties, a specific classification flow chart for these substances is given in 2.3.5.*

2.2.43.1.6 If substances of Class 4.3, as a result of admixtures, come into different categories of risk from those to which the substances mentioned by name in Table A of Chapter 3.2 belong, these mixtures shall be assigned to the entries to which they belong on the basis of their actual degree of danger.

NOTE: *For the classification of solutions and mixtures (such as preparations and wastes) see also 2.1.3.*

2.2.43.1.7 On the basis of the test procedures in accordance with the Manual of Tests and Criteria, Part III, Section 33.4, and the criteria set out in paragraph 2.2.43.1.5, it may also be determined whether the nature of a substance mentioned by name is such that the substance is not subject to the provisions for this Class.

Assignment of packing groups

2.2.43.1.8 Substances and articles classified under the various entries in Table A of Chapter 3.2 shall be assigned to packing groups I, II or III on the basis of test procedures of the Manual of Tests and Criteria, Part III, section 33.4, in accordance with the following criteria:

(a) Packing group I shall be assigned to any substance which reacts vigorously with water at ambient temperature and generally demonstrates a tendency for the gas produced to ignite spontaneously, or one which reacts readily with water at ambient temperatures such that the rate of evolution of flammable gas is equal to or greater than 10 litres per kilogram of substance over any one minute period;

(b) Packing group II shall be assigned to any substance which reacts readily with water at ambient temperature such that the maximum rate of evolution of flammable gas is equal to or greater than 20 litres per kilogram of substance per hour, and which does not meet the criteria of packing group I;

(c) Packing group III shall be assigned to any substance which reacts slowly with water at ambient temperature such that the maximum rate of evolution of flammable gas is greater than 1 litre per kilogram of substance per hour, and which does not meet the criteria of packing groups I or II.

2.2.43.2 ***Substances not accepted for carriage***

Water-reactive solids, oxidizing, assigned to UN No. 3133, shall not be accepted for carriage unless they meet the requirements for Class 1 (see also 2.1.3.7).

2.2.43.3 *List of collective entries*

Substances which, in contact with water, emit flammable gases

Subsidiary risk	Physical state	Classification code	UN No.	Name of substance or article
Without subsidiary risk W	**liquid**	**W1**	1389	ALKALI METAL AMALGAM, LIQUID
			1391	ALKALI METAL DISPERSION having a flash-point above 60 °C or
			1391	ALKALINE EARTH METAL DISPERSION having a flash-point above 60 °C
			1392	ALKALINE EARTH METAL AMALGAM, LIQUID
			1420	POTASSIUM METAL ALLOYS, LIQUID
			1421	ALKALI METAL ALLOY, LIQUID, N.O.S.
			1422	POTASSIUM SODIUM ALLOYS, LIQUID
			3398	ORGANOMETALLIC SUBSTANCE, LIQUID, WATER-REACTIVE
			3148	WATER-REACTIVE LIQUID, N.O.S.
	solid	**W2**[a]	1390	ALKALI METAL AMIDES
			3401	ALKALI METAL AMALGAM, SOLID
			3402	ALKALINE EARTH METAL AMALGAM, SOLID
			3170	ALUMINIUM SMELTING BY-PRODUCTS or
			3170	ALUMINIUM REMELTING BY-PRODUCTS
			3403	POTASSIUM METAL ALLOYS, SOLID
			3404	POTASSIUM SODIUM ALLOYS, SOLID
			1393	ALKALINE EARTH METAL ALLOY, N.O.S.
			1409	METAL HYDRIDES, WATER-REACTIVE, N.O.S.
			3208	METALLIC SUBSTANCE, WATER-REACTIVE, N.O.S.
			3395	ORGANOMETALLIC SUBSTANCE, SOLID, WATER-REACTIVE
			2813	WATER-REACTIVE SOLID, N.O.S.
	articles	**W3**	3292	BATTERIES, CONTAINING SODIUM or
			3292	CELLS, CONTAINING SODIUM
Liquid, flammable		**WF1**	1391	ALKALI METAL DISPERSION having a flash-point of not more than 60 °C or
			1391	ALKALINE EARTH METAL DISPERSION having a flash-point of not more than 60 °C
			3399	ORGANOMETALLIC SUBSTANCE, LIQUID, WATER-REACTIVE, FLAMMABLE
Solid, flammable		**WF2**	3396	ORGANOMETALLIC SUBSTANCE, SOLID, WATER-REACTIVE, FLAMMABLE
			3132	WATER-REACTIVE SOLID, FLAMMABLE, N.O.S.
Solid, self-heating		**WS**[b]	3397	ORGANOMETALLIC SUBSTANCE, SOLID, WATER-REACTIVE, SELF-HEATING
			3209	METALLIC SUBSTANCE, WATER-REACTIVE, SELF-HEATING, N.O.S.
			3135	WATER-REACTIVE SOLID, SELF-HEATING, N.O.S.
Solid, oxidizing		**WO**	3133	WATER-REACTIVE SOLID, OXIDIZING, N.O.S. (not allowed, see 2.2.43.2)
Toxic WT	**liquid**	**WT1**	3130	WATER-REACTIVE LIQUID, TOXIC, N.O.S.
	solid	**WT2**	3134	WATER-REACTIVE SOLID, TOXIC, N.O.S.
Corrosive WC	**liquid**	**WC1**	3129	WATER-REACTIVE LIQUID, CORROSIVE, N.O.S.
	solid	**WC2**	3131	WATER-REACTIVE SOLID, CORROSIVE, N.O.S.
Flammable, corrosive		**WFC**[c]	2988	CHLOROSILANES, WATER-REACTIVE, FLAMMABLE, CORROSIVE, NO.S.
				(No other collective entry with this classification code available, if need be, classification under a collective entry with a classification code to be determined according to the table of precedence of hazard in 2.1.3.10.)

---

[a] *Metals and metal alloys which, in contact with water, do not emit flammable gases and are not pyrophoric or self-heating, but which are readily flammable, are substances of Class 4.1. Alkaline-earth metals and alkaline-earth metal alloys in pyrophoric form are substances of Class 4.2. Dust and powders of metals in pyrophoric form are substances of Class 4.2. Metals and metal alloys in pyrophoric form are substances of Class 4.2. Compounds of phosphorus with heavy metals such as iron, copper, etc. are not subject to the provisions of ADN.*

[b] *Metals and metal alloys in pyrophoric form are substances of Class 4.2.*

[c] *Chlorosilanes, having a flash-point of less than 23 °C and which, in contact with water, do not emit flammable gases, are substances of Class 3. Chlorosilanes, having a flash-point equal to or greater than 23 °C and which, in contact with water, do not emit flammable gases, are substances of Class 8.*

2.2.51 Class 5.1 Oxidizing substances

2.2.51.1 *Criteria*

2.2.51.1.1 The heading of Class 5.1 covers substances which, while in themselves not necessarily combustible, may, generally by yielding oxygen, cause or contribute to the combustion of other material, and articles containing such substances.

2.2.51.1.2 The substances of Class 5.1 and articles containing such substances are subdivided as follows:

O Oxidizing substances without subsidiary risk or articles containing such substances:

O1 Liquid;
O2 Solid;
O3 Articles;

OF Oxidizing substances, solid, flammable;

OS Oxidizing substances, solid, self-heating;

OW Oxidizing substances, solid which, in contact with water, emit flammable gases;

OT Oxidizing substances, toxic:

OT1 Liquid;
OT2 Solid;

OC Oxidizing substances, corrosive:

OC1 Liquid;
OC2 Solid;

OTC Oxidizing substances, toxic, corrosive.

2.2.51.1.3 Substances and articles classified in Class 5.1 are listed in Table A of Chapter 3.2. The assignment of substances and articles not mentioned by name in Table A of Chapter 3.2 to the relevant entry of 2.2.51.3 in accordance with the provisions of Chapter 2.1 can be based on the tests, methods and criteria in paragraphs 2.2.51.1.6 to 2.2.51.1.9 below and the Manual of Tests and Criteria, Part III, Section 34.4. In the event of divergence between test results and known experience, judgement based on known experience shall take precedence over test results.

2.2.51.1.4 If substances of Class 5.1, as a result of admixtures, come into different categories of risk from those to which the substances mentioned by name in Table A of Chapter 3.2 belong, these mixtures or solutions shall be assigned to the entries to which they belong on the basis of their actual degree of danger.

NOTE: *For the classification of solutions and mixtures (such as preparations and wastes), see also Section 2.1.3.*

2.2.51.1.5 On the basis of the test procedures in the Manual of Tests and Criteria, Part III, Section 34.4 and the criteria set out in 2.2.51.1.6-2.2.51.1.9 it may also be determined whether the nature of a substance mentioned by name in Table A of Chapter 3.2 is such that the substance is not subject to the provisions for this class.

Oxidizing solids

Classification

2.2.51.1.6 When oxidizing solid substances not mentioned by name in Table A of Chapter 3.2 are assigned to one of the entries listed in 2.2.51.3 on the basis of the test procedure in accordance with the Manual of Tests and Criteria, Part III, sub-section 34.4.1, the following criteria shall apply:

A solid substance shall be assigned to Class 5.1 if, in the 4:1 or the 1:1 sample-to-cellulose ratio (by mass) tested, it ignites or burns or exhibits mean burning times equal to or less than that of a 3:7 mixture (by mass) of potassium bromate and cellulose.

Assignment of packing groups

2.2.51.1.7 Oxidizing solids classified under the various entries in Table A of Chapter 3.2 shall be assigned to packing groups I, II or III on the basis of test procedures of the Manual of Tests and Criteria, Part III, sub-section 34.4.1, in accordance with the following criteria:

(a) Packing group I: any substance which, in the 4:1 or 1:1 sample-to-cellulose ratio (by mass) tested, exhibits a mean burning time less than the mean burning time of a 3:2 mixture, by mass, of potassium bromate and cellulose;

(b) Packing group II: any substance which, in the 4:1 or 1:1 sample-to-cellulose ratio (by mass) tested, exhibits a mean burning time equal to or less than the mean burning time of a 2:3 mixture (by mass) of potassium bromate and cellulose and the criteria for packing group I are not met;

(c) Packing group III: any substance which, in the 4:1 or 1:1 sample-to-cellulose ratio (by mass) tested, exhibits a mean burning time equal to or less than the mean burning time of a 3:7 mixture (by mass) of potassium bromate and cellulose and the criteria for packing groups I and II are not met.

Oxidizing liquids

Classification

2.2.51.1.8 When oxidizing liquid substances not mentioned by name in Table A of Chapter 3.2 are assigned to one of the entries listed in sub-section 2.2.51.3 on the basis of the test procedure in accordance with the Manual of Tests and Criteria, Part III, sub-section 34.4.2, the following criteria shall apply:

A liquid substance shall be assigned to Class 5.1 if, in the 1:1 mixture, by mass, of substance and cellulose tested, it exhibits a pressure rise of 2070 kPa gauge or more and a mean pressure rise time equal to or less than the mean pressure rise time of a 1:1 mixture, by mass, of 65% aqueous nitric acid and cellulose.

Assignment of packing groups

2.2.51.1.9 Oxidizing liquids classified under the various entries in Table A of Chapter 3.2 shall be assigned to packing groups I, II or III on the basis of test procedures of the Manual of Tests and Criteria, Part III, section 34.4.2, in accordance with the following criteria:

(a) Packing group I: any substance which, in the 1:1 mixture, by mass, of substance and cellulose tested, spontaneously ignites; or the mean pressure rise time of a 1:1 mixture, by mass, of substance and cellulose is less than that of a 1:1 mixture, by mass, of 50% perchloric acid and cellulose;

(b) Packing group II: any substance which, in the 1:1 mixture, by mass, of substance and cellulose tested, exhibits a mean pressure rise time less than or equal to the mean pressure rise time of a 1:1 mixture, by mass, of 40% aqueous sodium chlorate solution and cellulose; and the criteria for packing group I are not met;

(c) Packing group III: any substance which, in the 1:1 mixture, by mass, of substance and cellulose tested, exhibits a mean pressure rise time less than or equal to the mean pressure rise time of a 1:1 mixture, by mass, of 65% aqueous nitric acid and cellulose; and the criteria for packing groups I and II are not met.

2.2.51.2 ***Substances not accepted for carriage***

2.2.51.2.1 The chemically unstable substances of Class 5.1 shall not be accepted for carriage unless the necessary steps have been taken to prevent their dangerous decomposition or polymerization during carriage. To this end it shall in particular be ensured that receptacles and tanks do not contain any material liable to promote these reactions.

2.2.51.2.2 The following substances and mixtures shall not be accepted for carriage:

- oxidizing solids, self-heating, assigned to UN No. 3100, oxidizing solids, water-reactive, assigned to UN No. 3121 and oxidizing solids, flammable, assigned to UN No. 3137, unless they meet the requirements for Class 1 (see also 2.1.3.7);

- hydrogen peroxide, not stabilized or hydrogen peroxide, aqueous solutions, not stabilized containing more than 60 % hydrogen peroxide;

- tetranitromethane not free from combustible impurities;

- perchloric acid solutions containing more than 72 % (mass) acid, or mixtures of perchloric acid with any liquid other than water;

- chloric acid solution containing more than 10 % chloric acid or mixtures of chloric acid with any liquid other than water;

- halogenated fluor compounds other than UN Nos. 1745 BROMINE PENTAFLUORIDE; 1746 BROMINE TRIFLUORIDE and 2495 IODINE PENTAFLUORIDE of Class 5.1 as well as UN Nos. 1749 CHLORINE TRIFLUORIDE and 2548 CHLORINE PENTAFLUORIDE of Class 2;

- ammonium chlorate and its aqueous solutions and mixtures of a chlorate with an ammonium salt;

- ammonium chlorite and its aqueous solutions and mixtures of a chlorite with an ammonium salt;

- mixtures of a hypochlorite with an ammonium salt;

- ammonium bromate and its aqueous solutions and mixtures of a bromate with an ammonium salt;

- ammonium permanganate and its aqueous solutions and mixtures of a permanganate with an ammonium salt;
- ammonium nitrate containing more than 0.2 % combustible substances (including any organic substance calculated as carbon) unless it is a constituent of a substance or article of Class 1;
- fertilizers having an ammonium nitrate content (in determining the ammonium nitrate content, all nitrate ions for which a molecular equivalent of ammonium ions is present in the mixture shall be calculated as ammonium nitrate) or a content in combustible substances exceeding the values specified in special provision 307 except under the conditions applicable to Class 1;
- ammonium nitrite and its aqueous solutions and mixtures of an inorganic nitrite with an ammonium salt;
- mixtures of potassium nitrate, sodium nitrite and an ammonium salt.

2.2.51.3 ***List of collective entries***

Classification	Form	Code	Collective entries
Oxidizing substances / **Without subsidiary risk O**	**liquid**	**O1**	3210 CHLORATES, INORGANIC, AQUEOUS SOLUTION, N.O.S. 3211 PERCHLORATES, INORGANIC, AQUEOUS SOLUTION, N.O.S. 3213 BROMATES, INORGANIC, AQUEOUS SOLUTION, N.O.S. 3214 PERMANGANATES, INORGANIC, AQUEOUS SOLUTION, N.O.S. 3216 PERSULPHATES, INORGANIC, AQUEOUS SOLUTION, N.O.S. 3218 NITRATES, INORGANIC, AQUEOUS SOLUTION, N.O.S. 3219 NITRITES, INORGANIC, AQUEOUS SOLUTION, N.O.S. 3139 OXIDIZING LIQUID, N.O.S.
	solid	**O2**	1450 BROMATES, INORGANIC, N.O.S 1461 CHLORATES, INORGANIC, N.O.S. 1462 CHLORITES, INORGANIC, N.O.S. 1477 NITRATES, INORGANIC, N.O.S 1481 PERCHLORATES, INORGANIC, N.O.S. 1482 PERMANGANATES, INORGANIC, N.O.S. 1483 PEROXIDES, INORGANIC, N.O.S 2627 NITRITES, INORGANIC, N.O.S. 3212 HYPOCHLORITES, INORGANIC, N.O.S. 3215 PERSULPHATES, INORGANIC, N.O.S. 1479 OXIDIZING SOLID, N.O.S.
	articles	**O3**	3356 OXYGEN GENERATOR, CHEMICAL
Solid, flammable		**OF**	3137 OXIDIZING SOLID, FLAMMABLE, N.O.S. (not allowed, see 2.2.51.2)
Solid, self-heating		**OS**	3100 OXIDIZING SOLID, SELF-HEATING, N.O.S. (not allowed, see 2.2.51.2)
Solid, water reactive		**OW**	3121 OXIDIZING SOLID, WATER REACTIVE, N.O.S. (not allowed, see 2.2.51.2)
Toxic OT	**liquid**	**OT1**	3099 OXIDIZING LIQUID, TOXIC, N.O.S.
	solid	**OT2**	3087 OXIDIZING SOLID, TOXIC, N.O.S.
Corrosive OC	**liquid**	**OC1**	3098 OXIDIZING LIQUID, CORROSIVE, N.O.S.
	solid	**OC2**	3085 OXIDIZING SOLID, CORROSIVE, N.O.S.
Toxic, corrosive		**OTC**	(No collective entry with this classification code available; if need be, classification under a collective entry with a classification code to be determined according to the table of precedence of hazard in 2.1.3.10.)

2.2.52 Class 5.2 Organic peroxides

2.2.52.1 *Criteria*

2.2.52.1.1 The heading of Class 5.2 covers organic peroxides and formulations of organic peroxides.

2.2.52.1.2 The substances of Class 5.2 are subdivided as follows:

P1 Organic peroxides, not requiring temperature control;
P2 Organic peroxides, requiring temperature control.

Definition

2.2.52.1.3 *Organic peroxides* are organic substances which contain the bivalent -O-O- structure and may be considered derivatives of hydrogen peroxide, where one or both of the hydrogen atoms have been replaced by organic radicals.

Properties

2.2.52.1.4 Organic peroxides are liable to exothermic decomposition at normal or elevated temperatures. The decomposition can be initiated by heat, contact with impurities (e.g. acids, heavy-metal compounds, amines), friction or impact. The rate of decomposition increases with temperature and varies with the organic peroxide formulation. Decomposition may result in the evolution of harmful, or flammable, gases or vapours. For certain organic peroxides the temperature shall be controlled during carriage. Some organic peroxides may decompose explosively, particularly if confined. This characteristic may be modified by the addition of diluents or by the use of appropriate packagings. Many organic peroxides burn vigorously. Contact of organic peroxides with the eyes is to be avoided. Some organic peroxides will cause serious injury to the cornea, even after brief contact, or will be corrosive to the skin.

***NOTE**: Test methods for determining the flammability of organic peroxides are set out in the Manual of Tests and Criteria, Part III, sub-section 32.4. Because organic peroxides may react vigorously when heated, it is recommended to determine their flash-point using small sample sizes such as described in ISO 3679:1983.*

Classification

2.2.52.1.5 Any organic peroxide shall be considered for classification in Class 5.2 unless the organic peroxide formulation contains:

(a) not more than 1.0 % available oxygen from the organic peroxides when containing not more than 1.0 % hydrogen peroxide;

(b) not more than 0.5 % available oxygen from the organic peroxides when containing more than 1.0 % but not more than 7.0 % hydrogen peroxide.

***NOTE:** The available oxygen content (%) of an organic peroxide formulation is given by the formula*

$$16 \times 3 (n_i \times c_i / m_i)$$

where:

n_i = *number of peroxygen groups per molecule of organic peroxide i;*
c_i = *concentration (mass %) of organic peroxide i; and*
m_i = *molecular mass of organic peroxide i.*

2.2.52.1.6 Organic peroxides are classified into seven types according to the degree of danger they present. The types of organic peroxide range from type A, which is not accepted for carriage in the packaging in which it is tested, to type G, which is not subject to the provisions of Class 5.2. The classification of types B to F is directly related to the maximum quantity allowed in one packaging. The principles to be applied to the classification of substances not listed in 2.2.52.4 are set out in the Manual of Tests and Criteria, Part II.

2.2.52.1.7 Organic peroxides which have already been classified and are already permitted for carriage in packagings are listed in 2.2.52.4, those already permitted for carriage in IBCs are listed in 4.1.4.2 of ADR, packing instruction IBC520 and those already permitted for carriage in tanks in accordance with Chapters 4.2 and 4.3 of ADR are listed in 4.2.5.2 of ADR, portable tank instruction T23. Each permitted substance listed is assigned to a generic entry of Table A of Chapter 3.2 (UN Nos. 3101 to 3120) and appropriate subsidiary risks and remarks providing relevant transport information are given.

These generic entries specify:

- the type (B to F) of organic peroxide (see 2.2.52.1.6 above);
- physical state (liquid/solid); and
- temperature control (when required), see 2.2.52.1.15 to 2.2.52.1.18.

Mixtures of these formulations may be classified as the same type of organic peroxide as that of the most dangerous component and be carried under the conditions of carriage given for this type. However, as two stable components can form a thermally less stable mixture, the self-accelerating decomposition temperature (SADT) of the mixture shall be determined and, if necessary, the control and emergency temperatures derived from the SADT in accordance paragraph 2.2.52.1.16.

2.2.52.1.8 Classification of organic peroxides, formulations or mixtures of organic peroxides not listed in 2.2.52.4, 4.1.4.2 of ADR, packing instruction IBC520 or 4.2.5.2 of ADR, portable tank instruction T23, and assignment to a collective entry shall be made by the competent authority of the country of origin. The statement of approval shall contain the classification and the relevant conditions of carriage. If the country of origin is not a Contracting Party to ADN, the classification and conditions of carriage shall be recognized by the competent authority of the first country Contracting Party to ADN reached by the consignment.

2.2.52.1.9 Samples of organic peroxides or formulations of organic peroxides not listed in 2.2.52.4, for which a complete set of test results is not available and which are to be carried for further testing or evaluation, shall be assigned to one of the appropriate entries for organic peroxides type C provided the following conditions are met:

- the available data indicate that the sample would be no more dangerous than organic peroxides type B;
- the sample is packaged in accordance with packing method OP2 of 4.1.4.1 of ADR and the quantity per cargo transport unit is limited to 10 kg;
- the available data indicate that the control temperature, if any, is sufficiently low to prevent any dangerous decomposition and sufficiently high to prevent any dangerous phase separation.

Desensitization of organic peroxides

2.2.52.1.10 In order to ensure safety during carriage, organic peroxides are in many cases desensitized by organic liquids or solids, inorganic solids or water. Where a percentage of a substance is stipulated, this refers to the percentage by mass, rounded to the nearest whole number. In general, desensitization shall be such that, in case of spillage, the organic peroxide will not concentrate to a dangerous extent.

2.2.52.1.11 Unless otherwise stated for the individual organic peroxide formulation, the following definition(s) shall apply to diluents used for desensitization:

- diluents type A are organic liquids which are compatible with the organic peroxide and which have a boiling point of not less than 150 °C. Type A diluents may be used for desensitizing all organic peroxides.

- diluents type B are organic liquids which are compatible with the organic peroxide and which have a boiling point of less than 150 °C but not less than 60 °C and a flash-point of not less than 5 °C.

Type B diluents may be used for desensitization of all organic peroxides provided that the boiling point of the liquid is at least 60 °C higher than the SADT in a 50 kg package.

2.1.52.1.12 Diluents, other than type A or type B, may be added to organic peroxide formulations as listed in 2.2.52.4 provided that they are compatible. However, replacement of all or part of a type A or type B diluent by another diluent with differing properties requires that the organic peroxide formulation be re-assessed in accordance with the normal acceptance procedure for Class 5.2.

2.2.52.1.13 Water may only be used for the desensitization of organic peroxides which are listed in 2.2.52.4 or in the competent authority decision according to 2.2.52.1.8 as being "with water" or "as a stable dispersion in water". Samples of organic peroxides or formulations of organic peroxides not listed in 2.2.52.4 may also be desensitized with water provided the requirements of 2.2.52.1.9 are met.

2.2.52.1.14 Organic and inorganic solids may be used for desensitization of organic peroxides provided that they are compatible. Compatible liquids and solids are those which have no detrimental influence on the thermal stability and hazard type of the organic peroxide formulation.

Temperature control requirements

2.2.52.1.15 Certain organic peroxides may only be carried under temperature-controlled conditions. The control temperature is the maximum temperature at which the organic peroxide can be safely carried. It is assumed that the temperature of the immediate surroundings of a package only exceeds 55 °C during carriage for a relatively short time in a 24 hour period. In the event of loss of temperature control, it may be necessary to implement emergency procedures. The emergency temperature is the temperature at which such procedures shall be implemented.

2.2.52.1.16 The control and emergency temperatures are derived from the SADT which is defined as the lowest temperature at which self-accelerating decomposition may occur with a substance in the packaging as used during carriage (see Table 1). The SADT shall be determined in order to decide whether a substance shall be subjected to temperature control during carriage. Provisions for the determination of the SADT are given in the Manual of Tests and Criteria, Part II, Sections 20 and 28.4.

Table 1: Derivation of control and emergency temperatures

Type of receptacle	SADT [a]	Control temperature	Emergency temperature
Single packagings and IBCs	20 °C or less	20 °C below SADT	10 °C below SADT
	over 20 °C to 35 °C	15 °C below SADT	10 °C below SADT
	over 35 °C	10 °C below SADT	5 °C below SADT
Tanks	Not greater than 50 °C	10 °C below SADT	5 °C below SADT

[a] *SADT of the substance as packaged for carriage.*

2.2.52.1.17 The following organic peroxides shall be subject to temperature control during carriage:

- organic peroxides types B and C with an SADT ≤ 50 °C;
- organic peroxides type D showing a medium effect when heated under confinement with an SADT ≤ 50 °C or showing a low or no effect when heated under confinement with an SADT ≤ 45 °C; and
- organic peroxides types E and F with an SADT ≤ 45 °C.

***NOTE**: Provisions for the determination of the effects of heating under confinement are given in the Manual of Tests and Criteria, Part II, Section 20 and Sub-section 28.4.*

2.2.52.1.18 Where applicable, control and emergency temperatures are listed in 2.2.52.4. The actual temperature during carriage may be lower than the control temperature but shall be selected so as to avoid dangerous separation of phases.

2.2.52.2 ***Substances not accepted for carriage***

Organic peroxides, type A, shall not be accepted for carriage under the provisions of Class 5.2 (see Manual of Tests and Criteria, Part II, paragraph 20.4.3 (a)).

2.2.52.3 *List of collective entries*

Organic peroxides			ORGANIC PEROXIDE TYPE A, LIQUID	Not accepted for carriage, see 2.2.52.2
			ORGANIC PEROXIDE TYPE A, SOLID	
		3101	ORGANIC PEROXIDE TYPE B, LIQUID	
		3102	ORGANIC PEROXIDE TYPE B, SOLID	
		3103	ORGANIC PEROXIDE TYPE C, LIQUID	
		3104	ORGANIC PEROXIDE TYPE C, SOLID	
Not requiring temperature control	**P1**	3105	ORGANIC PEROXIDE TYPE D, LIQUID	
		3106	ORGANIC PEROXIDE TYPE D, SOLID	
		3107	ORGANIC PEROXIDE TYPE E, LIQUID	
		3108	ORGANIC PEROXIDE TYPE E, SOLID	
		3109	ORGANIC PEROXIDE TYPE F, LIQUID	
		3110	ORGANIC PEROXIDE TYPE F, SOLID	
			ORGANIC PEROXIDE TYPE G, LIQUID	Not subject to the provisions applicable to Class 5.2, see 2.2.52.1.6
			ORGANIC PEROXIDE TYPE G, SOLID	
		3111	ORGANIC PEROXIDE TYPE B, LIQUID, TEMPERATURE CONTROLLED	
		3112	ORGANIC PEROXIDE TYPE B, SOLID, TEMPERATURE CONTROLLED	
		3113	ORGANIC PEROXIDE TYPE C, LIQUID, TEMPERATURE CONTROLLED	
		3114	ORGANIC PEROXIDE TYPE C, SOLID, TEMPERATURE CONTROLLED	
Requiring temperature control	**P2**	3115	ORGANIC PEROXIDE TYPE D, LIQUID, TEMPERATURE CONTROLLED	
		3116	ORGANIC PEROXIDE TYPE D, SOLID, TEMPERATURE CONTROLLED	
		3117	ORGANIC PEROXIDE TYPE E, LIQUID, TEMPERATURE CONTROLLED	
		3118	ORGANIC PEROXIDE TYPE E, SOLID, TEMPERATURE CONTROLLED	
		3119	ORGANIC PEROXIDE TYPE F, LIQUID, TEMPERATURE CONTROLLED	
		3120	ORGANIC PEROXIDE TYPE F, SOLID, TEMPERATURE CONTROLLED	

2.2.52.4 *List of currently assigned organic peroxides in packagings*

In the column "Packing Method", codes "OP1" to "OP8" refer to packing methods in 4.1.4.1 of ADR packing instruction P520 (see also 4.1.7.1 of ADR). Organic peroxides to be carried shall fulfill the classification and he control and emergency temperatures (derived from the SADT) as listed. For substances permitted in IBCs, see 4.1.4.2 of ADR, packing instruction IBC520 and, for those permitted in tanks according to Chapters 4.2 and 4.3 of ADR, see 4.2.5.2 of ADR, portable tank instruction T23.

ORGANIC PEROXIDE	Concentration (%)	Diluent type A (%)	Diluent type B (%) 1)	Inert solid (%)	Water	Packing Method	Control temperature (°C)	Emergency temperature (°C)	Number (Generic entry)	Subsidiary risks and remarks
ACETYL ACETONE PEROXIDE	≤ 42	≥ 48			≥8	OP7			3105	2)
"	≤ 32 as a paste					OP7			3106	20)
ACETYL CYCLOHEXANESULPHONYL PEROXIDE	≤ 82				≥ 12	OP4	-10	0	3112	3)
"	≤ 32		≥ 68			OP7	-10	0	3115	
tert-AMYL HYDROPEROXIDE	≤ 88	≥ 6			≥ 6	OP8			3107	
tert-AMYL PEROXYACETATE	≤ 62	≥ 38				OP7			3105	
tert-AMYL PEROXYBENZOATE	≤ 100					OP5			3103	
tert-AMYL PEROXY-2-ETHYLHEXANOATE	≤ 100					OP7	+20	+25	3115	
tert-AMYL PEROXY-2-ETHYLHEXYL CARBONATE	≤ 100					OP7			3105	
tert-AMYL PEROXY ISOPROPYL CARBONATE	≤ 77	≥ 23				OP5			3103	
tert-AMYL PEROXYNEODECANOATE	≤ 77		≥ 23			OP7	0	+10	3115	
"	≤ 47	≥ 53				OP8	0	+ 10	3119	
tert-AMYL PEROXYPIVALATE	≤ 77		≥ 23			OP5	+10	+15	3113	
tert-AMYLPEROXY-3,5,5-TRIMETHYLHEXANOATE	≤ 100					OP7			3105	3)
tert-BUTYL CUMYL PEROXIDE	> 42 – 100					OP8			3107	
"	≤ 52			≥ 48		OP8			3108	
n-BUTYL-4,4-DI-(tert-BUTYLPEROXY)VALERATE	> 52 – 100					OP5			3103	
"	≤ 52			≥ 48		OP8			3108	
tert-BUTYL HYDROPEROXIDE	>79 – 90				≥ 10	OP5			3103	13)
"	≤ 80	≥ 20				OP7			3105	4) 13)
"	≤ 79				> 14	OP8			3107	13) 23)
"	≤ 72				≥ 28	OP8			3109	13)
tert-BUTYL HYDROPEROXIDE + DI-tert-BUTYLPEROXIDE	< 82 + >9				≥ 7	OP5			3103	13)
tert-BUTYL MONOPEROXYMALEATE	> 52 – 100					OP5			3102	3)
"	≤ 52	≥ 48				OP6			3103	
"	≤ 52			≥ 48		OP8			3108	
"	≤ 52 as a paste					OP8			3108	
tert-BUTYL PEROXYACETATE	> 52 – 77	≥ 23				OP5			3101	3)
"	> 32 – 52	≥ 48				OP6			3103	
"	≤ 32		≥ 68			OP8			3109	
tert-BUTYL PEROXYBENZOATE	> 77 – 100					OP5			3103	
"	> 52 - 77	≥23				OP7			3105	
"	≤ 52			≥ 48		OP7			3106	
tert-BUTYL PEROXYBUTYL FUMARATE	≤ 52	≥ 48				OP7			3105	

ORGANIC PEROXIDE	Concentration (%)	Diluent type A (%)	Diluent type B (%) 1)	Inert solid (%)	Water	Packing Method	Control temperature (°C)	Emergency temperature (°C)	Number (Generic entry)	Subsidiary risks and remarks
tert-BUTYL PEROXYCROTONATE	≤ 77	≥ 23				OP7			3105	
tert-BUTYL PEROXYDIETHYLACETATE	≤ 100					OP5	+20	+25	3113	
tert-BUTYL PEROXY-2-ETHYLHEXANOATE	> 52 – 100					OP6	+20	+25	3113	
"	> 32 – 52		≥ 48			OP8	+30	+35	3117	
"	≤ 52			≥ 48		OP8	+20	+25	3118	
"	≤ 32		≥ 68			OP8	+40	+45	3119	
tert-BUTYL PEROXY-2-ETHYLHEXANOATE + 2,2-DI-(tert-BUTYLPEROXY)BUTANE	≤ 12 + ≤ 14	≥ 14		≥ 60		OP7			3106	
"	≤ 31 + ≤ 36		≥ 33			OP7	+35	+40	3115	
tert-BUTYL PEROXY-2-ETHYLHEXYLCARBONATE	≤ 100					OP7			3105	
tert-BUTYL PEROXYISOBUTYRATE	> 52 – 77		≥ 23			OP5	+15	+20	3111	3)
"	≤ 52		≥ 48			OP7	+15	+20	3115	
tert-BUTYLPEROXY ISOPROPYLCARBONATE	≤ 77	≥ 23				OP5			3103	
1-(2-tert-BUTYLPEROXY ISOPROPYL)-3-ISOPROPENYLBENZENE	≤ 77	≥ 23				OP7			3105	
"	≤ 42			≥ 58		OP8			3108	
tert-BUTYL PEROXY-2-METHYLBENZOATE	≤ 100					OP5			3103	
tert-BUTYL PEROXYNEODECANOATE	> 77 – 100					OP7	-5	+5	3115	
"	≤ 77		≥ 23			OP7	0	+10	3115	
"	≤ 52 as a stable dispersion in water					OP8	0	+10	3119	
"	≤ 42 as a stable dispersion in water (frozen)					OP8	0	+10	3118	
"	≤ 32	≥ 68				OP8	0	+10	3119	
tert-BUTYL PEROXYNEOHEPTANOATE	≤ 77	≥ 23				OP7	0	+10	3115	
"	≤ 42 as a stable dispersion in water					OP8	0	+10	3117	
tert-BUTYL PEROXYPIVALATE	> 67 - 77	≥ 23				OP5	0	+10	3113	
"	> 27 – 67		≥ 33			OP7	0	+10	3115	
"	≤ 27		≥ 73			OP8	+30	+35	3119	
tert-BUTYLPEROXY STEARYLCARBONATE	≤ 100					OP7			3106	
tert-BUTYL PEROXY-3,5,5-TRIMETHYLHEXANOATE	> 32 – 100					OP7			3105	
"	≤ 42			≥ 58		OP7			3106	
"	≤ 32		≥ 68			OP8			3109	

ORGANIC PEROXIDE	Concentration (%)	Diluent type A (%)	Diluent type B (%) 1)	Inert solid (%)	Water	Packing Method	Control temperature (°C)	Emergency temperature (°C)	Number (Generic entry)	Subsidiary risks and remarks
3-CHLOROPEROXYBENZOIC ACID	> 57 – 86			≥ 14		OP1			3102	3)
"	≤ 57			≥ 3	≥ 40	OP7			3106	
"	≤ 77			≥ 6	≥ 17	OP7			3106	
CUMYL HYDROPEROXIDE	> 90 – 98	≤10				OP8			3107	13)
"	≤ 90	≥ 10				OP8			3109	13) 18)
CUMYL PEROXYNEODECANOATE	≤ 87	≥ 13				OP7	- 10	0	3115	
"	≤ 77		≥ 23			OP7	-10	0	3115	
"	≤ 52 as a stable dispersion in water					OP8	-10	0	3119	
CUMYL PEROXYNEOHEPTANOATE	≤ 77	≥ 23				OP7	-10	0	3115	
CUMYL PEROXYPIVALATE	≤ 77		≥ 23			OP7	-5	+5	3115	
CYCLOHEXANONE PEROXIDE(S)	≤ 91				≥ 9	OP6			3104	13)
"	≤ 72	≥ 28				OP7			3105	5)
"	≤ 72 as a paste					OP7			3106	5) 20)
"	≤ 32			≥ 68					Exempt	29)
DIACETONE ALCOHOL PEROXIDES	≤ 57		≥ 26		≥ 8	OP7	+40	+45	3115	6)
DIACETYL PEROXIDE	≤ 27		≥ 73			OP7	+20	+25	3115	7) 13)
DI-tert-AMYL PEROXIDE	≤ 100					OP8			3107	
2,2-DI-(tert-AMYLPEROXY)BUTANE	≤ 57	≥ 43				OP7			3105	
1,1-DI-(tert-AMYLPEROXY)CYCLOHEXANE	≤ 82	≥ 18				OP6			3103	
DIBENZOYL PEROXIDE	> 51 - 100			≤ 48		OP2			3102	3)
"	> 77 - 94				≥ 6	OP4			3102	3)
"	≤ 77				≥ 23	OP6			3104	
"	≤ 62			≥ 28	≥ 10	OP7			3106	
"	> 52 – 62 as a paste					OP7			3106	20)
"	> 35 – 52			≥ 48		OP7			3106	
"	> 36 – 42	≥ 18			≤ 40	OP8			3107	
"	≤ 56.5 as a paste				≥ 15	OP8			3108	
"	≤ 52 as a paste					OP8			3108	20)
"	≤ 42 as a stable dispersion in water					OP8			3109	
"	≤ 35			≥ 65					Exempt	29)
DI-(4-tert-BUTYLCYCLOHEXYL) PEROXYDICARBONATE	≤ 100					OP6	+30	+35	3114	
"	≤ 42 as a stable dispersion in water					OP8	+30	+35	3119	

ORGANIC PEROXIDE	Concentration (%)	Diluent type A (%)	Diluent type B (%) 1)	Inert solid (%)	Water	Packing Method	Control temperature (°C)	Emergency temperature (°C)	Number (Generic entry)	Subsidiary risks and remarks
DI-tert-BUTYL PEROXIDE	> 52 – 100					OP8			3107	
"	≤ 52		≥ 48			OP8			3109	25)
DI-tert-BUTYL PEROXYAZELATE	≤ 52	≥ 48				OP7			3105	
2,2-DI-(tert-BUTYLPEROXY)BUTANE	≤ 52	≥ 48				OP6			3103	
1,6-Di-(tert-BUTYLPEROXYCARBONYLOXY) HEXANE	≤ 72	≥ 28				OP5			3103	
1,1-DI-(tert-BUTYLPEROXY) CYCLOHEXANE	> 80 - 100					OP5			3101	3)
"	≤ 72		≥ 28			OP5			3103	30)
"	> 52 - 80	≥ 20				OP5			3103	
"	> 42 – 52	≥ 48				OP7			3105	
"	≤ 42	≥ 13		≥ 45		OP7			3106	
"	≤ 42	≥ 58				OP8			3109	
"	≤ 27	≥ 25				OP8			3107	21)
"	≤ 13	≥ 13	≥ 74			OP8			3109	
1,1-DI-(tert-BUTYLPEROXY) CYCLOHEXANE + tert-BUTYL PEROXY-2-ETHYLHEXANOATE	≤ 43 + ≤ 16	≥ 41				OP 7			3105	
DI-n-BUTYL PEROXYDICARBONATE	> 27 - 52		≥ 48			OP7	-15	-5	3115	
"	≤ 27		≥ 73			OP8	-10	0	3117	
"	≤ 42 as a stable dispersion in water (frozen)					OP8	-15	-5	3118	
DI-sec-BUTYL PEROXYDICARBONATE	> 52 - 100					OP4	-20	-10	3113	
"	≤ 52		≥ 48			OP7	-15	-5	3115	
DI-(2-tert-BUTYLPEROXYISOPROPYL)BENZENE(S)	> 42 - 100			≤ 57		OP7			3106	
"	≤ 42			≥ 58					Exempt	29)
DI-(tert-BUTYLPEROXY) PHTHALATE	> 42 - 52	≥ 48				OP7			3105	
"	≤ 52 as a paste					OP7			3106	20)
"	≤ 42	≥ 58				OP8			3107	
2,2-DI-(tert-BUTYLPEROXY)PROPANE	≤ 52	≥ 48				OP7			3105	
"	≤ 42	≥ 13		≥ 45		OP7			3106	

ORGANIC PEROXIDE	Concentration (%)	Diluent type A (%)	Diluent type B (%) 1)	Inert solid (%)	Water	Packing Method	Control temperature (°C)	Emergency temperature (°C)	Number (Generic entry)	Subsidiary risks and remarks
1,1-DI-(tert-BUTYLPEROXY)-3,3,5-TRIMETHYLCYCLOHEXANE	> 90 - 100					OP5			3101	3)
"	≤ 90		≥ 10			OP5			3103	30)
"	> 57 – 90	≥ 10				OP5			3103	
"	≤ 77		≥ 23			OP5			3103	
"	≤ 57			≥ 43		OP8			3110	
"	≤ 57	≥ 43				OP8			3107	
"	≤ 32	≥ 26	≥ 42			OP8			3107	
DICETYL PEROXYDICARBONATE	≤ 100					OP7	+30	+35	3116	
"	≤ 42 as a stable dispersion in water					OP8	+30	+35	3119	
DI-4-CHLOROBENZOYL PEROXIDE	≤ 77				≥ 23	OP5			3102	3)
"	≤ 52 as a paste					OP7			3106	20)
"	≤ 32			≥ 68					Exempt	29)
DICUMYL PEROXIDE	> 52 - 100					OP8			3110	12)
"	≤ 52			≥ 48					Exempt	29)
DICYCLOHEXYL PEROXYDICARBONATE	> 91 - 100					OP3	+10	+15	3112	3)
"	≤ 91				≥ 9	OP5	+10	+15	3114	
"	≤ 42 as a stable dispersion in water					OP8	+15	+20	3119	
DIDECANOYL PEROXIDE	≤ 100					OP6	+30	+35	3114	
2,2-DI-(4,4-DI (tert-BUTYLPEROXY) CYCLOHEXYL) PROPANE	≤ 42			≥ 58		OP7			3106	
"	≤ 22		≥ 78			OP8			3107	
DI-2,4-DICHLOROBENZOYL PEROXIDE	≤ 77				≥ 23	OP5			3102	3)
"	≤ 52 as a paste					OP8	+ 20	+ 25	3118	
"	≤ 52 as a paste with silicon oil					OP7			3106	
DI-(2-ETHOXYETHYL) PEROXYDICARBONATE	≤ 52		≥ 48			OP7	-10	0	3115	
DI-(2-ETHYLHEXYL) PEROXYDICARBONATE	> 77 – 100					OP5	-20	-10	3113	
"	≤ 77		≥ 23			OP7	-15	-5	3115	
"	≤ 62 as a stable dispersion in water					OP8	-15	-5	3119	
"	≤ 52 as a stable dispersion in water (frozen)					OP8	-15	-5	3120	
2,2-DIHYDROPEROXYPROPANE	≤ 27			≥ 73		OP5			3102	3)

ORGANIC PEROXIDE	Concentration (%)	Diluent type A (%)	Diluent type B (%) 1)	Inert solid (%)	Water	Packing Method	Control temperature (°C)	Emergency temperature (°C)	Number (Generic entry)	Subsidiary risks and remarks
DI-(1-HYDROXYCYCLOHEXYL) PEROXIDE	≤ 100					OP7			3106	
DIISOBUTYRYL PEROXIDE	> 32 – 52		≥ 48			OP5	-20	-10	3111	3)
"	≤ 32		≥ 68			OP7	-20	-10	3115	
DIISOPROPYLBENZENE DIHYDROPEROXIDE	≤ 82	≥ 5			≥ 5	OP7			3106	24)
DIISOPROPYL PEROXYDICARBONATE	> 52-100					OP2	-15	-5	3112	3)
"	≤ 52		≥ 48			OP7	-20	-10	3115	
"	≤ 28	≥ 72				OP7	-15	-5	3115	
DILAUROYL PEROXIDE	≤ 100					OP7			3106	
"	≤ 42 as a stable dispersion in water					OP8			3109	
DI-(3-METHOXYBUTYL) PEROXYDICARBONATE	≤ 52		≥ 48			OP7	-5	+5	3115	
DI-(2-METHYLBENZOYL) PEROXIDE	≤ 87				≥ 13	OP5	+30	+35	3112	3)
DI-(3-METHYLBENZOYL) PEROXIDE + BENZOYL (3-METHYLBENZOYL) PEROXIDE + DIBENZOYL PEROXIDE	≤ 20 + ≤ 18 + ≤ 4		≥ 58			OP7	+35	+40	3115	
DI-(4-METHYLBENZOYL) PEROXIDE	≤ 52 as a paste with silicon oil					OP7			3106	
2,5-DIMETHYL-2,5-DI-(BENZOYLPEROXY)HEXANE	> 82-100					OP5			3102	3)
"	≤ 82			≥ 18		OP7			3106	
"	≤ 82				≥ 18	OP5			3104	
2,5-DIMETHYL-2,5-DI-(tert-BUTYLPEROXY)HEXANE	> 52 – 100					OP7			3105	
"	≤ 77			≥ 23		OP8			3108	
"	≤ 52	≥ 48				OP8			3109	
"	≤ 47 as a paste					OP8			3108	
2,5-DIMETHYL-2,5-DI-(tert-BUTYLPEROXY)HEXYNE-3	> 86-100					OP5			3101	3)
"	>52-86	≥ 14				OP5			3103	26)
"	≤ 52			≥ 48		OP7			3106	
2,5-DIMETHYL-2,5-DI-(2-ETHYLHEXANOYLPEROXY)HEXANE	≤ 100					OP5	+20	+25	3113	
2,5-DIMETHYL-2,5-DIHYDROPEROXYHEXANE	≤ 82				≥ 18	OP6			3104	
2,5-DIMETHYL-2,5-DI-(3,5,5-TRIMETHYLHEXANOYLPEROXY)HEXANE	≤ 77	≥ 23				OP7			3105	
1,1-DIMETHYL-3-HYDROXYBUTYL PEROXYNEOHEPTANOATE	≤ 52	≥ 48				OP8	0	+10	3117	

ORGANIC PEROXIDE	Concentration (%)	Diluent type A (%)	Diluent type B (%) 1)	Inert solid (%)	Water	Packing Method	Control temperature (°C)	Emergency temperature (°C)	Number (Generic entry)	Subsidiary risks and remarks
DIMYRISTYL PEROXYDICARBONATE	≤ 100					OP7	+20	+25	3116	
"	≤ 42 as a stable dispersion in water					OP8	+20	+25	3119	
DI-(2-NEODECANOYLPEROXYISOPROPYL) BENZENE	≤ 52	≥ 48				OP7	-10	0	3115	
DI-n-NONANOYL PEROXIDE	≤ 100					OP7	0	+10	3116	
DI-n-OCTANOYL PEROXIDE	≤ 100					OP5	+10	+15	3114	
DI-(2-PHENOXYETHYL) PEROXYDICARBONATE	>85 – 100					OP5			3102	3)
"	≤ 85				≥ 15	OP7			3106	
DIPROPIONYL PEROXIDE	≤ 27		≥ 73			OP8	+15	+20	3117	
DI-n-PROPYL PEROXYDICARBONATE	≤ 100					OP3	-25	-15	3113	
"	≤ 77		≥ 23			OP5	-20	-10	3113	
DISUCCINIC ACID PEROXIDE	> 72 – 100					OP4			3102	3) 17)
"	≤ 72				≥ 28	OP7	+10	+15	3116	
DI-(3,5,5-TRIMETHYLHEXANOYL) PEROXIDE	> 38-82	≥ 18				OP7	0	+10	3115	
"	≤ 52 as a stable dispersion in water					OP8	+10	+15	3119	
"	≤ 38	≥ 62				OP8	+20	+25	3119	
ETHYL 3,3-DI-(tert-AMYLPEROXY)BUTYRATE	≤ 67	≥ 33				OP7			3105	
ETHYL 3,3-DI-(tert-BUTYLPEROXY)BUTYRATE	> 77 - 100					OP5			3103	
"	≤ 77	≥ 23				OP7			3105	
"	≤ 52			≥ 48		OP7			3106	
1-(2-ETHYLHEXANOYLPEROXY)-1,3-DIMETHYLBUTYL PEROXYPIVALATE	≤ 52	≥ 45	≥ 10			OP7	-20	-10	3115	
tert-HEXYL PEROXYNEODECANOATE	≤ 71	≥ 29				OP7	0	+10	3115	
tert-HEXYL PEROXYPIVALATE	≤ 72		≥ 28			OP7	+10	+15	3115	
3-HYDROXY-1,1-DIMETHYLBUTYL PEROXYNEODECANOATE	≤ 77	≥ 23				OP 7	- 5	+ 5	3115	
"	≤ 52	≥ 48				OP 8	- 5	+ 5	3117	
"	≤ 52 as a stable dispersion in water					OP 8	- 5	+ 5	3119	
ISOPROPYL sec-BUTYL PEROXYDICARBONATE +DI-sec-BUTYL PEROXYDICARBONATE +DI-ISOPROPYL PEROXYDICARBONATE	≤ 32 + ≤ 15 – 18 ≤ 12 – 15	≥ 38				OP7	-20	-10	3115	
"	≤ 52 + ≤ 28 + ≤ 22					OP5	-20	-10	3111	3)
ISOPROPYLCUMYL HYDROPEROXIDE	≤ 72	≥ 28				OP8			3109	13)

ORGANIC PEROXIDE	Concentration (%)	Diluent type A (%)	Diluent type B (%) 1)	Inert solid (%)	Water	Packing Method	Control temperature (°C)	Emergency temperature (°C)	Number (Generic entry)	Subsidiary risks and remarks
p-MENTHYL HYDROPEROXIDE	> 72 - 100					OP7			3105	13)
"	≤ 72	≥ 28				OP8			3109	27)
METHYLCYCLOHEXANONE PEROXIDE(S)	≤ 67		≥ 33			OP7	+35	+40	3115	
METHYL ETHYL KETONE PEROXIDE(S)	see remark 8)	≥ 48				OP5			3101	3) 8) 13)
"	see remark 9)	≥ 55				OP7			3105	9)
"	see remark 10)	≥ 60				OP8			3107	10)
METHYL ISOBUTYL KETONE PEROXIDE(S)	≤ 62	≥ 19				OP7			3105	22)
METHYL ISOPROPYL KETONE PEROXIDE(S)	See remark 31)	≥ 70				OP8			3109	31)
ORGANIC PEROXIDE, LIQUID, SAMPLE						OP2			3103	11)
ORGANIC PEROXIDE, LIQUID, SAMPLE, TEMPERATURE CONTROLLED						OP2			3113	11)
ORGANIC PEROXIDE, SOLID, SAMPLE						OP2			3104	11)
ORGANIC PEROXIDE, SOLID, SAMPLE, TEMPERATURE CONTROLLED						OP2			3114	11)
3,3,5,7,7-PENTAMETHYL-1,2,4-TRIOXEPANE	≤ 100					OP8			3107	
PEROXYACETIC ACID, TYPE D, stabilized	≤ 43					OP7			3105	13) 14) 19)
PEROXYACETIC ACID, TYPE E, stabilized	≤ 43					OP8			3107	13) 15) 19)
PEROXYACETIC ACID, TYPE F, stabilized	≤ 43					OP8			3109	13) 16) 19)
PEROXYLAURIC ACID	≤ 100					OP8	+35	+40	3118	
PINANYL HYDROPEROXIDE	> 56 – 100					OP7			3105	13)
"	≤ 56	≥ 44				OP8			3109	
POLYETHER POLY-tert-BUTYLPEROXY-CARBONATE	≤ 52		≥ 48			OP8			3107	
1,1,3,3-TETRAMETHYLBUTYL HYDROPEROXIDE	≤ 100					OP7			3105	
1,1,3,3-TETRAMETHYLBUTYL PEROXY-2-ETHYLHEXANOATE	≤ 100					OP7	+15	+20	3115	
1,1,3,3- TETRAMETHYLBUTYL PEROXYNEODECANOATE	≤ 72		≥ 28			OP7	-5	+5	3115	
"	≤ 52 as a stable dispersion in water					OP8	-5	+5	3119	
1,1,3,3-TETRAMETHYLBUTYL PEROXYPIVALATE	≤ 77	≥ 23				OP7	0	+10	3115	
3,6,9-TRIETHYL-3,6,9-TRIMETHYL -1,4,7 TRIPEROXONANE	≤ 42	≥ 58				OP7			3105	28)

Remarks (refer to the last column of the Table in 2.2.52.4):

1) *Diluent type B may always be replaced by diluent type A. The boiling point of diluent type B shall be at least 60°C higher than the SADT of the organic peroxide.*

2) *Available oxygen ≤ 4.7%.*

3) *"EXPLOSIVE" subsidiary risk label required (Model No.1, see 5.2.2.2.2).*

4) *Diluent may be replaced by di-tert-butyl peroxide.*

5) *Available oxygen ≤ 9%.*

6) *With ≤ 9% hydrogen peroxide; available oxygen ≤ 10%.*

7) *Only non-metallic packagings allowed.*

8) *Available oxygen > 10% and ≤ 10.7%, with or without water.*

9) *Available oxygen ≤ 10%, with or without water.*

10) *Available oxygen ≤ 8.2%, with or without water.*

11) *See 2.2.52.1.9.*

12) *Up to 2000 kg per receptacle assigned to ORGANIC PEROXIDE TYPE F on the basis of large scale trials.*

13) *"CORROSIVE" subsidiary risk label required (Model No.8, see 5.2.2.2.2).*

14) *Peroxyacetic acid formulations which fulfil the criteria of the Manual of Tests and Criteria, paragraph 20.4.3 (d).*

15) *Peroxyacetic acid formulations which fulfil the criteria of the Manual of Tests and Criteria, paragraph 20.4.3 (e).*

16) *Peroxyacetic acid formulations which fulfil the criteria of the Manual of Tests and Criteria, paragraph 20.4.3 (f).*

17) *Addition of water to this organic peroxide will decrease its thermal stability.*

18) *No "CORROSIVE" subsidiary risk label (Model No.8, see 5.2.2.2.2) required for concentrations below 80%.*

19) *Mixtures with hydrogen peroxide, water and acid(s).*

20) *With diluent type A, with or without water.*

21) *With ≥ 25% diluent type A by mass, and in addition ethylbenzene.*

22) *With ≥ 19% diluent type A by mass, and in addition methyl isobutyl ketone.*

23) *With < 6% di-tert-butyl peroxide.*

24) *With ≤ 8% 1-isopropylhydroperoxy-4-isopropylhydroxybenzene.*

25) *Diluent type B with boiling point > 110 °C.*

26) *With < 0.5% hydroperoxides content.*

27) *For concentrations more than 56%, "CORROSIVE" subsidiary risk label required (Model No.8, see 5.2.2.2.2).*

28) *Available active oxygen ≤ 7.6% in diluent type A having a 95% boil-off point in the range of 200 - 260 °C.*

29) *Not subject to the requirements of ADN for Class 5.2.*

30) *Diluent type B with boiling point > 130 °C.*

31) *Active oxygen ≤ 6.7%.*

2.2.61 Class 6.1 Toxic substances

2.2.61.1 *Criteria*

2.2.61.1.1 The heading of Class 6.1 covers substances of which it is known by experience or regarding which it is presumed from experiments on animals that in relatively small quantities they are able by a single action or by action of short duration to cause damage to human health, or death, by inhalation, by cutaneous absorption or by ingestion.

2.2.61.1.2 Substances of Class 6.1 are subdivided as follows:

T Toxic substances without subsidiary risk:

T1 Organic, liquid;
T2 Organic, solid;
T3 Organometallic substances;
T4 Inorganic, liquid;
T5 Inorganic, solid;
T6 Liquid, used as pesticides;
T7 Solid, used as pesticides;
T8 Samples;
T9 Other toxic substances;

TF Toxic substances, flammable:

TF1 Liquid;
TF2 Liquid, used as pesticides;
TF3 Solid;

TS Toxic substances, self-heating, solid;

TW Toxic substances, which, in contact with water, emit flammable gases:

TW1 Liquid;
TW2 Solid;

TO Toxic substances, oxidizing:

TO1 Liquid;
TO2 Solid;

TC Toxic substances, corrosive:

TC1 Organic, liquid;
TC2 Organic, solid;
TC3 Inorganic, liquid;
TC4 Inorganic, solid;

TFC Toxic substances, flammable, corrosive.

Definitions

2.2.61.1.3 For the purposes of ADN:

LD_{50} (median lethal dose) for acute oral toxicity is the statistically derived single dose of a substance that can be expected to cause death within 14 days in 50 per cent of young adult

albino rats when administered by the oral route. The LD_{50} value is expressed in terms of mass of test substance per mass of test animal (mg/kg);

LD_{50} for acute dermal toxicity is that dose of the substance which, administered by continuous contact for 24 hours with the bare skin of albino rabbits, is most likely to cause death within 14 days in one half of the animals tested. The number of animals tested shall be sufficient to give a statistically significant result and be in conformity with good pharmacological practice. The result is expressed in milligrams per kg body mass;

LC_{50} for acute toxicity on inhalation is that concentration of vapour, mist or dust which, administered by continuous inhalation to both male and female young adult albino rats for one hour, is most likely to cause death within 14 days in one half of the animals tested. A solid substance shall be tested if at least 10% (by mass) of its total mass is likely to be dust in a respirable range, e.g. the aerodynamic diameter of that particle-fraction is 10 μm or less. A liquid substance shall be tested if a mist is likely to be generated in a leakage of the transport containment. Both for solid and liquid substances more than 90% (by mass) of a specimen prepared for inhalation toxicity shall be in the respirable range as defined above. The result is expressed in milligrams per litre of air for dusts and mists or in millilitres per cubic metre of air (parts per million) for vapours.

Classification and assignment of packing groups

2.2.61.1.4 Substances of Class 6.1 shall be classified in three packing groups according to the degree of danger they present for carriage, as follows:

Packing group I:	highly toxic substances
Packing group II:	toxic substances
Packing group III:	slightly toxic substances.

2.2.61.1.5 Substances, mixtures, solutions and articles classified in Class 6.1 are listed in Table A of Chapter 3.2. The assignment of substances, mixtures and solutions not mentioned by name in Table A of Chapter 3.2 to the relevant entry of sub-section 2.2.61.3 and to the relevant packing group in accordance with the provisions of Chapter 2.1, shall be made according to the following criteria in 2.2.61.1.6 to 2.2.61.1.11.

2.2.61.1.6 To assess the degree of toxicity, account shall be taken of human experience of instances of accidental poisoning, as well as special properties possessed by any individual substances: liquid state, high volatility, any special likelihood of cutaneous absorption, and special biological effects.

2.2.61.1.7 In the absence of observations on humans, the degree of toxicity shall be assessed using the available data from animal experiments in accordance with the table below:

	Packing group	Oral toxicity LD_{50} (mg/kg)	Dermal toxicity LD_{50} (mg/kg)	Inhalation toxicity by dusts and mists LC_{50} (mg/l)
Highly toxic	I	≤ 5.0	≤ 50	≤ 0.2
Toxic	II	> 5.0 and ≤ 50	> 50 and ≤ 200	> 0.2 and ≤ 2.0
Slightly toxic	III [a]	> 50 and ≤ 300	> 200 and $\leq$ 1 000	> 2.0 and ≤ 4.0

[a] *Tear gas substances shall be included in packing group II even if data concerning their toxicity correspond to packing group III criteria.*

2.2.61.1.7.1 Where a substance exhibits different degrees of toxicity for two or more kinds of exposure, it shall be classified under the highest such degree of toxicity.

2.2.61.1.7.2 Substances meeting the criteria of Class 8 and with an inhalation toxicity of dusts and mists (LC_{50}) leading to packing group I shall only be accepted for an allocation to Class 6.1 if the toxicity through oral ingestion or dermal contact is at least in the range of packing groups I or II. Otherwise an assignment to Class 8 shall be made if appropriate (see footnote [6] in 2.2.8.1.4).

2.2.61.1.7.3 The criteria for inhalation toxicity of dusts and mists are based on LC_{50} data relating to 1-hour exposure, and where such information is available it shall be used. However, where only LC_{50} data relating to 4-hour exposure are available, such figures can be multiplied by four and the product substituted in the above criteria, i.e. LC_{50} value multiplied by four (4 hour) is considered the equivalent of LC_{50} (1 hour).

Inhalation toxicity of vapours

2.2.61.1.8 Liquids giving off toxic vapours shall be classified into the following groups where "V" is the saturated vapour concentration (in ml/m^3 of air) (volatility) at 20 °C and standard atmospheric pressure:

	Packing group	
Highly toxic	I	Where $V \geq 10\ LC_{50}$ and $LC_{50} \leq 1\ 000$ ml/m^3
Toxic	II	Where $V \geq LC_{50}$ and $LC_{50} \leq 3\ 000$ ml/m^3 and the criteria for packing group I are not met
Slightly toxic	III[a]	Where $V \geq 1/5\ LC_{50}$ and $LC_{50} \leq 5\ 000$ ml/m^3 and the criteria for packing groups I and II are not met

[a] *Tear gas substances shall be included in packing group II even if data concerning their toxicity correspond to packing group III criteria.*

These criteria for inhalation toxicity of vapours are based on LC_{50} data relating to 1-hour exposure, and where such information is available, it shall be used.

However, where only LC_{50} data relating to 4-hour exposure to the vapours are available, such figures can be multiplied by two and the product substituted in the above criteria, i.e. LC_{50} (4 hour) × 2 is considered the equivalent of LC_{50} (1 hour).

Group borderlines inhalation toxicity of vapours

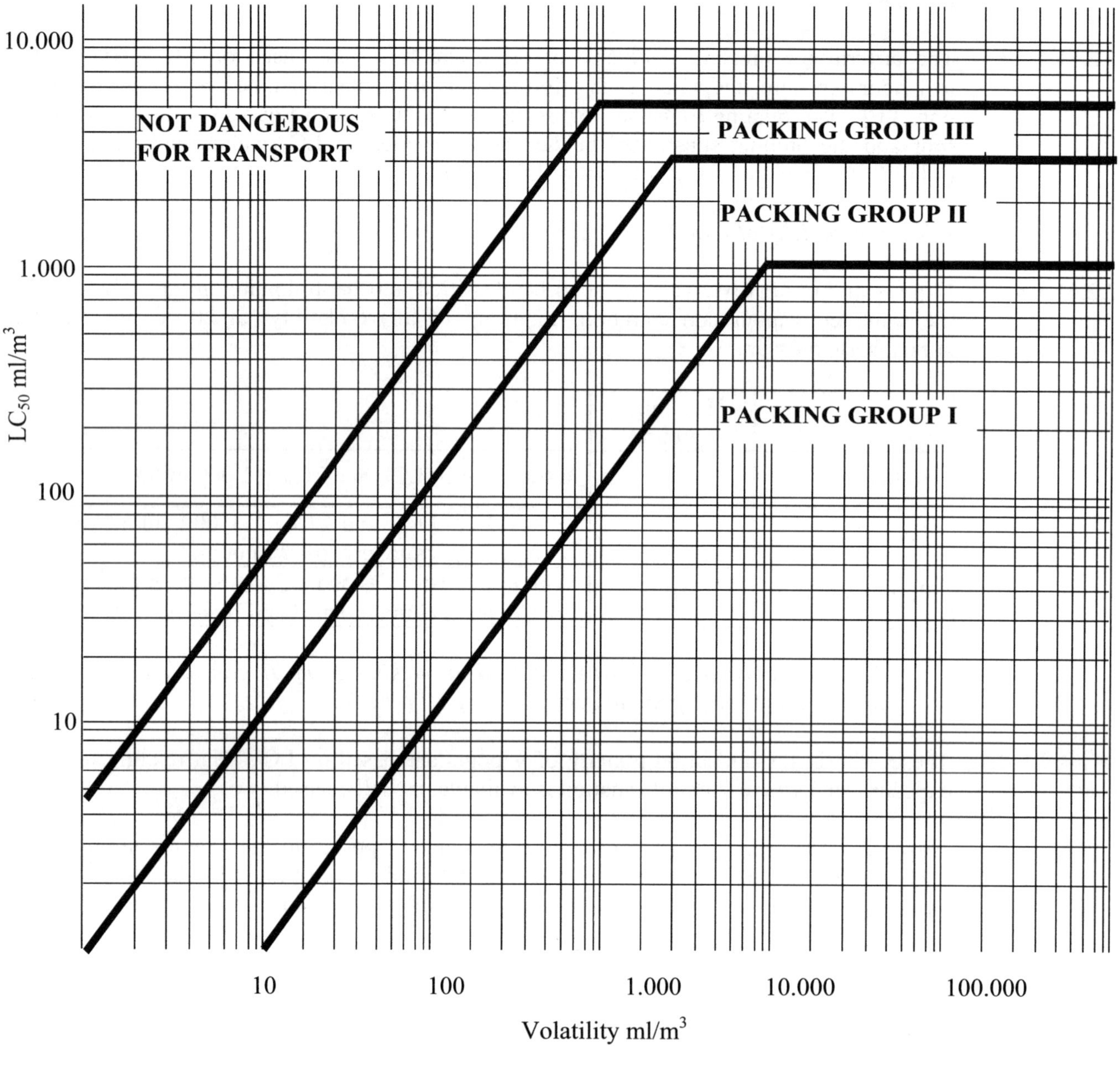

In this figure, the criteria are expressed in graphical form, as an aid to easy classification. However, due to approximations inherent in the use of graphs, substances falling on or near group borderlines shall be checked using numerical criteria.

Mixtures of liquids

2.2.61.1.9 Mixtures of liquids which are toxic on inhalation shall be assigned to packing groups according to the following criteria:

2.2.61.1.9.1 If LC_{50} is known for each of the toxic substances constituting the mixture, the packing group may be determined as follows:

(a) calculation of the LC_{50} of the mixture:

$$LC_{50}\ (\text{mixture}) = \frac{1}{\sum_{i=1}^{1} \frac{f_i}{LC_{50i}}}$$

where f_i = molar fraction of constituent i of the mixture;

LC_{50i} = average lethal concentration of constituent i in ml/m^3.

(b) calculation of volatility of each mixture constituent:

$$V_i = P_i \times \frac{10^6}{101.3}\ (ml/m^3)$$

where P_i = partial pressure of constituent i in kPa at 20 °C and at standard atmospheric pressure.

(c) calculation of the ratio of volatility to LC_{50}:

$$R = \sum_{i=1}^{n} \frac{V_i}{LC_{50i}}$$

(d) the values calculated for LC_{50} (mixture) and R are then used to determine the packing group of the mixture:

Packing group I: $R \geq 10$ and LC_{50} (mixture) $\leq$ 1 000 ml/m^3;

Packing group II: $R \geq 1$ and LC_{50} (mixture) $\leq$ 3 000 ml/m^3, if the mixture does not meet the criteria for packing group I;

Packing group III: $R \geq 1/5$ and LC_{50} (mixture) $\leq$ 5 000 ml/m^3, if the mixture does not meet the criteria of packing groups I or II.

2.2.61.1.9.2 In the absence of LC_{50} data on the toxic constituent substances, the mixture may be assigned to a group based on the following simplified threshold toxicity tests. When these threshold tests are used, the most restrictive group shall be determined and used for carrying the mixture.

2.2.61.1.9.3 A mixture is assigned to packing group I only if it meets both of the following criteria:

(a) A sample of the liquid mixture is vaporized and diluted with air to create a test atmosphere of 1000 ml/m^3 vaporized mixture in air. Ten albino rats (5 male and 5 female) are exposed to the test atmosphere for 1 hour and observed for 14 days. If five or more of the animals die within the 14-day observation period, the mixture is presumed to have an LC_{50} equal to or less than 1000 ml/m^3;

(b) A sample of vapour in equilibrium with the liquid mixture is diluted with 9 equal volumes of air to form a test atmosphere. Ten albino rats (5 male and 5 female) are exposed to the test atmosphere for 1 hour and observed for 14 days. If five or more of the animals die within the 14-day observation period, the mixture is presumed to have a volatility equal to or greater than 10 times the mixture LC_{50}.

2.2.61.1.9.4 A mixture is assigned to packing group II only if it meets both of the following criteria, and does not meet the criteria for packing group I:

(a) A sample of the liquid mixture is vaporized and diluted with air to create a test atmosphere of 3000 ml/m^3 vaporized mixture in air. Ten albino rats (5 male and 5 female) are exposed to the test atmosphere for 1 hour and observed for 14 days. If five or more of the animals die within the 14-day observation period, the mixture is presumed to have an LC_{50} equal to or less than 3000 ml/m^3;

(b) A sample of the vapour in equilibrium with the liquid mixture is used to form a test atmosphere. Ten albino rats (5 male and 5 female) are exposed to the test atmosphere for 1 hour and observed for 14 days. If five or more of the animals die within the 14-day observation period, the mixture is presumed to have a volatility equal to or greater than the mixture LC_{50}.

2.2.61.1.9.5 A mixture is assigned to packing group III only if it meets both of the following criteria, and does not meet the criteria for packing groups I or II:

(a) A sample of the liquid mixture is vaporized and diluted with air to create a test atmosphere of 5000 ml/m^3 vaporized mixture in air. Ten albino rats (5 male and 5 female) are exposed to the test atmosphere for 1 hour and observed for 14 days. If five or more of the animals die within the 14-day observation period, the mixture is presumed to have an LC_{50} equal to or less than 5000 ml/m^3;

(b) The vapour concentration (volatility) of the liquid mixture is measured and if the vapour concentration is equal to or greater than 1000 ml/m^3, the mixture is presumed to have a volatility equal to or greater than 1/5 the mixture LC_{50}.

Methods for determining oral and dermal toxicity of mixtures

2.2.61.1.10 When classifying and assigning the appropriate packing group to mixtures in Class 6.1 in accordance with the oral and dermal toxicity criteria (see 2.2.61.1.3), it is necessary to determine the acute LD_{50} of the mixture.

2.2.61.1.10.1 If a mixture contains only one active substance, and the LD_{50} of that constituent is known, in the absence of reliable acute oral and dermal toxicity data on the actual mixture to be carried, the oral or dermal LD_{50} may be obtained by the following method:

$$LD_{50} \text{ value of preparation} = \frac{LD_{50} \text{ value of active substance} \times 100}{\text{percentage of active substance by mass}}$$

2.2.61.1.10.2 If a mixture contains more than one active constituent, there are three possible approaches that may be used to determine the oral or dermal LD_{50} of the mixture. The preferred method is to obtain reliable acute oral and dermal toxicity data on the actual mixture to be carried. If reliable, accurate data are not available, then either of the following methods may be performed:

(a) Classify the formulation according to the most hazardous constituent of the mixture as if that constituent were present in the same concentration as the total concentration of all active constituents; or

(b) Apply the formula:

$$\frac{C_A}{T_A} + \frac{C_B}{T_B} + \ldots + \frac{C_Z}{T_Z} = \frac{100}{T_M}$$

where:

C = the percentage concentration of constituent A, B, ... Z in the mixture;

T = the oral LD_{50} values of constituent A, B, ... Z;

T_M = the oral LD_{50} value of the mixture.

***NOTE:** This formula can also be used for dermal toxicities provided that this information is available on the same species for all constituents. The use of this formula does not take into account any potentiation or protective phenomena.*

Classification of pesticides

2.2.61.1.11 All active pesticide substances and their preparations for which the LC_{50} and/or LD_{50} values are known and which are classified in Class 6.1 shall be classified under appropriate packing groups in accordance with the criteria given in 2.2.61.1.6 to 2.2.61.1.9. Substances and preparations which are characterized by subsidiary risks shall be classified according to the precedence of hazard Table in 2.1.3.10 with the assignment of appropriate packing groups.

2.2.61.1.11.1 If the oral or dermal LD_{50} value for a pesticide preparation is not known, but the LD_{50} value of its active substance(s) is known, the LD_{50} value for the preparation may be obtained by applying the procedures in 2.2.61.1.10.

***NOTE**: LD_{50} toxicity data for a number of common pesticides may be obtained from the most current edition of the document "The WHO Recommended Classification of Pesticides by Hazard and Guidelines to Classification" available from the International Programme on Chemical Safety, World Health Organisation (WHO), 1211 Geneva 27, Switzerland. While that document may be used as a source of LD_{50} data for pesticides, its classification system shall not be used for purposes of transport classification of, or assignment of packing groups to, pesticides, which shall be in accordance with the requirements of ADN.*

2.2.61.1.11.2 The proper shipping name used in the carriage of the pesticide shall be selected on the basis of the active ingredient, of the physical state of the pesticide and any subsidiary risks it may exhibit (see 3.1.2).

2.2.61.1.12 If substances of Class 6.1, as a result of admixtures, come into categories of risk different from those to which the substances mentioned by name in Table A of Chapter 3.2 belong, these mixtures or solutions shall be assigned to the entries to which they belong on the basis of their actual degree of danger.

NOTE: *For the classification of solutions and mixtures (such as preparations and wastes), see also 2.1.3.*

2.2.61.1.13 On the basis of the criteria of 2.2.61.1.6 to 2.2.61.1.11, it may also be determined whether the nature of a solution or mixture mentioned by name or containing a substance mentioned by name is such that the solution or mixture is not subject to the requirements for this Class.

2.2.61.1.14 Substances, solutions and mixtures, with the exception of substances and preparations used as pesticides, which do not meet the criteria of Directives 67/548/EEC[3] or 1999/45/EC[4] as amended and which are not therefore classified as highly toxic, toxic or harmful according to these directives, as amended, may be considered as substances not belonging to Class 6.1.

2.2.61.2 ***Substances not accepted for carriage***

2.2.61.2.1 Chemically unstable substances of Class 6.1 shall not be accepted for carriage unless the necessary steps have been taken to prevent their dangerous decomposition or polymerization during carriage. To this end, it shall in particular be ensured that receptacles and tanks do not contain any substance(s) likely to cause such a reaction.

2.2.61.2.2 The following substances and mixtures shall not be accepted for carriage:

- Hydrogen cyanide, anhydrous or in solution, which do not meet the descriptions of UN Nos. 1051, 1613, 1614 and 3294;
- Metal carbonyls, having a flash-point below 23 °C, other than UN Nos. 1259 NICKEL CARBONYL and 1994 IRON PENTACARBONYL;
- 2,3,7,8-TETRACHLORODIBENZO-p-DIOXINE (TCDD) in concentrations considered highly toxic in accordance with the criteria in 2.2.61.1.7;
- UN No. 2249 DICHLORODIMETHYL ETHER, SYMMETRICAL;
- Preparations of phosphides without additives inhibiting the emission of toxic flammable gases.

---

[3] *Council Directive 67/548/EEC of 27 June 1967 on the approximation of laws, regulations and administrative provisions relating to the classification, packaging and labelling of dangerous substances (Official Journal of the European Communities No. L 196 of 16.08.1967, page 1).*

[4] *Directive 1999/45/EC of the European Parliament and of the Council of 31 May 1999 on the approximation of laws, regulations and administrative provisions of the Member States relating to the classification, packaging and labelling of dangerous preparations (Official Journal of the European Communities No. L 200 of 30 July 1999, pages 1 to 68).*

2.2.61.3 ***List of collective entries***

Toxic substances <u>without</u> subsidiary risk(s)

Organic

liquid [a] **T1**

1583	CHLOROPICRIN MIXTURE, N.O.S.
1602	DYE, LIQUID, TOXIC, N.O.S., or
1602	DYE INTERMEDIATE, LIQUID, TOXIC, N.O.S.
1693	TEAR GAS SUBSTANCE, LIQUID, N.O.S.
1851	MEDICINE, LIQUID, TOXIC, N.O.S.
2206	ISOCYANATES, TOXIC, N.O.S. or
2206	ISOCYANATE SOLUTION, TOXIC, N.O.S.
3140	ALKALOIDS, LIQUID, N.O.S. or
3140	ALKALOID SALTS, LIQUID, N.O.S.
3142	DISINFECTANT, LIQUID, TOXIC, N.O.S.
3144	NICOTINE COMPOUND, LIQUID, N.O.S. or
3144	NICOTINE PREPARATION, LIQUID, N.O.S.
3172	TOXINS, EXTRACTED FROM LIVING SOURCES, LIQUID, N.O.S.
3276	NITRILES, TOXIC, LIQUID, N.O.S
3278	ORGANOPHOSPHORUS COMPOUND, TOXIC, LIQUID, N.O.S.
3381	TOXIC BY INHALATION LIQUID, N.O.S. with an inhalation toxicity lower than or equal to 200 ml/m^3 and saturated vapour concentration greater than or equal to 500 LC_{50}
3382	TOXIC BY INHALATION LIQUID, N.O.S. with an inhalation toxicity lower than or equal to 1000 ml/m^3 and saturated vapour concentration greater than or equal to 10 LC_{50}
2810	TOXIC LIQUID, ORGANIC, N.O.S.

solid [a, b] **T2**

1544	ALKALOIDS, SOLID, N.O.S. or
1544	ALKALOID SALTS, SOLID, N.O.S.
1601	DISINFECTANT, SOLID, TOXIC, N.O.S.
1655	NICOTINE COMPOUND, SOLID, N.O.S., or
1655	NICOTINE PREPARATION, SOLID, N.O.S.
3448	TEAR GAS SUBSTANCE, SOLID, N.O.S.
3143	DYE, SOLID, TOXIC, N.O.S. or
3143	DYE INTERMEDIATE, SOLID, TOXIC, N.O.S.
3462	TOXINS, EXTRACTED FROM LIVING SOURCES, SOLID, N.O.S.
3249	MEDICINE, SOLID, TOXIC, N.O.S.
3464	ORGANOPHOSPHORUS COMPOUND, TOXIC, SOLID, N.O.S.
3439	NITRILES, TOXIC, SOLID, N.O.S.
2811	TOXIC SOLID, ORGANIC, N.O.S.

Organometallic [c, d] **T3**

2026	PHENYLMERCURIC COMPOUND, N.O.S.
2788	ORGANOTIN COMPOUND, LIQUID, N.O.S.
3146	ORGANOTIN COMPOUND, SOLID, N.O.S.
3280	ORGANOARSENIC COMPOUND, LIQUID, N.O.S.
3465	ORGANOARSENIC COMPOUND, SOLID, N.O.S.
3281	METAL CARBONYLS, LIQUID, N.O.S.
3466	METAL CARBONYLS, SOLID, N.O.S.
3282	ORGANOMETALLIC COMPOUND, TOXIC, LIQUID, N.O.S.
3467	ORGANOMETALLIC COMPOUND, TOXIC, SOLID, N.O.S.

(cont'd on next page)

[a] *Substances and preparations containing alkaloids or nicotine used as pesticides shall be classified under UN No. 2588 PESTICIDES, SOLID, TOXIC, N.O.S., UN No. 2902 PESTICIDES, LIQUID, TOXIC, N.O.S. or UN No. 2903 PESTICIDES, LIQUID, TOXIC, FLAMMABLE, N.O.S.*

[b] *Active substances and triturations or mixtures of substances intended for laboratories and experiments and for the manufacture of pharmaceutical products with other substances shall be classified according to their toxicity (see 2.2.61.1.7 to 2.2.61.1.11).*

[c] *Self-heating substances, slightly toxic and spontaneously combustible organometallic compounds, are substances of Class 4.2.*

[d] *Water-reactive substances, slightly toxic, and water-reactive organometallic compounds, are substances of Class 4.3.*

2.2.61.3 *List of collective entries (cont'd)*

Toxic substances without subsidiary risk(s) *(cont'd)*

Inorganic

liquid [e] **T4**

1556	ARSENIC COMPOUND, LIQUID, N.O.S., inorganic including: Arsenates, n.o.s., Arsenites, n.o.s.; and Arsenic sulphides, n.o.s.
1935	CYANIDE SOLUTION, N.O.S.
2024	MERCURY COMPOUND, LIQUID, N.O.S.
3141	ANTIMONY COMPOUND, INORGANIC, LIQUID, N.O.S.
3440	SELENIUM COMPOUND, LIQUID, N.O.S.
3381	TOXIC BY INHALATION LIQUID, N.O.S. with an inhalation toxicity lower than or equal to 200 ml/m^3 and saturated vapour concentration greater than or equal to 500 LC_{50}
3382	TOXIC BY INHALATION LIQUID, N.O.S. with an inhalation toxicity lower than or equal to 1000 ml/m^3 and saturated vapour concentration greater than or equal to 10 LC_{50}
3287	TOXIC LIQUID, INORGANIC, N.O.S.

solids [f, g] **T5**

1549	ANTIMONY COMPOUND, INORGANIC, SOLID, N.O.S
1557	ARSENIC COMPOUND, SOLID, N.O.S., including: Arsenates, n.o.s.; Arsenites, n.o.s.; and Arsenic sulphides, n.o.s.
1564	BARIUM COMPOUND, N.O.S.
1566	BERYLLIUM COMPOUND, N.O.S.
1588	CYANIDES, INORGANIC, SOLID, N.O.S.
1707	THALLIUM COMPOUND, N.O.S.
2025	MERCURY COMPOUND, SOLID, N.O.S.
2291	LEAD COMPOUND, SOLUBLE, N.O.S.
2570	CADMIUM COMPOUND
2630	SELENATES or
2630	SELENITES
2856	FLUOROSILICATES, N.O.S.
3283	SELENIUM COMPOUND, SOLID, N.O.S.
3284	TELLURIUM COMPOUND, N.O.S.
3285	VANADIUM COMPOUND, N.O.S.
3288	TOXIC SOLID, INORGANIC, N.O.S.

Pesticides

liquid [h] **T6**

2992	CARBAMATE PESTICIDE, LIQUID, TOXIC
2994	ARSENICAL PESTICIDE, LIQUID, TOXIC
2996	ORGANOCHLORINE PESTICIDE, LIQUID, TOXIC
2998	TRIAZINE PESTICIDE, LIQUID, TOXIC
3006	THIOCARBAMATE PESTICIDE, LIQUID, TOXIC
3010	COPPER BASED PESTICIDE, LIQUID, TOXIC
3012	MERCURY BASED PESTICIDE, LIQUID, TOXIC
3014	SUBSTITUTED NITROPHENOL PESTICIDE, LIQUID, TOXIC
3016	BIPYRIDILIUM PESTICIDE, LIQUID, TOXIC
3018	ORGANOPHOSPHORUS PESTICIDE, LIQUID, TOXIC
3020	ORGANOTIN PESTICIDE, LIQUID, TOXIC
3026	COUMARIN DERIVATIVE PESTICIDE, LIQUID, TOXIC
3348	PHENOXYACETIC ACID DERIVATIVE PESTICIDE, LIQUID, TOXIC
3352	PYRETHROID PESTICIDE, LIQUID, TOXIC

(cont'd on next page)

---

[e] *Mercury fulminate, wetted with not less than 20% water, or mixture of alcohol and water by mass is a substance of Class 1, UN No. 0135.*

[f] *Ferricyanides, ferrocyanides, alkaline thiocyanates and ammonium thiocyanates are not subject to the provisions of ADN.*

[g] *Lead salts and lead pigments which, when mixed in a ratio of 1:1,000 with 0.07M hydrochloric acid and stirred for one hour at a temperature of 23 °C ± 2 °C, exhibit a solubility of 5% or less, are not subject to the provisions of ADN.*

[h] *Articles impregnated with this pesticide, such as fibreboard plates, paper strips, cotton-wool balls, sheets of plastics material, in hermetically closed wrappings, are not subject to the provisions of ADN.*

2.2.61.3 *List of collective entries (cont'd)*

Toxic substances without subsidiary risk(s) ***(cont'd)***

Group	Subgroup	Code	UN No.	Name
Pesticides ***(cont'd)***			2902	PESTICIDE, LIQUID, TOXIC, N.O.S.
	Solid[h]	**T7**	2757	CARBAMATE PESTICIDE, SOLID, TOXIC
			2759	ARSENICAL PESTICIDE, SOLID, TOXIC
			2761	ORGANOCHLORINE PESTICIDE, SOLID, TOXIC
			2763	TRIAZINE PESTICIDE, SOLID, TOXIC
			2771	THIOCARBAMATE PESTICIDE, SOLID, TOXIC
			2775	COPPER BASED PESTICIDE, SOLID, TOXIC
			2777	MERCURY BASED PESTICIDE, SOLID, TOXIC
			2779	SUBSTITUTED NITROPHENOL PESTICIDE, SOLID, TOXIC
			2781	BIPYRIDILIUM PESTICIDE, SOLID, TOXIC
			2783	ORGANOPHOSPHORUS PESTICIDE, SOLID, TOXIC
			2786	ORGANOTIN PESTICIDE, SOLID, TOXIC
			3027	COUMARIN DERIVATIVE PESTICIDE, SOLID, TOXIC
			3048	ALUMINIUM PHOSPHIDE PESTICIDE
			3345	PHENOXYACETIC ACID DERIVATIVE PESTICIDE, SOLID, TOXIC
			3349	PYRETHROID PESTICIDE, SOLID, TOXIC
			2588	PESTICIDE, SOLID, TOXIC, N.O.S.
Samples		**T8**	3315	CHEMICAL SAMPLE, TOXIC
Other toxic substances [i]		**T9**	3243	SOLIDS CONTAINING TOXIC LIQUID, N.O.S.

Toxic substances with subsidiary risk(s)

Group	Subgroup	Code	UN No.	Name
Flammable **TF**	**Liquid** [j, k]	**TF1**	3071	MERCAPTANS, LIQUID, TOXIC, FLAMMABLE, N.O.S. or
			3071	MERCAPTAN MIXTURE, LIQUID, TOXIC, FLAMMABLE, N.O.S.
			3080	ISOCYANATES, TOXIC, FLAMMABLE, N.O.S. or
			3080	ISOCYANATE SOLUTION, TOXIC, FLAMMABLE, N.O.S.
			3275	NITRILES, TOXIC, FLAMMABLE, N.O.S.
			3279	ORGANOPHOSPHORUS COMPOUND, TOXIC, FLAMMABLE, N.O.S.
			3383	TOXIC BY INHALATION LIQUID, FLAMMABLE, N.O.S. with an inhalation toxicity lower than or equal to 200 ml/m^3 and saturated vapour concentration greater than or equal to 500 LC_{50}
			3384	TOXIC BY INHALATION LIQUID, FLAMMABLE, N.O.S. with an inhalation toxicity lower than or equal to 1000 ml/m^3 and saturated vapour concentration greater than or equal to 10 LC_{50}
			2929	TOXIC LIQUID, FLAMMABLE, ORGANIC, N.O.S.

(cont'd on next page)

h *Articles impregnated with this pesticide, such as fibreboard plates, paper strips, cotton-wool balls, sheets of plastics material, in hermetically closed wrappings, are not subject to the provisions of ADN.*

i *Mixtures of solids which are not subject to the provisions of ADN and of toxic liquids may be carried under UN No. 3243 without first applying the classification criteria of Class 6.1, provided there is no free liquid visible at the time the substance is loaded or at the time the packaging, container or cargo transport unit is closed. Each packaging shall correspond to a design type that has passed a leakproofness test at the packing group II level. This entry shall not be used for solids containing a packing group I liquid.*

j *Highly toxic or toxic, flammable liquids having a flash-point below 23 °C excluding substances which are highly toxic on inhalation, i.e. UN Nos. 1051, 1092, 1098, 1143, 1163, 1182, 1185, 1238, 1239, 1244, 1251, 1259, 1613, 1614, 1695, 1994, 2334, 2382, 2407, 2438, 2480, 2482, 2484, 2485, 2606, 2929, 3279 and 3294 are substances of Class 3.*

k *Flammable liquids, slightly toxic, with the exception of substances and preparations used as pesticides, having a flash-point between 23 °C and 60 °C inclusive, are substances of Class 3.*

2.2.61.3 ***List of collective entries (cont'd)***

Toxic substances with subsidiary risk(s) (cont'd)

Subsidiary risk	Sub-group	Classification code	UN No.	Name and description
Flammable TF *(cont'd)*	**pesticides, liquid (flash-point not less than 23 °C)**	**TF2**	2991	CARBAMATE PESTICIDE, LIQUID, TOXIC, FLAMMABLE
			2993	ARSENICAL PESTICIDE, LIQUID, TOXIC, FLAMMABLE
			2995	ORGANOCHLORINE PESTICIDE, LIQUID, TOXIC, FLAMMABLE
			2997	TRIAZINE PESTICIDE, LIQUID, TOXIC, FLAMMABLE
			3005	THIOCARBAMATE PESTICIDE, LIQUID, TOXIC, FLAMMABLE
			3009	COPPER BASED PESTICIDE, LIQUID, TOXIC, FLAMMABLE
			3011	MERCURY BASED PESTICIDE, LIQUID, TOXIC, FLAMMABLE
			3013	SUBSTITUTED NITROPHENOL PESTICIDE, LIQUID, TOXIC, FLAMMABLE
			3015	BIPYRIDILIUM PESTICIDE, LIQUID, TOXIC, FLAMMABLE
			3017	ORGANOPHOSPHORUS PESTICIDE, LIQUID, TOXIC, FLAMMABLE
			3019	ORGANOTIN PESTICIDE, LIQUID, TOXIC, FLAMMABLE
			3025	COUMARIN DERIVATIVE PESTICIDE, LIQUID, TOXIC, FLAMMABLE
			3347	PHENOXYACETIC ACID DERIVATIVE PESTICIDE, LIQUID, TOXIC, FLAMMABLE
			3351	PYRETHROID PESTICIDE, LIQUID, TOXIC, FLAMMABLE
			2903	PESTICIDE, LIQUID, TOXIC, FLAMMABLE, N.O.S.
	solid	**TF3**	1700	TEAR GAS CANDLES
			2930	TOXIC SOLID, FLAMMABLE, ORGANIC, N.O.S.
Solid, self-heating [c] **TS**			3124	TOXIC SOLID, SELF-HEATING, N.O.S.
Water-reactive [d] **TW**	**liquid**	**TW1**	3385	TOXIC BY INHALATION LIQUID, WATER-REACTIVE, N.O.S. with an inhalation toxicity lower than or equal to 200 ml/m^3 and saturated vapour concentration greater than or equal to 500 LC_{50}
			3386	TOXIC BY INHALATION LIQUID, WATER-REACTIVE, N.O.S. with an inhalation toxicity lower than or equal to 1000 ml/m^3 and saturated vapour concentration greater than or equal to 10 LC_{50}
			3123	TOXIC LIQUID, WATER-REACTIVE, N.O.S.
	solid [n]	**TW2**	3125	TOXIC SOLID, WATER-REACTIVE, N.O.S.
Oxidizing [l] **TO**	**liquid**	**TO1**	3387	TOXIC BY INHALATION LIQUID, OXIDIZING, N.O.S. with an inhalation toxicity lower than or equal to 200 ml/m^3 and saturated vapour concentration greater than or equal to 500 LC_{50}
			3388	TOXIC BY INHALATION LIQUID, OXIDIZING, N.O.S. with an inhalation toxicity lower than or equal to 1000 ml/m^3 and saturated vapour concentration greater than or equal to 10 LC_{50}
			3122	TOXIC LIQUID, OXIDIZING, N.O.S.
	solid	**TO2**	3086	TOXIC SOLID, OXIDIZING, N.O.S.
Corrosive [m] **TC** — **organic**	**liquid**	**TC1**	3277	CHLOROFORMATES, TOXIC, CORROSIVE, N.O.S.
			3361	CHLOROSILANES, TOXIC, CORROSIVE, N.O.S.
			3389	TOXIC BY INHALATION LIQUID, CORROSIVE, N.O.S. with an inhalation toxicity lower than or equal to 200 ml/m^3 and saturated vapour concentration greater than or equal to 500 LC_{50}
			3390	TOXIC BY INHALATION LIQUID, CORROSIVE, N.O.S. with an inhalation toxicity lower than or equal to 1000 ml/m^3 and saturated vapour concentration greater than or equal to 10 LC_{50}
			2927	TOXIC LIQUID, CORROSIVE, ORGANIC, N.O.S.
	solid	**TC2**	2928	TOXIC SOLID, CORROSIVE, ORGANIC, N.O.S.

(cont'd on next page)

---

c *Self-heating substances, slightly toxic and spontaneously combustible organometallic compounds, are substances of Class 4.2.*

d *Water-reactive substances, slightly toxic, and water-reactive organometallic compounds, are substances of Class 4.3.*

l *Oxidizing substances, slightly toxic, are substances of Class 5.1.*

m *Substances slightly toxic and slightly corrosive, are substances of Class 8.*

n *Metal phosphides assigned to UN Nos. 1360, 1397, 1432, 1714, 2011 and 2013 are substances of Class 4.3.*

2.2.61.3 *List of collective entries (cont'd)*

Toxic substances with subsidiary risk(s) (cont'd)

Corrosive [m] **TC** *(cont'd)*	**organic** *(cont'd)*	**solid**	**TC2**	2928	TOXIC SOLID, CORROSIVE, ORGANIC, N.O.S.
	inorganic	**liquid**	**TC3**	3389	TOXIC BY INHALATION LIQUID, CORROSIVE, N.O.S. with an inhalation toxicity lower than or equal to 200 ml/m^3 and saturated vapour concentration greater than or equal to 500 LC_{50}
				3390	TOXIC BY INHALATION LIQUID, CORROSIVE, N.O.S. with an inhalation toxicity lower than or equal to 1000 ml/m^3 and saturated vapour concentration greater than or equal to 10 LC_{50}
				3289	TOXIC LIQUID, CORROSIVE, INORGANIC, N.O.S.
		solid	**TC4**	3290	TOXIC SOLID, CORROSIVE, INORGANIC, N.O.S.
Flammable, corrosive TFC				2742	CHLOROFORMATES, TOXIC, CORROSIVE, FLAMMABLE, N.O.S.
				3362	CHLOROSILANES, TOXIC, CORROSIVE, FLAMMABLE, N.O.S.
					(No other collective entry available; if need be, classification under a collective entry with a classification code to be determined according to the table of precedence of hazards in 2.1.3.10)

[m] *Substances slightly toxic and slightly corrosive, are substances of Class 8.*

2.2.62 **Class 6.2 Infectious substances**

2.2.62.1 ***Criteria***

2.2.62.1.1 The heading of Class 6.2 covers infectious substances. For the purposes of ADN, infectious substances are substances which are known or are reasonably expected to contain pathogens. Pathogens are defined as micro-organisms (including bacteria, viruses, rickettsiae, parasites, fungi) and other agents such as prions, which can cause disease in humans or animals.

NOTE 1: *Genetically modified micro-organisms and organisms, biological products, diagnostic specimens and infected live animals shall be assigned to this Class if they meet the conditions for this Class.*

NOTE 2: *Toxins from plant, animal or bacterial sources which do not contain any infectious substances or organisms or which are not contained in them are substances of Class 6.1, UN No. 3172 or 3462.*

2.2.62.1.2 Substances of Class 6.2 are subdivided as follows:

I1 Infectious substances affecting humans;

I2 Infectious substances affecting animals only;

I3 Clinical waste;

I4 Biological substances, category B.

Definitions

2.2.62.1.3 For the purposes of ADN,

"Biological products" are those products derived from living organisms which are manufactured and distributed in accordance with the requirements of appropriate national authorities, which may have special licensing requirements, and are used either for prevention, treatment, or diagnosis of disease in humans or animals, or for development, experimental or investigational purposes related thereto. They include, but are not limited to, finished or unfinished products such as vaccines;

"*Cultures*" are the result of a process by which pathogens are intentionally propagated. This definition does not include human or animal patient specimens as defined in this paragraph;

"*Genetically modified micro-organisms and organisms*" are micro-organisms and organisms in which genetic material has been purposely altered through genetic engineering in a way that does not occur naturally;

"*Medical or clinical wastes*" are wastes derived from the medical treatment of animals or humans or from bio-research;

"Patient specimens" are human or animal materials, collected directly from humans or animals, including, but not limited to, excreta, secreta, blood and its components, tissue and tissue fluid swabs, and body parts being carried for purposes such as research, diagnosis, investigational activities, disease treatment and prevention.

Classification

2.2.62.1.4 Infectious substances shall be classified in Class 6.2 and assigned to UN Nos 2814, 2900, 3291 or 3373, as appropriate.

Infectious substances are divided into the following categories:

2.2.62.1.4.1 Category A: An infectious substance which is carried in a form that, when exposure to it occurs, is capable of causing permanent disability, life-threatening or fatal disease in otherwise healthy humans or animals. Indicative examples of substances that meet these criteria are given in the table in this paragraph.

NOTE *: An exposure occurs when an infectious substance is released outside of the protective packaging, resulting in physical contact with humans or animals.*

(a) Infectious substances meeting these criteria which cause disease in humans or both in humans and animals shall be assigned to UN No. 2814. Infectious substances which cause disease only in animals shall be assigned to UN No. 2900;

(b) Assignment to UN No. 2814 or UN No. 2900 shall be based on the known medical history and symptoms of the source human or animal, endemic local conditions, or professional judgement concerning individual circumstances of the source human or animal.

NOTE 1*: The proper shipping name for UN No. 2814 is "INFECTIOUS SUBSTANCE, AFFECTING HUMANS". The proper shipping name for UN No. 2900 is "INFECTIOUS SUBSTANCE, AFFECTING ANIMALS only".*

NOTE 2**:** *The following table is not exhaustive. Infectious substances, including new or emerging pathogens, which do not appear in the table but which meet the same criteria shall be assigned to Category A. In addition, if there is doubt as to whether or not a substance meets the criteria it shall be included in Category A.*

NOTE 3: *In the following table, the micro-organisms written in italics are bacteria, mycoplasmas, rickettsia or fungi.*

INDICATIVE EXAMPLES OF INFECTIOUS SUBSTANCES INCLUDED IN CATEGORY A IN ANY FORM UNLESS OTHERWISE INDICATED (2.2.62.1.4.1)	
UN Number and name	**Microorganism**
UN No. 2814 Infectious substances affecting humans	*Bacillus anthracis (cultures only)* *Brucella abortus (cultures only)* *Brucella melitensis (cultures only)* *Brucella suis (cultures only)* *Burkholderia mallei - Pseudomonas mallei* – Glanders *(cultures only)* *Burkholderia pseudomallei – Pseudomonas pseudomallei (cultures only)* *Chlamydia psittaci* - avian strains (cultures only) *Clostridium botulinum (cultures only)* *Coccidioides immitis (cultures only)* *Coxiella burnetii (cultures only)* Crimean-Congo haemorrhagic fever virus Dengue virus (cultures only) Eastern equine encephalitis virus (cultures only) *Escherichia coli*, verotoxigenic (cultures only) [a] Ebola virus Flexal virus *Francisella tularensis (cultures only)* Guanarito virus Hantaan virus Hantavirus causing haemorrhagic fever with renal syndrome Hendra virus Hepatitis B virus (cultures only) Herpes B virus (cultures only) Human immunodeficiency virus (cultures only) Highly pathogenic avian influenza virus (cultures only) Japanese Encephalitis virus (cultures only) Junin virus Kyasanur Forest disease virus Lassa virus Machupo virus Marburg virus Monkeypox virus *Mycobacterium tuberculosis (cultures only)* [a] Nipah virus Omsk haemorrhagic fever virus Poliovirus (cultures only) Rabies virus (cultures only) *Rickettsia prowazekii (cultures only)* *Rickettsia rickettsii* (cultures only) Rift Valley fever virus (cultures only) Russian spring-summer encephalitis virus (cultures only) Sabia virus *Shigella dysenteriae type 1 (cultures only)* [a] Tick-borne encephalitis virus (cultures only) Variola virus Venezuelan equine encephalitis virus (cultures only) West Nile virus (cultures only) Yellow fever virus (cultures only) *Yersinia pestis* (cultures only)

[a] *Nevertheless, when the cultures are intended for diagnostic or clinical purposes, they may be classified as infectious substances of Category B.*

INDICATIVE EXAMPLES OF INFECTIOUS SUBSTANCES INCLUDED IN CATEGORY A IN ANY FORM UNLESS OTHERWISE INDICATED (2.2.62.1.4.1)	
UN Number and name	**Microorganism**
UN No. 2900 Infectious substances affecting animals only	African swine fever virus (cultures only) Avian paramyxovirus Type 1 - Velogenic Newcastle disease virus (cultures only) Classical swine fever virus (cultures only) Foot and mouth disease virus (cultures only) Lumpy skin disease virus (cultures only) *Mycoplasma mycoides* - Contagious bovine pleuropneumonia (cultures only) Peste des petits ruminants virus (cultures only) Rinderpest virus (cultures only) Sheep-pox virus (cultures only) Goatpox virus (cultures only) Swine vesicular disease virus (cultures only) Vesicular stomatitis virus (cultures only)

2.2.62.1.4.2 Category B: An infectious substance which does not meet the criteria for inclusion in Category A. Infectious substances in Category B shall be assigned to UN No. 3373.

NOTE: *The proper shipping name of UN No. 3373 is "BIOLOGICAL SUBSTANCE, CATEGORY B".*

2.2.62.1.5 *Exemptions*

2.2.62.1.5.1 Substances which do not contain infectious substances or substances which are unlikely to cause disease in humans or animals are not subject to the provisions of ADN unless they meet the criteria for inclusion in another class.

2.2.62.1.5.2 Substances containing microorganisms which are non-pathogenic to humans or animals are not subject to ADN unless they meet the criteria for inclusion in another class.

2.2.62.1.5.3 Substances in a form that any present pathogens have been neutralized or inactivated such that they no longer pose a health risk are not subject to ADN unless they meet the criteria for inclusion in another class.

2.2.62.1.5.4 Substances where the concentration of pathogens is at a level naturally encountered (including foodstuff and water samples) and which are not considered to pose a significant risk of infection are not subject to ADN unless they meet the criteria for inclusion in another class.

2.2.62.1.5.5 Dried blood spots, collected by applying a drop of blood onto absorbent material, or faecal occult blood screening tests and blood or blood components which have been collected for the purposes of transfusion or for the preparation of blood products to be used for transfusion or transplantation and any tissues or organs intended for use in transplantation are not subject to the provisions of ADN.

2.2.62.1.5.6 Human or animal specimens for which there is minimal likelihood that pathogens are present are not subject to ADN if the specimen is carried in a packaging which will prevent any leakage and which is marked with the words "Exempt human specimen" or "Exempt animal specimen", as appropriate.

The packaging is deemed to comply with the above requirements if it meets the following conditions:

(a) The packaging consists of three components:

(i) a leak-proof primary receptacle(s);

(ii) a leak-proof secondary packaging; and

(iii) an outer packaging of adequate strength for its capacity, mass and intended use, and with at least one surface having minimum dimensions of 100 mm × 100 mm;

(b) For liquids, absorbent material in sufficient quantity to absorb the entire contents is placed between the primary receptacle(s) and the secondary packaging so that, during carriage, any release or leak of a liquid substance will not reach the outer packaging and will not compromise the integrity of the cushioning material;

(c) When multiple fragile primary receptacles are placed in a single secondary packaging, they are either individually wrapped or separated to prevent contact between them.

NOTE 1: *An element of professional judgement is required to determine if a substance is exempt under this paragraph. That judgement should be based on the known medical history, symptoms and individual circumstances of the source, human or animal, and endemic local conditions. Examples of specimens which may be carried under this paragraph include blood or urine tests to monitor cholesterol levels, blood glucose levels, hormone levels, or prostate specific antibodies (PSA); those required to monitor organ function such as heart, liver or kidney function for humans or animals with non-infectious diseases, or for therapeutic drug monitoring; those conducted for insurance or employment purposes and intended to determine the presence of drugs or alcohol; pregnancy tests; biopsies to detect cancer; and antibody detection in humans or animals in the absence of any concern for infection (e.g. evaluation of vaccine induced immunity, diagnosis of autoimmune disease, etc.).*

NOTE 2: For air transport, packagings for specimens exempted under this paragraph shall meet the conditions in (a) to (c).

2.2.62.1.6-2.2.62.1.8 *(Reserved)*

2.2.62.1.9 *Biological products*

For the purposes of ADN, biological products are divided into the following groups:

(a) those which are manufactured and packaged in accordance with the requirements of appropriate national authorities and carried for the purposes of final packaging or distribution, and use for personal health care by medical professionals or individuals. Substances in this group are not subject to the provisions of ADN;

(b) those which do not fall under paragraph (a) and are known or reasonably believed to contain infectious substances and which meet the criteria for inclusion in Category A or Category B. Substances in this group shall be assigned to UN No. 2814, UN No. 2900 or UN No. 3373, as appropriate.

***NOTE**: Some licensed biological products may present a biohazard only in certain parts of the world. In that case, competent authorities may require these biological products to be in compliance with local requirements for infectious substances or may impose other restrictions.*

2.2.62.1.10 *Genetically modified micro-organisms and organisms*

Genetically modified micro-organisms not meeting the definition of infectious substance shall be classified according to section 2.2.9.

2.2.62.1.11 *Medical or clinical wastes*

2.2.62.1.11.1 Medical or clinical wastes containing Category A infectious substances shall be assigned to UN No. 2814 or UN No. 2900 as appropriate. Medical or clinical wastes containing infectious substances in Category B shall be assigned to UN No. 3291.

***NOTE**: Medical or clinical wastes assigned to number 18 01 03 (Wastes from human or animal health care and/or related research – wastes from natal care, diagnosis, treatment or prevention of disease in humans – wastes whose collection and disposal is subject to special requirement in order to prevent infection) or 18 02 02 (Wastes from human or animal health care and/or related research – wastes from research, diagnosis, treatment or prevention of disease involving animals – wastes whose collection and disposal is subject to special requirements in order to prevent infection) according to the list of wastes annexed to the Commission Decision 2000/532/EC*[5] *as amended, shall be classified according to the provisions set out in this paragraph, based on the medical or veterinary diagnosis concerning the patient or the animal.*

2.2.62.1.11.2 Medical or clinical wastes which are reasonably believed to have a low probability of containing infectious substances shall be assigned to UN No. 3291. For the assignment, international, regional or national waste catalogues may be taken into account.

NOTE 1: *The proper shipping name for UN No. 3291 is "CLINICAL WASTE, UNSPECIFIED, N.O.S." or "(BIO) MEDICAL WASTE, N.O.S". or "REGULATED MEDICAL WASTE, N.O.S.".*

***NOTE 2**: Notwithstanding the classification criteria set out above, medical or clinical wastes assigned to number 18 01 04 (Wastes from human or animal health care and/or related research – wastes from natal care, diagnosis, treatment or prevention of disease in humans – wastes whose collection and disposal is not subject to special requirements in order to prevent infection) or 18 02 03 (Wastes from human or animal health care and/or related research – wastes from research, diagnosis, treatment or prevention of disease involving animals – wastes whose collection and disposal is not subject to special requirements in order to prevent infection) according to the list of wastes annexed to the Commission Decision 2000/532/EC*[5] *as amended, are not subject to the provisions of ADN.*

2.2.62.1.11.3 Decontaminated medical or clinical wastes which previously contained infectious substances are not subject to the provisions of ADN unless they meet the criteria for inclusion in another class.

2.2.62.1.11.4 Medical or clinical wastes assigned to UN No. 3291 are assigned to packing group II.

---

[5] *Commission Decision 2000/532/EC of 3 May 2000 replacing Decision 94/3/EC establishing a list of wastes pursuant to Article 1(a) of Council Directive 75/442/EEC on waste (replaced by the Directive 2006/12/EC of the European Parliament and of the Council (Official Journal of the European Communities No. L 114 of 27 April 2006, page 9)) and Council Decision 94/904/EC establishing a list of hazardous waste pursuant to Article 1(4) of Council Directive 91/689/EEC on hazardous waste (Official Journal of the European Communities No. L 226 of 6 September 2000, page 3).*

2.2.62.1.12 *Infected animals*

2.2.62.1.12.1 Unless an infectious substance cannot be consigned by any other means, live animals shall not be used to consign such a substance. A live animal which has been intentionally infected and is known or suspected to contain an infectious substance shall only be carried under terms and conditions approved by the competent authority[6].

2.2.62.1.12.2 Animal material affected by pathogens of Category A or by pathogens which would be assigned to Category A in cultures only, shall be assigned to UN 2814 or UN 2900 as appropriate. Animal material affected by pathogens of Category B, other than those which would be assigned to Category A if they were in cultures, shall be assigned to UN 3373.

2.2.62.2 ***Substances not accepted for carriage***

Live vertebrate or invertebrate animals shall not be used to carry an infectious agent unless the agent cannot be carried by other means or unless this carriage has been approved by the competent authority (see 2.2.62.1.12.1).

2.2.62.3 ***List of collective entries***

Effects on humans	**I1**	2814 INFECTIOUS SUBSTANCE, AFFECTING HUMANS
Effects on animals only	**I2**	2900 INFECTIOUS SUBSTANCE, AFFECTING ANIMALS only
Clinical waste	**I3**	3291 CLINICAL WASTE, UNSPECIFIED, N.O.S. or 3291 (BIO)MEDICAL WASTE, N.O.S. or 3291 REGULATED MEDICAL WASTE, N.O.S.
Biological substances	**I4**	3373 BIOLOGICAL SUBSTANCE, CATEGORY B

[6] *Such regulations are contained in, e.g. Directive 91/628/EEC (Official Journal of the European Communities No. L 340 of 11 December 1991, p. 17) and in the Recommendations of the Council of Europe (Ministerial Committee) on the carriage of certain animal species.*

2.2.7 **Class 7** **Radioactive material**

2.2.7.1 ***Definitions***

2.2.7.1.1 *Radioactive material* means any material containing radionuclides where both the activity concentration and the total activity in the consignment exceed the values specified in 2.2.7.2.2.1 to 2.2.7.2.2.6.

2.2.7.1.2 *Contamination*

Contamination means the presence of a radioactive substance on a surface in quantities in excess of 0.4 Bq/cm^2 for beta and gamma emitters and low toxicity alpha emitters, or 0.04 Bq/cm^2 for all other alpha emitters.

Non-fixed contamination means contamination that can be removed from a surface during routine conditions of carriage.

Fixed contamination means contamination other than non-fixed contamination.

2.2.7.1.3 *Definitions of specific terms*

A_1 and A_2

A_1 means the activity value of special form radioactive material which is listed in the Table in 2.2.7.2.2.1 or derived in 2.2.7.2.2.2 and is used to determine the activity limits for the requirements of ADN.

A_2 means the activity value of radioactive material, other than special form radioactive material, which is listed in the Table in 2.2.7.2.2.1 or derived in 2.2.7.2.2.2 and is used to determine the activity limits for the requirements of ADN.

Fissile material means uranium-233, uranium-235, plutonium-239, plutonium-241, or any combination of these radionuclides. Excepted from this definition is:

(a) Natural uranium or depleted uranium which is unirradiated; and

(b) Natural uranium or depleted uranium which has been irradiated in thermal reactors only.

Low dispersible radioactive material means either a solid radioactive material or a solid radioactive material in a sealed capsule, that has limited dispersibility and is not in powder form.

Low specific activity (LSA) material means radioactive material which by its nature has a limited specific activity, or radioactive material for which limits of estimated average specific activity apply. External shielding materials surrounding the LSA material shall not be considered in determining the estimated average specific activity.

Low toxicity alpha emitters are: natural uranium; depleted uranium; natural thorium; uranium-235 or uranium-238; thorium-232; thorium-228 and thorium-230 when contained in ores or physical and chemical concentrates; or alpha emitters with a half-life of less than 10 days.

Specific activity of a radionuclide means the activity per unit mass of that nuclide. The specific activity of a material shall mean the activity per unit mass of the material in which the radionuclides are essentially uniformly distributed.

Special form radioactive material means either:

(a) An indispersible solid radioactive material; or

(b) A sealed capsule containing radioactive material.

Surface contaminated object (SCO) means a solid object which is not itself radioactive but which has radioactive material distributed on its surfaces.

Unirradiated thorium means thorium containing not more than 10^{-7} g of uranium-233 per gram of thorium-232.

Unirradiated uranium means uranium containing not more than 2×10^3 Bq of plutonium per gram of uranium-235, not more than 9×10^6 Bq of fission products per gram of uranium-235 and not more than 5×10^{-3} g of uranium-236 per gram of uranium-235.

Uranium - natural, depleted, enriched means the following:

Natural uranium means uranium (which may be chemically separated) containing the naturally occurring distribution of uranium isotopes (approximately 99.28% uranium-238, and 0.72% uranium-235 by mass).

Depleted uranium means uranium containing a lesser mass percentage of uranium-235 than in natural uranium.

Enriched uranium means uranium containing a greater mass percentage of uranium-235 than 0.72%.

In all cases, a very small mass percentage of uranium-234 is present.

2.2.7.2 ***Classification***

2.2.7.2.1 *General provisions*

2.2.7.2.1.1 Radioactive material shall be assigned to one of the UN number specified in Table 2.2.7.2.1.1 depending on the activity level of the radionuclides contained in a package, the fissile or non-fissile properties of these radionuclides, the type of package to be presented for carriage, and the nature or form of the contents of the package, or special arrangements governing the carriage operation, in accordance with the provisions laid down in 2.2.7.2.2 to 2.2.7.2.5.

Table 2.2.7.2.1.1 Assignment of UN numbers

Excepted packages (1.7.1.5)	
UN 2908	RADIOACTIVE MATERIAL, EXCEPTED PACKAGE - EMPTY PACKAGING
UN 2909	RADIOACTIVE MATERIAL, EXCEPTED PACKAGE – ARTICLES MANUFACTURED FROM NATURAL URANIUM or DEPLETED URANIUM or NATURAL THORIUM
UN 2910	RADIOACTIVE MATERIAL, EXCEPTED PACKAGE - LIMITED QUANTITY OF MATERIAL
UN 2911	RADIOACTIVE MATERIAL, EXCEPTED PACKAGE - INSTRUMENTS or ARTICLES

Low specific activity radioactive material (2.2.7.2.3.1)		
	UN 2912	RADIOACTIVE MATERIAL, LOW SPECIFIC ACTIVITY (LSA-I), non-fissile or fissile-excepted
	UN 3321	RADIOACTIVE MATERIAL, LOW SPECIFIC ACTIVITY (LSA-II), non fissile or fissile-excepted
	UN 3322	RADIOACTIVE MATERIAL, LOW SPECIFIC ACTIVITY (LSA-III), non fissile or fissile-excepted
	UN 3324	RADIOACTIVE MATERIAL, LOW SPECIFIC ACTIVITY (LSA-II), FISSILE
	UN 3325	RADIOACTIVE MATERIAL, LOW SPECIFIC ACTIVITY (LSA-III), FISSILE
Surface contaminated objects (2.2.7.2.3.2)		
	UN 2913	RADIOACTIVE MATERIAL, SURFACE CONTAMINATED OBJECTS (SCO-I or SCO-II), non-fissile or fissile-excepted
	UN 3326	RADIOACTIVE MATERIAL, SURFACE CONTAMINATED OBJECTS (SCO-I or SCO-II), FISSILE
Type A packages (2.2.7.2.4.4)		
	UN 2915	RADIOACTIVE MATERIAL, TYPE A PACKAGE, non-special form, non-fissile or fissile-excepted
	UN 3327	RADIOACTIVE MATERIAL, TYPE A PACKAGE, FISSILE, non-special form
	UN 3332	RADIOACTIVE MATERIAL, TYPE A PACKAGE, SPECIAL FORM, non fissile or fissile-excepted
	UN 3333	RADIOACTIVE MATERIAL, TYPE A PACKAGE, SPECIAL FORM, FISSILE
Type B(U) packages (2.2.7.2.4.6)		
	UN 2916	RADIOACTIVE MATERIAL, TYPE B(U) PACKAGE, non-fissile or fissile-excepted
	UN 3328	RADIOACTIVE MATERIAL, TYPE B(U) PACKAGE, FISSILE
Type B(M) packages (2.2.7.2.4.6)		
	UN 2917	RADIOACTIVE MATERIAL, TYPE B(M) PACKAGE, non-fissile or fissile-excepted
	UN 3329	RADIOACTIVE MATERIAL, TYPE B(M) PACKAGE, FISSILE
Type C packages (2.2.7.2.4.6)		
	UN 3323	RADIOACTIVE MATERIAL, TYPE C PACKAGE, non fissile or fissile-excepted
	UN 3330	RADIOACTIVE MATERIAL, TYPE C PACKAGE, FISSILE
Special arrangement (2.2.7.2.5)		
	UN 2919	RADIOACTIVE MATERIAL, TRANSPORTED UNDER SPECIAL ARRANGEMENT, non-fissile or fissile-excepted
	UN 3331	RADIOACTIVE MATERIAL, TRANSPORTED UNDER SPECIAL ARRANGEMENT, FISSILE
Uranium hexafluoride (2.2.7.2.4.5)		
	UN 2977	RADIOACTIVE MATERIAL, URANIUM HEXAFLUORIDE, FISSILE
	UN 2978	RADIOACTIVE MATERIAL, URANIUM HEXAFLUORIDE, non-fissile or fissile-excepted

2.2.7.2.2 *Determination of activity level*

2.2.7.2.2.1 The following basic values for individual radionuclides are given in Table 2.2.7.2.2.1:

(a) A_1 and A_2 in TBq;

(b) Activity concentration for exempt material in Bq/g; and

(c) Activity limits for exempt consignments in Bq.

Table 2.2.7.2.2.1: Basic radionuclides values for individual radionuclides

Radionuclide (atomic number)	A_1 (TBq)	A_2 (TBq)	Activity concentration for exempt material (Bq/g)	Activity limit for an exempt consignment (Bq)
Actinium (89)				
Ac-225 (a)	8×10^{-1}	6×10^{-3}	1×10^{1}	1×10^{4}
Ac-227 (a)	9×10^{-1}	9×10^{-5}	1×10^{-1}	1×10^{3}
Ac-228	6×10^{-1}	5×10^{-1}	1×10^{1}	1×10^{6}
Silver (47)				
Ag-105	2×10^{0}	2×10^{0}	1×10^{2}	1×10^{6}
Ag-108m (a)	7×10^{-1}	7×10^{-1}	1×10^{1} (b)	1×10^{6} (b)
Ag-110m (a)	4×10^{-1}	4×10^{-1}	1×10^{1}	1×10^{6}
Ag-111	2×10^{0}	6×10^{-1}	1×10^{3}	1×10^{6}
Aluminium (13)				
Al-26	1×10^{-1}	1×10^{-1}	1×10^{1}	1×10^{5}
Americium (95)				
Am-241	1×10^{1}	1×10^{-3}	1×10^{0}	1×10^{4}
Am-242m (a)	1×10^{1}	1×10^{-3}	1×10^{0} (b)	1×10^{4} (b)
Am-243 (a)	5×10^{0}	1×10^{-3}	1×10^{0} (b)	1×10^{3} (b)
Argon (18)				
Ar-37	4×10^{1}	4×10^{1}	1×10^{6}	1×10^{8}
Ar-39	4×10^{1}	2×10^{1}	1×10^{7}	1×10^{4}
Ar-41	3×10^{-1}	3×10^{-1}	1×10^{2}	1×10^{9}
Arsenic (33)				
As-72	3×10^{-1}	3×10^{-1}	1×10^{1}	1×10^{5}
As-73	4×10^{1}	4×10^{1}	1×10^{3}	1×10^{7}
As-74	1×10^{0}	9×10^{-1}	1×10^{1}	1×10^{6}
As-76	3×10^{-1}	3×10^{-1}	1×10^{2}	1×10^{5}
As-77	2×10^{1}	7×10^{-1}	1×10^{3}	1×10^{6}
Astatine (85)				
At-211 (a)	2×10^{1}	5×10^{-1}	1×10^{3}	1×10^{7}
Gold (79)				
Au-193	7×10^{0}	2×10^{0}	1×10^{2}	1×10^{7}

Radionuclide (atomic number)	A_1 (TBq)	A_2 (TBq)	Activity concentration for exempt material (Bq/g)	Activity limit for an exempt consignment (Bq)
Au-194	1×10^0	1×10^0	1×10^1	1×10^6
Au-195	1×10^1	6×10^0	1×10^2	1×10^7
Au-198	1×10^0	6×10^{-1}	1×10^2	1×10^6
Au-199	1×10^1	6×10^{-1}	1×10^2	1×10^6
Barium (56)				
Ba-131 (a)	2×10^0	2×10^0	1×10^2	1×10^6
Ba-133	3×10^0	3×10^0	1×10^2	1×10^6
Ba-133m	2×10^1	6×10^{-1}	1×10^2	1×10^6
Ba-140 (a)	5×10^{-1}	3×10^{-1}	1×10^1 (b)	1×10^5 (b)
Beryllium (4)				
Be-7	2×10^1	2×10^1	1×10^3	1×10^7
Be-10	4×10^1	6×10^{-1}	1×10^4	1×10^6
Bismuth (83)				
Bi-205	7×10^{-1}	7×10^{-1}	1×10^1	1×10^6
Bi-206	3×10^{-1}	3×10^{-1}	1×10^1	1×10^5
Bi-207	7×10^{-1}	7×10^{-1}	1×10^1	1×10^6
Bi-210	1×10^0	6×10^{-1}	1×10^3	1×10^6
Bi-210m (a)	6×10^{-1}	2×10^{-2}	1×10^1	1×10^5
Bi-212 (a)	7×10^{-1}	6×10^{-1}	1×10^1 (b)	1×10^5 (b)
Berkelium (97)				
Bk-247	8×10^0	8×10^{-4}	1×10^0	1×10^4
Bk-249 (a)	4×10^1	3×10^{-1}	1×10^3	1×10^6
Bromine (35)				
Br-76	4×10^{-1}	4×10^{-1}	1×10^1	1×10^5
Br-77	3×10^0	3×10^0	1×10^2	1×10^6
Br-82	4×10^{-1}	4×10^{-1}	1×10^1	1×10^6
Carbon (6)				
C-11	1×10^0	6×10^{-1}	1×10^1	1×10^6
C-14	4×10^1	3×10^0	1×10^4	1×10^7
Calcium (20)				
Ca-41	Unlimited	Unlimited	1×10^5	1×10^7

Radionuclide (atomic number)	A_1 (TBq)	A_2 (TBq)	Activity concentration for exempt material (Bq/g)	Activity limit for an exempt consignment (Bq)
Ca-45	4×10^1	1×10^0	1×10^4	1×10^7
Ca-47 (a)	3×10^0	3×10^{-1}	1×10^1	1×10^6
Cadmium (48)				
Cd-109	3×10^1	2×10^0	1×10^4	1×10^6
Cd-113m	4×10^1	5×10^{-1}	1×10^3	1×10^6
Cd-115 (a)	3×10^0	4×10^{-1}	1×10^2	1×10^6
Cd-115m	5×10^{-1}	5×10^{-1}	1×10^3	1×10^6
Cerium (58)				
Ce-139	7×10^0	2×10^0	1×10^2	1×10^6
Ce-141	2×10^1	6×10^{-1}	1×10^2	1×10^7
Ce-143	9×10^{-1}	6×10^{-1}	1×10^2	1×10^6
Ce-144 (a)	2×10^{-1}	2×10^{-1}	1×10^2 (b)	1×10^5 (b)
Californium (98)				
Cf-248	4×10^1	6×10^{-3}	1×10^1	1×10^4
Cf-249	3×10^0	8×10^{-4}	1×10^0	1×10^3
Cf-250	2×10^1	2×10^{-3}	1×10^1	1×10^4
Cf-251	7×10^0	7×10^{-4}	1×10^0	1×10^3
Cf-252	1×10^{-1}	3×10^{-3}	1×10^1	1×10^4
Cf-253 (a)	4×10^1	4×10^{-2}	1×10^2	1×10^5
Cf-254	1×10^{-3}	1×10^{-3}	1×10^0	1×10^3
Chlorine (17)				
Cl-36	1×10^1	6×10^{-1}	1×10^4	1×10^6
Cl-38	2×10^{-1}	2×10^{-1}	1×10^1	1×10^5
Curium (96)				
Cm-240	4×10^1	2×10^{-2}	1×10^2	1×10^5
Cm-241	2×10^0	1×10^0	1×10^2	1×10^6
Cm-242	4×10^1	1×10^{-2}	1×10^2	1×10^5
Cm-243	9×10^0	1×10^{-3}	1×10^0	1×10^4
Cm-244	2×10^1	2×10^{-3}	1×10^1	1×10^4
Cm-245	9×10^0	9×10^{-4}	1×10^0	1×10^3
Cm-246	9×10^0	9×10^{-4}	1×10^0	1×10^3

Radionuclide (atomic number)	***A***$_1$ **(TBq)**	***A***$_2$ **(TBq)**	**Activity concentration for exempt material (Bq/g)**	**Activity limit for an exempt consignment (Bq)**
Cm-247 (a)	3×10^{0}	1×10^{-3}	1×10^{0}	1×10^{4}
Cm-248	2×10^{-2}	3×10^{-4}	1×10^{0}	1×10^{3}
Cobalt (27)				
Co-55	5×10^{-1}	5×10^{-1}	1×10^{1}	1×10^{6}
Co-56	3×10^{-1}	3×10^{-1}	1×10^{1}	1×10^{5}
Co-57	1×10^{1}	1×10^{1}	1×10^{2}	1×10^{6}
Co-58	1×10^{0}	1×10^{0}	1×10^{1}	1×10^{6}
Co-58m	4×10^{1}	4×10^{1}	1×10^{4}	1×10^{7}
Co-60	4×10^{-1}	4×10^{-1}	1×10^{1}	1×10^{5}
Chromium (24)				
Cr-51	3×10^{1}	3×10^{1}	1×10^{3}	1×10^{7}
Caesium (55)				
Cs-129	4×10^{0}	4×10^{0}	1×10^{2}	1×10^{5}
Cs-131	3×10^{1}	3×10^{1}	1×10^{3}	1×10^{6}
Cs-132	1×10^{0}	1×10^{0}	1×10^{1}	1×10^{5}
Cs-134	7×10^{-1}	7×10^{-1}	1×10^{1}	1×10^{4}
Cs-134m	4×10^{1}	6×10^{-1}	1×10^{3}	1×10^{5}
Cs-135	4×10^{1}	1×10^{0}	1×10^{4}	1×10^{7}
Cs-136	5×10^{-1}	5×10^{-1}	1×10^{1}	1×10^{5}
Cs-137 (a)	2×10^{0}	6×10^{-1}	1×10^{1} (b)	1×10^{4} (b)
Copper (29)				
Cu-64	6×10^{0}	1×10^{0}	1×10^{2}	1×10^{6}
Cu-67	1×10^{1}	7×10^{-1}	1×10^{2}	1×10^{6}
Dysprosium (66)				
Dy-159	2×10^{1}	2×10^{1}	1×10^{3}	1×10^{7}
Dy-165	9×10^{-1}	6×10^{-1}	1×10^{3}	1×10^{6}
Dy-166 (a)	9×10^{-1}	3×10^{-1}	1×10^{3}	1×10^{6}
Erbium (68)				
Er-169	4×10^{1}	1×10^{0}	1×10^{4}	1×10^{7}
Er-171	8×10^{-1}	5×10^{-1}	1×10^{2}	1×10^{6}
Europium (63)				

Radionuclide (atomic number)	A_1 (TBq)	A_2 (TBq)	Activity concentration for exempt material (Bq/g)	Activity limit for an exempt consignment (Bq)
Eu-147	2×10^0	2×10^0	1×10^2	1×10^6
Eu-148	5×10^{-1}	5×10^{-1}	1×10^1	1×10^6
Eu-149	2×10^1	2×10^1	1×10^2	1×10^7
Eu-150(short lived)	2×10^0	7×10^{-1}	1×10^3	1×10^6
Eu-150(long lived)	7×10^{-1}	7×10^{-1}	1×10^1	1×10^6
Eu-152	1×10^0	1×10^0	1×10^1	1×10^6
Eu-152m	8×10^{-1}	8×10^{-1}	1×10^2	1×10^6
Eu-154	9×10^{-1}	6×10^{-1}	1×10^1	1×10^6
Eu-155	2×10^1	3×10^0	1×10^2	1×10^7
Eu-156	7×10^{-1}	7×10^{-1}	1×10^1	1×10^6
Fluorine (9)				
F-18	1×10^0	6×10^{-1}	1×10^1	1×10^6
Iron (26)				
Fe-52 (a)	3×10^{-1}	3×10^{-1}	1×10^1	1×10^6
Fe-55	4×10^1	4×10^1	1×10^4	1×10^6
Fe-59	9×10^{-1}	9×10^{-1}	1×10^1	1×10^6
Fe-60 (a)	4×10^1	2×10^{-1}	1×10^2	1×10^5
Gallium (31)				
Ga-67	7×10^0	3×10^0	1×10^2	1×10^6
Ga-68	5×10^{-1}	5×10^{-1}	1×10^1	1×10^5
Ga-72	4×10^{-1}	4×10^{-1}	1×10^1	1×10^5
Gadolinium (64)				
Gd-146 (a)	5×10^{-1}	5×10^{-1}	1×10^1	1×10^6
Gd-148	2×10^1	2×10^{-3}	1×10^1	1×10^4
Gd-153	1×10^1	9×10^0	1×10^2	1×10^7
Gd-159	3×10^0	6×10^{-1}	1×10^3	1×10^6
Germanium (32)				
Ge-68 (a)	5×10^{-1}	5×10^{-1}	1×10^1	1×10^5
Ge-71	4×10^1	4×10^1	1×10^4	1×10^8
Ge-77	3×10^{-1}	3×10^{-1}	1×10^1	1×10^5
Hafnium (72)				

Radionuclide (atomic number)	A_1 (TBq)	A_2 (TBq)	Activity concentration for exempt material (Bq/g)	Activity limit for an exempt consignment (Bq)
Hf-172 (a)	6×10^{-1}	6×10^{-1}	1×10^{1}	1×10^{6}
Hf-175	3×10^{0}	3×10^{0}	1×10^{2}	1×10^{6}
Hf-181	2×10^{0}	5×10^{-1}	1×10^{1}	1×10^{6}
Hf-182	Unlimited	Unlimited	1×10^{2}	1×10^{6}
Mercury (80)				
Hg-194 (a)	1×10^{0}	1×10^{0}	1×10^{1}	1×10^{6}
Hg-195m (a)	3×10^{0}	7×10^{-1}	1×10^{2}	1×10^{6}
Hg-197	2×10^{1}	1×10^{1}	1×10^{2}	1×10^{7}
Hg-197m	1×10^{1}	4×10^{-1}	1×10^{2}	1×10^{6}
Hg-203	5×10^{0}	1×10^{0}	1×10^{2}	1×10^{5}
Holmium (67)				
Ho-166	4×10^{-1}	4×10^{-1}	1×10^{3}	1×10^{5}
Ho-166m	6×10^{-1}	5×10^{-1}	1×10^{1}	1×10^{6}
Iodine (53)				
I-123	6×10^{0}	3×10^{0}	1×10^{2}	1×10^{7}
I-124	1×10^{0}	1×10^{0}	1×10^{1}	1×10^{6}
I-125	2×10^{1}	3×10^{0}	1×10^{3}	1×10^{6}
I-126	2×10^{0}	1×10^{0}	1×10^{2}	1×10^{6}
I-129	Unlimited	Unlimited	1×10^{2}	1×10^{5}
I-131	3×10^{0}	7×10^{-1}	1×10^{2}	1×10^{6}
I-132	4×10^{-1}	4×10^{-1}	1×10^{1}	1×10^{5}
I-133	7×10^{-1}	6×10^{-1}	1×10^{1}	1×10^{6}
I-134	3×10^{-1}	3×10^{-1}	1×10^{1}	1×10^{5}
I-135 (a)	6×10^{-1}	6×10^{-1}	1×10^{1}	1×10^{6}
Indium (49)				
In-111	3×10^{0}	3×10^{0}	1×10^{2}	1×10^{6}
In-113m	4×10^{0}	2×10^{0}	1×10^{2}	1×10^{6}
In-114m (a)	1×10^{1}	5×10^{-1}	1×10^{2}	1×10^{6}
In-115m	7×10^{0}	1×10^{0}	1×10^{2}	1×10^{6}
Iridium (77)				
Ir-189 (a)	1×10^{1}	1×10^{1}	1×10^{2}	1×10^{7}

Radionuclide (atomic number)	A_1 (TBq)	A_2 (TBq)	Activity concentration for exempt material (Bq/g)	Activity limit for an exempt consignment (Bq)
Ir-190	7×10^{-1}	7×10^{-1}	1×10^{1}	1×10^{6}
Ir-192	1×10^{0}(c)	6×10^{-1}	1×10^{1}	1×10^{4}
Ir-194	3×10^{-1}	3×10^{-1}	1×10^{2}	1×10^{5}
Potassium (19)				
K-40	9×10^{-1}	9×10^{-1}	1×10^{2}	1×10^{6}
K-42	2×10^{-1}	2×10^{-1}	1×10^{2}	1×10^{6}
K-43	7×10^{-1}	6×10^{-1}	1×10^{1}	1×10^{6}
Krypton (36)				
Kr-79	$4 x 10^{0}$	$1 x 10^{0}$	$1 x 10^{3}$	$1 x 10^{5}$
Kr-81	4×10^{1}	4×10^{1}	1×10^{4}	1×10^{7}
Kr-85	1×10^{1}	1×10^{1}	1×10^{5}	1×10^{4}
Kr-85m	8×10^{0}	3×10^{0}	1×10^{3}	1×10^{10}
Kr-87	2×10^{-1}	2×10^{-1}	1×10^{2}	1×10^{9}
Lanthanum (57)				
La-137	3×10^{1}	6×10^{0}	1×10^{3}	1×10^{7}
La-140	4×10^{-1}	4×10^{-1}	1×10^{1}	1×10^{5}
Lutetium (71)				
Lu-172	6×10^{-1}	6×10^{-1}	1×10^{1}	1×10^{6}
Lu-173	8×10^{0}	8×10^{0}	1×10^{2}	1×10^{7}
Lu-174	9×10^{0}	9×10^{0}	1×10^{2}	1×10^{7}
Lu-174m	2×10^{1}	1×10^{1}	1×10^{2}	1×10^{7}
Lu-177	3×10^{1}	7×10^{-1}	1×10^{3}	1×10^{7}
Magnesium (12)				
Mg-28 (a)	3×10^{-1}	3×10^{-1}	1×10^{1}	1×10^{5}
Manganese (25)				
Mn-52	3×10^{-1}	3×10^{-1}	1×10^{1}	1×10^{5}
Mn-53	Unlimited	Unlimited	1×10^{4}	1×10^{9}
Mn-54	1×10^{0}	1×10^{0}	1×10^{1}	1×10^{6}
Mn-56	3×10^{-1}	3×10^{-1}	1×10^{1}	1×10^{5}
Molybdenum (42)				
Mo-93	4×10^{1}	2×10^{1}	1×10^{3}	1×10^{8}

Radionuclide (atomic number)	A_1 (TBq)	A_2 (TBq)	Activity concentration for exempt material (Bq/g)	Activity limit for an exempt consignment (Bq)
Mo-99 (a)	1×10^{0}	6×10^{-1}	1×10^{2}	1×10^{6}
Nitrogen (7)				
N-13	9×10^{-1}	6×10^{-1}	1×10^{2}	1×10^{9}
Sodium (11)				
Na-22	5×10^{-1}	5×10^{-1}	1×10^{1}	1×10^{6}
Na-24	2×10^{-1}	2×10^{-1}	1×10^{1}	1×10^{5}
Niobium (41)				
Nb-93m	4×10^{1}	3×10^{1}	1×10^{4}	1×10^{7}
Nb-94	7×10^{-1}	7×10^{-1}	1×10^{1}	1×10^{6}
Nb-95	1×10^{0}	1×10^{0}	1×10^{1}	1×10^{6}
Nb-97	9×10^{-1}	6×10^{-1}	1×10^{1}	1×10^{6}
Neodymium (60)				
Nd-147	6×10^{0}	6×10^{-1}	1×10^{2}	1×10^{6}
Nd-149	6×10^{-1}	5×10^{-1}	1×10^{2}	1×10^{6}
Nickel (28)				
Ni-59	Unlimited	Unlimited	1×10^{4}	1×10^{8}
Ni-63	4×10^{1}	3×10^{1}	1×10^{5}	1×10^{8}
Ni-65	4×10^{-1}	4×10^{-1}	1×10^{1}	1×10^{6}
Neptunium (93)				
Np-235	4×10^{1}	4×10^{1}	1×10^{3}	1×10^{7}
Np-236(short-lived)	2×10^{1}	2×10^{0}	1×10^{3}	1×10^{7}
Np-236(long-lived)	9×10^{0}	2×10^{-2}	1×10^{2}	1×10^{5}
Np-237	2×10^{1}	2×10^{-3}	1×10^{0} (b)	1×10^{3} (b)
Np-239	7×10^{0}	4×10^{-1}	1×10^{2}	1×10^{7}
Osmium (76)				
Os-185	1×10^{0}	1×10^{0}	1×10^{1}	1×10^{6}
Os-191	1×10^{1}	2×10^{0}	1×10^{2}	1×10^{7}
Os-191m	4×10^{1}	3×10^{1}	1×10^{3}	1×10^{7}
Os-193	2×10^{0}	6×10^{-1}	1×10^{2}	1×10^{6}
Os-194 (a)	3×10^{-1}	3×10^{-1}	1×10^{2}	1×10^{5}
Phosphorus (15)				

Radionuclide (atomic number)	A_1 (TBq)	A_2 (TBq)	Activity concentration for exempt material (Bq/g)	Activity limit for an exempt consignment (Bq)
P-32	5×10^{-1}	5×10^{-1}	1×10^{3}	1×10^{5}
P-33	4×10^{1}	1×10^{0}	1×10^{5}	1×10^{8}
Protactinium (91)				
Pa-230 (a)	2×10^{0}	7×10^{-2}	1×10^{1}	1×10^{6}
Pa-231	4×10^{0}	4×10^{-4}	1×10^{0}	1×10^{3}
Pa-233	5×10^{0}	7×10^{-1}	1×10^{2}	1×10^{7}
Lead (82)				
Pb-201	1×10^{0}	1×10^{0}	1×10^{1}	1×10^{6}
Pb-202	4×10^{1}	2×10^{1}	1×10^{3}	1×10^{6}
Pb-203	4×10^{0}	3×10^{0}	1×10^{2}	1×10^{6}
Pb-205	Unlimited	Unlimited	1×10^{4}	1×10^{7}
Pb-210 (a)	1×10^{0}	5×10^{-2}	1×10^{1} (b)	1×10^{4} (b)
Pb-212 (a)	7×10^{-1}	2×10^{-1}	1×10^{1} (b)	1×10^{5} (b)
Palladium (46)				
Pd-103 (a)	4×10^{1}	4×10^{1}	1×10^{3}	1×10^{8}
Pd-107	Unlimited	Unlimited	1×10^{5}	1×10^{8}
Pd-109	2×10^{0}	5×10^{-1}	1×10^{3}	1×10^{6}
Promethium (61)				
Pm-143	3×10^{0}	3×10^{0}	1×10^{2}	1×10^{6}
Pm-144	7×10^{-1}	7×10^{-1}	1×10^{1}	1×10^{6}
Pm-145	3×10^{1}	1×10^{1}	1×10^{3}	1×10^{7}
Pm-147	4×10^{1}	2×10^{0}	1×10^{4}	1×10^{7}
Pm-148m (a)	8×10^{-1}	7×10^{-1}	1×10^{1}	1×10^{6}
Pm-149	2×10^{0}	6×10^{-1}	1×10^{3}	1×10^{6}
Pm-151	2×10^{0}	6×10^{-1}	1×10^{2}	1×10^{6}
Polonium (84)				
Po-210	4×10^{1}	2×10^{-2}	1×10^{1}	1×10^{4}
Praseodymium (59)				
Pr-142	4×10^{-1}	4×10^{-1}	1×10^{2}	1×10^{5}
Pr-143	3×10^{0}	6×10^{-1}	1×10^{4}	1×10^{6}
Platinum (78)				

Radionuclide (atomic number)	A_1 (TBq)	A_2 (TBq)	Activity concentration for exempt material (Bq/g)	Activity limit for an exempt consignment (Bq)
Pt-188 (a)	1×10^{0}	8×10^{-1}	1×10^{1}	1×10^{6}
Pt-191	4×10^{0}	3×10^{0}	1×10^{2}	1×10^{6}
Pt-193	4×10^{1}	4×10^{1}	1×10^{4}	1×10^{7}
Pt-193m	4×10^{1}	5×10^{-1}	1×10^{3}	1×10^{7}
Pt-195m	1×10^{1}	5×10^{-1}	1×10^{2}	1×10^{6}
Pt-197	2×10^{1}	6×10^{-1}	1×10^{3}	1×10^{6}
Pt-197m	1×10^{1}	6×10^{-1}	1×10^{2}	1×10^{6}
Plutonium (94)				
Pu-236	3×10^{1}	3×10^{-3}	1×10^{1}	1×10^{4}
Pu-237	2×10^{1}	2×10^{1}	1×10^{3}	1×10^{7}
Pu-238	1×10^{1}	1×10^{-3}	1×10^{0}	1×10^{4}
Pu-239	1×10^{1}	1×10^{-3}	1×10^{0}	1×10^{4}
Pu-240	1×10^{1}	1×10^{-3}	1×10^{0}	1×10^{3}
Pu-241 (a)	4×10^{1}	6×10^{-2}	1×10^{2}	1×10^{5}
Pu-242	1×10^{1}	1×10^{-3}	1×10^{0}	1×10^{4}
Pu-244 (a)	4×10^{-1}	1×10^{-3}	1×10^{0}	1×10^{4}
Radium (88)				
Ra-223 (a)	4×10^{-1}	7×10^{-3}	1×10^{2} (b)	1×10^{5} (b)
Ra-224 (a)	4×10^{-1}	2×10^{-2}	1×10^{1} (b)	1×10^{5} (b)
Ra-225 (a)	2×10^{-1}	4×10^{-3}	1×10^{2}	1×10^{5}
Ra-226 (a)	2×10^{-1}	3×10^{-3}	1×10^{1} (b)	1×10^{4} (b)
Ra-228 (a)	6×10^{-1}	2×10^{-2}	1×10^{1} (b)	1×10^{5} (b)
Rubidium (37)				
Rb-81	2×10^{0}	8×10^{-1}	1×10^{1}	1×10^{6}
Rb-83 (a)	2×10^{0}	2×10^{0}	1×10^{2}	1×10^{6}
Rb-84	1×10^{0}	1×10^{0}	1×10^{1}	1×10^{6}
Rb-86	5×10^{-1}	5×10^{-1}	1×10^{2}	1×10^{5}
Rb-87	Unlimited	Unlimited	1×10^{4}	1×10^{7}
Rb(nat)	Unlimited	Unlimited	1×10^{4}	1×10^{7}
Rhenium (75)				
Re-184	1×10^{0}	1×10^{0}	1×10^{1}	1×10^{6}

Radionuclide (atomic number)	A_1 (TBq)	A_2 (TBq)	Activity concentration for exempt material (Bq/g)	Activity limit for an exempt consignment (Bq)
Re-184m	3×10^{0}	1×10^{0}	1×10^{2}	1×10^{6}
Re-186	2×10^{0}	6×10^{-1}	1×10^{3}	1×10^{6}
Re-187	Unlimited	Unlimited	1×10^{6}	1×10^{9}
Re-188	4×10^{-1}	4×10^{-1}	1×10^{2}	1×10^{5}
Re-189 (a)	3×10^{0}	6×10^{-1}	1×10^{2}	1×10^{6}
Re(nat)	Unlimited	Unlimited	1×10^{6}	1×10^{9}
Rhodium (45)				
Rh-99	2×10^{0}	2×10^{0}	1×10^{1}	1×10^{6}
Rh-101	4×10^{0}	3×10^{0}	1×10^{2}	1×10^{7}
Rh-102	5×10^{-1}	5×10^{-1}	1×10^{1}	1×10^{6}
Rh-102m	2×10^{0}	2×10^{0}	1×10^{2}	1×10^{6}
Rh-103m	4×10^{1}	4×10^{1}	1×10^{4}	1×10^{8}
Rh-105	1×10^{1}	8×10^{-1}	1×10^{2}	1×10^{7}
Radon (86)				
Rn-222 (a)	3×10^{-1}	4×10^{-3}	1×10^{1} (b)	1×10^{8} (b)
Ruthenium (44)				
Ru-97	5×10^{0}	5×10^{0}	1×10^{2}	1×10^{7}
Ru-103 (a)	2×10^{0}	2×10^{0}	1×10^{2}	1×10^{6}
Ru-105	1×10^{0}	6×10^{-1}	1×10^{1}	1×10^{6}
Ru-106 (a)	2×10^{-1}	2×10^{-1}	1×10^{2} (b)	1×10^{5} (b)
Sulphur (16)				
S-35	4×10^{1}	3×10^{0}	1×10^{5}	1×10^{8}
Antimony (51)				
Sb-122	4×10^{-1}	4×10^{-1}	1×10^{2}	1×10^{4}
Sb-124	6×10^{-1}	6×10^{-1}	1×10^{1}	1×10^{6}
Sb-125	2×10^{0}	1×10^{0}	1×10^{2}	1×10^{6}
Sb-126	4×10^{-1}	4×10^{-1}	1×10^{1}	1×10^{5}
Scandium (21)				
Sc-44	5×10^{-1}	5×10^{-1}	1×10^{1}	1×10^{5}
Sc-46	5×10^{-1}	5×10^{-1}	1×10^{1}	1×10^{6}
Sc-47	1×10^{1}	7×10^{-1}	1×10^{2}	1×10^{6}

Radionuclide (atomic number)	A_1 (TBq)	A_2 (TBq)	Activity concentration for exempt material (Bq/g)	Activity limit for an exempt consignment (Bq)
Sc-48	3×10^{-1}	3×10^{-1}	1×10^{1}	1×10^{5}
Selenium (34)				
Se-75	3×10^{0}	3×10^{0}	1×10^{2}	1×10^{6}
Se-79	4×10^{1}	2×10^{0}	1×10^{4}	1×10^{7}
Silicon (14)				
Si-31	6×10^{-1}	6×10^{-1}	1×10^{3}	1×10^{6}
Si-32	4×10^{1}	5×10^{-1}	1×10^{3}	1×10^{6}
Samarium (62)				
Sm-145	1×10^{1}	1×10^{1}	1×10^{2}	1×10^{7}
Sm-147	Unlimited	Unlimited	1×10^{1}	1×10^{4}
Sm-151	4×10^{1}	1×10^{1}	1×10^{4}	1×10^{8}
Sm-153	9×10^{0}	6×10^{-1}	1×10^{2}	1×10^{6}
Tin (50)				
Sn-113 (a)	4×10^{0}	2×10^{0}	1×10^{3}	1×10^{7}
Sn-117m	7×10^{0}	4×10^{-1}	1×10^{2}	1×10^{6}
Sn-119m	4×10^{1}	3×10^{1}	1×10^{3}	1×10^{7}
Sn-121m (a)	4×10^{1}	9×10^{-1}	1×10^{3}	1×10^{7}
Sn-123	8×10^{-1}	6×10^{-1}	1×10^{3}	1×10^{6}
Sn-125	4×10^{-1}	4×10^{-1}	1×10^{2}	1×10^{5}
Sn-126 (a)	6×10^{-1}	4×10^{-1}	1×10^{1}	1×10^{5}
Strontium (38)				
Sr-82 (a)	2×10^{-1}	2×10^{-1}	1×10^{1}	1×10^{5}
Sr-85	2×10^{0}	2×10^{0}	1×10^{2}	1×10^{6}
Sr-85m	5×10^{0}	5×10^{0}	1×10^{2}	1×10^{7}
Sr-87m	3×10^{0}	3×10^{0}	1×10^{2}	1×10^{6}
Sr-89	6×10^{-1}	6×10^{-1}	1×10^{3}	1×10^{6}
Sr-90 (a)	3×10^{-1}	3×10^{-1}	1×10^{2} (b)	1×10^{4} (b)
Sr-91 (a)	3×10^{-1}	3×10^{-1}	1×10^{1}	1×10^{5}
Sr-92 (a)	1×10^{0}	3×10^{-1}	1×10^{1}	1×10^{6}
Tritium (1)				
T(H-3)	4×10^{1}	4×10^{1}	1×10^{6}	1×10^{9}

Radionuclide (atomic number)	A_1 (TBq)	A_2 (TBq)	Activity concentration for exempt material (Bq/g)	Activity limit for an exempt consignment (Bq)
Tantalum (73)				
Ta-178(long-lived)	1×10^0	8×10^{-1}	1×10^1	1×10^6
Ta-179	3×10^1	3×10^1	1×10^3	1×10^7
Ta-182	9×10^{-1}	5×10^{-1}	1×10^1	1×10^4
Terbium (65)				
Tb-157	4×10^1	4×10^1	1×10^4	1×10^7
Tb-158	1×10^0	1×10^0	1×10^1	1×10^6
Tb-160	1×10^0	6×10^{-1}	1×10^1	1×10^6
Technetium (43)				
Tc-95m (a)	2×10^0	2×10^0	1×10^1	1×10^6
Tc-96	4×10^{-1}	4×10^{-1}	1×10^1	1×10^6
Tc-96m (a)	4×10^{-1}	4×10^{-1}	1×10^3	1×10^7
Tc-97	Unlimited	Unlimited	1×10^3	1×10^8
Tc-97m	4×10^1	1×10^0	1×10^3	1×10^7
Tc-98	8×10^{-1}	7×10^{-1}	1×10^1	1×10^6
Tc-99	4×10^1	9×10^{-1}	1×10^4	1×10^7
Tc-99m	1×10^1	4×10^0	1×10^2	1×10^7
Tellurium (52)				
Te-121	2×10^0	2×10^0	1×10^1	1×10^6
Te-121m	5×10^0	3×10^0	1×10^2	1×10^6
Te-123m	8×10^0	1×10^0	1×10^2	1×10^7
Te-125m	2×10^1	9×10^{-1}	1×10^3	1×10^7
Te-127	2×10^1	7×10^{-1}	1×10^3	1×10^6
Te-127m (a)	2×10^1	5×10^{-1}	1×10^3	1×10^7
Te-129	7×10^{-1}	6×10^{-1}	1×10^2	1×10^6
Te-129m (a)	8×10^{-1}	4×10^{-1}	1×10^3	1×10^6
Te-131m (a)	7×10^{-1}	5×10^{-1}	1×10^1	1×10^6
Te-132 (a)	5×10^{-1}	4×10^{-1}	1×10^2	1×10^7
Thorium (90)				
Th-227	1×10^1	5×10^{-3}	1×10^1	1×10^4
Th-228 (a)	5×10^{-1}	1×10^{-3}	1×10^0 (b)	1×10^4 (b)

Radionuclide (atomic number)	A_1 (TBq)	A_2 (TBq)	Activity concentration for exempt material (Bq/g)	Activity limit for an exempt consignment (Bq)
Th-229	5×10^{0}	5×10^{-4}	1×10^{0} (b)	1×10^{3} (b)
Th-230	1×10^{1}	1×10^{-3}	1×10^{0}	1×10^{4}
Th-231	4×10^{1}	2×10^{-2}	1×10^{3}	1×10^{7}
Th-232	Unlimited	Unlimited	1×10^{1}	1×10^{4}
Th-234 (a)	3×10^{-1}	3×10^{-1}	1×10^{3} (b)	1×10^{5} (b)
Th(nat)	Unlimited	Unlimited	1×10^{0} (b)	1×10^{3} (b)
Titanium (22)				
Ti-44 (a)	5×10^{-1}	4×10^{-1}	1×10^{1}	1×10^{5}
Thallium (81)				
Tl-200	9×10^{-1}	9×10^{-1}	1×10^{1}	1×10^{6}
Tl-201	1×10^{1}	4×10^{0}	1×10^{2}	1×10^{6}
Tl-202	2×10^{0}	2×10^{0}	1×10^{2}	1×10^{6}
Tl-204	1×10^{1}	7×10^{-1}	1×10^{4}	1×10^{4}
Thulium (69)				
Tm-167	7×10^{0}	8×10^{-1}	1×10^{2}	1×10^{6}
Tm-170	3×10^{0}	6×10^{-1}	1×10^{3}	1×10^{6}
Tm-171	4×10^{1}	4×10^{1}	1×10^{4}	1×10^{8}
Uranium (92)				
U-230 (fast lung absorption) (a)(d)	4×10^{1}	1×10^{-1}	1×10^{1} (b)	1×10^{5} (b)
U-230 (medium lung absorption) (a)(e)	4×10^{1}	4×10^{-3}	1×10^{1}	1×10^{4}
U-230 (slow lung absorption) (a)(f)	3×10^{1}	3×10^{-3}	1×10^{1}	1×10^{4}
U-232 (fast lung absorption) (d)	4×10^{1}	1×10^{-2}	1×10^{0} (b)	1×10^{3} (b)
U-232 (medium lung absorption) (e)	4×10^{1}	7×10^{-3}	1×10^{1}	1×10^{4}
U-232 (slow lung absorption) (f)	1×10^{1}	1×10^{-3}	1×10^{1}	1×10^{4}
U-233 (fast lung absorption) (d)	4×10^{1}	9×10^{-2}	1×10^{1}	1×10^{4}
U-233 (medium lung absorption) (e)	4×10^{1}	2×10^{-2}	1×10^{2}	1×10^{5}
U-233 (slow lung absorption) (f)	4×10^{1}	6×10^{-3}	1×10^{1}	1×10^{5}
U-234 (fast lung absorption) (d)	4×10^{1}	9×10^{-2}	1×10^{1}	1×10^{4}
U-234 (medium lung absorption) (e)	4×10^{1}	2×10^{-2}	1×10^{2}	1×10^{5}
U-234 (slow lung absorption) (f)	4×10^{1}	6×10^{-3}	1×10^{1}	1×10^{5}
U-235 (all lung absorption types)	Unlimited	Unlimited	1×10^{1} (b)	1×10^{4} (b)

Radionuclide (atomic number)	A_1 (TBq)	A_2 (TBq)	Activity concentration for exempt material (Bq/g)	Activity limit for an exempt consignment (Bq)
(a)(d)(e)(f)				
U-236 (fast lung absorption) (d)	Unlimited	Unlimited	1×10^{1}	1×10^{4}
U-236 (medium lung absorption) (e)	4×10^{1}	2×10^{-2}	1×10^{2}	1×10^{5}
U-236 (slow lung absorption) (f)	4×10^{1}	6×10^{-3}	1×10^{1}	1×10^{4}
U-238 (all lung absorption types) (d)(e)(f)	Unlimited	Unlimited	1×10^{1} (b)	1×10^{4} (b)
U (nat)	Unlimited	Unlimited	1×10^{0} (b)	1×10^{3} (b)
U (enriched to 20% or less) (g)	Unlimited	Unlimited	1×10^{0}	1×10^{3}
U (dep)	Unlimited	Unlimited	1×10^{0}	1×10^{3}
Vanadium (23)				
V-48	4×10^{-1}	4×10^{-1}	1×10^{1}	1×10^{5}
V-49	4×10^{1}	4×10^{1}	1×10^{4}	1×10^{7}
Tungsten (74)				
W-178 (a)	9×10^{0}	5×10^{0}	1×10^{1}	1×10^{6}
W-181	3×10^{1}	3×10^{1}	1×10^{3}	1×10^{7}
W-185	4×10^{1}	8×10^{-1}	1×10^{4}	1×10^{7}
W-187	2×10^{0}	6×10^{-1}	1×10^{2}	1×10^{6}
W-188 (a)	4×10^{-1}	3×10^{-1}	1×10^{2}	1×10^{5}
Xenon (54)				
Xe-122 (a)	4×10^{-1}	4×10^{-1}	1×10^{2}	1×10^{9}
Xe-123	2×10^{0}	7×10^{-1}	1×10^{2}	1×10^{9}
Xe-127	4×10^{0}	2×10^{0}	1×10^{3}	1×10^{5}
Xe-131m	4×10^{1}	4×10^{1}	1×10^{4}	1×10^{4}
Xe-133	2×10^{1}	1×10^{1}	1×10^{3}	1×10^{4}
Xe-135	3×10^{0}	2×10^{0}	1×10^{3}	1×10^{10}
Yttrium (39)				
Y-87 (a)	1×10^{0}	1×10^{0}	1×10^{1}	1×10^{6}
Y-88	4×10^{-1}	4×10^{-1}	1×10^{1}	1×10^{6}
Y-90	3×10^{-1}	3×10^{-1}	1×10^{3}	1×10^{5}
Y-91	6×10^{-1}	6×10^{-1}	1×10^{3}	1×10^{6}
Y-91m	2×10^{0}	2×10^{0}	1×10^{2}	1×10^{6}

Radionuclide (atomic number)	A_1 (TBq)	A_2 (TBq)	Activity concentration for exempt material (Bq/g)	Activity limit for an exempt consignment (Bq)
Y-92	2×10^{-1}	2×10^{-1}	1×10^{2}	1×10^{5}
Y-93	3×10^{-1}	3×10^{-1}	1×10^{2}	1×10^{5}
Ytterbium (70)				
Yb-169	4×10^{0}	1×10^{0}	1×10^{2}	1×10^{7}
Yb-175	3×10^{1}	9×10^{-1}	1×10^{3}	1×10^{7}
Zinc (30)				
Zn-65	2×10^{0}	2×10^{0}	1×10^{1}	1×10^{6}
Zn-69	3×10^{0}	6×10^{-1}	1×10^{4}	1×10^{6}
Zn-69m (a)	3×10^{0}	6×10^{-1}	1×10^{2}	1×10^{6}
Zirconium (40)				
Zr-88	3×10^{0}	3×10^{0}	1×10^{2}	1×10^{6}
Zr-93	Unlimited	Unlimited	1×10^{3} (b)	1×10^{7} (b)
Zr-95 (a)	2×10^{0}	8×10^{-1}	1×10^{1}	1×10^{6}
Zr-97 (a)	4×10^{-1}	4×10^{-1}	1×10^{1} (b)	1×10^{5} (b)

(a) A_1 and/or A_2 values for these parent radionuclides include contributions from daughter radionuclides with half-lives less than 10 days, as listed in the following:

Mg-28	Al-28
Ar-42	K-42
Ca-47	Sc-47
Ti-44	Sc-44
Fe-52	Mn-52m
Fe-60	Co-60m
Zn-69m	Zn-69
Ge-68	Ga-68
Rb-83	Kr-83m
Sr-82	Rb-82
Sr-90	Y-90
Sr-91	Y-91m
Sr-92	Y-92
Y-87	Sr-87m
Zr-95	Nb-95m
Zr-97	Nb-97m, Nb-97
Mo-99	Tc-99m
Tc-95m	Tc-95
Tc-96m	Tc-96
Ru-103	Rh-103m
Ru-106	Rh-106
Pd-103	Rh-103m
Ag-108m	Ag-108
Ag-110m	Ag-110

Cd-115	In-115m
In-114m	In-114
Sn-113	In-113m
Sn-121m	Sn-121
Sn-126	Sb-126m
Te-118	Sb-118
Te-127m	Te-127
Te-129m	Te-129
Te-131m	Te-131
Te-132	I-132
I-135	Xe-135m
Xe-122	I-122
Cs-137	Ba-137m
Ba-131	Cs-131
Ba-140	La-140
Ce-144	Pr-144m, Pr-144
Pm-148m	Pm-148
Gd-146	Eu-146
Dy-166	Ho-166
Hf-172	Lu-172
W-178	Ta-178
W-188	Re-188
Re-189	Os-189m
Os-194	Ir-194
Ir-189	Os-189m
Pt-188	Ir-188
Hg-194	Au-194
Hg-195m	Hg-195
Pb-210	Bi-210
Pb-212	Bi-212, Tl-208, Po-212
Bi-210m	Tl-206
Bi-212	Tl-208, Po-212
At-211	Po-211
Rn-222	Po-218, Pb-214, At-218, Bi-214, Po-214
Ra-223	Rn-219, Po-215, Pb-211, Bi-211, Po-211, Tl-207
Ra-224	Rn-220, Po-216, Pb-212, Bi-212, Tl-208, Po-212
Ra-225	Ac-225, Fr-221, At-217, Bi-213, Tl-209, Po-213, Pb-209
Ra-226	Rn-222, Po-218, Pb-214, At-218, Bi-214, Po-214
Ra-228	Ac-228
Ac-225	Fr-221, At-217, Bi-213, Tl-209, Po-213, Pb-209
Ac-227	Fr-223
Th-228	Ra-224, Rn-220, Po-216, Pb-212, Bi-212, Tl-208, Po-212
Th-234	Pa-234m, Pa-234
Pa-230	Ac-226, Th-226, Fr-222, Ra-222, Rn-218, Po-214
U-230	Th-226, Ra-222, Rn-218, Po-214
U-235	Th-231
Pu-241	U-237
Pu-244	U-240, Np-240m
Am-242m	Am-242, Np-238
Am-243	Np-239
Cm-247	Pu-243
Bk-249	Am-245
Cf-253	Cm-249

(b) Parent nuclides and their progeny included in secular equilibrium are listed in the following:

Sr-90	Y-90
Zr-93	Nb-93m
Zr-97	Nb-97
Ru-106	Rh-106
Ag-108m	Ag-108
Cs-137	Ba-137m
Ce-144	Pr-144
Ba-140	La-140
Bi-212	Tl-208 (0.36), Po-212 (0.64)
Pb-210	Bi-210, Po-210
Pb-212	Bi-212, Tl-208 (0.36), Po-212 (0.64)
Rn-222	Po-218, Pb-214, Bi-214, Po-214
Ra-223	Rn-219, Po-215, Pb-211, Bi-211, Tl-207
Ra-224	Rn-220, Po-216, Pb-212, Bi-212, Tl-208 (0.36), Po-212 (0.64)
Ra-226	Rn-222, Po-218, Pb-214, Bi-214, Po-214, Pb-210, Bi-210, Po-210
Ra-228	Ac-228
Th-228	Ra-224, Rn-220, Po-216, Pb212, Bi-212, Tl208 (0.36), Po-212 (0.64)
Th-229	Ra-225, Ac-225, Fr-221, At-217, Bi-213, Po-213, Pb-209
Th-nat	Ra-228, Ac-228, Th-228, Ra-224, Rn-220, Po-216, Pb-212, Bi-212, l208(0.36), Po-212 (0.64)
Th-234	Pa-234m
U-230	Th-226, Ra-222, Rn-218, Po-214
U-232	Th-228, Ra-224, Rn-220, Po-216, Pb-212, Bi-212, Tl-208 (0.36), Po-212 (0.64)
U-235	Th-231
U-238	Th-234, Pa-234m
U-nat	Th-234, Pa-234m, U-234, Th-230, Ra-226, Rn-222, Po-218, Pb-214, Bi-214, Po-214, Pb-210, Bi-210, Po-210
	Np-237 Pa-233
Am-242m	Am-242
Am-243	Np-239

(c) The quantity may be determined from a measurement of the rate of decay or a measurement of the radiation level at a prescribed distance from the source.

(d) These values apply only to compounds of uranium that take the chemical form of UF_6, UO_2F_2 and $UO_2(NO_3)_2$ in both normal and accident conditions of carriage.

(e) These values apply only to compounds of uranium that take the chemical form of UO_3, UF_4, UCl_4 and hexavalent compounds in both normal and accident conditions of carriage.

(f) These values apply to all compounds of uranium other than those specified in (d) and (e) above.

(g) These values apply to unirradiated uranium only.

2.2.7.2.2.2 For individual radionuclides which are not listed in Table 2.2.7.2.2.1 the determination of the basic radionuclide values referred to in 2.2.7.2.2.1 shall require multilateral approval. It is permissible to use an A_2 value calculated using a dose coefficient for the appropriate lung absorption type as recommended by the International Commission on Radiological Protection, if the chemical forms of each radionuclide under both normal and accident conditions of carriage are taken into consideration. Alternatively, the radionuclide values in Table 2.2.7.2.2.2 may be used without obtaining competent authority approval.

Table 2.2.7.2.2.2: Basic radionuclide values for unknown radionuclides or mixtures

Radioactive contents	A_1 (TBq)	A_2 (TBq)	Activity concentration for exempt material (Bq/g)	Activity limit for exempt consignments (Bq)
Only beta or gamma emitting nuclides are known to be present	0.1	0.02	1×10^1	1×10^4
Alpha emitting nuclides but no neutron emitters are known to be present	0.2	9×10^{-5}	1×10^{-1}	1×10^3
Neutron emitting nuclides are known to be present or no relevant data are available	0.001	9×10^{-5}	1×10^{-1}	1×10^3

2.2.7.2.2.3 In the calculations of A_1 and A_2 for a radionuclide not in Table 2.2.7.2.2.1, a single radioactive decay chain in which the radionuclides are present in their naturally occurring proportions, and in which no daughter nuclide has a half-life either longer than 10 days or longer than that of the parent nuclide, shall be considered as a single radionuclide; and the activity to be taken into account and the A_1 or A_2 value to be applied shall be those corresponding to the parent nuclide of that chain. In the case of radioactive decay chains in which any daughter nuclide has a half-life either longer than 10 days or greater than that of the parent nuclide, the parent and such daughter nuclides shall be considered as mixtures of different nuclides.

2.2.7.2.2.4 For mixtures of radionuclides, the determination of the basic radionuclide values referred to in 2.2.7.2.2.1 may be determined as follows:

$$X_m = \frac{1}{\Sigma_i \frac{f(i)}{X(i)}}$$

where,

f(i) is the fraction of activity or activity concentration of radionuclide i in the mixture;

X(i) is the appropriate value of A_1 or A_2, or the activity concentration for exempt material or the activity limit for an exempt consignment as appropriate for the radionuclide i; and

X_m is the derived value of A_1 or A_2, or the activity concentration for exempt material or the activity limit for an exempt consignment in the case of a mixture.

2.2.7.2.2.5 When the identity of each radionuclide is known but the individual activities of some of the radionuclides are not known, the radionuclides may be grouped and the lowest radionuclide value, as appropriate, for the radionuclides in each group may be used in applying the formulas in 2.2.7.2.2.4 and 2.2.7.2.4.4. Groups may be based on the total alpha activity and the total beta/gamma activity when these are known, using the lowest radionuclide values for the alpha emitters or beta/gamma emitters, respectively.

2.2.7.2.2.6 For individual radionuclides or for mixtures of radionuclides for which relevant data are not available, the values shown in Table 2.2.7.2.2.2 shall be used.

2.2.7.2.3 *Determination of other material characteristics*

2.2.7.2.3.1 Low specific activity (LSA) material

2.2.7.2.3.1.1 (*Reserved*)

2.2.7.2.3.1.2 LSA material shall be in one of three groups:

(a) LSA-I

(i) uranium and thorium ores and concentrates of such ores, and other ores containing naturally occurring radionuclides which are intended to be processed for the use of these radionuclides;

(ii) natural uranium, depleted uranium, natural thorium or their compounds or mixtures, providing they are unirradiated and in solid or liquid form;

(iii) radioactive material for which the A_2 value is unlimited, excluding material classified as fissile according to 2.2.7.2.3.5; or

(iv) other radioactive material in which the activity is distributed throughout and the estimated average specific activity does not exceed 30 times the values for activity concentration specified in 2.2.7.2.2.1 to 2.2.7.2.2.6, excluding material classified as fissile according to 2.2.7.2.3.5;

(b) LSA-II

(i) water with tritium concentration up to 0.8 TBq/l; or

(ii) other material in which the activity is distributed throughout and the estimated average specific activity does not exceed 10^{-4} A_2/g for solids and gases, and 10^{-5} A_2/g for liquids;

(c) LSA-III - Solids (e.g. consolidated wastes, activated materials), excluding powders, in which:

(i) the radioactive material is distributed throughout a solid or a collection of solid objects, or is essentially uniformly distributed in a solid compact binding agent (such as concrete, bitumen, ceramic, etc.);

(ii) the radioactive material is relatively insoluble, or it is intrinsically contained in a relatively insoluble matrix, so that, even under loss of packaging, the loss of radioactive material per package by leaching when placed in water for seven days would not exceed 0.1 A_2; and

(iii) the estimated average specific activity of the solid, excluding any shielding material, does not exceed 2 $\times 10^{-3}$ A_2/g.

2.2.7.2.3.1.3 LSA-III material shall be a solid of such a nature that if the entire contents of a package were subjected to the test specified in 2.2.7.2.3.1.4 the activity in the water would not exceed 0.1 A_2.

2.2.7.2.3.1.4 LSA-III material shall be tested as follows:

A solid material sample representing the entire contents of the package shall be immersed for 7 days in water at ambient temperature. The volume of water to be used in the test shall be sufficient to ensure that at the end of the 7 day test period the free volume of the unabsorbed and unreacted water remaining shall be at least 10% of the volume of the solid test sample itself. The water shall have an initial pH of 6-8 and a maximum conductivity of 1 mS/m at 20 °C. The total activity of the free volume of water shall be measured following the 7 day immersion of the test sample.

2.2.7.2.3.1.5 Demonstration of compliance with the performance standards in 2.2.7.2.3.1.4 shall be in accordance with 6.4.12.1 and 6.4.12.2 of ADR.

2.2.7.2.3.2 Surface contaminated object (SCO)

SCO is classified in one of two groups:

(a) SCO-I: A solid object on which:

(i) the non-fixed contamination on the accessible surface averaged over 300 cm^2 (or the area of the surface if less than 300 cm^2) does not exceed 4 Bq/cm^2 for beta and gamma emitters and low toxicity alpha emitters, or 0.4 Bq/cm^2 for all other alpha emitters; and

(ii) the fixed contamination on the accessible surface averaged over 300 cm^2 (or the area of the surface if less than 300 cm^2) does not exceed 4×10^4 Bq/cm^2 for beta and gamma emitters and low toxicity alpha emitters, or 4×10^3 Bq/cm^2 for all other alpha emitters; and

(iii) the non-fixed contamination plus the fixed contamination on the inaccessible surface averaged over 300 cm^2 (or the area of the surface if less than 300 cm^2)

does not exceed 4×10^4 Bq/cm^2 for beta and gamma emitters and low toxicity alpha emitters, or 4×10^3 Bq/cm^2 for all other alpha emitters;

(b) SCO-II: A solid object on which either the fixed or non-fixed contamination on the surface exceeds the applicable limits specified for SCO-I in (a) above and on which:

(i) the non-fixed contamination on the accessible surface averaged over 300 cm^2 (or the area of the surface if less than 300 cm^2) does not exceed 400 Bq/cm^2 for beta and gamma emitters and low toxicity alpha emitters, or 40 Bq/cm^2 for all other alpha emitters; and

(ii) the fixed contamination on the accessible surface, averaged over 300 cm^2 (or the area of the surface if less than 300 cm^2) does not exceed 8×10^5 Bq/cm^2 for beta and gamma emitters and low toxicity alpha emitters, or 8×10^4 Bq/cm^2 for all other alpha emitters; and

(iii) the non-fixed contamination plus the fixed contamination on the inaccessible surface averaged over 300 cm^2 (or the area of the surface if less than 300 cm^2) does not exceed 8×10^5 Bq/cm^2 for beta and gamma emitters and low toxicity alpha emitters, or 8×10^4 Bq/cm^2 for all other alpha emitters.

2.2.7.2.3.3 Special form radioactive material

2.2.7.2.3.3.1 Special form radioactive material shall have at least one dimension not less than 5 mm. When a sealed capsule constitutes part of the special form radioactive material, the capsule shall be so manufactured that it can be opened only by destroying it. The design for special form radioactive material requires unilateral approval.

2.2.7.2.3.3.2 Special form radioactive material shall be of such a nature or shall be so designed that if it is subjected to the tests specified in 2.2.7.2.3.3.4 to 2.2.7.2.3.3.8, it shall meet the following requirements:

(a) It would not break or shatter under the impact, percussion and bending tests 2.2.7.2.3.3.5 (a), (b), (c), 2.2.7.2.3.3.6 (a) as applicable;

(b) It would not melt or disperse in the applicable heat test 2.2.7.2.3.3.5 (d) or 2.2.7.2.3.3.6 (b) as applicable; and

(c) The activity in the water from the leaching tests specified in 2.2.7.2.3.3.7 and 2.2.7.2.3.3.8 would not exceed 2 kBq; or alternatively for sealed sources, the leakage rate for the volumetric leakage assessment test specified in ISO 9978:1992 "Radiation Protection - Sealed Radioactive Sources - Leakage Test Methods", would not exceed the applicable acceptance threshold acceptable to the competent authority.

2.2.7.2.3.3.3 Demonstration of compliance with the performance standards in 2.2.7.2.3.3.2 shall be in accordance with 6.4.12.1 and 6.4.12.2 of ADR.

2.2.7.2.3.3.4 Specimens that comprise or simulate special form radioactive material shall be subjected to the impact test, the percussion test, the bending test, and the heat test specified in 2.2.7.2.3.3.5 or alternative tests as authorized in 2.2.7.2.3.3.6. A different specimen may be used for each of the tests. Following each test, a leaching assessment or volumetric leakage test shall be performed on the specimen by a method no less sensitive than the methods given in 2.2.7.2.3.3.7 for indispersible solid material or 2.2.7.2.3.3.8 for encapsulated material.

2.2.7.2.3.3.5 The relevant test methods are:

(a) Impact test: The specimen shall drop onto the target from a height of 9 m. The target shall be as defined in 6.4.14 of ADR;

(b) Percussion test: The specimen shall be placed on a sheet of lead which is supported by a smooth solid surface and struck by the flat face of a mild steel bar so as to cause an impact equivalent to that resulting from a free drop of 1.4 kg through 1 m. The lower part of the bar shall be 25 mm in diameter with the edges rounded off to a radius of (3.0 ± 0.3) mm. The lead, of hardness number 3.5 to 4.5 on the Vickers scale and not more than 25 mm thick, shall cover an area greater than that covered by the specimen. A fresh surface of lead shall be used for each impact. The bar shall strike the specimen so as to cause maximum damage;

(c) Bending test: The test shall apply only to long, slender sources with both a minimum length of 10 cm and a length to minimum width ratio of not less than 10. The specimen shall be rigidly clamped in a horizontal position so that one half of its length protrudes from the face of the clamp. The orientation of the specimen shall be such that the specimen will suffer maximum damage when its free end is struck by the flat face of a steel bar. The bar shall strike the specimen so as to cause an impact equivalent to that resulting from a free vertical drop of 1.4 kg through 1 m. The lower part of the bar shall be 25 mm in diameter with the edges rounded off to a radius of (3.0 ± 0.3) mm;

(d) Heat test: The specimen shall be heated in air to a temperature of 800 °C and held at that temperature for a period of 10 minutes and shall then be allowed to cool.

2.2.7.2.3.3.6 Specimens that comprise or simulate radioactive material enclosed in a sealed capsule may be excepted from:

(a) The tests prescribed in 2.2.7.2.3.3.5 (a) and (b) provided the mass of the special form radioactive material:

(i) is less than 200 g and they are alternatively subjected to the Class 4 impact test prescribed in ISO 2919:1999 "Radiation protection - Sealed radioactive sources - General requirements and classification"; or

(ii) is less than 500 g and they are alternatively subjected to the Class 5 impact test prescribed in ISO 2919:1999 "Radiation protection - Sealed radioactive sources General requirements and classification"; and

(b) The test prescribed in 2.2.7.2.3.3.5 (d) provided they are alternatively subjected to the Class 6 temperature test specified in ISO 2919:1999 "Radiation protection - Sealed radioactive sources - General requirements and classification".

2.2.7.2.3.3.7 For specimens which comprise or simulate indispersible solid material, a leaching assessment shall be performed as follows:

(a) The specimen shall be immersed for 7 days in water at ambient temperature. The volume of water to be used in the test shall be sufficient to ensure that at the end of the 7 day test period the free volume of the unabsorbed and unreacted water remaining shall be at least 10% of the volume of the solid test sample itself. The water shall have an initial pH of 6-8 and a maximum conductivity of 1 mS/m at 20 °C;

(b) The water with specimen shall then be heated to a temperature of (50 ± 5) °C and maintained at this temperature for 4 hours;

(c) The activity of the water shall then be determined;

(d) The specimen shall then be kept for at least 7 days in still air at not less than 30 °C and relative humidity not less than 90%;

(e) The specimen shall then be immersed in water of the same specification as in (a) above and the water with the specimen heated to (50 ± 5) °C and maintained at this temperature for 4 hours;

(f) The activity of the water shall then be determined.

2.2.7.2.3.3.8 For specimens which comprise or simulate radioactive material enclosed in a sealed capsule, either a leaching assessment or a volumetric leakage assessment shall be performed as follows:

(a) The leaching assessment shall consist of the following steps:

(i) the specimen shall be immersed in water at ambient temperature. The water shall have an initial pH of 6-8 with a maximum conductivity of 1 mS/m at 20 C;

(ii) the water and specimen shall be heated to a temperature of (50 ± 5) °C and maintained at this temperature for 4 hours;

(iii) the activity of the water shall then be determined;

(iv) the specimen shall then be kept for at least 7 days in still air at not less than 30°C and relative humidity of not less than 90%;

(v) the process in (i), (ii) and (iii) shall be repeated;

(b) The alternative volumetric leakage assessment shall comprise any of the tests prescribed in ISO 9978:1992 "Radiation Protection - Sealed radioactive sources - Leakage test methods", which are acceptable to the competent authority.

2.2.7.2.3.4 Low dispersible radioactive material

2.2.7.2.3.4.1 The design for low dispersible radioactive material shall require multilateral approval. Low dispersible radioactive material shall be such that the total amount of this radioactive material in a package shall meet the following requirements:

(a) The radiation level at 3 m from the unshielded radioactive material does not exceed 10 mSv/h;

(b) If subjected to the tests specified in 6.4.20.3 and 6.4.20.4 of ADR, the airborne release in gaseous and particulate forms of up to 100 μm aerodynamic equivalent diameter would not exceed 100 A_2. A separate specimen may be used for each test; and

(c) If subjected to the test specified in 2.2.7.2.3.1.4 the activity in the water would not exceed 100 A_2. In the application of this test, the damaging effects of the tests specified in (b) above shall be taken into account.

2.2.7.2.3.4.2 Low dispersible radioactive material shall be tested as follows:

A specimen that comprises or simulates low dispersible radioactive material shall be subjected to the enhanced thermal test specified in 6.4.20.3 of ADR and the impact test specified in 6.4.20.4 of ADR. A different specimen may be used for each of the tests. Following each test, the specimen shall be subjected to the leach test specified in 2.2.7.2.3.1.4. After each test it shall be determined if the applicable requirements of 2.2.7.2.3.4.1 have been met.

2.2.7.2.3.4.3 Demonstration of compliance with the performance standards in 2.2.7.2.3.4.1 and 2.2.7.2.3.4.2 shall be in accordance with 6.4.12.1 and 6.4.12.2 of ADR.

2.2.7.2.3.5 Fissile material

Packages containing fissile radionuclides shall be classified under the relevant entry of table 2.2.7.2.1.1 for fissile material unless one of the conditions (a) to (d) of this paragraph is met. Only one type of exception is allowed per consignment.

(a) A mass limit per consignment such that:

$$\frac{\text{mass of uranium} - 235\ (\text{g})}{\text{X}} + \frac{\text{mass of other fissile material (g)}}{\text{Y}} < 1$$

where X and Y are the mass limits defined in Table 2.2.7.2.3.5, provided that the smallest external dimension of each package is not less than 10 cm and that either:

(i) each individual package contains not more than 15 g of fissile material; for unpackaged material, this quantity limitation shall apply to the consignment being carried in or on the wagon/vehicle/conveyance; or

(ii) the fissile material is a homogeneous hydrogenous solution or mixture where the ratio of fissile nuclides to hydrogen is less than 5% by mass; or

(iii) there are not more than 5 g of fissile material in any 10 litre volume of material.

Neither beryllium nor deuterium shall be present in quantities exceeding 1% of the applicable consignment mass limits provided in Table 2.2.7.2.3.5, except for deuterium in natural concentration in hydrogen;

(b) Uranium enriched in uranium-235 to a maximum of 1% by mass, and with a total plutonium and uranium-233 content not exceeding 1% of the mass of uranium-235, provided that the fissile material is distributed essentially homogeneously throughout the material. In addition, if uranium-235 is present in metallic, oxide or carbide forms, it shall not form a lattice arrangement;

(c) Liquid solutions of uranyl nitrate enriched in uranium-235 to a maximum of 2% by mass, with a total plutonium and uranium-233 content not exceeding 0.002% of the mass of uranium, and with a minimum nitrogen to uranium atomic ratio (N/U) of 2;

(d) Packages containing, individually, a total plutonium mass not more than 1 kg, of which not more than 20% by mass may consist of plutonium-239, plutonium-241 or any combination of those radionuclides.

Table 2.2.7.2.3.5: Consignment mass limits for exceptions from the requirements for packages containing fissile material

Fissile material	Fissile material mass (g) mixed with substances having an average hydrogen density less than or equal to water	Fissile material mass (g) mixed with substances having an average hydrogen density greater than water
Uranium-235 (X)	400	290
Other fissile material (Y)	250	180

2.2.7.2.4 *Classification of packages or unpacked material*

The quantity of radioactive material in a package shall not exceed the relevant limits for the package type as specified below.

2.2.7.2.4.1 Classification as excepted package

2.2.7.2.4.1.1 Packages may be classified as excepted packages if:

(a) They are empty packagings having contained radioactive material;

(b) They contain instruments or articles in limited quantities;

(c) They contain articles manufactured of natural uranium, depleted uranium or natural thorium; or

(d) They contain radioactive material in limited quantities.

2.2.7.2.4.1.2 A package containing radioactive material may be classified as an excepted package provided that the radiation level at any point on its external surface does not exceed 5 μSv/h.

Table 2.2.7.2.4.1.2: Activity limits for excepted packages

Physical state of contents	Instruments or article		Materials
	Item limits [a]	Package limits [a]	Package limits [a]
(1)	(2)	(3)	(4)
Solids			
special form	10^{-2} A_1	A_1	10^{-3} A_1
other form	10^{-2} A_2	A_2	10^{-3} A_2
Liquids	10^{-3} A_2	10^{-1} A_2	10^{-4} A_2
Gases			
Tritium	2×10^{-2} A_2	2×10^{-1} A_2	2×10^{-2} A_2
special form	10^{-3} A_1	10^{-2} A_1	10^{-3} A_1
other forms	10^{-3} A_2	10^{-2} A_2	10^{-3} A_2

[a] *For mixtures of radionuclides, see 2.2.7.2.2.4 to 2.2.7.2.2.6.*

2.2.7.2.4.1.3 Radioactive material which is enclosed in or is included as a component part of an instrument or other manufactured article may be classified under UN No. 2911 RADIOACTIVE MATERIAL, EXCEPTED PACKAGE - INSTRUMENTS or ARTICLES provided that:

(a) The radiation level at 10 cm from any point on the external surface of any unpackaged instrument or article is not greater than 0.1 mSv/h; and

(b) Each instrument or manufactured article bears the marking "RADIOACTIVE" except:

(i) radioluminescent time-pieces or devices;

(ii) consumer products that either have received regulatory approval according to 1.7.1.4 (d) or do not individually exceed the activity limit for an exempt consignment in Table 2.2.7.2.2.1 (column 5), provided such products are carried in a package that bears the marking "RADIOACTIVE" on an internal surface in such a manner that warning of the presence of radioactive material is visible on opening the package; and

(c) The active material is completely enclosed by non-active components (a device performing the sole function of containing radioactive material shall not be considered to be an instrument or manufactured article); and

(d) The limits specified in columns 2 and 3 of Table 2.2.7.2.4.1.2 are met for each individual item and each package, respectively.

2.2.7.2.4.1.4 Radioactive material with an activity not exceeding the limit specified in column 4 of Table 2.2.7.2.4.1.2, may be classified under UN No. 2910 RADIOACTIVE MATERIAL, EXCEPTED PACKAGE - LIMITED QUANTITY OF MATERIAL provided that:

(a) The package retains its radioactive contents under routine conditions of carriage; and

(b) The package bears the marking "RADIOACTIVE" on an internal surface in such a manner that a warning of the presence of radioactive material is visible on opening the package.

2.2.7.2.4.1.5 An empty packaging which had previously contained radioactive material with an activity not exceeding the limit specified in column 4 of Table 2.2.7.2.4.1.2 may be classified under UN No. 2908 RADIOACTIVE MATERIAL, EXCEPTED PACKAGE - EMPTY PACKAGING, provided that:

(a) It is in a well-maintained condition and securely closed;

(b) The outer surface of any uranium or thorium in its structure is covered with an inactive sheath made of metal or some other substantial material;

(c) The level of internal non-fixed contamination, when averaged over any 300 cm^2, does not exceed:

(i) 400 Bq/cm^2 for beta and gamma emitters and low toxicity alpha emitters; and

(ii) 40 Bq/cm^2 for all other alpha emitters; and

(d) Any labels which may have been displayed on it in conformity with 5.2.2.1.11.1 are no longer visible.

2.2.7.2.4.1.6 Articles manufactured of natural uranium, depleted uranium or natural thorium and articles in which the sole radioactive material is unirradiated natural uranium, unirradiated depleted uranium or unirradiated natural thorium may be classified under UN No. 2909 RADIOACTIVE MATERIAL, EXCEPTED PACKAGE - ARTICLES MANUFACTURED FROM NATURAL URANIUM or DEPLETED URANIUM or NATURAL THORIUM, provided that the outer surface of the uranium or thorium is enclosed in an inactive sheath made of metal or some other substantial material.

2.2.7.2.4.2 Classification as Low specific activity (LSA) material

Radioactive material may only be classified as LSA material if the conditions of 2.2.7.2.3.1 and 4.1.9.2 of ADR are met.

2.2.7.2.4.3 Classification as Surface contaminated object (SCO)

Radioactive material may be classified as SCO if the conditions of 2.2.7.2.3.2 and 4.1.9.2 of ADR are met.

2.2.7.2.4.4 Classification as Type A package

Packages containing radioactive material may be classified as Type A packages provided that the following conditions are met:

Type A packages shall not contain activities greater than the following:

(a) For special form radioactive material - A_1; or

(b) For all other radioactive material - A_2.

For mixtures of radionuclides whose identities and respective activities are known, the following condition shall apply to the radioactive contents of a Type A package:

$$\Sigma_i \frac{B(i)}{A_1(i)} + \Sigma_j \frac{C(j)}{A_2(j)} \leq 1$$

where B(i) is the activity of radionuclide i as special form radioactive material;

$A_1(i)$ is the A_1 value for radionuclide i;

C (j) is the activity of radionuclide j as other than special form radioactive material; and

A_2 (j) is the A_2 value for radionuclide j.

2.2.7.2.4.5 Classification of Uranium hexafluoride

Uranium hexafluoride shall only be assigned to UN Nos. 2977 RADIOACTIVE MATERIAL, URANIUM HEXAFLUORIDE, FISSILE, or 2978 RADIOACTIVE MATERIAL, URANIUM HEXAFLUORIDE, non-fissile or fissile-excepted.

2.2.7.2.4.5.1 Packages containing uranium hexafluoride shall not contain:

(a) A mass of uranium hexafluoride different from that authorized for the package design;

(b) A mass of uranium hexafluoride greater than a value that would lead to an ullage smaller than 5% at the maximum temperature of the package as specified for the plant systems where the package shall be used; or

(c) Uranium hexafluoride other than in solid form or at an internal pressure above atmospheric pressure when presented for carriage.

2.2.7.2.4.6 Classification as Type B(U), Type B(M) or Type C packages

2.2.7.2.4.6.1 Packages not otherwise classified in 2.2.7.2.4 (2.2.7.2.4.1 to 2.2.7.2.4.5) shall be classified in accordance with the competent authority approval certificate for the package issued by the country of origin of design.

2.2.7.2.4.6.2 A package may only be classified as a Type B(U) if it does not contain:

(a) Activities greater than those authorized for the package design;

(b) Radionuclides different from those authorized for the package design; or

(c) Contents in a form, or a physical or chemical state different from those authorized for the package design;

as specified in the certificate of approval.

2.2.7.2.4.6.3 A package may only be classified as a Type B(M) if it does not contain:

(a) Activities greater than those authorized for the package design;

(b) Radionuclides different from those authorized for the package design; or

(c) Contents in a form, or a physical or chemical state different from those authorized for the package design;

as specified in the certificate of approval.

2.2.7.2.4.6.4 A package may only be classified as a Type C if it does not contain:

(a) Activities greater than those authorized for the package design;

(b) Radionuclides different from those authorized for the package design; or

(c) Contents in a form, or physical or chemical state different from those authorized for the package design;

as specified in the certificate of approval.

2.2.7.2.5 *Special arrangements*

Radioactive material shall be classified as transported under special arrangement when it is intended to be carried in accordance with 1.7.4.

2.2.8 Class 8 Corrosive substances

2.2.8.1 *Criteria*

2.2.8.1.1 The heading of Class 8 covers substances and articles containing substances of this class which by chemical action attack epithelial tissue - of skin or mucous membranes - with which they are in contact, or which in the event of leakage are capable of damaging or destroying other goods, or means of transport. The heading of this class also covers other substances which form a corrosive liquid only in the presence of water, or which produce corrosive vapour or mist in the presence of natural moisture of the air.

2.2.8.1.2 Substances and articles of Class 8 are subdivided as follows:

C1-C10 Corrosive substances without subsidiary risk:

C1-C4 Acid substances:
- C1 Inorganic, liquid;
- C2 Inorganic, solid;
- C3 Organic, liquid;
- C4 Organic, solid;

C5-C8 Basic substances:
- C5 Inorganic, liquid;
- C6 Inorganic, solid;
- C7 Organic, liquid;
- C8 Organic, solid;

C9-C10 Other corrosive substances:
- C9 Liquid;
- C10 Solid;

C11 Articles;

CF Corrosive substances, flammable:
- CF1 Liquid;
- CF2 Solid;

CS Corrosive substances, self-heating:
- CS1 Liquid;
- CS2 Solid;

CW Corrosive substances which, in contact with water, emit flammable gases:
- CW1 Liquid;
- CW2 Solid;

CO Corrosive substances, oxidizing:
- CO1 Liquid;
- CO2 Solid;

CT Corrosive substances, toxic:

- CT1 Liquid;
- CT2 Solid;

CFT Corrosive substances, flammable, liquid, toxic;
COT Corrosive substances, oxidizing, toxic.

Classification and assignment of packing groups

2.2.8.1.3 Substances of Class 8 shall be classified in three packing groups according to the degree of danger they present for carriage, as follows:

Packing group I:	highly corrosive substances
Packing group II:	corrosive substances
Packing group III:	slightly corrosive substances.

2.2.8.1.4 Substances and articles classified in Class 8 are listed in Table A of Chapter 3.2. Allocation of substances to packing groups I, II and III has been made on the basis of experience taking into account such additional factors as inhalation risk (see 2.2.8.1.5) and reactivity with water (including the formation of dangerous decomposition products).

2.2.8.1.5 A substance or preparation meeting the criteria of Class 8 having an inhalation toxicity of dusts and mists (LC_{50}) in the range of packing group I, but toxicity through oral ingestion or dermal contact only in the range of packing group III or less, shall be allocated to Class 8.

2.2.8.1.6 Substances, including mixtures, not mentioned by name in Table A of Chapter 3.2 can be assigned to the relevant entry of sub-section 2.2.8.3, and to the relevant packing group on the basis of the length of time of contact necessary to produce full thickness destruction of human skin in accordance with the criteria of (a) to (c) below.

Liquids, and solids which may become liquid during carriage, which are judged not to cause full thickness destruction of human skin shall still be considered for their potential to cause corrosion to certain metal surfaces. In assigning the packing group, account shall be taken of human experience in instances of accidental exposure. In the absence of human experience, the grouping shall be based on data obtained from experiments in accordance with OECD Guideline 404[8].

(a) Packing group I is assigned to substances that cause full thickness destruction of intact skin tissue within an observation period up to 60 minutes starting after the exposure time of 3 minutes or less.

(b) Packing group II is assigned to substances that cause full thickness destruction of intact skin tissue within an observation period up to 14 days starting after the exposure time of more than 3 minutes but not more than 60 minutes.

(c) Packing group III is assigned to substances that:

– cause full thickness destruction of intact skin tissue within an observation period up to 14 days starting after the exposure time of more than 60 minutes but not more than 4 hours; or

– are judged not to cause full thickness destruction of intact skin tissue, but which exhibit a corrosion rate on either steel or aluminium surfaces exceeding 6.25 mm a year at a test temperature of 55 °C when tested on both

[8] *OECD guidelines for Testing of Chemicals, No. 404 "Acute Dermal Irritation/Corrosion" (1992).*

materials. For the purposes of testing steel, type S235JR+CR (1.0037 resp. St 37-2), S275J2G3+CR (1.0144 resp. St 44-3), ISO 3574, Unified Numbering System (UNS) G10200 or SAE 1020, and for testing aluminium, non-clad, types 7075-T6 or AZ5GU-T6 shall be used. An acceptable test is prescribed in the Manual of Tests and Criteria, Part III, Section 37.

NOTE: *Where an initial test on either steel or aluminium indicates the substance being tested is corrosive the follow up test on the other metal is not required.*

2.2.8.1.7 If substances of Class 8, as a result of admixtures, come into categories of risk different from those to which the substances mentioned by name in Table A of Chapter 3.2 belong, these mixtures or solutions shall be assigned to the entries to which they belong, on the basis of their actual degree of danger.

NOTE: *For the classification of solutions and mixtures (such as preparations and wastes), see also 2.1.3.*

2.2.8.1.8 On the basis of the criteria set out in paragraph 2.2.8.1.6, it may also be determined whether the nature of a solution or mixture mentioned by name or containing a substance mentioned by name is such that the solution or mixture is not subject to the provisions for this class.

2.2.8.1.9 Substances, solutions and mixtures, which

– do not meet the criteria of Directives 67/548/EEC[3] or 1999/45/EC[4] as amended and therefore are not classified as corrosive according to these directives, as amended; and

– do not exhibit a corrosive effect on steel or aluminium,

may be considered as substances not belonging to Class 8.

NOTE*: UN No. 1910 calcium oxide and UN No. 2812 sodium aluminate, listed in the UN Model Regulations, are not subject to the provisions of ADN.*

2.2.8.2 ***Substances not accepted for carriage***

2.2.8.2.1 The chemically unstable substances of Class 8 shall not be accepted for carriage unless the necessary steps have been taken to prevent their dangerous decomposition or polymerization during carriage. To this end it shall in particular be ensured that receptacles and tanks do not contain any substance liable to promote these reactions.

2.2.8.2.2 The following substances shall not be accepted for carriage:

---

3 Council Directive 67/548/EEC of 27 June 1967 on the approximation of laws, regulations and administrative provisions relating to the classification, packaging and labelling of dangerous substances (Official Journal of the European Communities No. L 196 of 16.08.1967, page 1).

4 Directive 1999/45/EC of the European Parliament and of the Council of 31 May 1999 on the approximation of laws, regulations and administrative provisions of the Member States relating to the classification, packaging and labelling of dangerous preparations (Official Journal of the European Communities No. L 200 of 30 July 1999, pages 1 to 68).

- UN No. 1798 NITROHYDROCHLORIC ACID;
- chemically unstable mixtures of spent sulphuric acid;
- chemically unstable mixtures of nitrating acid or mixtures of residual sulphuric and nitric acids, not denitrated;
- perchloric acid aqueous solution with more than 72 % pure acid, by mass, or mixtures of perchloric acid with any liquid other than water.

2.2.8.3 *List of collective entries*

Corrosive substances <u>without</u> subsidiary risk

Group	Subgroup	State	Code	UN No.	Name and description
Acid C1-C4	inorganic	liquid	C1	2584	ALKYLSULPHONIC ACIDS, LIQUID with more than 5% free sulphuric acid or
				2584	ARYLSULPHONIC ACIDS, LIQUID with more than 5% free sulphuric acid
				2693	BISULPHITES, AQUEOUS SOLUTION, N.O.S.
				2837	BISULPHATES, AQUEOUS SOLUTION
				3264	CORROSIVE LIQUID, ACIDIC, INORGANIC, N.O.S.
		solid	C2	1740	HYDROGENDIFLUORIDES, SOLID, N.O.S.
				2583	ALKYLSULPHONIC ACIDS, SOLID with more than 5% free sulphuric acid or
				2583	ARYLSULPHONIC ACIDS, SOLID with more than 5% free sulphuric acid
				3260	CORROSIVE SOLID, ACIDIC, INORGANIC, N.O.S.
	organic	liquid	C3	2586	ALKYLSULPHONIC ACIDS, LIQUID with not more than 5% free sulphuric acid or
				2586	ARYLSULPHONIC ACIDS, LIQUID with not more than 5% free sulphuric acid
				2987	CHLOROSILANES, CORROSIVE, N.O.S.
				3145	ALKYLPHENOLS, LIQUID, N.O.S. (including C_2-C_{12} homologues)
				3265	CORROSIVE LIQUID, ACIDIC, ORGANIC, N.O.S
		solid	C4	2430	ALKYLPHENOLS, SOLID, N.O.S. (including C_2-C_{12} homologues)
				2585	ALKYLSULPHONIC ACIDS, SOLID with not more than 5% free sulphuric acid or
				2585	ARYLSULPHONIC ACIDS, SOLID with not more than 5% free sulphuric acid
				3261	CORROSIVE SOLID, ACIDIC, ORGANIC, N.O.S.
Basic C5-C8	inorganic	liquid	C5	1719	CAUSTIC ALKALI LIQUID, N.O.S.
				2797	BATTERY FLUID, ALKALI
				3266	CORROSIVE LIQUID, BASIC, INORGANIC, N.O.S.
		solid	C6	3262	CORROSIVE SOLID, BASIC, INORGANIC, N.O.S.
	organic	liquid	C7	2735	AMINES, LIQUID, CORROSIVE, N.O.S. or
				2735	POLYAMINES, LIQUID, CORROSIVE, N.O.S.
				3267	CORROSIVE LIQUID, BASIC, ORGANIC, N.O.S.
		solid	C8	3259	AMINES, SOLID, CORROSIVE, N.O.S., or
				3259	POLYAMINES, SOLID, CORROSIVE, N.O.S.
				3263	CORROSIVE SOLID, BASIC, ORGANIC, N.O.S.
Other corrosive substances C9-C10		liquid	C9	1903	DISINFECTANT, LIQUID, CORROSIVE, N.O.S
				2801	DYE, LIQUID, CORROSIVE, N.O.S. or
				2801	DYE INTERMEDIATE, LIQUID, CORROSIVE, N.O.S.
				3066	PAINT (including paint, enamel, stain, shellac, varnish, polish, liquid filler and lacquer base) or
				3066	PAINT RELATED MATERIAL (including paint thinning or reducing compound)
				1760	CORROSIVE LIQUID, N.O.S.
		solid [a]	C10	3147	DYE, SOLID, CORROSIVE, N.O.S. or
				3147	DYE INTERMEDIATE, SOLID, CORROSIVE, N.O.S.
				3244	SOLIDS CONTAINING CORROSIVE LIQUID, N.O.S.
				1759	CORROSIVE SOLID, N.O.S.
Articles			C11	2794	BATTERIES, WET, FILLED WITH ACID, electric storage
				2795	BATTERIES, WET, FILLED WITH ALKALI, electric storage
				2800	BATTERIES, WET, NON-SPILLABLE, electric storage
				3028	BATTERIES, DRY, CONTAINING POTASSIUM HYDROXIDE SOLID, electric storage

(cont'd on next page)

[a] *Mixtures of solids which are not subject to the provisions of ADN and of corrosive liquids may be carried under UN No. 3244 without being subject to the classification criteria of Class 8, provided there is no free liquid visible at the time the substance is loaded or at the time the packaging or cargo transport unit is closed. Each packaging shall correspond to a design type which has passed the leakproofness test for Packing group II level.*

Corrosive substances with subsidiary risk(s)

Subsidiary risk	State	Code	UN No. and name of collective entry
(cont'd) Flammable[b] CF	liquid	CF1	3470 PAINT, CORROSIVE, FLAMMABLE (including paint, enamel, stain, shellac, varnish, polish, liquid filler and lacquer base) or 3470 PAINT RELATED MATERIAL, CORROSIVE, FLAMMABLE (including paint thinning or reducing compound) 2734 AMINES, LIQUID, CORROSIVE, FLAMMABLE, N.O.S. or 2734 POLYAMINES, LIQUID, CORROSIVE, FLAMMABLE, N.O.S. 2986 CHLOROSILANES, CORROSIVE, FLAMMABLE, N.O.S. 2920 CORROSIVE LIQUID, FLAMMABLE, N.O.S.
	solid	CF2	2921 CORROSIVE SOLID, FLAMMABLE, N.O.S.
Self-heating CS	liquid	CS1	3301 CORROSIVE LIQUID, SELF-HEATING, N.O.S.
	solid	CS2	3095 CORROSIVE SOLID, SELF-HEATING, N.O.S.
Water-reactive CW	liquid[b]	CW1	3094 CORROSIVE LIQUID, WATER-REACTIVE, N.O.S.
	solid	CW2	3096 CORROSIVE SOLID, WATER-REACTIVE, N.O.S.
Oxidizing CO	liquid	CO1	3093 CORROSIVE LIQUID, OXIDIZING, N.O.S.
	solid	CO2	3084 CORROSIVE SOLID, OXIDIZING, N.O.S.
Toxic[d] CT	liquid[c]	CT1	3471 HYDROGENDIFLUORIDES SOLUTION, N.O.S. 2922 CORROSIVE LIQUID, TOXIC, N.O.S.
	solid[e]	CT2	2923 CORROSIVE SOLID, TOXIC, N.O.S.
Flammable, liquid, toxic[d]		CFT	No collective entry with this classification code available; if need be, classification under a collective entry with a classification code to be determined according to table of precedence of hazard in 2.1.3.10.
Oxidizing, toxic[d, e]		COT	No collective entry with this classification code available; if need be, classification under a collective entry with a classification code to be determined according to table of precedence of hazard in 2.1.3.10.

---

[b] *Chlorosilanes which, in contact with water or moist air, emit flammable gases, are substances of Class 4.3.*

[c] *Chloroformates having predominantly toxic properties are substances of Class 6.1.*

[d] *Corrosive substances which are highly toxic by inhalation, as defined in 2.2.61.1.4 to 2.2.61.1.9 are substances of Class 6.1.*

[e] *UN No. 2505 AMMONIUM FLUORIDE, UN No. 1812 POTASSIUM FLUORIDE, UN No. 1690 SODIUM FLUORIDE, SOLD, UN No. 2674 SODIUM FLUOROSILICATE, UN No. 2856 FLUOROSILICATES, N.O.S. , UN No. 3415 SODIUM FLUORIDE SOLUTION and UN No. 3422 POTASSIUM FLUORIDE SOLUTION are substances of Class 6.1.*

2.2.9 Class 9 Miscellaneous dangerous substances and articles

2.2.9.1 *Criteria*

2.2.9.1.1 The heading of Class 9 covers substances and articles which, during carriage, present a danger not covered by the heading of other classes.

2.2.9.1.2 The substances and articles of Class 9 are subdivided as follows:

M1 Substances which, on inhalation as fine dust, may endanger health;

M2 Substances and apparatus which, in the event of fire, may form dioxins;

M3 Substances evolving flammable vapour;

M4 Lithium batteries;

M5 Life-saving appliances;

M6-M8 Environmentally hazardous substances:

M6 Pollutant to the aquatic environment, liquid;
M7 Pollutant to the aquatic environment, solid;
M8 Genetically modified micro-organisms and organisms;

M9-M10 Elevated temperature substances:

M9 Liquid;
M10 Solid;

M11 Other substances presenting a danger during carriage, but not meeting the definitions of another class.

Definitions and classification

2.2.9.1.3 Substances and articles classified in Class 9 are listed in Table A of Chapter 3.2. The assignment of substances and articles not mentioned by name in Table A of Chapter 3.2 to the relevant entry of that Table or of sub-section 2.2.9.3 shall be done in accordance with 2.2.9.1.4 to 2.2.9.1.14 below.

Substances which, on inhalation as fine dust, may endanger health

2.2.9.1.4 Substances which, on inhalation as fine dust, may endanger health include asbestos and mixtures containing asbestos.

Substances and apparatus which, in the event of fire, may form dioxins

2.2.9.1.5 Substances and apparatus which, in the event of fire, may form dioxins include polychlorinated biphenyls (PCBs) and terphenyls (PCTs) and polyhalogenated biphenyls and terphenyls and mixtures containing these substances, as well as apparatus such as transformers, condensers and apparatus containing those substances or mixtures.

NOTE: *Mixtures with a PCB or PCT content of not more than 50 mg/kg are not subject to the provisions of ADN.*

Substances evolving flammable vapour

2.2.9.1.6 Substances evolving flammable vapour include polymers containing flammable liquids with a flash-point not exceeding 55 °C.

Lithium batteries

2.2.9.1.7 The term "lithium battery" covers all cells and batteries containing lithium in any form. They may be assigned to Class 9 if they meet the requirements of special provision 230 of Chapter 3.3. They are not subject to the provisions of ADN if they meet the requirements of special provision 188 of Chapter 3.3. They shall be classified in accordance with the procedures of Section 38.3 of the Manual of Tests and Criteria.

Life-saving appliances

2.2.9.1.8 Life-saving appliances include life-saving appliances and motor vehicle components which meet the descriptions of special provisions 235 or 296 of Chapter 3.3.

Environmentally hazardous substances

2.2.9.1.9 *(Deleted)*

Pollutants to the aquatic environment

2.2.9.1.10

2.2.9.1.10.1 For carriage in packages or in bulk, substances, solutions and mixtures meeting the criteria for <u>Acute 1</u>, <u>Chronic 1</u> or <u>Chronic 2</u> in Chapter 2.4 (see also 2.1.3.8) shall be considered to be environmentally hazardous (aquatic environment). Substances which cannot be assigned to other classes in ADN or to other Class 9 entries and which meet these criteria shall be assigned to UN Nos. 3077, ENVIRONMENTALLY HAZARDOUS SUBSTANCE, SOLID, N.O.S., or 3082, ENVIRONMENTALLY HAZARDOUS SUBSTANCE, LIQUID, N.O.S, and to packing group III.

2.2.9.1.10.2 For carriage in tank vessels, the substances, solutions and mixtures referred to in 2.2.9.1.10.1 and those meeting the criteria for <u>Acute 2</u>, <u>Acute 3</u> or <u>Chronic 3</u> in Chapter 2.4 shall be considered to be environmentally hazardous.

Substances classified as environmentally hazardous which meet the criteria for Acute or Chronic Category 1 shall be assigned to group 'N1'.

Substances classified as environmentally hazardous which meet the criteria for Chronic Categories 2 or 3 shall be assigned to group 'N2'.

Substances classified as environmentally hazardous which meet the criteria for Acute Categories 2 or 3 shall be assigned to group 'N3'.

Substances which meet the criteria of 2.2.9.1.10 shall be assigned to UN Nos. 3082, ENVIRONMENTALLY HAZARDOUS SUBSTANCE, LIQUID, N.O.S, or 3077, ENVIRONMENTALLY HAZARDOUS SUBSTANCE, SOLID, N.O.S., MOLTEN. Those that meet the additional criteria in this paragraph shall be assigned to identification Nos. 9005, ENVIRONMENTALLY HAZARDOUS SUBSTANCE, SOLID, N.O.S, MOLTEN, or 9006, ENVIRONMENTALLY HAZARDOUS SUBSTANCE, LIQUID, N.O.S.

2.2.9.1.10.3 Notwithstanding the provisions of 2.2.9.1.10.1,

(a) Substances which cannot be assigned to entries other than UN Nos. 3077 and 3082 in Class 9 or to other entries in Classes 1 to 8, but which are identified in Council Directive 67/548/EEC of 27 June 1967 on the approximation of laws, regulations and administrative provisions relating to the classification, packaging and labelling of dangerous substances,[9] as amended, as substances to which letter N 'Environmentally hazardous' (R50; R50/53; R51/53) has been allocated; and

(b) Solutions and mixtures (such as preparations and wastes) of substances to which letter N 'Environmentally hazardous' (R50; R50/53; R51/53) has been allocated in Directive 67/548/EEC, as amended, and which, according to Directive 1999/45/EC of the European Parliament and of the Council of 31 May 1999 concerning the approximation of the laws, regulations and administrative provisions of the Member States relating to the classification, packaging and labelling of dangerous preparations,[10] as amended, are also allocated letter N 'Environmentally hazardous' (R50; R50/53; R51/53) and which cannot be assigned to entries other than UN Nos. 3077 and 3082 in Class 9 or to other entries in Classes 1 to 8;

Shall be assigned to UN Nos. 3077 and 3082, as appropriate.

2.2.9.1.10.4 Classification flowchart for environmentally hazardous substances

---

[9] *Official Journal of the European Communities No. 196, of 16 August 1967, pp. 1-5.*
[10] *Official Journal of the European Communities No. L 200, of 30 July 1999, pp. 1-68.*

Classification flowchart for environmentally hazardous substances (aquatic environment)

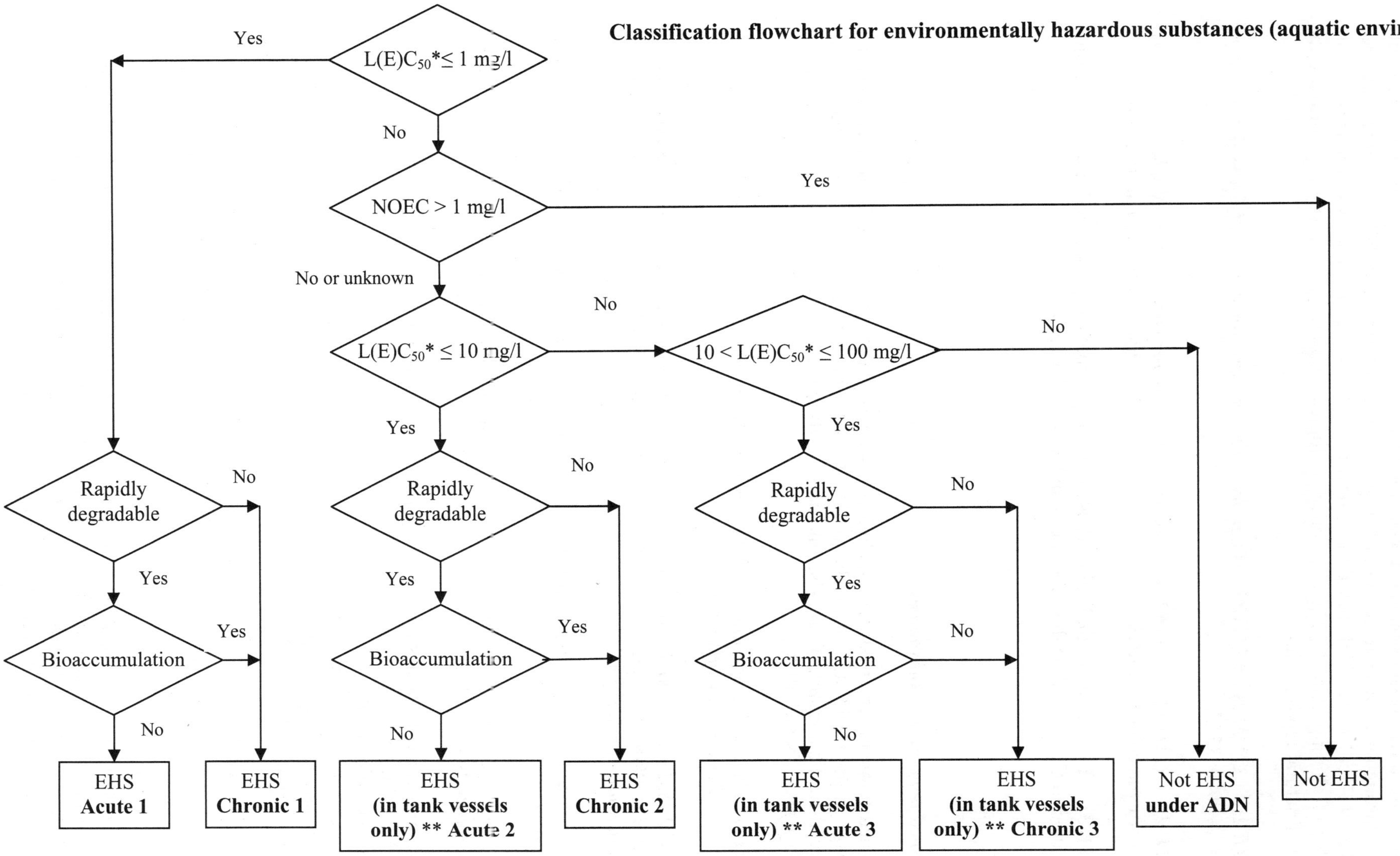

EHS = Environmentally hazardous substance (aquatic environment).

* Lowest value of 96-hour LC_{50}, 48-hour EC_{50} or 72- or 92-hour ErC_{50} as appropriate.

** Substances not considered dangerous for the environment when transported in packages.

Genetically modified micro-organisms or organisms

2.2.9.1.11 Genetically modified micro-organisms (GMMOs) and genetically modified organisms (GMOs) are micro-organisms and organisms in which genetic material has been purposely altered through genetic engineering in a way that does not occur naturally. They are assigned to Class 9 (UN No. 3245) if they do not meet the definition of infectious substances, but are capable of altering animals, plants or microbiological substances in a way not normally the result of natural reproduction.

***NOTE 1**: GMMOs and GMOs which are infectious are substances of Class 6.2, UN Nos. 2814, 2900 or 3373).*

***NOTE 2:** GMMOs or GMOs are not subject to the provisions of ADN when authorized for use by the competent authorities of the countries of origin, transit and destination*[11].

NOTE 3: Live animals shall not be used to carry genetically modified micro-organisms classified in Class 9 unless the substance can be carried no other way.

2.2.9.1.12 (*Deleted*)

Elevated temperature substances

2.2.9.1.13 Elevated temperature substances include substances which are carried or handed over for carriage in the liquid state at or above 100 °C and, in the case of those with a flash-point, below their flash-point. They also include solids which are carried or handed over for carriage at or above 240 °C.

***NOTE 1:** Elevated temperature substances may be assigned to Class 9 only if they do not meet the criteria of any other class.*

***NOTE 2:** Substances having a flash-point above 60 °C which are carried or handed over for carriage within a range of 15 K below the flash-point are substances of Class 3, identification number 9001.*

Other substances presenting a danger during carriage but not meeting the definitions of another class.

2.2.9.1.14 The following other miscellaneous substances not meeting the definitions of another class are assigned to Class 9:

Solid ammonia compounds having a flash-point below 60 °C
Low hazard dithionites
Highly volatile liquids
Substances emitting noxious fumes
Substances containing allergens
Chemical kits and first aid kits

The following miscellaneous substances not meeting the definition of another class are assigned to Class 9 when they are carried in bulk or in tank vessels:

---

[11] *See in particular Part C of Directive 2001/18/EC of the European Parliament and of the Council on the deliberate release into the environment of genetically modified organisms and repealing Council Directive 90/220/EEC (Official Journal of the European Communities, No. L 106, of 17 April 2001, pp. 8-14), which sets out the authorization procedures for the European Community.*

- UN 2071 AMMONIUM NITRATE FERTILIZERS: uniform non-segregating mixtures of nitrogen phosphate or nitrogen potash types or complete fertilizers of nitrogen phosphate potash type, with not more than 70% ammonium nitrate and not more than 0.4% total combustible material, or with not more than 45% ammonium nitrate irrespective of their content of combustible material;

  NOTE 1: *In determining the ammonium nitrate content, all nitrate ions for which a molecular equivalent of ammonium ions is present in the mixture shall be calculated as ammonium nitrate mass.*

  NOTE 2: *Ammonium nitrate fertilizers of Class 9 are not subject to ADN if:*

  - *the results of the trough test (see Manual of Tests and Criteria, Part III, sub-section 38.2) show that they are not capable of undergoing self-sustaining decomposition; and*

  - *the calculation referred to in NOTE 1 does not give an excess of nitrate greater than 10% by mass, calculated in KNO_3.*

- UN 2216 FISH MEAL, STABILIZED (humidity between 5% by mass and 12% by mass with not more than 15% fat by mass); or

- UN 2216 FISH SCRAP, STABILIZED (humidity between 5% by mass and 12% by mass with not more than 15% fat by mass);

- Identification No. 9003 SUBSTANCES HAVING A FLASH-POINT ABOVE 60° C AND NOT MORE THAN 100° C which cannot be assigned to another class or another entry of Class 9;

- Identification No. 9004, 4,4'-DIPHENYLMETHANE DIISOCYANATE.;

- Identification No. 9005, ENVIRONMENTALLY HAZARDOUS SUBSTANCE, SOLID, N.O.S, MOLTEN, which cannot be assigned to UN No. 3077;

- Identification No. 9006, ENVIRONMENTALLY HAZARDOUS SUBSTANCE, LIQUID, N.O.S., which cannot be assigned to UN No. 3082.

NOTE: *UN No. 1845 carbon dioxide, solid (dry ice), UN No. 2807 magnetized material, UN No. 3166 engine, internal combustion or vehicle, flammable gas powered or vehicle, flammable liquid powered, UN No. 3171 battery-powered vehicle or 3171 battery-powered equipment (wet battery), UN No. 3334 aviation regulated liquid, n.o.s., UN No. 3335 aviation regulated solid, n.o.s. and UN No. 3363 dangerous goods in machinery or dangerous goods in apparatus, listed in the UN Model Regulations, are not subject to the provisions of ADN.*

Assignment of the packing groups

2.2.9.1.15 When indicated in column 4 of Table A of Chapter 3.2, substances and articles of Class 9 are assigned to one of the following packing groups according to their degree of danger:

Packing group II: substances presenting medium danger;
Packing group III: substances presenting low danger.

2.2.9.2 ***Substances and articles not accepted for carriage***

The following substances and articles shall not be accepted for carriage:

- Lithium batteries which do not meet the relevant conditions of special provisions 188, 230 or 636 of Chapter 3.3;

- Uncleaned empty containment vessels for apparatus such as transformers, condensers and hydraulic apparatus containing substances assigned to UN Nos. 2315, 3151, 3152 or 3432.

2.2.9.3 *List of collective entries*

Substances which, on inhalation as fine dust, may endanger health		**M1**	2212 BLUE ASBESTOS (crocidolite) or 2212 BROWN ASBESTOS (amosite, mysorite) 2590 WHITE ASBESTOS (chrysotile, actinolite, anthophyllite, tremolite)
Substances and apparatus which, in the event of fire, may form dioxins		**M2**	2315 POLYCHLORINATED BIPHENYLS, LIQUID 3432 POLYCHLORINATED BIPHENYLS, SOLID 3151 POLYHALOGENATED BIPHENYLS, LIQUID or 3151 POLYHALOGENATED TERPHENYLS, LIQUID 3152 POLYHALOGENATED BIPHENYLS, SOLID or 3152 POLYHALOGENATED TERPHENYLS, SOLID
Substances evolving flammable vapour		**M3**	2211 POLYMERIC BEADS, EXPANDABLE, evolving flammable vapour 3314 PLASTICS MOULDING COMPOUND in dough, sheet or extruded rope form evolving flammable vapour
Lithium batteries		**M4**	3090 LITHIUM BATTERIES (including lithium alloy batteries) 3091 LITHIUM BATTERIES CONTAINED IN EQUIPMENT (including lithium alloy batteries) or 3091 LITHIUM BATTERIES PACKED WITH EQUIPMENT (including lithium alloy batteries) 3480 LITHIUM ION BATTERIES (including lithium ion polymer batteries) 3481 LITHIUM ION BATTERIES CONTAINED IN EQUIPMENT (including lithium ion polymer batteries) or 3481 LITHIUM ION BATTERIES PACKED WITH EQUIPMENT (including lithium ion polymer batteries)
Live-saving appliances		**M5**	2990 LIFE-SAVING APPLIANCES, SELF-INFLATING 3072 LIFE-SAVING APPLIANCES NOT SELF-INFLATING containing dangerous goods as equipment 3268 AIR BAG INFLATORS or 3268 AIR BAG MODULES or 3268 SEAT-BELT PRETENSIONERS
Environmentally hazardous substances	**pollutant to the aquatic environment, liquid**	**M6**	3082 ENVIRONMENTALLY HAZARDOUS SUBSTANCE, LIQUID, N.O.S.
	pollutant to the aquatic environment, solid	**M7**	3077 ENVIRONMENTALLY HAZARDOUS SUBSTANCE, SOLID, N.O.S.
	genetically modified micro-organisms and organisms	**M8**	3245 GENETICALLY MODIFIED MICROORGANISMS or 3245 GENETICALLY MODIFIED ORGANISMS
Elevated temperature substances	**liquid**	**M9**	3257 ELEVATED TEMPERATURE LIQUID, N.O.S., at or above 100 °C and below its flash-point (including molten metal, molten salts, etc.)
	solid	**M10**	3258 ELEVATED TEMPERATURE SOLID, N.O.S., at or above 240 °C

(cont'd on next page)

2.2.9.3 *List of collective entries (cont'd)*

Other substances or articles presenting a danger during carriage, but not meeting the definitions of another class	**M11**	No collective entry available. Only substances listed in Table A of Chapter 3.2 are subject to the provisions for Class 9 under this classification code, as follows: 1841 ACETALDEHYDE AMMONIA 1931 ZINC DITHIONITE (ZINC HYDROSULPHITE) 1941 DIBROMODIFLUOROMETHANE 1990 BENZALDEHYDE 2969 CASTOR BEANS, or 2969 CASTOR MEAL, or 2969 CASTOR POMACE, or 2969 CASTOR FLAKE 3316 CHEMICAL KIT, or 3316 FIRST AID KIT 3359 FUMIGATED UNIT

CHAPTER 2.3

TEST METHODS

2.3.0 General

Unless otherwise provided for in Chapter 2.2 or in this Chapter, the test methods to be used for the classification of dangerous goods are those described in the Manual of Tests and Criteria.

2.3.1 Exudation test for blasting explosives of Type A

2.3.1.1 Blasting explosives of type A (UN No. 0081) shall, if they contain more than 40 % liquid nitric ester, in addition to the testing specified in the Manual of Tests and Criteria, satisfy the following exudation test.

2.3.1.2 The apparatus for testing blasting explosive for exudation (figs. 1 to 3) consists of a hollow bronze cylinder. This cylinder, which is closed at one end by a plate of the same metal, has an internal diameter of 15.7 mm and a depth of 40 mm.

It is pierced by 20 holes 0.5 mm in diameter (four sets of five holes) on the circumference. A bronze piston, cylindrically fashioned over a length of 48 mm and having a total length of 52 mm, slides into the vertically placed cylinder.

The piston, whose diameter is 15.6 mm, is loaded with a mass of 2 220 g so that a pressure of 120 kPa (1.20 bar) is exerted on the base of the cylinder.

2.3.1.3 A small plug of blasting explosive weighing 5 to 8 g, 30 mm long and 15 mm in diameter, is wrapped in very fine gauze and placed in the cylinder; the piston and its loading mass are then placed on it so that the blasting explosive is subjected to a pressure of 120 kPa (1.20 bar). The time taken for the appearance of the first signs of oily droplets (nitroglycerine) at the outer orifices of the cylinder holes is noted.

2.3.1.4 The blasting explosive is considered satisfactory if the time elapsing before the appearance of the liquid exudations is more than five minutes, the test having been carried out at a temperature of 15 °C to 25 °C.

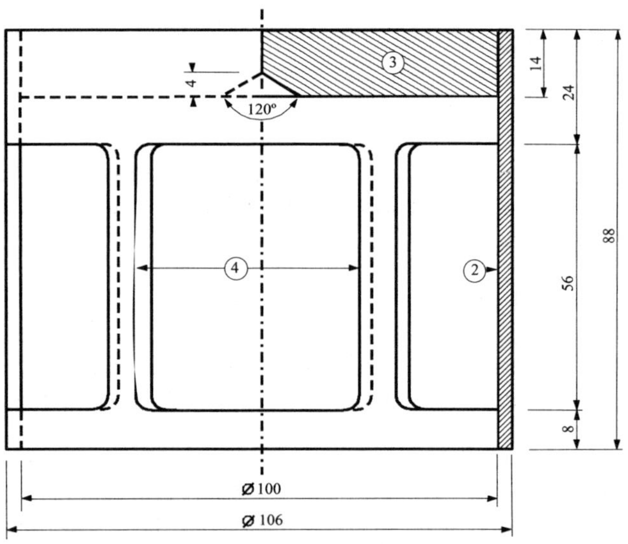

Fig.1: Bell-form charge, mass 2220 g, capable of being suspended from a bronze piston

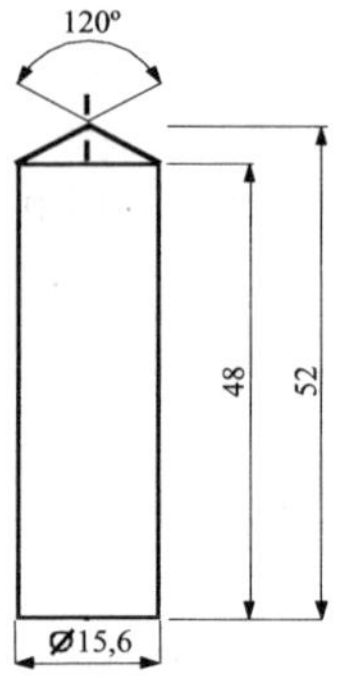

Fig.2: Cylindrical bronze piston, dimensions in mm

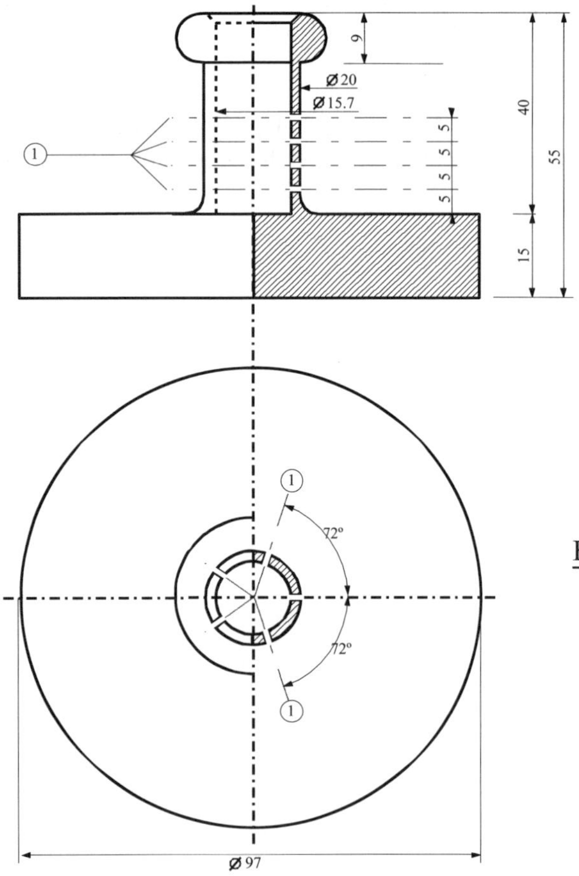

Fig.3: Hollow bronze cylinder, closed at one end; Plan and cut dimensions in mm

Fig. 1 to 3

(1) 4 series of 5 holes at 0.5 Ø
(2) copper
(3) iron plate with centre cone at the inferior face
(4) 4 openings, approximately 46x56, set at even intervals on the periphery

2.3.2 Tests relating to nitrated cellulose mixtures of Class 4.1

2.3.2.1 Nitrocellulose heated for half an hour at 132 °C shall not give off visible yellowish-brown nitrous fumes (nitrous gases). The ignition temperature shall be above 180 °C. See 2.3.2.3 to 2.3.2.8, 2.3.2.9 (a) and 2.3.2.10 below.

2.3.2.2 3 g of plasticized nitrocellulose, heated for one hour at 132 °C, shall not give off visible yellowish-brown nitrous fumes (nitrous gases). The ignition temperature shall be above 170 °C. See 2.3.2.3 to 2.3.2.8, 2.3.2.9 (b) and 2.3.2.10 below.

2.3.2.3 The test procedures set out below are to be applied when differences of opinion arise as to the acceptability of substances for carriage by road.

2.3.2.4 If other methods or test procedures are used to verify the conditions of stability prescribed above in this section, those methods shall lead to the same findings as could be reached by the methods specified below.

2.3.2.5 In carrying out the stability tests by heating described below, the temperature of the oven containing the sample under test shall not deviate by more than 2 °C from the prescribed temperature; the prescribed duration of a 30-minute or 60-minute test shall be observed to within two minutes. The oven shall be such that the required temperature is restored not more than five minutes after insertion of the sample.

2.3.2.6 Before undergoing the tests in 2.3.2.9 and 2.3.2.10, the samples shall be dried for not less than 15 hours at the ambient temperature in a vacuum desiccator containing fused and granulated calcium chloride, the sample substance being spread in a thin layer; for this purpose, substances which are neither in powder form nor fibrous shall be ground, or grated, or cut into small pieces. The pressure in the desiccator shall be brought below 6.5 kPa (0.065 bar).

2.3.2.7 Before being dried as prescribed in 2.3.2.6 above, substances conforming to 2.3.2.2 shall undergo preliminary drying in a well-ventilated oven, with its temperature set at 70 °C, until the loss of mass per quarter-hour is less than 0.3 % of the original mass.

2.3.2.8 Weakly nitrated nitrocellulose conforming to 2.3.2.1 shall first undergo preliminary drying as prescribed in 2.3.2.7 above; drying shall then be completed by keeping the nitrocellulose for at least 15 hours over concentrated sulphuric acid in a desiccator.

2.3.2.9 ***Test of chemical stability under heat***

(a) *Test of the substance listed in paragraph 2.3.2.1 above.*

(i) In each of two glass test tubes having the following dimensions:

length	350	mm
internal diameter	16	mm
thickness of wall	1.5	mm

is placed 1 g of substance dried over calcium chloride (if necessary the drying shall be carried out after reducing the substance to pieces weighing not more than 0.05 g each).

Both test tubes, completely covered with loose-fitting closures, are then so placed in an oven that at least four-fifths of their length is visible, and are kept at a constant temperature of 132 °C for 30 minutes. It is observed whether nitrous gases in the form of yellowish-brown fumes clearly visible against a white background are given off during this time.

(ii) In the absence of such fumes the substance is deemed to be stable.

(b) *Test of plasticized nitrocellulose (see 2.3.2.2)*

(i) 3 g of plasticized nitrocellulose are placed in glass test tubes, similar to those referred to in (a), which are then placed in an oven kept at a constant temperature of 132 °C.

(ii) The test tubes containing the plasticized nitrocellulose are kept in the oven for one hour. During this time no yellowish-brown nitrous fumes (nitrous gases) shall be visible. Observation and appraisal as in (a).

2.3.2.10 ***Ignition temperature (see 2.3.2.1 and 2.3.2.2)***

(a) The ignition temperature is determined by heating 0.2 g of substance enclosed in a glass test tube immersed in a Wood's alloy bath. The test tube is placed in the bath when the latter has reached 100 °C. The temperature of the bath is then progressively increased by 5 °C per minute;

(b) The test tubes must have the following dimensions:

length	125 mm
internal diameter	15 mm
thickness of wall	0.5 mm

and shall be immersed to a depth of 20 mm;

(c) The test shall be repeated three times, the temperature at which ignition of the substance occurs, i.e., slow or rapid combustion, deflagration or detonation, being noted each time;

(d) The lowest temperature recorded in the three tests is the ignition temperature.

2.3.3 **Tests relating to flammable liquids of Classes 3, 6.1 and 8**

2.3.3.1 ***Test for determining flash-point***

2.3.3.1.1 The flash-point shall be determined by means of one of the following types of apparatus:

(a) Abel;

(b) Abel-Pensky;

(c) Tag;

(d) Pensky-Martens;

(e) Apparatus in accordance with ISO 3679:1983 or ISO 3680:1983.

2.3.3.1.2 To determine the flash-point of paints, gums and similar viscous products containing solvents, only apparatus and test methods suitable for determining the flash-point for viscous liquids shall be used, in accordance with the following standards:

(a) International Standard ISO 3679:1983;

(b) International Standard ISO 3680:1983;

(c) International Standard ISO 1523:1983;

(d) German Standard DIN 53213:1978, Part 1.

2.3.3.1.3 The test procedure shall be either according to an equilibrium method or according to a non-equilibrium method.

2.3.3.1.4 For the procedure according to an equilibrium method, see:

(a) International Standard ISO 1516:1981;

(b) International Standard ISO 3680:1983;

(c) International Standard ISO 1523:1983;

(d) International Standard ISO 3679:1983.

2.3.3.1.5 The procedure according to a non-equilibrium method shall be:

(a) for the Abel apparatus, see:

(i) British Standard BS 2000 Part 170:1995;

(ii) French Standard NF MO7-011:1988;

(iii) French Standard NF T66-009:1969;

(b) for the Abel-Pensky apparatus, see:

(i) German Standard DIN 51755, Part 1:1974 (for temperatures from 5 °C to 65 °C);

(ii) German Standard DIN 51755, Part 2:1978 (for temperatures below 5 °C);

(iii) French Standard NF MO7-036:1984;

(c) for the Tag apparatus, see American Standard ASTM D 56:1993;

(d) for the Pensky-Martens apparatus, see:

(i) International Standard ISO 2719:1988;

(ii) European Standard EN 22719 in each of its national versions (e.g. BS 2000, part 404/EN 22719):1994;

(iii) American Standard ASTM D 93:1994;

(iv) Institute of Petroleum Standard IP 34:1988.

2.3.3.1.6 The test methods listed in 2.3.3.1.4 and 2.3.3.1.5 shall only be used for flash-point ranges which are specified in the individual methods. The possibility of chemical reactions between the substance and the sample holder shall be considered when selecting the method to be used. The apparatus shall, as far as is consistent with safety, be placed in a draught-free position. For safety, a method utilizing a small sample size, around 2 ml, shall be used for organic peroxides and self-reactive substances (also known as "energetic" substances), or for toxic substances.

2.3.3.1.7 When the flash-point, determined by a non-equilibrium method in accordance with 2.3.3.1.5 is found to be 23 ± 2 °C or 60 ± 2 °C, it shall be confirmed for each temperature range by an equilibrium method in accordance with 2.3.3.1.4.

2.3.3.1.8 In the event of a dispute as to the classification of a flammable liquid, the classification proposed by the consignor shall be accepted if a check-test of the flash-point, yields a result not differing by more than 2 °C from the limits (23 °C and 60 °C respectively) stated in 2.2.3.1. If the difference is more than 2 °C, a second check-test shall be carried out, and the lowest figure of the flash-points obtained in either check-test shall be adopted.

2.3.3.2 ***Test for determining peroxide content***

To determine the peroxide content of a liquid, the procedure is as follows:

A quantity p (about 5 g, weighed to the nearest 0.01 g) of the liquid to be titrated is placed in an Erlenmeyer flask; 20 cm^3 of acetic anhydride and about 1 g of powdered solid potassium iodide are added; the flask is shaken and, after 10 minutes, heated for 3 minutes to about 60 °C. When it has been left to cool for 5 minutes, 25 cm^3 of water are added. After this, it is left standing for half an hour, then the liberated iodine is titrated with a decinormal solution of sodium thiosulphate, no indicator being added; complete discoloration indicates the end of the reaction. If n is the number of cm^3 of thiosulphate solution required, the percentage of peroxide (calculated as H_2O_2) present in the sample is obtained by the formula:

$$\frac{17n}{100p}$$

2.3.4 **Test for determining fluidity**

To determine the fluidity of liquid, viscous or pasty substances and mixtures, the following test method shall be used.

2.3.4.1 ***Test apparatus***

Commercial penetrometer conforming to ISO 2137:1985, with a guide rod of 47.5 g ± 0.05 g; sieve disc of duralumin with conical bores and a mass of 102.5 g ± 0.05 g (see Figure 1); penetration vessel with an inside diameter of 72 mm to 80 mm for reception of the sample.

2.3.4.2 ***Test procedure***

The sample is poured into the penetration vessel not less than half an hour before the measurement. The vessel is then hermetically closed and left standing until the measurement. The sample in the hermetically closed penetration vessel is heated to 35 °C ± 0.5 °C and is placed on the penetrometer table immediately prior to measurement (not more than two minutes). The point S of the sieve disc is then brought into contact with the surface of the liquid and the rate of penetration is measured.

2.3.4.3 ***Evaluation of test results***

A substance is pasty if, after the centre S has been brought into contact with the surface of the sample, the penetration indicated by the dial gauge:

(a) after a loading time of 5 s ± 0.1 s, is less than 15.0 mm ± 0.3 mm; or

(b) after a loading time of 5 s ± 0.1 s, is greater than 15.0 mm ± 0.3 mm, but the additional penetration after another 55 s ± 0.5 s is less than 5.0 mm ± 0.5 mm.

***NOTE**: In the case of samples having a flow point, it is often impossible to produce a steady level surface in the penetration vessel and, hence, to establish satisfactory initial measuring conditions for the contact of the point S. Furthermore, with some samples, the impact of the sieve disc can cause an elastic deformation of the surface and, in the first few seconds, simulate a deeper penetration. In all these cases, it may be appropriate to make the evaluation in paragraph (b) above.*

Figure 1 – Penetrometer

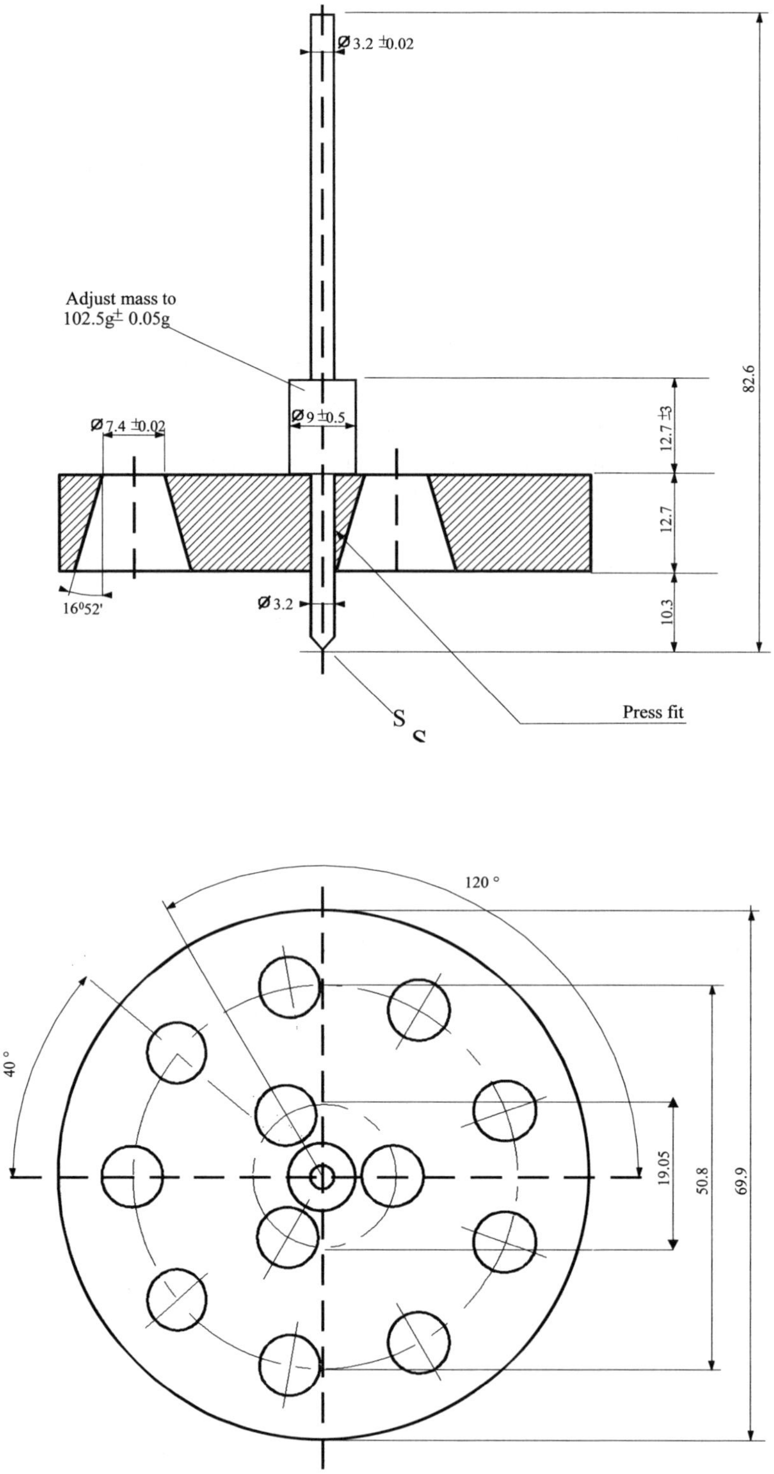

Tolerances not specified are ± 0.1 mm.

2.3.5 Classification of organometallic substances in Classes 4.2 and 4.3

Depending on their properties as determined in accordance with tests N.1 to N.5 of the Manual of Tests and Criteria, Part III, section 33, organometallic substances may be classified in Classes 4.2 or 4.3, as appropriate, in accordance with the flowchart scheme given in Figure 2.3.5.

NOTE 1: *Depending on their other properties and on the precedence of hazard table (see 2.1.3.10), organometallic substances may have to be classified in other classes as appropriate.*

NOTE 2*:* *Flammable solutions with organometallic compounds in concentrations which are not liable to spontaneous combustion or, in contact with water, do not emit flammable gases in dangerous quantities, are substances of Class 3.*

Figure 2.3.5 Flowchart scheme for the classification of organometallic substances in Classes 4.2 and 4.3 [b]

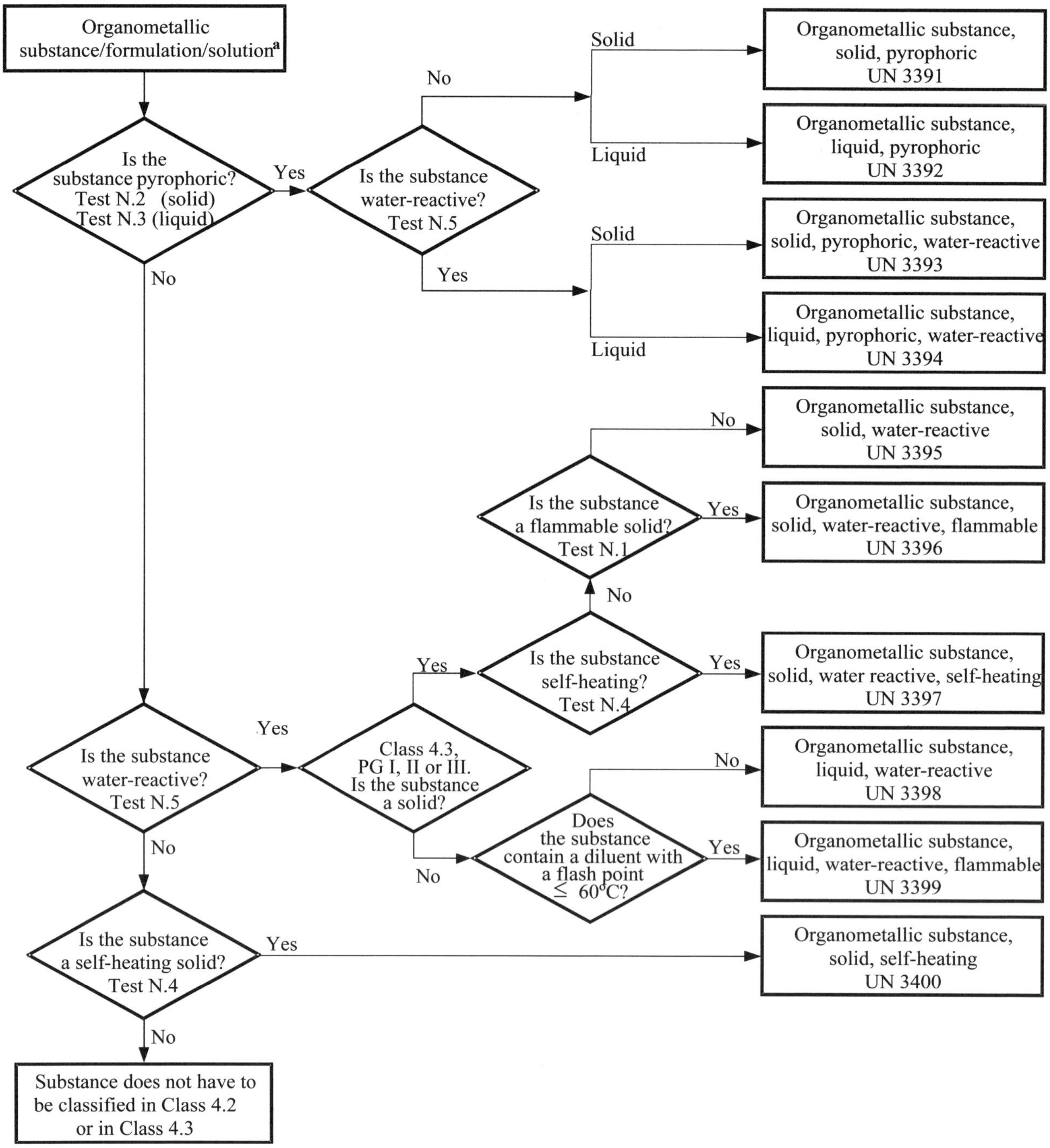

[a] *If applicable and testing is relevant, taking into account reactivity properties, class 6.1 and 8 properties should be considered according to the precedence of hazard table of 2.1.3.10.*

[b] *Test methods N.1 to N.5 can be found in the Manual of Tests and Criteria, Part III, Section 33.*

CHAPTER 2.4

CRITERIA FOR SUBSTANCES HAZARDOUS TO THE AQUATIC ENVIRONMENT

2.4.1 General definitions

2.4.1.1 Environmentally hazardous substances include, inter alia, liquid or solid substances pollutant to the aquatic environment and solutions and mixtures of such substances (such as preparations and wastes). For the purposes of this Chapter, 'substance' means chemical elements and their compounds in the natural state or obtained by any production process, including any additive necessary to preserve the stability of the product and any impurities deriving from the process used, but excluding any solvent which may be separated without affecting the stability of the substance or changing its composition.

2.4.1.2 The aquatic environment may be considered in terms of the aquatic organisms that live in the water, and the aquatic ecosystem of which they are part.[1] The basis, therefore, of the identification of hazard is the aquatic toxicity of the substance or mixture, although this may be modified by further information on the degradation and bioaccumulation behaviour.

2.4.1.3 While the following classification procedure is intended to apply to all substances and mixtures, it is recognized that in some cases, e.g. metals or poorly soluble inorganic compounds, special guidance will be necessary.[2]

2.4.1.4 The following definitions apply for acronyms or terms used in this section:

- BCF: Bioconcentration Factor;
- BOD: Biochemical Oxygen Demand;
- COD: Chemical Oxygen Demand;
- GLP: Good Laboratory Practices;
- EC_{50}: the effective concentration of substance that causes 50% of the maximum response;
- ErC_{50}: EC_{50} in terms of reduction of growth;
- K_{ow}: octanol/water partition coefficient;
- LC_{50} (50% lethal concentration): the concentration of a substance in water which causes the death of 50% (one half) in a group of test animals;
- $L(E)C_{50}$: LC_{50} or EC_{50};
- NOEC: No Observed Effect Concentration;
- OECD Test Guidelines: test guidelines published by the Organisation for Economic Co-operation and Development (OECD).

---

[1] *This does not address aquatic pollutants for which there may be a need to consider effects beyond the aquatic environment such as the impacts on human health, etc*
[2] *See annex 10 of GHS.*

2.4.2 Definitions and data requirements

2.4.2.1 The basic elements for classification of environmentally hazardous substances (aquatic environment) are as follows:

- Acute aquatic toxicity;
- Potential for or actual bioaccumulation;
- Degradation (biotic or abiotic) for organic chemicals; and
- Chronic aquatic toxicity.

2.4.2.2 While data from internationally harmonized test methods are preferred, in practice, data from national methods may also be used where they are considered as equivalent. In general, it has been agreed that freshwater and marine species toxicity data can be considered as equivalent data and are preferably to be derived using OECD Test Guidelines or equivalent according to the principles of Good Laboratory Practices (GLP). Where such data are not available, classification should be based on the best available data.

2.4.2.3 **Acute aquatic toxicity** shall normally be determined using a fish 96-hour LC_{50} (OECD Test Guideline 203 or equivalent), a crustacea species 48-hour EC_{50} (OECD Test Guideline 202 or equivalent) and/or an algal species 72- or 96-hour EC_{50} (OECD Test Guideline 201 or equivalent). These species are considered as surrogate for all aquatic organisms, and data on other species such as Lemna may also be considered if the test methodology is suitable.

2.4.2.4 **Bioaccumulation** means net result of uptake, transformation and elimination of a substance in an organism due to all routes of exposure (i.e. air, water, sediment/soil and food).

The **potential for bioaccumulation** shall normally be determined by using the octanol/water partition coefficient, usually reported as a log K_{ow} determined by OECD Test Guideline 107 or 117. While this represents a potential to bioaccumulate, an experimentally determined Bioconcentration Factor (BCF) provides a better measure and should be used in preference when available. A BCF should be determined according to OECD Test Guideline 305.

2.4.2.5 **Environmental degradation** may be biotic or abiotic (e.g. hydrolysis) and the criteria reflect this fact. Ready biodegradation can most easily be defined using the OECD biodegradability tests, OECD Test Guideline 301 (A - F). A pass level in these tests can be considered as indicative of rapid degradation in most environments. These are freshwater tests and thus the use of the results from OECD Test Guideline 306, which is more suitable for marine environments, has also been included. Where such data are not available, a BOD_5 (5 days)/COD ratio ≥ 0.5 is considered as indicative of rapid degradation. Abiotic degradation such as hydrolysis, primary degradation, both abiotic and biotic, degradation in non-aquatic media and proven rapid degradation in the environment may all be considered in defining rapid degradability.[3]

Substances shall be considered rapidly degradable in the environment if the following criteria are met:

(a) In 28-day ready biodegradation studies, the following levels of degradation are achieved:

(i) Tests based on dissolved organic carbon: 70%;

---

[3] *Special guidance on data interpretation is provided in Chapter 4.1 and Annex 9 to GHS.*

(ii) Tests based on oxygen depletion or carbon dioxide generation: 60% of theoretical maxima;

These levels of biodegradation shall be achieved within 10 days of the start of degradation, which point is taken as the time when 10% of the substance has been degraded; or

(b) In those cases where only BOD and COD data are available, when the ratio of BOD_5/COD is ≥ 0.5; or

(c) If other convincing scientific evidence is available to demonstrate that the substance or mixture can be degraded (biotically and/or abiotically) in the aquatic environment to a level above 70% within a 28-day period.

2.4.2.6 **Chronic toxicity** data are less available than acute data and the range of testing procedures less standardized. Data generated according to OECD Test Guidelines 210 (Fish Early Life Stage) or 211 (Daphnia Reproduction) and 201 (Algal Growth Inhibition) can be accepted. Other validated and internationally accepted tests could also be used. The No Observed Effect Concentrations (NOECs) or other equivalent L(E)Cx should be used.

2.4.3 Classification categories and criteria for substances

***NOTE**: Chronic Category 4 of Chapter 4.1 of GHS is reproduced in this section for information, although it is not relevant in the context of ADN.*

2.4.3.1 The following substances shall be considered to be environmentally hazardous (aquatic environment):

(a) For carriage in packages, substances which meet the criteria for Acute 1, Chronic 1 or Chronic 2, according to the tables below; and

(b) For carriage in tank vessels, substances which meet the criteria for Acute 1, Acute 2 or Acute 3, or Chronic 1, Chronic 2 or Chronic 3, according to the tables below.

Acute toxicity

Category: Acute 1	
96 hr LC_{50} (for fish)	≤1 mg/l and/or
48 hr EC_{50} (for crustacea)	≤1 mg/l and/or
72 or 96 hr ErC_{50} (for algae or other aquatic plants)	≤1 mg/l

Category: Acute 2	
96 hr LC_{50} (for fish)	>1 - ≤10 mg/l and/or
48 hr EC_{50} (for crustacea)	>1 - ≤10 mg/l and/or
72 or 96 hr ErC_{50} (for algae or other aquatic plants)	>1 - ≤10 mg/l

Category: Acute 3	
96 hr LC_{50} (for fish)	>10 - ≤100 mg/l and/or
48 hr EC_{50} (for crustacea)	>10 - ≤100 mg/l and/or
72 or 96 hr ErC_{50} (for algae or other aquatic plants)	>10 - ≤100 mg/l

Chronic toxicity

Category: Chronic 1	
96 hr LC_{50} (for fish)	$\leq$ 1 mg/l and/or
48 hr EC_{50} (for crustacea)	$\leq$ 1 mg/l and/or
72 or 96 hr ErC_{50} (for algae or other aquatic plants)	$\leq$ 1 mg/l
and the substance is not rapidly degradable and/or the log $K_{ow} \geq 4$ (unless the experimentally determined BCF <500).	

Category: Chronic 2	
96 hr LC_{50} (for fish)	> 1 to $\leq$ 10 mg/l and/or
48 hr EC_{50} (for crustacea)	> 1 to $\leq$ 10 mg/l and/or
72 or 96 hr ErC_{50} (for algae or other aquatic plants)	> 1 to $\leq$ 10 mg/l
and the substance is not rapidly degradable and/or the log $K_{ow} \geq 4$ (unless the experimentally determined BCF <500), unless the chronic toxicity NOECs are > 1 mg/l.	

Category: Chronic 3	
96 hr LC_{50} (for fish)	> 10 to $\leq$ 100 mg/l and/or
48 hr EC_{50} (for crustacea)	> 10 to $\leq$ 100 mg/l and/or
72 or 96 hr ErC_{50} (for algae or other aquatic plants)	$\geq$ 10 to $\leq$ 100 mg/l
and the substance is not rapidly degradable and/or the log $K_{ow} \geq 4$ (unless the experimentally determined BCF <500), unless the chronic toxicity NOECs are > 1 mg/l.	

Category: Chronic 4
Poorly soluble substances for which no acute toxicity is recorded at levels up to the water solubility, and which are not rapidly degradable and have a log $K_{ow} \geq 4$, indicating a potential to bioaccumulate, will be classified in this category unless other scientific evidence exists showing classification to be unnecessary. Such evidence would include an experimentally determined BCF < 500, or chronic toxicity NOECs > 1 mg/l, or evidence of rapid degradation in the environment.
Substances which come under Chronic Category 4 alone are not considered to be environmentally hazardous in the sense of ADN.

2.4.4 Classification categories and criteria for mixtures

NOTE: *Chronic Category 4 of Chapter 4.1 of GHS is reproduced in this section for information, although it is not relevant in the context of ADN.*

2.4.4.1 The classification system for mixtures covers all classification categories which are used for substances, meaning Acute Categories 1 to 3 and Chronic Categories 1 to 4. In order to make use of all available data for purposes of classifying the aquatic environmental hazards of the mixture, the following assumption has been made and is applied where appropriate.

The 'relevant ingredients' of a mixture are those which are present in a concentration of 1% (mass) or greater, unless there is a presumption (e.g. in the case of highly toxic ingredients) that an ingredient present at less than 1% can still be relevant for classifying the mixture for aquatic environmental hazards.

2.4.4.2 The approach for classification of aquatic environmental hazards is tiered and is dependent upon the type of information available for the mixture itself and for its ingredients. Elements of the tiered approach include:

(a) Classification based on tested mixtures;

(b) Classification based on bridging principles,

(c) Use of 'summation of classified ingredients' and/or an 'additivity formula'.

Figure 2.4.4.2 outlines the process to be followed.

Figure 2.4.4.2: Tiered approach to classification of mixtures for acute and chronic environmental hazards

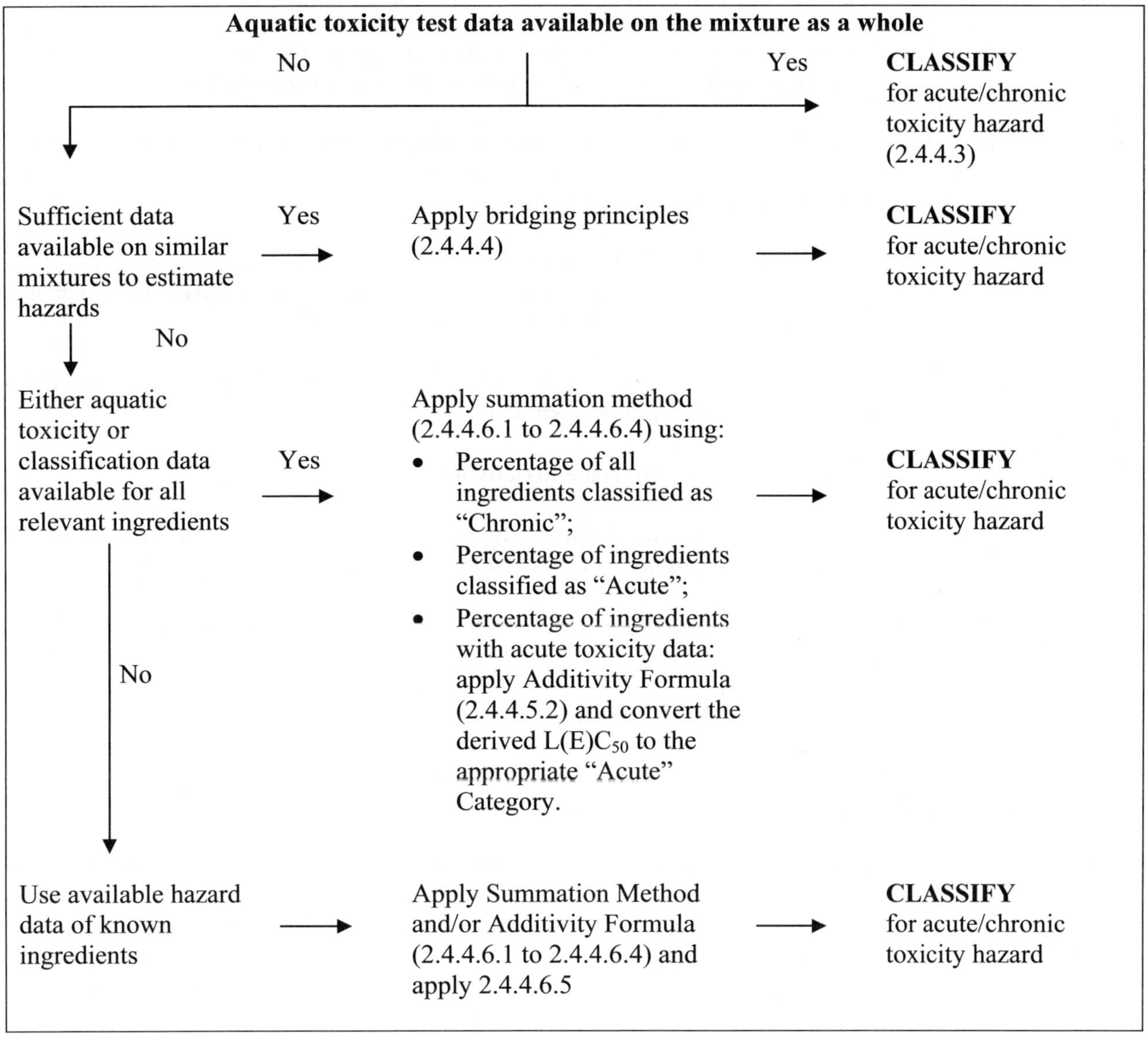

2.4.4.3 ***Classification of mixtures when data are available for the complete mixture***

2.4.4.3.1 When the mixture as a whole has been tested to determine its aquatic toxicity, it can be classified according to the criteria that have been agreed for substances, but only for acute toxicity. The classification shall be based on the data for fish, crustacea, algae and plants. Classification of mixtures by using LC_{50}, EC_{50} or ErC_{50} data for the mixture as a whole is not possible for chronic categories since both toxicity data and environmental fate data are needed, and there are no degradability and bioaccumulation data for mixtures as a whole. It is not possible to apply the criteria for chronic classification because the data from degradability and bioaccumulation tests of mixtures cannot be interpreted; they are meaningful only for single substances.

2.4.4.3.2 When there are acute toxicity test data (LC_{50} or EC_{50}) available for the mixture as a whole, these data as well as information with respect to the classification of ingredients for chronic toxicity shall be used to complete the classification for tested mixtures as follows. When chronic (long-term) toxicity data (NOEC) are also available, these should be used as well.

(a) $L(E)C_{50}$ (LC_{50} or EC_{50}) of the tested mixture $\leq$ 100 mg/l and NOEC of the tested mixture $\leq$ 1.0 mg/l or unknown:

(i) Classify mixture as Category Acute 1, 2 or 3;

(ii) Apply summation of classified ingredients approach (see 2.4.4.6) for chronic classification (Chronic 1 to 4 or no need for chronic classification);

(b) $L(E)C_{50}$ of the tested mixture $\leq$ 100 mg/l and NOEC of the tested mixture > 1.0 mg/l:

(i) Classify mixture as Category Acute 1, 2 or 3;

(ii) Apply summation of classified ingredients approach (see 2.4.4.6) for classification as Category Chronic 1. If the mixture is not classified as Category Chronic 1, then there is no need for chronic classification;

(c) $L(E)C_{50}$ of the tested mixture > 100 mg/l, or above the water solubility, and NOEC of the tested mixture $\leq$ 1.0 mg/l or unknown:

(i) No need to classify for acute toxicity hazard;

(ii) Apply summation of classified ingredients approach (see 2.4.4.6) for chronic classification (Chronic 4 or no need for chronic classification);

(d) $L(E)C_{50}$ of the tested mixture >100 mg/l, or above the water solubility, and NOEC of the tested mixture > 1.0 mg/l:

– No need to classify for acute or chronic toxicity hazard.

2.4.4.4 ***Classification of mixtures when data are not available for the complete mixture: bridging principles***

2.4.4.4.1 Where the mixture itself has not been tested to determine its aquatic environmental hazard, but there are sufficient data on the individual ingredients and similar tested mixtures to adequately characterize the hazards of the mixture, these data shall be used in accordance with the following agreed bridging rules. This ensures that the classification process uses the available data to the greatest extent possible in characterizing the hazards of the mixture without the necessity for additional testing in animals.

2.4.4.4.2 *Dilution*

2.4.4.4.2.1 If a mixture is formed by diluting another classified mixture or a substance with a diluent which has an equivalent or lower aquatic hazard classification than the least toxic original ingredient and which is not expected to affect the aquatic hazards of other ingredients, then the mixture will be classified as equivalent to the original mixture or substance.

2.4.4.4.2.2 If a mixture is formed by diluting another classified mixture or a substance with water or other totally non-toxic material, the toxicity of the mixture shall be calculated from the original mixture or substance.

2.4.4.4.3 *Batching*

The aquatic hazard classification of one production batch of a complex mixture can be assumed to be substantially equivalent to that of another production batch of the same commercial product and produced by or under the control of the same manufacturer, unless there is reason to believe there is significant variation such that the aquatic hazard classification of the batch has changed. If the latter occurs, new classification is necessary.

2.4.4.4.4 *Concentration of mixtures which are classified with the most severe classification categories (Chronic 1 and Acute 1)*

If a mixture is classified as Chronic 1 and/or Acute 1, and ingredients of the mixture which are classified as Chronic 1 and/or Acute 1 are further concentrated, the more concentrated mixture shall be classified with the same classification category as the original mixture without additional testing.

2.4.4.4.5 *Interpolation within one toxicity category*

If mixtures A and B are in the same classification category and mixture C is made in which the toxicologically active ingredients have concentrations intermediate to those in mixtures A and B, then mixture C shall be in the same category as A and B. Note that the identity of the ingredients is the same in all three mixtures.

2.4.4.4.6 *Substantially similar mixtures*

Given the following:

(a) Two mixtures:

(i) A + B;

(ii) C + B;

(b) The concentration of ingredient B is the same in both mixtures;

(c) The concentration of ingredient A in mixture (i) equals that of ingredient C in mixture (ii);

(d) Classifications for A and C are available and are the same, i.e. they are in the same hazard category and are not expected to affect the aquatic toxicity of B.

Then there is no need to test mixture (ii) if mixture (i) is already characterized by testing and both mixtures are classified in the same category.

2.4.4.5 ***Classification of mixtures when data are available for all ingredients or only for some ingredients of the mixture***

2.4.4.5.1 The classification of a mixture is based on summation of the concentrations of its classified ingredients. The percentage of ingredients classified as 'Acute' or 'Chronic' will feed straight into the summation method. Details of the summation method are described in 2.4.4.6.1 to 2.4.4.6.4.

2.4.4.5.2 Mixtures can be made of a combination of both ingredients that are classified (as Acute 1 to 3 and/or Chronic 1 to 4) and those for which adequate test data are available. When adequate toxicity data are available for more than one ingredient in the mixture, the combined toxicity of those ingredients may be calculated using the following additivity formula, and the calculated toxicity may be used to assign that portion of the mixture an acute hazard category, which is then subsequently used in applying the summation method.

$$\frac{\sum C_i}{L(E)C_{50m}} = \sum_n \frac{C_i}{L(E)C_{50i}}$$

where:

Ci	= concentration of ingredient i (weight percentage)
$L(E)C_{50i}$	= LC_{50} or EC_{50} (in mg/l) for ingredient i
N	= number of ingredients, and i is running from 1 to n
$L(E)C_{50m}$	= $L(E)C_{50}$ of the part of the mixture with test data

2.4.4.5.3 When applying the additivity formula for part of the mixture, it is preferable to calculate the toxicity of this part of the mixture using for each substance toxicity values that relate to the same species (i.e. fish, daphnia or algae) and then to use the highest toxicity (lowest value) obtained (i.e. use the most sensitive of the three species). However, when toxicity data for each ingredient are not available in the same species, the toxicity value of each ingredient shall be selected in the same manner that toxicity values are selected for the classification of substances, i.e. the highest toxicity (from the most sensitive test organism) is used. The calculated acute toxicity may then be used to classify this part of the mixture as Acute 1, 2 or 3 using the same criteria described for substances.

2.4.4.5.4 If a mixture is classified in more than one way, the method yielding the more conservative result shall be used.

2.4.4.6 ***Summation method***

2.4.4.6.1 *Classification procedures*

In general, a more severe classification for mixtures overrides a less severe classification, e.g. a classification with Chronic 1 overrides a classification with Chronic 2. As a consequence, the classification procedure is already completed if the result of the classification is Chronic 1. A more severe classification than Chronic 1 is not possible; therefore, it is not necessary to pursue the classification procedure further.

2.4.4.6.2 *Classification for Acute Categories 1, 2 and 3*

2.4.4.6.2.1 First, all ingredients classified as Acute 1 shall be considered. If the sum of these ingredients is ≥ 25%, the whole mixture is classified as Acute 1. If the result of the calculation is a classification of the mixture as Acute 1, the classification process is completed.

2.4.4.6.2.2 In cases where the mixture is not classified as Acute 1, classification of the mixture as Acute 2 shall be considered. A mixture is classified as Acute 2 if 10 times the sum of all ingredients classified as Acute 1 plus the sum of all ingredients classified as Acute 2 is ≥ 25%. If the result of the calculation is classification of the mixture as Acute 2, the classification process is completed.

2.4.4.6.2.3 In cases where the mixture is not classified either as Acute 1 or Acute 2, classification of the mixture as Acute 3 shall be considered. A mixture is classified as Acute 3 if 100 times the sum of all ingredients classified as Acute 1 plus 10 times the sum of all ingredients classified as Acute 2 plus the sum of all ingredients classified as Acute 3 is ≥ 25%.

2.4.4.6.2.4 The classification of mixtures for acute hazards based on this summation of classified ingredients is summarized in Table 2.4.4.6.2.4.

Table 2.4.4.6.2.4: Classification of a mixture for acute hazards based on summation of classified ingredients

Sum of ingredients classified as:	Mixture is classified as:
Acute 1 × M[a] ≥ 25%	Acute 1
(M ×10 x Acute 1) + Acute 2 ≥ 25%	Acute 2
(M × 100 × Acute 1) + (10 × Acute 2) + Acute 3 ≥ 25%	Acute 3

[a] For explanation of the M factor, see 2.4.4.6.4.

2.4.4.6.3 *Classification for Chronic Categories 1, 2, 3 and 4*

2.4.4.6.3.1 First, all ingredients classified as Chronic 1 are considered. If the sum of these ingredients is ≥ 25%, the mixture shall be classified as Chronic 1. If the result of the calculation is a classification of the mixture as Chronic 1, the classification procedure is completed.

2.4.4.6.3.2 In cases where the mixture is not classified as Chronic 1, classification of the mixture as Chronic 2 shall be considered. A mixture is classified as Chronic 2 if 10 times the sum of all ingredients classified as Chronic 1 plus the sum of all ingredients classified as Chronic 2 is ≥ 25%. If the result of the calculation is classification of the mixture as Chronic 2, the classification process is completed.

2.4.4.6.3.3 In cases where the mixture is not classified either as Chronic 1 or Chronic 2, classification of the mixture as Chronic 3 shall be considered. A mixture is classified as Chronic 3 if 100 times the sum of all ingredients classified as Chronic 1 plus 10 times the sum of all ingredients classified as Chronic 2 plus the sum of all ingredients classified as Chronic 3 is ≥ 25%.

2.4.4.6.3.4 If the mixture is still not classified in Category Chronic 1, 2 or 3, classification of the mixture as Chronic 4 need not be considered for the purposes of ADN. A mixture is classified as Chronic 4 if the sum of the percentages of ingredients classified as Chronic 1, 2, 3 and 4 is ≥ 25%.

2.4.4.6.3.5 The classification of mixtures for chronic hazards based on this summation of classified ingredients is summarized in Table 2.4.4.6.3.4.

Table 2.4.4.6.3.4: Classification of a mixture for chronic hazards based on summation of classified ingredients

Sum of ingredients classified as:	Mixture is classified as:
Chronic 1 × M[a] ≥ 25%	Chronic 1
(M × 10 × Chronic 1) + Chronic 2 ≥ 25%	Chronic 2
(M × 100 × Chronic 1) + (10 × Chronic 2) + Chronic 3 ≥ 25%	Chronic 3
Chronic 1 + Chronic 2 + Chronic 3 + Chronic 4 ≥ 25 %	Chronic 4

[a] For explanation of the M factor, see 2.4.4.6.4.

2.4.4.6.4 *Mixtures with highly toxic ingredients*

Acute Category 1 ingredients with toxicities well below 1 mg/l may influence the toxicity of the mixture and shall be given increased weight in applying the summation method. When a mixture contains ingredients classified as Acute or Chronic 1, the tiered approach described in 2.4.4.6.2 and 2.4.4.6.3 shall be applied using a weighted sum by multiplying the concentrations of Acute 1 ingredients by a factor, instead of merely adding up the percentages. This means that the concentration of "Acute 1" in the left column of Table 2.4.4.6.2.4 and the concentration of "Chronic 1" in the left column of Table 2.4.4.6.3.4 are multiplied by the appropriate multiplying factor. The multiplying factors to be applied to these ingredients are defined using the toxicity value, as summarized in Table 2.4.4.6.4 below. Therefore, in order to classify a mixture containing Acute/Chronic 1 ingredients, the classifier needs to be informed of the value of the M factor in order to apply the summation method. Alternatively, the additivity formula (see 2.4.4.5.2) may be used when toxicity data are available for all highly toxic ingredients in the mixture and there is convincing evidence that all other ingredients, including those for which specific acute toxicity data are not available, are of low or no toxicity and do not significantly contribute to the environmental hazard of the mixture.

Table 2.4.4.6.4 Multiplying factors for highly toxic ingredients of mixtures

$L(E)C_{50}$ value	Multiplying factor (M)
$0.1 < L(E)C_{50} \leq 1$	1
$0.01 < L(E)C_{50} \leq 0.1$	10
$0.001 < L(E)C_{50} \leq 0.01$	100
$0.0001 < L(E)C_{50} \leq 0.001$	1 000
$0.00001 < L(E)C_{50} \leq 0.0001$	10 000
(continue in factor 10 intervals)	

2.4.4.6.5 *Classification of mixtures with ingredients without any useable information*

In the event that no useable information on acute and/or chronic aquatic hazard is available for one or more relevant ingredients, it is concluded that the mixture cannot be attributed (a) definitive hazard category(ies). In this situation, the mixture shall be classified based on the known ingredients only, with the additional statement that: "×% of the mixture consists of (a) ingredient(s) of unknown hazards to the aquatic environment".

PART 3

Dangerous goods list, special provisions and exemptions related to limited and excepted quantities

CHAPTER 3.1

GENERAL

3.1.1 Introduction

In addition to the provisions referred to or given in the tables of this Part, the general requirements of each Part, Chapter and/or Section are to be observed. These general requirements are not given in the tables. When a general requirement is contradictory to a special provision, the special provision prevails.

3.1.2 Proper shipping name

NOTE: *For proper shipping names used for the carriage of samples, see 2.1.4.1.*

3.1.2.1 The proper shipping name is that portion of the entry most accurately describing the goods in Table A or Table C in Chapter 3.2, which is shown in upper case characters (plus any numbers, Greek letters, "sec", "tert", and the letters "m", "n", "o", "p", which form an integral part of the name). Particulars concerning the vapour pressure (vp) and the boiling point (bp) in column (2) of Table C in chapter 3.2 are part of the proper shipping name. An alternative proper shipping name may be shown in brackets following the main proper shipping name. In Table A, it is shown in upper case characters (e.g., ETHANOL (ETHYL ALCOHOL)). In Table C, it is shown in lower case characters (e.g. ACETONITRILE (methyl cyanide)). Portions of an entry appearing in lower case need not be considered as part of the proper shipping name unless otherwise stated above.

3.1.2.2 When conjunctions such as "and" or "or" are in lower case or when segments of the name are punctuated by commas, the entire name of the entry need not necessarily be shown in the transport document or package markings. This is the case particularly when a combination of several distinct entries are listed under a single UN Number. Examples illustrating the selection of the proper shipping name for such entries are:

(a) UN 1057 LIGHTERS or LIGHTER REFILLS - The proper shipping name is the most appropriate of the following possible combinations:

LIGHTERS
LIGHTER REFILLS;

(b) UN 2793 FERROUS METAL BORINGS, SHAVINGS, TURNINGS or CUTTINGS in a form liable to self-heating. The proper shipping name is the most appropriate of the following combinations:

FERROUS METAL BORINGS

FERROUS METAL SHAVINGS

FERROUS METAL TURNINGS

FERROUS METAL CUTTINGS.

3.1.2.3 Proper shipping names may be used in the singular or plural as appropriate. In addition, when qualifying words are used as part of the proper shipping name, their sequence on documentation or package markings is optional. For instance, "DIMETHYLAMINE AQUEOUS SOLUTION" may alternatively be shown "AQUEOUS SOLUTION OF DIMETHYLAMINE". Commercial or military names for goods of Class 1 which contain the proper shipping name supplemented by additional descriptive text may be used.

3.1.2.4 Many substances have an entry for both the liquid and solid state (see definitions for liquid and solid in 1.2.1), or for the solid and solution. These are allocated separate UN numbers which are not necessarily adjacent to each other[1].

3.1.2.5 Unless it is already included in capital letters in the name indicated in Table A or Table C in Chapter 3.2, the qualifying word "MOLTEN" shall be added as part of the proper shipping name when a substance, which is a solid in accordance with the definition in 1.2.1, is offered for carriage in the molten state (e.g. ALKYLPHENOL, SOLID, N.O.S., MOLTEN).

3.1.2.6 Except for self-reactive substances and organic peroxides and unless it is already included in capital letters in the name indicated in Column (2) of Table A of Chapter 3.2, the word "STABILIZED" shall be added as part of the proper shipping name of a substance which without stabilization would be forbidden from carriage in accordance with paragraphs 2.2.X.2 due to it being liable to dangerously react under conditions normally encountered in carriage (e.g.: "TOXIC LIQUID, ORGANIC, N.O.S., STABILIZED").

When temperature control is used to stabilize such substances to prevent the development of any dangerous excess pressure, then:

(a) For liquids: (see 3.1.2.6 of ADR);

(b) For gases: the conditions of carriage shall be approved by the competent authority.

3.1.2.7 Hydrates may be carried under the proper shipping name for the anhydrous substance.

3.1.2.8 ***Generic or "not otherwise specified" (N.O.S.) names***

3.1.2.8.1 Generic and "not otherwise specified" proper shipping names that are assigned to special provision 274 in Column (6) of Table A in Chapter 3.2 or remark 27 in column (20) of Table C in Chapter 3.2 shall be supplemented with the technical name of the goods unless a national law or international convention prohibits its disclosure if it is a controlled substance. For explosive substances and articles of Class 1, the dangerous goods description may be supplemented by additional descriptive text to indicate commercial or military names. Technical names shall be entered in brackets immediately following the proper shipping name. An appropriate modifier, such as "contains" or "containing" or other qualifying words such as "mixture", "solution", etc. and the percentage of the technical constituent may also be used. For example: "UN 1993 FLAMMABLE LIQUID, N.O.S. (CONTAINS XYLENE AND BENZENE), 3, II".

3.1.2.8.1.1 The technical name shall be a recognized chemical name, if relevant a biological name, or other name currently used in scientific and technical handbooks, journals and texts. Trade names shall not be used for this purpose. In the case of pesticides, only ISO common name(s), other name(s) in the World Health Organization (WHO) Recommended Classification of Pesticides by Hazard and Guidelines to Classification, or the name(s) of the active substance(s) may be used.

3.1.2.8.1.2 When a mixture of dangerous goods is described by one of the "N.O.S." or "generic" entries to which special provision 274 has been allocated in Column (6) of Table A in Chapter 3.2, or remark 27 has been allocated in column (20) of Table C in Chapter 3.2, not more than the two constituents which most predominantly contribute to the hazard or hazards of a mixture need to be shown, excluding controlled substances when their disclosure is prohibited by

1 *Details are provided in the alphabetical index (Table B of Chapter 3.2), e.g.:*
NITROXYLENES, LIQUID 6.1 1665
NITROXYLENES, SOLID 6.1 3447

national law or international convention. If a package containing a mixture is labelled with any subsidiary risk label, one of the two technical names shown in parentheses shall be the name of the constituent which compels the use of the subsidiary risk label.

***NOTE**: see 5.4.1.2.2.*

3.1.2.8.1.3 Examples illustrating the selection of the proper shipping name supplemented with the technical name of goods for such N.O.S. entries are:

UN 2902 PESTICIDE, LIQUID, TOXIC, N.O.S. (drazoxolon);

UN 3394 ORGANOMETALLIC SUBSTANCE, LIQUID, PYROPHORIC, WATER-REACTIVE (trimethylgallium).

3.1.2.8.1.4 Examples illustrating how the proper shipping name is supplemented in the indication of the vapour pressure or the boiling-point for N.O.S entries for carriage in tank vessels are:

UN 1268 PETROLEUM DISTILLATES, N.O.S. or PETROLEUM PRODUCTS, N.O.S., 110 kPa $<$ pv50 $\leq$ 150 kpa;

UN 1993 FLAMMABLE LIQUID, N.O.S. (ACETONE with more than 10% BENZENE), pv 50 $\leq$ 110 kPa, 85° C $<$ bp $\leq$ 115° C.

3.1.2.9 ***Mixtures and solutions containing one dangerous substance***

When mixtures and solutions have to be regarded as the dangerous substance mentioned by name in accordance with the classification requirements of 2.1.3.3, the qualifying word "SOLUTION" or "MIXTURE", as appropriate, shall be added as part of the proper shipping name, e.g. "ACETONE SOLUTION". In addition, the concentration of the solution or mixture may also be indicated, e.g. "ACETONE 75% SOLUTION".

CHAPTER 3.2

DANGEROUS GOODS LIST

3.2.1 Table A: List of dangerous goods in numerical order

Explanations concerning Table A:

As a rule, each row of Table A deals with the substance(s) or article(s) covered by a specific UN number or an identification number. However, when substances or articles belonging to the same UN number have different chemical properties, physical properties and/or carriage conditions, several consecutive rows may be used for that UN number or identification number.

Each column of Table A is dedicated to a specific subject as indicated in the explanatory notes below. The intersection of columns and rows (cell) contains information concerning the subject treated in that column, for the substance(s) or article(s) of that row:

– The first four cells identify the substance(s) or article(s) belonging to that row (additional information in that respect may be given by the special provisions referred to in Column (6));

– The following cells give the applicable special provisions, either in the form of complete information or in coded form. The codes cross-refer to detailed information that is to be found in the numbers indicated in the explanatory notes below. An empty cell means either that there is no special provision and that only the general requirements apply, or that the carriage restriction indicated in the explanatory notes is in force.

The applicable general requirements are not referred to in the corresponding cells.

Explanatory notes for each column:

Column (1) "UN number/identification number".

Contains the UN number or the identification number:

– of the dangerous substance or article if the substance or article has been assigned its own specific UN number or identification number, or

– of the generic or n.o.s. entry to which the dangerous substances or articles not mentioned by name shall be assigned in accordance with the criteria ("decision trees") of Part 2.

Column (2) "Name and description"

Contains, in upper case characters, the name of the substance or article, if the substance or article has been assigned its own specific UN number or identification number, or of the generic or n.o.s. entry to which it has been assigned in accordance with the criteria ("decision trees") of Part 2. This name shall be used as the proper shipping name or, when applicable, as part of the proper shipping name (see 3.1.2 for further details on the proper shipping name).

A descriptive text in lower case characters is added after the proper shipping name to clarify the scope of the entry if the classification and/or carriage conditions of the substance or article may be different under certain conditions.

Column (3a) "Class"

Contains the number of the Class, whose heading covers the dangerous substance or article. This Class number is assigned in accordance with the procedures and criteria of Part 2.

Column (3b) "Classification code"

Contains the classification code of the dangerous substance or article.

- For dangerous substances or articles of Class 1, the code consists of a division number and compatibility group letter, which are assigned in accordance with the procedures and criteria of 2.2.1.1.4.
- For dangerous substances or articles of Class 2, the code consists of a number and one or more letters representing the hazardous property group, which are explained in 2.2.2.1.2 and 2.2.2.1.3.
- For dangerous substances or articles of Classes 3, 4.1, 4.2, 4.3, 5.1, 5.2, 6.1, 6.2, 8 and 9, the codes are explained in 2.2.x.1.2. [1]
- Dangerous substances or articles of Class 7 do not have a classification code.

Column (4) "Packing group"

Contains the packing group number(s) (I, II or III) assigned to the dangerous substance. These packing group numbers are assigned on the basis of the procedures and criteria of Part 2. Certain articles and substances are not assigned to packing groups.

Column (5) "Labels"

Contains the model number of the labels/placards (see 5.2.2.2 and 5.3.1.1.7) that have to be affixed to packages, containers, tank-containers, portable tanks, MEGCs, vehicles and wagons. However:

- For substances or articles of Class 7, 7X means label model No. 7A, 7B or 7C as appropriate according to the category (see 5.1.5.3.4 and 5.2.2.1.11.1) or placard No. 7D (see 5.3.1.1.3 and 5.3.1.1.7.2).

The general provisions on labelling/placarding (e.g. number of labels, their location) are to be found in 5.2.2.1 for packages, and in 5.3.1, for containers, tank-containers, MEGCs, portable tanks, vehicles and wagons.

---

1 *x = the Class number of the dangerous substance or article, without dividing point if applicable.*

***NOTE**: Special provisions, indicated in Column (6), may change the above labelling provisions.*

Column (6) "Special provisions"

Contains the numeric codes of special provisions that have to be met. These provisions concern a wide array of subjects, mainly connected with the contents of Columns (1) to (5) (e.g. carriage prohibitions, exemptions from certain requirements, explanations concerning the classification of certain forms of the dangerous goods concerned and additional labelling or marking provisions), and are listed in Chapter 3.3 in numerical order. If Column (6) is empty, no special provisions apply to the contents of Columns (1) to (5) for the dangerous goods concerned. Special provisions specific to inland navigation begin at 800.

Column (7a) "Limited Quantities"

Contains an alphanumeric code with the following meaning:

- "LQ0" signifies that no exemption from the provisions of ADN exists for the dangerous goods packed in limited quantities;
- All the other alphanumeric codes starting with the letters "LQ" signify that the provisions of ADN are not applicable if the conditions indicated in Chapter 3.4 are fulfilled.

Column (7b) "Excepted Quantities"

Contains an alphanumeric code with the following meaning:

- "E0" signifies that no exemption from the provisions of ADN exists for the dangerous goods packed in excepted quantities;
- All the other alphanumerical codes starting with the letter "E" signify that the provisions of ADN are not applicable if the conditions indicated in Chapter 3.5 are fulfilled.

Column (8) "Carriage permitted"

This column contains the alphabetic codes concerning the permitted form of carriage in inland navigation vessels.

If column (8) is empty, the substance or article may only be carried in packages.

If column (8) contains code "B", carriage is permitted in packages or in bulk (see 7.1.1.11).

If column (8) contains code "T", carriage is permitted in packages and in tank vessels. In the event of carriage in tank vessels, the requirements of Table C are applicable (see 7.2.1.21).

If "carriage prohibited" appears in column (8), carriage is not permitted.

If "free" appears in column (8), the substance is not subject to the requirements of ADN.

Column (9) “Equipment required”

This column contains the alphanumeric codes for the equipment required for the carriage of the dangerous substance or article (see 8.1.5).

Column (10) “Ventilation”

This column contains the alphanumeric codes of the special requirements concerning ventilation applicable to carriage with the following meaning:

– alphanumeric codes starting with the letters “VE” mean that special additional conditions are applicable to carriage. These can be found in 7.1.6.12 and establish special requirements.

Column (11) “Provisions concerning loading, unloading and carriage”

This column contains the alphanumeric codes of the special requirements applicable to carriage with the following meaning:

– alphanumeric codes starting with the letters “CO”, “ST” and “RA” mean that special additional conditions are applicable to carriage in bulk. These can be found in 7.1.6.11 and establish special requirements.

– alphanumeric codes starting with the letters “LO” mean that special additional conditions are applicable prior to loading. These can be found in 7.1.6.13 and establish special requirements.

– alphanumeric codes starting with the letters “HA” mean that special additional conditions are applicable to the handling and stowage of the cargo. These can be found in 7.1.6.14 and establish special requirements.

– alphanumeric codes starting with the letters “IN” mean that special additional conditions are applicable to the inspection of holds during carriage. These can be found in 7.1.6.16 and establish special requirements.

Column (12) “Number of blue cones/lights”

This column contains the number of cones/lights which should constitute the marking of the vessel during the carriage of this dangerous substance or article (see 7.1.5).

Column (13) “Additional requirements/Remarks”

This column contains additional requirements or observations concerning the carriage of this dangerous substance or article

UN No. or ID No.	Name and description	Class	Classi-fication Code	Packing group	Labels	Special provi-sions	Limited and excepted quantities		Carriage permitted	Equipment required	Venti-lation	Provisions concerning loading, unloading and carriage		Number of blue cones/ lights	Remarks
	3.1.2	2.2	2.2	2.1.1.3	5.2.2	3.3	3.4.6	3.5.1.2	3.2.1	8.1.5	7.1.6	7.1.6		7.1.5	3.2.1
(1)	(2)	(3a)	(3b)	(4)	(5)	(6)	(7a)	(7b)	(8)	(9)	(10)	(11)		(12)	(13)
0004	AMMONIUM PICRATE dry or wetted with less than 10% water, by mass	1	1.1D		1		LQ0	E0		PP		LO01	HA01, HA02, HA03, HA04, HA05, HA06	3	
0005	CARTRIDGES FOR WEAPONS with bursting charge	1	1.1F		1		LQ0	E0		PP		LO01	HA01, HA02, HA03, HA04, HA05, HA06	3	
0006	CARTRIDGES FOR WEAPONS with bursting charge	1	1.1E		1		LQ0	E0		PP		LO01	HA01, HA02, HA03, HA04, HA05, HA06	3	
0007	CARTRIDGES FOR WEAPONS with bursting charge	1	1.2F		1		LQ0	E0		PP		LO01	HA01, HA02, HA03, HA04, HA05, HA06	3	
0009	AMMUNITION, INCENDIARY with or without burster, expelling charge or propelling charge	1	1.2G		1		LQ0	E0		PP		LO01	HA01, HA03, HA04, HA05, HA06	3	
0010	AMMUNITION, INCENDIARY with or without burster, expelling charge or propelling charge	1	1.3G		1		LQ0	E0		PP		LO01	HA01, HA03, HA04, HA05, HA06	3	
0012	CARTRIDGES FOR WEAPONS, INERT PROJECTILE or CARTRIDGES, SMALL ARMS	1	1.4S		1.4		LQ0	E0		PP		LO01	HA01, HA03, HA04, HA05, HA06	0	
0014	CARTRIDGES FOR WEAPONS, BLANK or CARTRIDGES, SMALL ARMS, BLANK	1	1.4S		1.4		LQ0	E0		PP		LO01	HA01, HA03, HA04, HA05, HA06	0	

UN No. or ID No.	Name and description 3.1.2	Class 2.2	Classi-fication Code 2.2	Packing group 2.1.1.3	Labels 5.2.2	Special provi-sions 3.3	Limited and excepted quantities 3.4.6	3.5.1.2	Carriage permitted 3.2.1	Equipment required 8.1.5	Venti-lation 7.1.6	Provisions concerning loading, unloading and carriage 7.1.6	Number of blue cones/ lights 7.1.5	Remarks 3.2.1
(1)	(2)	(3a)	(3b)	(4)	(5)	(6)	(7a)	(7b)	(8)	(9)	(10)	(11)	(12)	(13)
0015	AMMUNITION, SMOKE with or without burster, expelling charge or propelling charge	1	1.2G		1		LQ0	E0		PP		LO01 HA01, HA03, HA04, HA05, HA06	3	
0015	AMMUNITION, SMOKE with or without burster, expelling charge or propelling charge, containing corrosive substances	1	1.2G		1 +8		LQ0	E0		PP		LO01 HA01, HA03, HA04, HA05, HA06	3	
0016	AMMUNITION, SMOKE with or without burster, expelling charge or propelling charge	1	1.3G		1		LQ0	E0		PP		LO01 HA01, HA03, HA04, HA05, HA06	3	
0016	AMMUNITION, SMOKE with or without burster, expelling charge or propelling charge, containing corrosive substances	1	1.3G		1 +8		LQ0	E0		PP		LO01 HA01, HA03, HA04, HA05, HA06	3	
0018	AMMUNITION, TEAR-PRODUCING with burster, expelling charge or propelling charge	1	1.2G		1+6.1+8	802	LQ0	E0		PP		LO01 HA01, HA03, HA04, HA05, HA06	3	
0019	AMMUNITION, TEAR-PRODUCING with burster, expelling charge or propelling charge	1	1.3G		1+6.1+8	802	LQ0	E0		PP		LO01 HA01, HA03, HA04, HA05, HA06	3	
0020	AMMUNITION, TOXIC with burster, expelling charge or propelling charge	1	1.2K	CARRIAGE PROHIBITED										
0021	AMMUNITION, TOXIC with burster, expelling charge or propelling charge	1	1.3K	CARRIAGE PROHIBITED										
0027	BLACK POWDER (GUNPOWDER), granular or as a meal	1	1.1D		1		LQ0	E0		PP		LO01 HA01, HA02, HA03, HA04, HA05, HA06	3	
0028	BLACK POWDER (GUNPOWDER), COMPRESSED or BLACK POWDER (GUNPOWDER), IN PELLETS	1	1.1D		1		LQ0	E0		PP		LO01 HA01, HA02, HA03, HA04, HA05, HA06	3	

UN No. or ID No.	Name and description	Class	Classi-fication Code	Packing group	Labels	Special provi-sions	Limited and excepted quantities		Carriage permitted	Equipment required	Venti-lation	Provisions concerning loading, unloading and carriage		Number of blue cones/ lights	Remarks
	3.1.2	2.2	2.2	2.1.1.3	5.2.2	3.3	3.4.6	3.5.1.2	3.2.1	8.1.5	7.1.6	7.1.6		7.1.5	3.2.1
(1)	(2)	(3a)	(3b)	(4)	(5)	(6)	(7a)	(7b)	(8)	(9)	(10)	(11)		(12)	(13)
0029	DETONATORS, NON-ELECTRIC for blasting	1	1.1B		1		LQ0	E0		PP		LO01	HA01, HA02, HA03, HA04, HA05, HA06	3	
0030	DETONATORS, ELECTRIC for blasting	1	1.1B		1		LQ0	E0		PP		LO01	HA01, HA02, HA03, HA04, HA05, HA06	3	
0033	BOMBS with bursting charge	1	1.1F		1		LQ0	E0		PP		LO01	HA01, HA02, HA03, HA04, HA05, HA06	3	
0034	BOMBS with bursting charge	1	1.1D		1		LQ0	E0		PP		LO01	HA01, HA02, HA03, HA04, HA05, HA06	3	
0035	BOMBS with bursting charge	1	1.2D		1		LQ0	E0		PP		LO01	HA01, HA03, HA04, HA05, HA06	3	
0037	BOMBS, PHOTO-FLASH	1	1.1F		1		LQ0	E0		PP		LO01	HA01, HA02, HA03, HA04, HA05, HA06	3	
0038	BOMBS, PHOTO-FLASH	1	1.1D		1		LQ0	E0		PP		LO01	HA01, HA02, HA03, HA04, HA05, HA06	3	
0039	BOMBS, PHOTO-FLASH	1	1.2G		1		LQ0	E0		PP		LO01	HA01, HA03, HA04, HA05, HA06	3	

UN No. or ID No.	Name and description	Class	Classi-fication Code	Packing group	Labels	Special provi-sions	Limited and excepted quantities		Carriage permitted	Equipment required	Venti-lation	Provisions concerning loading, unloading and carriage		Number of blue cones/ lights	Remarks
	3.1.2	2.2	2.2	2.1.1.3	5.2.2	3.3	3.4.6	3.5.1.2	3.2.1	8.1.5	7.1.6	7.1.6		7.1.5	3.2.1
(1)	(2)	(3a)	(3b)	(4)	(5)	(6)	(7a)	(7b)	(8)	(9)	(10)	(11)		(12)	(13)
0042	BOOSTERS without detonator	1	1.1D		1		LQ0	E0		PP		LO01	HA01, HA02, HA03, HA04, HA05, HA06	3	
0043	BURSTERS, explosive	1	1.1D		1		LQ0	E0		PP		LO01	HA01, HA02, HA03, HA04, HA05, HA06	3	
0044	PRIMERS, CAP TYPE	1	1.4S		1.4		LQ0	E0		PP		LO01	HA01, HA03, HA04, HA05, HA06	0	
0048	CHARGES, DEMOLITION	1	1.1D		1		LQ0	E0		PP		LO01	HA01, HA02, HA03, HA04, HA05, HA06	3	
0049	CARTRIDGES, FLASH	1	1.1G		1		LQ0	E0		PP		LO01	HA01, HA02, HA03, HA04, HA05, HA06	3	
0050	CARTRIDGES, FLASH	1	1.3G		1		LQ0	E0		PP		LO01	HA01, HA03, HA04, HA05, HA06	3	
0054	CARTRIDGES, SIGNAL	1	1.3G		1		LQ0	E0		PP		LO01	HA01, HA03, HA04, HA05, HA06	3	
0055	CASES, CARTRIDGE, EMPTY, WITH PRIMER	1	1.4S		1.4		LQ0	E0		PP		LO01	HA01, HA03, HA04, HA05, HA06	0	

UN No. or ID No.	Name and description	Class	Classi-fication Code	Packing group	Labels	Special provi-sions	Limited and excepted quantities		Carriage permitted	Equipment required	Venti-lation	Provisions concerning loading, unloading and carriage		Number of blue cones/ lights	Remarks
	3.1.2	2.2	2.2	2.1.1.3	5.2.2	3.3	3.4.6	3.5.1.2	3.2.1	8.1.5	7.1.6	7.1.6		7.1.5	3.2.1
(1)	(2)	(3a)	(3b)	(4)	(5)	(6)	(7a)	(7b)	(8)	(9)	(10)	(11)		(12)	(13)
0056	CHARGES, DEPTH	1	1.1D		1		LQ0	E0		PP		LO01	HA01, HA02, HA03, HA04, HA05, HA06	3	
0059	CHARGES, SHAPED without detonator	1	1.1D		1		LQ0	E0		PP		LO01	HA01, HA02, HA03, HA04, HA05, HA06	3	
0060	CHARGES, SUPPLEMENTARY, EXPLOSIVE	1	1.1D		1		LQ0	E0		PP		LO01	HA01, HA02, HA03, HA04, HA05, HA06	3	
0065	CORD, DETONATING, flexible	1	1.1D		1		LQ0	E0		PP		LO01	HA01, HA02, HA03, HA04, HA05, HA06	3	
0066	CORD, IGNITER	1	1.4G		1.4		LQ0	E0		PP		LO01	HA01, HA03, HA04, HA05, HA06	1	
0070	CUTTERS, CABLE, EXPLOSIVE	1	1.4S		1.4		LQ0	E0		PP		LO01	HA01, HA03, HA04, HA05, HA06	0	
0072	CYCLOTRIMETHYLENE-TRINITRAMINE (CYCLONITE; HEXOGEN; RDX), WETTED with not less than 15% water, by mass	1	1.1D		1	266	LQ0	E0		PP		LO01	HA01, HA02, HA03, HA04, HA05, HA06	3	
0073	DETONATORS FOR AMMUNITION	1	1.1B		1		LQ0	E0		PP		LO01	HA01, HA02, HA03, HA04, HA05, HA06	3	

UN No. or ID No.	Name and description	Class	Classi-fication Code	Packing group	Labels	Special provi-sions	Limited and excepted quantities		Carriage permitted	Equipment required	Venti-lation	Provisions concerning loading, unloading and carriage		Number of blue cones/ lights	Remarks
	3.1.2	2.2	2.2	2.1.1.3	5.2.2	3.3	3.4.6	3.5.1.2	3.2.1	8.1.5	7.1.6	7.1.6		7.1.5	3.2.1
(1)	(2)	(3a)	(3b)	(4)	(5)	(6)	(7a)	(7b)	(8)	(9)	(10)	(11)		(12)	(13)
0074	DIAZODINITROPHENOL, WETTED with not less than 40% water, or mixture of alcohol and water, by mass	1	1.1A		1	266	LQ0	E0		PP		LO01	HA01, HA02, HA03, HA04, HA05, HA06	3	
0075	DIETHYLENEGLYCOL DINITRATE, DESENSITIZED with not less than 25% non-volatile, water-insoluble phlegmatizer, by mass	1	1.1D		1	266	LQ0	E0		PP		LO01	HA01, HA02, HA03, HA04, HA05, HA06	3	
0076	DINITROPHENOL, dry or wetted with less than 15% water, by mass	1	1.1D		1+6.1	802	LQ0	E0		PP		LO01	HA01, HA02, HA03, HA04, HA05, HA06	3	
0077	DINITROPHENOLATES, alkali metals, dry or wetted with less than 15% water, by mass	1	1.3C		1+6.1	802	LQ0	E0		PP		LO01	HA01, HA03, HA04, HA05, HA06	3	
0078	DINITRORESORCINOL, dry or wetted with less than 15% water, by mass	1	1.1D		1		LQ0	E0		PP		LO01	HA01, HA02, HA03, HA04, HA05, HA06	3	
0079	HEXANITRODIPHENYLAMINE (DIPICRYLAMINE; HEXYL)	1	1.1D		1		LQ0	E0		PP		LO01	HA01, HA02, HA03, HA04, HA05, HA06	3	
0081	EXPLOSIVE, BLASTING, TYPE A	1	1.1D		1	616 617	LQ0	E0		PP		LO01	HA01, HA02, HA03, HA04, HA05, HA06	3	

UN No. or ID No.	Name and description	Class	Classi-fication Code	Packing group	Labels	Special provi-sions	Limited and excepted quantities		Carriage permitted	Equipment required	Venti-lation	Provisions concerning loading, unloading and carriage		Number of blue cones/ lights	Remarks
	3.1.2	2.2	2.2	2.1.1.3	5.2.2	3.3	3.4.6	3.5.1.2	3.2.1	8.1.5	7.1.6	7.1.6		7.1.5	3.2.1
(1)	(2)	(3a)	(3b)	(4)	(5)	(6)	(7a)	(7b)	(8)	(9)	(10)	(11)		(12)	(13)
0082	EXPLOSIVE, BLASTING, TYPE B	1	1.1D		1	617	LQ0	E0		PP		LO01	HA01, HA02, HA03, HA04, HA05, HA06	3	
0083	EXPLOSIVE, BLASTING, TYPE C	1	1.1D		1	267 617	LQ0	E0		PP		LO01	HA01, HA02, HA03, HA04, HA05, HA06	3	
0084	EXPLOSIVE, BLASTING, TYPE D	1	1.1D		1	617	LQ0	E0		PP		LO01	HA01, HA02, HA03, HA04, HA05, HA06	3	
0092	FLARES, SURFACE	1	1.3G		1		LQ0	E0		PP		LO01	HA01, HA03, HA04, HA05, HA06	3	
0093	FLARES, AERIAL	1	1.3G		1		LQ0	E0		PP		LO01	HA01, HA03, HA04, HA05, HA06	3	
0094	FLASH POWDER	1	1.1G		1		LQ0	E0		PP		LO01	HA01, HA02, HA03, HA04, HA05, HA06	3	
0099	FRACTURING DEVICES, EXPLOSIVE without detonator, for oil wells	1	1.1D		1		LQ0	E0		PP		LO01	HA01, HA02, HA03, HA04, HA05, HA06	3	
0101	FUSE, NON-DETONATING	1	1.3G		1		LQ0	E0		PP		LO01	HA01, HA03, HA04, HA05, HA06	3	

UN No. or ID No.	Name and description	Class	Classi-fication Code	Packing group	Labels	Special provi-sions	Limited and excepted quantities		Carriage permitted	Equipment required	Venti-lation	Provisions concerning loading, unloading and carriage		Number of blue cones/ lights	Remarks
	3.1.2	2.2	2.2	2.1.1.3	5.2.2	3.3	3.4.6	3.5.1.2	3.2.1	8.1.5	7.1.6	7.1.6		7.1.5	3.2.1
(1)	(2)	(3a)	(3b)	(4)	(5)	(6)	(7a)	(7b)	(8)	(9)	(10)	(11)		(12)	(13)
0102	CORD (FUSE), DETONATING, metal clad	1	1.2D		1		LQ0	E0		PP		LO01	HA01, HA03, HA04, HA05, HA06	3	
0103	FUSE, IGNITER, tubular, metal clad	1	1.4G		1.4		LQ0	E0		PP		LO01	HA01, HA03, HA04, HA05, HA06	1	
0104	CORD (FUSE), DETONATING, MILD EFFECT, metal clad	1	1.4D		1.4		LQ0	E0		PP		LO01	HA01, HA03, HA04, HA05, HA06	1	
0105	FUSE, SAFETY	1	1.4S		1.4		LQ0	E0		PP		LO01	HA01, HA03, HA04, HA05, HA06	0	
0106	FUZES, DETONATING	1	1.1B		1		LQ0	E0		PP		LO01	HA01, HA02, HA03, HA04, HA05, HA06	3	
0107	FUZES, DETONATING	1	1.2B		1		LQ0	E0		PP		LO01	HA01, HA02, HA03, HA04, HA05, HA06	3	
0110	GRENADES, PRACTICE, hand or rifle	1	1.4S		1.4		LQ0	E0		PP		LO01	HA01, HA03, HA04, HA05, HA06	0	
0113	GUANYLNITROSAMINOGUANYLIDENE HYDRAZINE, WETTED with not less than 30% water, by mass	1	1.1A		1	266	LQ0	E0		PP		LO01	HA01, HA02, HA03, HA04, HA05, HA06	3	

UN No. or ID No.	Name and description	Class	Classi-fication Code	Packing group	Labels	Special provi-sions	Limited and excepted quantities		Carriage permitted	Equipment required	Venti-lation	Provisions concerning loading, unloading and carriage		Number of blue cones/ lights	Remarks
	3.1.2	2.2	2.2	2.1.1.3	5.2.2	3.3	3.4.6	3.5.1.2	3.2.1	8.1.5	7.1.6	7.1.6		7.1.5	3.2.1
(1)	(2)	(3a)	(3b)	(4)	(5)	(6)	(7a)	(7b)	(8)	(9)	(10)	(11)		(12)	(13)
0114	GUANYLNITROSAMINO-GUANYLTETRAZENE (TETRAZENE), WETTED with not less than 30% water, or mixture of alcohol and water, by mass	1	1.1A		1	266	LQ0	E0		PP		LO01	HA01, HA02, HA03, HA04, HA05, HA06	3	
0118	HEXOLITE (HEXOTOL), dry or wetted with less than 15% water, by mass	1	1.1D		1		LQ0	E0		PP		LO01	HA01, HA02, HA03, HA04, HA05, HA06	3	
0121	IGNITERS	1	1.1G		1		LQ0	E0		PP		LO01	HA01, HA02, HA03, HA04, HA05, HA06	3	
0124	JET PERFORATING GUNS, CHARGED, oil well, without detonator	1	1.1D		1		LQ0	E0		PP		LO01	HA01, HA02, HA03, HA04, HA05, HA06	3	
0129	LEAD AZIDE, WETTED with not less than 20% water, or mixture of alcohol and water, by mass	1	1.1A		1	266	LQ0	E0		PP		LO01	HA01, HA02, HA03, HA04, HA05, HA06	3	
0130	LEAD STYPHNATE (LEAD TRINITRORESORCINATE), WETTED with not less than 20% water, or mixture of alcohol and water, by mass	1	1.1A		1	266	LQ0	E0		PP		LO01	HA01, HA02, HA03, HA04, HA05, HA06	3	
0131	LIGHTERS, FUSE	1	1.4S		1.4		LQ0	E0		PP		LO01	HA01, HA03, HA04, HA05, HA06	0	
0132	DEFLAGRATING METAL SALTS OF AROMATIC NITRODERIVATIVES, N.O.S.	1	1.3C		1	274	LQ0	E0		PP		LO01	HA01, HA03, HA04, HA05, HA06	3	

UN No. or ID No.	Name and description	Class	Classi-fication Code	Packing group	Labels	Special provi-sions	Limited and excepted quantities		Carriage permitted	Equipment required	Venti-lation	Provisions concerning loading, unloading and carriage		Number of blue cones/ lights	Remarks
	3.1.2	2.2	2.2	2.1.1.3	5.2.2	3.3	3.4.6	3.5.1.2	3.2.1	8.1.5	7.1.6	7.1.6		7.1.5	3.2.1
(1)	(2)	(3a)	(3b)	(4)	(5)	(6)	(7a)	(7b)	(8)	(9)	(10)	(11)		(12)	(13)
0133	MANNITOL HEXANITRATE (NITROMANNITE), WETTED with not less than 40% water, or mixture of alcohol and water, by mass	1	1.1D		1	266	LQ0	E0		PP		LO01	HA01, HA02, HA03, HA04, HA05, HA06	3	
0135	MERCURY FULMINATE, WETTED with not less than 20% water, or mixture of alcohol and water, by mass	1	1.1A		1	266	LQ0	E0		PP		LO01	HA01, HA02, HA03, HA04, HA05, HA06	3	
0136	MINES with bursting charge	1	1.1F		1		LQ0	E0		PP		LO01	HA01, HA02, HA03, HA04, HA05, HA06	3	
0137	MINES with bursting charge	1	1.1D		1		LQ0	E0		PP		LO01	HA01, HA02, HA03, HA04, HA05, HA06	3	
0138	MINES with bursting charge	1	1.2D		1		LQ0	E0		PP		LO01	HA01, HA03, HA04, HA05, HA06	3	
0143	NITROGLYCERIN, DESENSITIZED with not less than 40% non-volatile water-insoluble phlegmatizer, by mass	1	1.1D		1+6.1	266 271 802	LQ0	E0		PP		LO01	HA01, HA02, HA03, HA04, HA05, HA06	3	
0144	NITROGLYCERIN SOLUTION IN ALCOHOL with more than 1% but not more than 10% nitroglycerin	1	1.1D		1	500	LQ0	E0		PP		LO01	HA01, HA02, HA03, HA04, HA05, HA06	3	

UN No. or ID No.	Name and description	Class	Classi-fication Code	Packing group	Labels	Special provi-sions	Limited and excepted quantities		Carriage permitted	Equipment required	Venti-lation	Provisions concerning loading, unloading and carriage		Number of blue cones/ lights	Remarks
	3.1.2	2.2	2.2	2.1.1.3	5.2.2	3.3	3.4.6	3.5.1.2	3.2.1	8.1.5	7.1.6	7.1.6		7.1.5	3.2.1
(1)	(2)	(3a)	(3b)	(4)	(5)	(6)	(7a)	(7b)	(8)	(9)	(10)	(11)		(12)	(13)
0146	NITROSTARCH, dry or wetted with less than 20% water, by mass	1	1.1D		1		LQ0	E0		PP		LO01	HA01, HA02, HA03, HA04, HA05, HA06	3	
0147	NITRO UREA	1	1.1D		1		LQ0	E0		PP		LO01	HA01, HA02, HA03, HA04, HA05, HA06	3	
0150	PENTAERYTHRITE TETRANITRATE (PENTAERYTHRITOL TETRANITRATE; PETN), WETTED with not less than 25% water, by mass, or DESENSITIZED with not less than 15% phlegmatizer, by mass	1	1.1D		1	266	LQ0	E0		PP		LO01	HA01, HA02, HA03, HA04, HA05, HA06	3	
0151	PENTOLITE, dry or wetted with less than 15% water, by mass	1	1.1D		1		LQ0	E0		PP		LO01	HA01, HA02, HA03, HA04, HA05, HA06	3	
0153	TRINITROANILINE (PICRAMIDE)	1	1.1D		1		LQ0	E0		PP		LO01	HA01, HA02, HA03, HA04, HA05, HA06	3	
0154	TRINITROPHENOL (PICRIC ACID), dry or wetted with less than 30% water, by mass	1	1.1D		1		LQ0	E0		PP		LO01	HA01, HA02, HA03, HA04, HA05, HA06	1	
0155	TRINITROCHLOROBENZENE (PICRYL CHLORIDE)	1	1.1D		1		LQ0	E0		PP		LO01	HA01, HA02, HA03, HA04, HA05, HA06	3	

UN No. or ID No.	Name and description	Class	Classi-fication Code	Packing group	Labels	Special provi-sions	Limited and excepted quantities		Carriage permitted	Equipment required	Venti-lation	Provisions concerning loading, unloading and carriage		Number of blue cones/ lights	Remarks
	3.1.2	2.2	2.2	2.1.1.3	5.2.2	3.3	3.4.6	3.5.1.2	3.2.1	8.1.5	7.1.6	7.1.6		7.1.5	3.2.1
(1)	(2)	(3a)	(3b)	(4)	(5)	(6)	(7a)	(7b)	(8)	(9)	(10)	(11)		(12)	(13)
0159	POWDER CAKE (POWDER PASTE), WETTED with not less than 25% water, by mass	1	1.3C		1	266	LQ0	E0		PP		LO01	HA01, HA03, HA04, HA05, HA06	3	
0160	POWDER, SMOKELESS	1	1.1C		1		LQ0	E0		PP		LO01	HA01, HA02, HA03, HA04, HA05, HA06	3	
0161	POWDER, SMOKELESS	1	1.3C		1		LQ0	E0		PP		LO01	HA01, HA03, HA04, HA05, HA06	3	
0167	PROJECTILES with bursting charge	1	1.1F		1		LQ0	E0		PP		LO01	HA01, HA02, HA03, HA04, HA05, HA06	3	
0168	PROJECTILES with bursting charge	1	1.1D		1		LQ0	E0		PP		LO01	HA01, HA02, HA03, HA04, HA05, HA06	3	
0169	PROJECTILES with bursting charge	1	1.2D		1		LQ0	E0		PP		LO01	HA01, HA03, HA04, HA05, HA06	3	
0171	AMMUNITION, ILLUMINATING with or without burster, expelling charge or propelling charge	1	1.2G		1		LQ0	E0		PP		LO01	HA01, HA03, HA04, HA05, HA06	3	
0173	RELEASE DEVICES, EXPLOSIVE	1	1.4S		1.4		LQ0	E0		PP		LO01	HA01, HA03, HA04, HA05, HA06	0	

UN No. or ID No.	Name and description	Class	Classi-fication Code	Packing group	Labels	Special provi-sions	Limited and excepted quantities		Carriage permitted	Equipment required	Venti-lation	Provisions concerning loading, unloading and carriage		Number of blue cones/ lights	Remarks
	3.1.2	2.2	2.2	2.1.1.3	5.2.2	3.3	3.4.6	3.5.1.2	3.2.1	8.1.5	7.1.6	7.1.6		7.1.5	3.2.1
(1)	(2)	(3a)	(3b)	(4)	(5)	(6)	(7a)	(7b)	(8)	(9)	(10)	(11)		(12)	(13)
0174	RIVETS, EXPLOSIVE	1	1.4S		1.4		LQ0	E0		PP		LO01	HA01, HA03, HA04, HA05, HA06	0	
0180	ROCKETS with bursting charge	1	1.1F		1		LQ0	E0		PP		LO01	HA01, HA02, HA03, HA04, HA05, HA06	3	
0181	ROCKETS with bursting charge	1	1.1E		1		LQ0	E0		PP		LO01	HA01, HA02, HA03, HA04, HA05, HA06	3	
0182	ROCKETS with bursting charge	1	1.2E		1		LQ0	E0		PP		LO01	HA01, HA03, HA04, HA05, HA06	3	
0183	ROCKETS with inert head	1	1.3C		1		LQ0	E0		PP		LO01	HA01, HA03, HA04, HA05, HA06	3	
0186	ROCKET MOTORS	1	1.3C		1		LQ0	E0		PP		LO01	HA01, HA03, HA04, HA05, HA06	3	
0190	SAMPLES, EXPLOSIVE, other than initiating explosive	1				16 274	LQ0	E0		PP		LO01	HA01, HA02, HA03, HA04, HA05, HA06	3	
0191	SIGNAL DEVICES, HAND	1	1.4G		1.4		LQ0	E0		PP		LO01	HA01, HA03, HA04, HA05, HA06	1	

UN No. or ID No.	Name and description	Class	Classi-fication Code	Packing group	Labels	Special provi-sions	Limited and excepted quantities		Carriage permitted	Equipment required	Venti-lation	Provisions concerning loading, unloading and carriage		Number of blue cones/ lights	Remarks
	3.1.2	2.2	2.2	2.1.1.3	5.2.2	3.3	3.4.6	3.5.1.2	3.2.1	8.1.5	7.1.6	7.1.6		7.1.5	3.2.1
(1)	(2)	(3a)	(3b)	(4)	(5)	(6)	(7a)	(7b)	(8)	(9)	(10)	(11)		(12)	(13)
0192	SIGNALS, RAILWAY TRACK, EXPLOSIVE	1	1.1G		1		LQ0	E0		PP		LO01	HA01, HA02, HA03, HA04, HA05, HA06	3	
0193	SIGNALS, RAILWAY TRACK, EXPLOSIVE	1	1.4S		1.4		LQ0	E0		PP		LO01	HA01, HA03, HA04, HA05, HA06	0	
0194	SIGNALS, DISTRESS, ship	1	1.1G		1		LQ0	E0		PP		LO01	HA01, HA02, HA03, HA04, HA05, HA06	3	
0195	SIGNALS, DISTRESS, ship	1	1.3G		1		LQ0	E0		PP		LO01	HA01, HA03, HA04, HA05, HA06	3	
0196	SIGNALS, SMOKE	1	1.1G		1		LQ0	E0		PP		LO01	HA01, HA02, HA03, HA04, HA05, HA06	3	
0197	SIGNALS, SMOKE	1	1.4G		1.4		LQ0	E0		PP		LO01	HA01, HA03, HA04, HA05, HA06	1	
0204	SOUNDING DEVICES, EXPLOSIVE	1	1.2F		1		LQ0	E0		PP		LO01	HA01, HA02, HA03, HA04, HA05, HA06	3	
0207	TETRANITROANILINE	1	1.1D		1		LQ0	E0		PP		LO01	HA01, HA02, HA03, HA04, HA05, HA06	3	

UN No. or ID No.	Name and description	Class	Classi-fication Code	Packing group	Labels	Special provi-sions	Limited and excepted quantities		Carriage permitted	Equipment required	Venti-lation	Provisions concerning loading, unloading and carriage		Number of blue cones/ lights	Remarks
	3.1.2	2.2	2.2	2.1.1.3	5.2.2	3.3	3.4.6	3.5.1.2	3.2.1	8.1.5	7.1.6	7.1.6		7.1.5	3.2.1
(1)	(2)	(3a)	(3b)	(4)	(5)	(6)	(7a)	(7b)	(8)	(9)	(10)	(11)		(12)	(13)
0208	TRINITROPHENYLMETHYLNITRAMINE (TETRYL)	1	1.1D		1		LQ0	E0		PP		LO01	HA01, HA02, HA03, HA04, HA05, HA06	3	
0209	TRINITROTOLUENE (TNT), dry or wetted with less than 30% water, by mass	1	1.1D		1		LQ0	E0		PP		LO01	HA01, HA02, HA03, HA04, HA05, HA06	3	
0212	TRACERS FOR AMMUNITION	1	1.3G		1		LQ0	E0		PP		LO01	HA01, HA03, HA04, HA05, HA06	3	
0213	TRINITROANISOLE	1	1.1D		1		LQ0	E0		PP		LO01	HA01, HA02, HA03, HA04, HA05, HA06	3	
0214	TRINITROBENZENE, dry or wetted with less than 30% water, by mass	1	1.1D		1		LQ0	E0		PP		LO01	HA01, HA02, HA03, HA04, HA05, HA06	3	
0215	TRINITROBENZOIC ACID, dry or wetted with less than 30% water, by mass	1	1.1D		1		LQ0	E0		PP		LO01	HA01, HA02, HA03, HA04, HA05, HA06	3	
0216	TRINITRO-m-CRESOL	1	1.1D		1		LQ0	E0		PP		LO01	HA01, HA02, HA03, HA04, HA05, HA06	3	

UN No. or ID No.	Name and description	Class	Classi-fication Code	Packing group	Labels	Special provi-sions	Limited and excepted quantities		Carriage permitted	Equipment required	Venti-lation	Provisions concerning loading, unloading and carriage		Number of blue cones/ lights	Remarks
	3.1.2	2.2	2.2	2.1.1.3	5.2.2	3.3	3.4.6	3.5.1.2	3.2.1	8.1.5	7.1.6	7.1.6		7.1.5	3.2.1
(1)	(2)	(3a)	(3b)	(4)	(5)	(6)	(7a)	(7b)	(8)	(9)	(10)	(11)		(12)	(13)
0217	TRINITRONAPHTHALENE	1	1.1D		1		LQ0	E0		PP		LO01	HA01, HA02, HA03, HA04, HA05, HA06	3	
0218	TRINITROPHENETOLE	1	1.1D		1		LQ0	E0		PP		LO01	HA01, HA02, HA03, HA04, HA05, HA06	3	
0219	TRINITRORESORCINOL (STYPHNIC ACID), dry or wetted with less than 20% water, or mixture of alcohol and water, by mass	1	1.1D		1		LQ0	E0		PP		LO01	HA01, HA02, HA03, HA04, HA05, HA06	3	
0220	UREA NITRATE, dry or wetted with less than 20% water, by mass	1	1.1D		1		LQ0	E0		PP		LO01	HA01, HA02, HA03, HA04, HA05, HA06	3	
0221	WARHEADS, TORPEDO with bursting charge	1	1.1D		1		LQ0	E0		PP		LO01	HA01, HA02, HA03, HA04, HA05, HA06	3	
0222	AMMONIUM NITRATE with more than 0.2% combustible substances, including any organic substance calculated as carbon, to the exclusion of any other added substance	1	1.1D		1		LQ0	E0		PP		LO01	HA01, HA02, HA03, HA04, HA05, HA06	3	
0224	BARIUM AZIDE, dry or wetted with less than 50% water, by mass	1	1.1A		1+6.1	802	LQ0	E0		PP		LO01	HA01, HA02, HA03, HA04, HA05, HA06	3	

UN No. or ID No.	Name and description	Class	Classi-fication Code	Packing group	Labels	Special provi-sions	Limited and excepted quantities		Carriage permitted	Equipment required	Venti-lation	Provisions concerning loading, unloading and carriage		Number of blue cones/ lights	Remarks
	3.1.2	2.2	2.2	2.1.1.3	5.2.2	3.3	3.4.6	3.5.1.2	3.2.1	8.1.5	7.1.6	7.1.6		7.1.5	3.2.1
(1)	(2)	(3a)	(3b)	(4)	(5)	(6)	(7a)	(7b)	(8)	(9)	(10)	(11)		(12)	(13)
0225	BOOSTERS WITH DETONATOR	1	1.1B		1		LQ0	E0		PP		LO01	HA01, HA02, HA03, HA04, HA05, HA06	3	
0226	CYCLOTETRAMETHYLENE-TETRANITRAMINE (HMX; OCTOGEN), WETTED with not less than 15% water, by mass	1	1.1D		1	266	LQ0	E0		PP		LO01	HA01, HA02, HA03, HA04, HA05, HA06	3	
0234	SODIUM DINITRO-o-CRESOLATE, dry or wetted with less than 15% water, by mass	1	1.3C		1		LQ0	E0		PP		LO01	HA01, HA03, HA04, HA05, HA06	3	
0235	SODIUM PICRAMATE, dry or wetted with less than 20% water, by mass	1	1.3C		1		LQ0	E0		PP		LO01	HA01, HA03, HA04, HA05, HA06	3	
0236	ZIRCONIUM PICRAMATE, dry or wetted with less than 20% water, by mass	1	1.3C		1		LQ0	E0		PP		LO01	HA01, HA03, HA04, HA05, HA06	3	
0237	CHARGES, SHAPED, FLEXIBLE, LINEAR	1	1.4D		1.4		LQ0	E0		PP		LO01	HA01, HA03, HA04, HA05, HA06	1	
0238	ROCKETS, LINE-THROWING	1	1.2G		1		LQ0	E0		PP		LO01	HA01, HA03, HA04, HA05, HA06	3	
0240	ROCKETS, LINE-THROWING	1	1.3G		1		LQ0	E0		PP		LO01	HA01, HA03, HA04, HA05, HA06	3	

UN No. or ID No.	Name and description	Class	Classi-fication Code	Packing group	Labels	Special provi-sions	Limited and excepted quantities		Carriage permitted	Equipment required	Venti-lation	Provisions concerning loading, unloading and carriage		Number of blue cones/ lights	Remarks
	3.1.2	2.2	2.2	2.1.1.3	5.2.2	3.3	3.4.6	3.5.1.2	3.2.1	8.1.5	7.1.6	7.1.6		7.1.5	3.2.1
(1)	(2)	(3a)	(3b)	(4)	(5)	(6)	(7a)	(7b)	(8)	(9)	(10)	(11)		(12)	(13)
0241	EXPLOSIVE, BLASTING, TYPE E	1	1.1D		1	617	LQ0	E0		PP		LO01	HA01, HA02, HA03, HA04, HA05, HA06	3	
0242	CHARGES, PROPELLING, FOR CANNON	1	1.3C		1		LQ0	E0		PP		LO01	HA01, HA03, HA04, HA05, HA06	3	
0243	AMMUNITION, INCENDIARY, WHITE PHOSPHORUS with burster, expelling charge or propelling charge	1	1.2H		1		LQ0	E0		PP		LO01	HA01, HA03, HA04, HA05, HA06	3	
0244	AMMUNITION, INCENDIARY, WHITE PHOSPHORUS with burster, expelling charge or propelling charge	1	1.3H		1		LQ0	E0		PP		LO01	HA01, HA03, HA04, HA05, HA06	3	
0245	AMMUNITION, SMOKE, WHITE PHOSPHORUS with burster, expelling charge or propelling charge	1	1.2H		1		LQ0	E0		PP		LO01	HA01, HA03, HA04, HA05, HA06	3	
0246	AMMUNITION, SMOKE, WHITE PHOSPHORUS with burster, expelling charge or propelling charge	1	1.3H		1		LQ0	E0		PP		LO01	HA01, HA03, HA04, HA05, HA06	3	
0247	AMMUNITION, INCENDIARY, liquid or gel, with burster, expelling charge or propelling charge	1	1.3J		1		LQ0	E0		PP		LO01	HA01, HA03, HA04, HA05, HA06	3	
0248	CONTRIVANCES, WATER-ACTIVATED with burster, expelling charge or propelling charge	1	1.2L		1	274	LQ0	E0		PP		LO01	HA01, HA03, HA04, HA05, HA06	3	
0249	CONTRIVANCES, WATER-ACTIVATED with burster, expelling charge or propelling charge	1	1.3L		1	274	LQ0	E0		PP		LO01	HA01, HA03, HA04, HA05, HA06	3	

UN No. or ID No.	Name and description	Class	Classi-fication Code	Packing group	Labels	Special provi-sions	Limited and excepted quantities		Carriage permitted	Equipment required	Venti-lation	Provisions concerning loading, unloading and carriage		Number of blue cones/ lights	Remarks
	3.1.2	2.2	2.2	2.1.1.3	5.2.2	3.3	3.4.6	3.5.1.2	3.2.1	8.1.5	7.1.6	7.1.6		7.1.5	3.2.1
(1)	(2)	(3a)	(3b)	(4)	(5)	(6)	(7a)	(7b)	(8)	(9)	(10)	(11)		(12)	(13)
0250	ROCKET MOTORS WITH HYPERGOLIC LIQUIDS with or without expelling charge	1	1.3L		1		LQ0	E0		PP		LO01	HA01, HA03, HA04, HA05, HA06	3	
0254	AMMUNITION, ILLUMINATING with or without burster, expelling charge or propelling charge	1	1.3G		1		LQ0	E0		PP		LO01	HA01, HA03, HA04, HA05, HA06	3	
0255	DETONATORS, ELECTRIC for blasting	1	1.4B		1.4		LQ0	E0		PP		LO01	HA01, HA02, HA03, HA04, HA05, HA06	1	
0257	FUZES, DETONATING	1	1.4B		1.4		LQ0	E0		PP		LO01	HA01, HA02, HA03, HA04, HA05, HA06	1	
0266	OCTOLITE (OCTOL), dry or wetted with less than 15% water, by mass	1	1.1D		1		LQ0	E0		PP		LO01	HA01, HA02, HA03, HA04, HA05, HA06	3	
0267	DETONATORS, NON-ELECTRIC for blasting	1	1.4B		1.4		LQ0	E0		PP		LO01	HA01, HA02, HA03, HA04, HA05, HA06	1	
0268	BOOSTERS WITH DETONATOR	1	1.2B		1		LQ0	E0		PP		LO01	HA01, HA02, HA03, HA04, HA05, HA06	3	
0271	CHARGES, PROPELLING	1	1.1C		1		LQ0	E0		PP		LO01	HA01, HA02, HA03, HA04, HA05, HA06	3	

UN No. or ID No.	Name and description	Class	Classi-fication Code	Packing group	Labels	Special provi-sions	Limited and excepted quantities		Carriage permitted	Equipment required	Venti-lation	Provisions concerning loading, unloading and carriage		Number of blue cones/ lights	Remarks
	3.1.2	2.2	2.2	2.1.1.3	5.2.2	3.3	3.4.6	3.5.1.2	3.2.1	8.1.5	7.1.6	7.1.6		7.1.5	3.2.1
(1)	(2)	(3a)	(3b)	(4)	(5)	(6)	(7a)	(7b)	(8)	(9)	(10)	(11)		(12)	(13)
0272	CHARGES, PROPELLING	1	1.3C		1		LQ0	E0		PP		LO01	HA01, HA03, HA04, HA05, HA06	3	
0275	CARTRIDGES, POWER DEVICE	1	1.3C		1		LQ0	E0		PP		LO01	HA01, HA03, HA04, HA05, HA06	3	
0276	CARTRIDGES, POWER DEVICE	1	1.4C		1.4		LQ0	E0		PP		LO01	HA01, HA03, HA04, HA05, HA06	1	
0277	CARTRIDGES, OIL WELL	1	1.3C		1		LQ0	E0		PP		LO01	HA01, HA03, HA04, HA05, HA06	3	
0278	CARTRIDGES, OIL WELL	1	1.4C		1.4		LQ0	E0		PP		LO01	HA01, HA03, HA04, HA05, HA06	1	
0279	CHARGES, PROPELLING, FOR CANNON	1	1.1C		1		LQ0	E0		PP		LO01	HA01, HA02, HA03, HA04, HA05, HA06	3	
0280	ROCKET MOTORS	1	1.1C		1		LQ0	E0		PP		LO01	HA01, HA02, HA03, HA04, HA05, HA06	3	
0281	ROCKET MOTORS	1	1.2C		1		LQ0	E0		PP		LO01	HA01, HA03, HA04, HA05, HA06	3	

UN No. or ID No.	Name and description	Class	Classi-fication Code	Packing group	Labels	Special provi-sions	Limited and excepted quantities		Carriage permitted	Equipment required	Venti-lation	Provisions concerning loading, unloading and carriage		Number of blue cones/ lights	Remarks
	3.1.2	2.2	2.2	2.1.1.3	5.2.2	3.3	3.4.6	3.5.1.2	3.2.1	8.1.5	7.1.6	7.1.6		7.1.5	3.2.1
(1)	(2)	(3a)	(3b)	(4)	(5)	(6)	(7a)	(7b)	(8)	(9)	(10)	(11)		(12)	(13)
0282	NITROGUANIDINE (PICRITE), dry or wetted with less than 20% water, by mass	1	1.1D		1		LQ0	E0		PP		LO01	HA01, HA02, HA03, HA04, HA05, HA06	3	
0283	BOOSTERS without detonator	1	1.2D		1		LQ0	E0		PP		LO01	HA01, HA03, HA04, HA05, HA06	3	
0284	GRENADES, hand or rifle, with bursting charge	1	1.1D		1		LQ0	E0		PP		LO01	HA01, HA02, HA03, HA04, HA05, HA06	3	
0285	GRENADES, hand or rifle, with bursting charge	1	1.2D		1		LQ0	E0		PP		LO01	HA01, HA03, HA04, HA05, HA06	3	
0286	WARHEADS, ROCKET with bursting charge	1	1.1D		1		LQ0	E0		PP		LO01	HA01, HA02, HA03, HA04, HA05, HA06	3	
0287	WARHEADS, ROCKET with bursting charge	1	1.2D		1		LQ0	E0		PP		LO01	HA01, HA03, HA04, HA05, HA06	3	
0288	CHARGES, SHAPED, FLEXIBLE, LINEAR	1	1.1D		1		LQ0	E0		PP		LO01	HA01, HA02, HA03, HA04, HA05, HA06	3	
0289	CORD, DETONATING, flexible	1	1.4D		1.4		LQ0	E0		PP		LO01	HA01, HA03, HA04, HA05, HA06	1	

UN No. or ID No.	Name and description	Class	Classi-fication Code	Packing group	Labels	Special provi-sions	Limited and excepted quantities		Carriage permitted	Equipment required	Venti-lation	Provisions concerning loading, unloading and carriage		Number of blue cones/ lights	Remarks
	3.1.2	2.2	2.2	2.1.1.3	5.2.2	3.3	3.4.6	3.5.1.2	3.2.1	8.1.5	7.1.6	7.1.6		7.1.5	3.2.1
(1)	(2)	(3a)	(3b)	(4)	(5)	(6)	(7a)	(7b)	(8)	(9)	(10)	(11)		(12)	(13)
0290	CORD (FUSE), DETONATING, metal clad	1	1.1D		1		LQ0	E0		PP		LO01	HA01, HA02, HA03, HA04, HA05, HA06	3	
0291	BOMBS with bursting charge	1	1.2F		1		LQ0	E0		PP		LO01	HA01, HA02, HA03, HA04, HA05, HA06	3	
0292	GRENADES, hand or rifle, with bursting charge	1	1.1F		1		LQ0	E0		PP		LO01	HA01, HA02, HA03, HA04, HA05, HA06	3	
0293	GRENADES, hand or rifle, with bursting charge	1	1.2F		1		LQ0	E0		PP		LO01	HA01, HA02, HA03, HA04, HA05, HA06	3	
0294	MINES with bursting charge	1	1.2F		1		LQ0	E0		PP		LO01	HA01, HA02, HA03, HA04, HA05, HA06	3	
0295	ROCKETS with bursting charge	1	1.2F		1		LQ0	E0		PP		LO01	HA01, HA02, HA03, HA04, HA05, HA06	3	
0296	SOUNDING DEVICES, EXPLOSIVE	1	1.1F		1		LQ0	E0		PP		LO01	HA01, HA02, HA03, HA04, HA05, HA06	3	

UN No. or ID No.	Name and description	Class	Classi-fication Code	Packing group	Labels	Special provi-sions	Limited and excepted quantities		Carriage permitted	Equipment required	Venti-lation	Provisions concerning loading, unloading and carriage		Number of blue cones/ lights	Remarks
	3.1.2	2.2	2.2	2.1.1.3	5.2.2	3.3	3.4.6	3.5.1.2	3.2.1	8.1.5	7.1.6	7.1.6		7.1.5	3.2.1
(1)	(2)	(3a)	(3b)	(4)	(5)	(6)	(7a)	(7b)	(8)	(9)	(10)	(11)		(12)	(13)
0297	AMMUNITION, ILLUMINATING with or without burster, expelling charge or propelling charge	1	1.4G		1.4		LQ0	E0		PP		LO01	HA01, HA03, HA04, HA05, HA06	1	
0299	BOMBS, PHOTO-FLASH	1	1.3G		1		LQ0	E0		PP		LO01	HA01, HA03, HA04, HA05, HA06	3	
0300	AMMUNITION, INCENDIARY with or without burster, expelling charge or propelling charge	1	1.4G		1.4		LQ0	E0		PP		LO01	HA01, HA03, HA04, HA05, HA06	1	
0301	AMMUNITION, TEAR-PRODUCING with burster, expelling charge or propelling charge	1	1.4G		1.4+6.1+8	802	LQ0	E0		PP		LO01	HA01, HA03, HA04, HA05, HA06	1	
0303	AMMUNITION, SMOKE with or without burster, expelling charge or propelling charge	1	1.4G		1.4		LQ0	E0		PP		LO01	HA01, HA03, HA04, HA05, HA06	1	
0303	AMMUNITION, SMOKE with or without burster, expelling charge or propelling charge, containing corrosive substances	1	1.4G		1.4 +8		LQ0	E0		PP		LO01	HA01, HA03, HA04, HA05, HA06	1	
0305	FLASH POWDER	1	1.3G		1		LQ0	E0		PP		LO01	HA01, HA03, HA04, HA05, HA06	3	
0306	TRACERS FOR AMMUNITION	1	1.4G		1.4		LQ0	E0		PP		LO01	HA01, HA03, HA04, HA05, HA06	1	
0312	CARTRIDGES, SIGNAL	1	1.4G		1.4		LQ0	E0		PP		LO01	HA01, HA03, HA04, HA05, HA06	1	

UN No. or ID No.	Name and description	Class	Classi-fication Code	Packing group	Labels	Special provi-sions	Limited and excepted quantities		Carriage permitted	Equipment required	Venti-lation	Provisions concerning loading, unloading and carriage		Number of blue cones/ lights	Remarks
	3.1.2	2.2	2.2	2.1.1.3	5.2.2	3.3	3.4.6	3.5.1.2	3.2.1	8.1.5	7.1.6	7.1.6		7.1.5	3.2.1
(1)	(2)	(3a)	(3b)	(4)	(5)	(6)	(7a)	(7b)	(8)	(9)	(10)	(11)		(12)	(13)
0313	SIGNALS, SMOKE	1	1.2G		1		LQ0	E0		PP		LO01	HA01, HA03, HA04, HA05, HA06	3	
0314	IGNITERS	1	1.2G		1		LQ0	E0		PP		LO01	HA01, HA03, HA04, HA05, HA06	3	
0315	IGNITERS	1	1.3G		1		LQ0	E0		PP		LO01	HA01, HA03, HA04, HA05, HA06	3	
0316	FUZES, IGNITING	1	1.3G		1		LQ0	E0		PP		LO01	HA01, HA03, HA04, HA05, HA06	3	
0317	FUZES, IGNITING	1	1.4G		1.4		LQ0	E0		PP		LO01	HA01, HA03, HA04, HA05, HA06	1	
0318	GRENADES, PRACTICE, hand or rifle	1	1.3G		1		LQ0	E0		PP		LO01	HA01, HA03, HA04, HA05, HA06	3	
0319	PRIMERS, TUBULAR	1	1.3G		1		LQ0	E0		PP		LO01	HA01, HA03, HA04, HA05, HA06	3	
0320	PRIMERS, TUBULAR	1	1.4G		1.4		LQ0	E0		PP		LO01	HA01, HA03, HA04, HA05, HA06	1	
0321	CARTRIDGES FOR WEAPONS with bursting charge	1	1.2E		1		LQ0	E0		PP		LO01	HA01, HA03, HA04, HA05, HA06	3	

UN No. or ID No.	Name and description	Class	Classi-fication Code	Packing group	Labels	Special provi-sions	Limited and excepted quantities		Carriage permitted	Equipment required	Venti-lation	Provisions concerning loading, unloading and carriage		Number of blue cones/ lights	Remarks
	3.1.2	2.2	2.2	2.1.1.3	5.2.2	3.3	3.4.6	3.5.1.2	3.2.1	8.1.5	7.1.6	7.1.6		7.1.5	3.2.1
(1)	(2)	(3a)	(3b)	(4)	(5)	(6)	(7a)	(7b)	(8)	(9)	(10)	(11)		(12)	(13)
0322	ROCKET MOTORS WITH HYPERGOLIC LIQUIDS with or without expelling charge	1	1.2L		1		LQ0	E0		PP		LO01	HA01, HA03, HA04, HA05, HA06	3	
0323	CARTRIDGES, POWER DEVICE	1	1.4S		1.4		LQ0	E0		PP		LO01	HA01, HA03, HA04, HA05, HA06	0	
0324	PROJECTILES with bursting charge	1	1.2F		1		LQ0	E0		PP		LO01	HA01, HA02, HA03, HA04, HA05, HA06	3	
0325	IGNITERS	1	1.4G		1.4		LQ0	E0		PP		LO01	HA01, HA03, HA04, HA05, HA06	1	
0326	CARTRIDGES FOR WEAPONS, BLANK	1	1.1C		1		LQ0	E0		PP		LO01	HA01, HA02, HA03, HA04, HA05, HA06	3	
0327	CARTRIDGES FOR WEAPONS, BLANK or CARTRIDGES, SMALL ARMS, BLANK	1	1.3C		1		LQ0	E0		PP		LO01	HA01, HA03, HA04, HA05, HA06	3	
0328	CARTRIDGES FOR WEAPONS, INERT PROJECTILE	1	1.2C		1		LQ0	E0		PP		LO01	HA01, HA03, HA04, HA05, HA06	3	
0329	TORPEDOES with bursting charge	1	1.1E		1		LQ0	E0		PP		LO01	HA01, HA02, HA03, HA04, HA05, HA06	3	

UN No. or ID No.	Name and description	Class	Classi-fication Code	Packing group	Labels	Special provi-sions	Limited and excepted quantities		Carriage permitted	Equipment required	Venti-lation	Provisions concerning loading, unloading and carriage		Number of blue cones/ lights	Remarks
	3.1.2	2.2	2.2	2.1.1.3	5.2.2	3.3	3.4.6	3.5.1.2	3.2.1	8.1.5	7.1.6	7.1.6		7.1.5	3.2.1
(1)	(2)	(3a)	(3b)	(4)	(5)	(6)	(7a)	(7b)	(8)	(9)	(10)	(11)		(12)	(13)
0330	TORPEDOES with bursting charge	1	1.1F		1		LQ0	E0		PP		LO01	HA01, HA02, HA03, HA04, HA05, HA06	3	
0331	EXPLOSIVE, BLASTING, TYPE B (AGENT, BLASTING, TYPE B)	1	1.5D		1.5	617	LQ0	E0		PP		LO01	HA01, HA03, HA04, HA05, HA06	3	
0332	EXPLOSIVE, BLASTING, TYPE E (AGENT, BLASTING, TYPE B)	1	1.5D		1.5	617	LQ0	E0		PP		LO01	HA01, HA03, HA04, HA05, HA06	3	
0333	FIREWORKS	1	1.1G		1	645	LQ0	E0		PP		LO01	HA01, HA02, HA03, HA04, HA05, HA06	3	
0334	FIREWORKS	1	1.2G		1	645	LQ0	E0		PP		LO01	HA01, HA03, HA04, HA05, HA06	3	
0335	FIREWORKS	1	1.3G		1	645	LQ0	E0		PP		LO01	HA01, HA03, HA04, HA05, HA06	3	
0336	FIREWORKS	1	1.4G		1.4	645 651	LQ0	E0		PP		LO01	HA01, HA03, HA04, HA05, HA06	1	
0337	FIREWORKS	1	1.4S		1.4	645	LQ0	E0		PP		LO01	HA01, HA03, HA04, HA05, HA06	0	

UN No. or ID No.	Name and description	Class	Classi-fication Code	Packing group	Labels	Special provi-sions	Limited and excepted quantities		Carriage permitted	Equipment required	Venti-lation	Provisions concerning loading, unloading and carriage		Number of blue cones/ lights	Remarks
	3.1.2	2.2	2.2	2.1.1.3	5.2.2	3.3	3.4.6	3.5.1.2	3.2.1	8.1.5	7.1.6	7.1.6		7.1.5	3.2.1
(1)	(2)	(3a)	(3b)	(4)	(5)	(6)	(7a)	(7b)	(8)	(9)	(10)	(11)		(12)	(13)
0338	CARTRIDGES FOR WEAPONS, BLANK or CARTRIDGES, SMALL ARMS, BLANK	1	1.4C		1.4		LQ0	E0		PP		LO01	HA01, HA03, HA04, HA05, HA06	1	
0339	CARTRIDGES FOR WEAPONS, INERT PROJECTILE or CARTRIDGES, SMALL ARMS	1	1.4C		1.4		LQ0	E0		PP		LO01	HA01, HA03, HA04, HA05, HA06	1	
0340	NITROCELLULOSE, dry or wetted with less than 25% water (or alcohol), by mass	1	1.1D		1		LQ0	E0		PP		LO01	HA01, HA02, HA03, HA04, HA05, HA06	3	
0341	NITROCELLULOSE, unmodified or plasticized with less than 18% plasticizing substance, by mass	1	1.1D		1		LQ0	E0		PP		LO01	HA01, HA02, HA03, HA04, HA05, HA06	3	
0342	NITROCELLULOSE, WETTED with not less than 25% alcohol, by mass	1	1.3C		1	105	LQ0	E0		PP		LO01	HA01, HA03, HA04, HA05, HA06	3	
0343	NITROCELLULOSE, PLASTICIZED with not less than 18% plasticizing substance, by mass	1	1.3C		1	105	LQ0	E0		PP		LO01	HA01, HA03, HA04, HA05, HA06	3	
0344	PROJECTILES with bursting charge	1	1.4D		1.4		LQ0	E0		PP		LO01	HA01, HA03, HA04, HA05, HA06	1	
0345	PROJECTILES, inert with tracer	1	1.4S		1.4		LQ0	E0		PP		LO01	HA01, HA03, HA04, HA05, HA06	0	

UN No. or ID No.	Name and description 3.1.2	Class 2.2	Classi-fication Code 2.2	Packing group 2.1.1.3	Labels 5.2.2	Special provi-sions 3.3	Limited and excepted quantities 3.4.6	 3.5.1.2	Carriage permitted 3.2.1	Equipment required 8.1.5	Venti-lation 7.1.6	Provisions concerning loading, unloading and carriage 7.1.6		Number of blue cones/ lights 7.1.5	Remarks 3.2.1
(1)	(2)	(3a)	(3b)	(4)	(5)	(6)	(7a)	(7b)	(8)	(9)	(10)	(11)		(12)	(13)
0346	PROJECTILES with burster or expelling charge	1	1.2D		1		LQ0	E0		PP		LO01	HA01, HA03, HA04, HA05, HA06	3	
0347	PROJECTILES with burster or expelling charge	1	1.4D		1.4		LQ0	E0		PP		LO01	HA01, HA03, HA04, HA05, HA06	1	
0348	CARTRIDGES FOR WEAPONS with bursting charge	1	1.4F		1.4		LQ0	E0		PP		LO01	HA01, HA02, HA03, HA04, HA05, HA06	1	
0349	ARTICLES, EXPLOSIVE, N.O.S.	1	1.4S		1.4	178 274	LQ0	E0		PP		LO01	HA01, HA03, HA04, HA05, HA06	0	
0350	ARTICLES, EXPLOSIVE, N.O.S.	1	1.4B		1.4	178 274	LQ0	E0		PP		LO01	HA01, HA02, HA03, HA04, HA05, HA06	1	
0351	ARTICLES, EXPLOSIVE, N.O.S.	1	1.4C		1.4	178 274	LQ0	E0		PP		LO01	HA01, HA03, HA04, HA05, HA06	1	
0352	ARTICLES, EXPLOSIVE, N.O.S.	1	1.4D		1.4	178 274	LQ0	E0		PP		LO01	HA01, HA03, HA04, HA05, HA06	1	
0353	ARTICLES, EXPLOSIVE, N.O.S.	1	1.4G		1.4	178 274	LQ0	E0		PP		LO01	HA01, HA03, HA04, HA05, HA06	1	

UN No. or ID No.	Name and description	Class	Classi-fication Code	Packing group	Labels	Special provi-sions	Limited and excepted quantities		Carriage permitted	Equipment required	Venti-lation	Provisions concerning loading, unloading and carriage		Number of blue cones/ lights	Remarks
	3.1.2	2.2	2.2	2.1.1.3	5.2.2	3.3	3.4.6	3.5.1.2	3.2.1	8.1.5	7.1.6	7.1.6		7.1.5	3.2.1
(1)	(2)	(3a)	(3b)	(4)	(5)	(6)	(7a)	(7b)	(8)	(9)	(10)	(11)		(12)	(13)
0354	ARTICLES, EXPLOSIVE, N.O.S.	1	1.1L		1	178 274	LQ0	E0		PP		LO01	HA01, HA02, HA03, HA04, HA05, HA06	3	
0355	ARTICLES, EXPLOSIVE, N.O.S.	1	1.2L		1	178 274	LQ0	E0		PP		LO01	HA01, HA03, HA04, HA05, HA06	3	
0356	ARTICLES, EXPLOSIVE, N.O.S.	1	1.3L		1	178 274	LQ0	E0		PP		LO01	HA01, HA03, HA04, HA05, HA06	3	
0357	SUBSTANCES, EXPLOSIVE, N.O.S.	1	1.1L		1	178 274	LQ0	E0		PP		LO01	HA01, HA02, HA03, HA04, HA05, HA06	3	
0358	SUBSTANCES, EXPLOSIVE, N.O.S.	1	1.2L		1	178 274	LQ0	E0		PP		LO01	HA01, HA03, HA04, HA05, HA06	3	
0359	SUBSTANCES, EXPLOSIVE, N.O.S.	1	1.3L		1	178 274	LQ0	E0		PP		LO01	HA01, HA03, HA04, HA05, HA06	3	
0360	DETONATOR ASSEMBLIES, NON-ELECTRIC for blasting	1	1.1B		1		LQ0	E0		PP		LO01	HA01, HA02, HA03, HA04, HA05, HA06	3	
0361	DETONATOR ASSEMBLIES, NON-ELECTRIC for blasting	1	1.4B		1.4		LQ0	E0		PP		LO01	HA01, HA02, HA03, HA04, HA05, HA06	1	

UN No. or ID No.	Name and description	Class	Classi-fication Code	Packing group	Labels	Special provi-sions	Limited and excepted quantities		Carriage permitted	Equipment required	Venti-lation	Provisions concerning loading, unloading and carriage			Number of blue cones/ lights	Remarks
	3.1.2	2.2	2.2	2.1.1.3	5.2.2	3.3	3.4.6	3.5.1.2	3.2.1	8.1.5	7.1.6	7.1.6			7.1.5	3.2.1
(1)	(2)	(3a)	(3b)	(4)	(5)	(6)	(7a)	(7b)	(8)	(9)	(10)	(11)			(12)	(13)
0362	AMMUNITION, PRACTICE	1	1.4G		1.4		LQ0	E0		PP		LO01	HA01, HA03, HA04, HA05, HA06		1	
0363	AMMUNITION, PROOF	1	1.4G		1.4		LQ0	E0		PP		LO01	HA01, HA03, HA04, HA05, HA06		1	
0364	DETONATORS FOR AMMUNITION	1	1.2B		1		LQ0	E0		PP		LO01	HA01, HA02, HA03, HA04, HA05, HA06		3	
0365	DETONATORS FOR AMMUNITION	1	1.4B		1.4		LQ0	E0		PP		LO01	HA01, HA02, HA03, HA04, HA05, HA06		1	
0366	DETONATORS FOR AMMUNITION	1	1.4S		1.4		LQ0	E0		PP		LO01	HA01, HA03, HA04, HA05, HA06		0	
0367	FUZES, DETONATING	1	1.4S		1.4		LQ0	E0		PP		LO01	HA01, HA03, HA04, HA05, HA06		0	
0368	FUZES, IGNITING	1	1.4S		1.4		LQ0	E0		PP		LO01	HA01, HA03, HA04, HA05, HA06		0	
0369	WARHEADS, ROCKET with bursting charge	1	1.1F		1		LQ0	E0		PP		LO01	HA01, HA02, HA03, HA04, HA05, HA06		3	

UN No. or ID No.	Name and description	Class	Classi-fication Code	Packing group	Labels	Special provi-sions	Limited and excepted quantities		Carriage permitted	Equipment required	Venti-lation	Provisions concerning loading, unloading and carriage		Number of blue cones/ lights	Remarks
	3.1.2	2.2	2.2	2.1.1.3	5.2.2	3.3	3.4.6	3.5.1.2	3.2.1	8.1.5	7.1.6	7.1.6		7.1.5	3.2.1
(1)	(2)	(3a)	(3b)	(4)	(5)	(6)	(7a)	(7b)	(8)	(9)	(10)	(11)		(12)	(13)
0370	WARHEADS, ROCKET with burster or expelling charge	1	1.4D		1.4		LQ0	E0		PP		LO01	HA01, HA03, HA04, HA05, HA06	1	
0371	WARHEADS, ROCKET with burster or expelling charge	1	1.4F		1.4		LQ0	E0		PP		LO01	HA01, HA02, HA03, HA04, HA05, HA06	1	
0372	GRENADES, PRACTICE, hand or rifle	1	1.2G		1		LQ0	E0		PP		LO01	HA01, HA03, HA04, HA05, HA06	3	
0373	SIGNAL DEVICES, HAND	1	1.4S		1.4		LQ0	E0		PP		LO01	HA01, HA03, HA04, HA05, HA06	0	
0374	SOUNDING DEVICES, EXPLOSIVE	1	1.1D		1		LQ0	E0		PP		LO01	HA01, HA02, HA03, HA04, HA05, HA06	3	
0375	SOUNDING DEVICES, EXPLOSIVE	1	1.2D		1		LQ0	E0		PP		LO01	HA01, HA03, HA04, HA05, HA06	3	
0376	PRIMERS, TUBULAR	1	1.4S		1.4		LQ0	E0		PP		LO01	HA01, HA03, HA04, HA05, HA06	0	
0377	PRIMERS, CAP TYPE	1	1.1B		1		LQ0	E0		PP		LO01	HA01, HA02, HA03, HA04, HA05, HA06	3	

UN No. or ID No.	Name and description	Class	Classi-fication Code	Packing group	Labels	Special provi-sions	Limited and excepted quantities		Carriage permitted	Equipment required	Venti-lation	Provisions concerning loading, unloading and carriage		Number of blue cones/ lights	Remarks
	3.1.2	2.2	2.2	2.1.1.3	5.2.2	3.3	3.4.6	3.5.1.2	3.2.1	8.1.5	7.1.6	7.1.6		7.1.5	3.2.1
(1)	(2)	(3a)	(3b)	(4)	(5)	(6)	(7a)	(7b)	(8)	(9)	(10)	(11)		(12)	(13)
0378	PRIMERS, CAP TYPE	1	1.4B		1.4		LQ0	E0		PP		LO01	HA01, HA02, HA03, HA04, HA05, HA06	1	
0379	CASES, CARTRIDGE, EMPTY, WITH PRIMER	1	1.4C		1.4		LQ0	E0		PP		LO01	HA01, HA03, HA04, HA05, HA06	1	
0380	ARTICLES, PYROPHORIC	1	1.2L		1		LQ0	E0		PP		LO01	HA01, HA03, HA04, HA05, HA06	3	
0381	CARTRIDGES, POWER DEVICE	1	1.2C		1		LQ0	E0		PP		LO01	HA01, HA03, HA04, HA05, HA06	3	
3820	COMPONENTS, EXPLOSIVE TRAIN, N.O.S.	1	1.2B		1	178 274	LQ0	E0		PP		LO01	HA01, HA02, HA03, HA04, HA05, HA06	3	
0383	COMPONENTS, EXPLOSIVE TRAIN, N.O.S.	1	1.4B		1.4	178 274	LQ0	E0		PP		LO01	HA01, HA02, HA03, HA04, HA05, HA06	1	
0384	COMPONENTS, EXPLOSIVE TRAIN, N.O.S.	1	1.4S		1.4	178 274	LQ0	E0		PP		LO01	HA01, HA03, HA04, HA05, HA06	0	
0385	5-NITROBENZOTRIAZOL	1	1.1D		1		LQ0	E0		PP		LO01	HA01, HA02, HA03, HA04, HA05, HA06	3	

UN No. or ID No.	Name and description	Class	Classi-fication Code	Packing group	Labels	Special provi-sions	Limited and excepted quantities		Carriage permitted	Equipment required	Venti-lation	Provisions concerning loading, unloading and carriage		Number of blue cones/ lights	Remarks
	3.1.2	2.2	2.2	2.1.1.3	5.2.2	3.3	3.4.6	3.5.1.2	3.2.1	8.1.5	7.1.6	7.1.6		7.1.5	3.2.1
(1)	(2)	(3a)	(3b)	(4)	(5)	(6)	(7a)	(7b)	(8)	(9)	(10)	(11)		(12)	(13)
0386	TRINITROBENZENESULPHONIC ACID	1	1.1D		1		LQ0	E0		PP		LO01	HA01, HA02, HA03, HA04, HA05, HA06	3	
0387	TRINITROFLUORENONE	1	1.1D		1		LQ0	E0		PP		LO01	HA01, HA02, HA03, HA04, HA05, HA06	3	
0388	TRINITROTOLUENE (TNT) AND TRINITROBENZENE MIXTURE or TRINITROTOLUENE (TNT) AND HEXANITROSTILBENE MIXTURE	1	1.1D		1		LQ0	E0		PP		LO01	HA01, HA02, HA03, HA04, HA05, HA06	3	
0389	TRINITROTOLUENE (TNT) MIXTURE CONTAINING TRINITROBENZENE AND HEXANITROSTILBENE	1	1.1D		1		LQ0	E0		PP		LO01	HA01, HA02, HA03, HA04, HA05, HA06	3	
0390	TRITONAL	1	1.1D		1		LQ0	E0		PP		LO01	HA01, HA02, HA03, HA04, HA05, HA06	3	
0391	CYCLOTRIMETHYLENETRINITRAMINE (CYCLONITE; HEXOGEN; RDX) AND CYCLOTETRAMETHYLENE-TETRANITRAMINE (HMX; OCTOGEN) MIXTURE, WETTED with not less than 15% water, by mass or DESENSITIZED with not less than 10% phlegmatiser by mass	1	1.1D		1	266	LQ0	E0		PP		LO01	HA01, HA02, HA03, HA04, HA05, HA06	3	
0392	HEXANITROSTILBENE	1	1.1D		1		LQ0	E0		PP		LO01	HA01, HA02, HA03, HA04, HA05, HA06	3	

UN No. or ID No.	Name and description 3.1.2	Class 2.2	Classi-fication Code 2.2	Packing group 2.1.1.3	Labels 5.2.2	Special provi-sions 3.3	Limited and excepted quantities 3.4.6	3.5.1.2	Carriage permitted 3.2.1	Equipment required 8.1.5	Venti-lation 7.1.6	Provisions concerning loading, unloading and carriage 7.1.6		Number of blue cones/ lights 7.1.5	Remarks 3.2.1
(1)	(2)	(3a)	(3b)	(4)	(5)	(6)	(7a)	(7b)	(8)	(9)	(10)	(11)		(12)	(13)
0393	HEXOTONAL	1	1.1D		1		LQ0	E0		PP		LO01	HA01, HA02, HA03, HA04, HA05, HA06	3	
0394	TRINITRORESORCINOL (STYPHNIC ACID), WETTED with not less than 20% water, or mixture of alcohol and water, by mass	1	1.1D		1		LQ0	E0		PP		LO01	HA01, HA02, HA03, HA04, HA05, HA06	3	
0395	ROCKET MOTORS, LIQUID FUELLED	1	1.2J		1		LQ0	E0		PP		LO01	HA01, HA03, HA04, HA05, HA06	3	
0396	ROCKET MOTORS, LIQUID FUELLED	1	1.3J		1		LQ0	E0		PP		LO01	HA01, HA03, HA04, HA05, HA06	3	
0397	ROCKETS, LIQUID FUELLED with bursting charge	1	1.1J		1		LQ0	E0		PP		LO01	HA01, HA02, HA03, HA04, HA05, HA06	3	
0398	ROCKETS, LIQUID FUELLED with bursting charge	1	1.2J		1		LQ0	E0		PP		LO01	HA01, HA03, HA04, HA05, HA06	3	
0399	BOMBS WITH FLAMMABLE LIQUID with bursting charge	1	1.1J		1		LQ0	E0		PP		LO01	HA01, HA02, HA03, HA04, HA05, HA06	3	
0400	BOMBS WITH FLAMMABLE LIQUID with bursting charge	1	1.2J		1		LQ0	E0		PP		LO01	HA01, HA03, HA04, HA05, HA06	3	

UN No. or ID No.	Name and description	Class	Classi-fication Code	Packing group	Labels	Special provi-sions	Limited and excepted quantities		Carriage permitted	Equipment required	Venti-lation	Provisions concerning loading, unloading and carriage		Number of blue cones/ lights	Remarks
	3.1.2	2.2	2.2	2.1.1.3	5.2.2	3.3	3.4.6	3.5.1.2	3.2.1	8.1.5	7.1.6	7.1.6		7.1.5	3.2.1
(1)	(2)	(3a)	(3b)	(4)	(5)	(6)	(7a)	(7b)	(8)	(9)	(10)	(11)		(12)	(13)
0401	DIPICRYL SULPHIDE, dry or wetted with less than 10% water, by mass	1	1.1D		1		LQ0	E0		PP		LO01	HA01, HA02, HA03, HA04, HA05, HA06	3	
0402	AMMONIUM PERCHLORATE	1	1.1D		1	152	LQ0	E0		PP		LO01	HA01, HA02, HA03, HA04, HA05, HA06	3	
0403	FLARES, AERIAL	1	1.4G		1.4		LQ0	E0		PP		LO01	HA01, HA03, HA04, HA05, HA06	1	
0404	FLARES, AERIAL	1	1.4S		1.4		LQ0	E0		PP		LO01	HA01, HA03, HA04, HA05, HA06	0	
0405	CARTRIDGES, SIGNAL	1	1.4S		1.4		LQ0	E0		PP		LO01	HA01, HA03, HA04, HA05, HA06	0	
0406	DINITROSOBENZENE	1	1.3C		1		LQ0	E0		PP		LO01	HA01, HA03, HA04, HA05, HA06	3	
0407	TETRAZOL-1-ACETIC ACID	1	1.4C		1.4		LQ0	E0		PP		LO01	HA01, HA03, HA04, HA05, HA06	1	
0408	FUZES, DETONATING with protective features	1	1.1D		1		LQ0	E0		PP		LO01	HA01, HA02, HA03, HA04, HA05, HA06	3	

UN No. or ID No.	Name and description	Class	Classi-fication Code	Packing group	Labels	Special provi-sions	Limited and excepted quantities		Carriage permitted	Equipment required	Venti-lation	Provisions concerning loading, unloading and carriage		Number of blue cones/ lights	Remarks
	3.1.2	2.2	2.2	2.1.1.3	5.2.2	3.3	3.4.6	3.5.1.2	3.2.1	8.1.5	7.1.6	7.1.6		7.1.5	3.2.1
(1)	(2)	(3a)	(3b)	(4)	(5)	(6)	(7a)	(7b)	(8)	(9)	(10)	(11)		(12)	(13)
0409	FUZES, DETONATING with protective features	1	1.2D		1		LQ0	E0		PP		LO01	HA01, HA03, HA04, HA05, HA06	3	
0410	FUZES, DETONATING with protective features	1	1.4D		1.4		LQ0	E0		PP		LO01	HA01, HA03, HA04, HA05, HA06	1	
0411	PENTAERYTHRITE TETRANITRATE (PENTAERYTHRITOL TETRANITRATE; PETN) with not less than 7% wax, by mass	1	1.1D		1	131	LQ0	E0		PP		LO01	HA01, HA02, HA03, HA04, HA05, HA06	3	
0412	CARTRIDGES FOR WEAPONS with bursting charge	1	1.4E		1.4		LQ0	E0		PP		LO01	HA01, HA03, HA04, HA05, HA06	1	
0413	CARTRIDGES FOR WEAPONS, BLANK	1	1.2C		1		LQ0	E0		PP		LO01	HA01, HA03, HA04, HA05, HA06	3	
0414	CHARGES, PROPELLING, FOR CANNON	1	1.2C		1		LQ0	E0		PP		LO01	HA01, HA03, HA04, HA05, HA06	3	
0415	CHARGES, PROPELLING	1	1.2C		1		LQ0	E0		PP		LO01	HA01, HA03, HA04, HA05, HA06	3	
0417	CARTRIDGES FOR WEAPONS, INERT PROJECTILE or CARTRIDGES, SMALL ARMS	1	1.3C		1		LQ0	E0		PP		LO01	HA01, HA03, HA04, HA05, HA06	3	

UN No. or ID No.	Name and description	Class	Classi-fication Code	Packing group	Labels	Special provi-sions	Limited and excepted quantities		Carriage permitted	Equipment required	Venti-lation	Provisions concerning loading, unloading and carriage		Number of blue cones/ lights	Remarks
	3.1.2	2.2	2.2	2.1.1.3	5.2.2	3.3	3.4.6	3.5.1.2	3.2.1	8.1.5	7.1.6	7.1.6		7.1.5	3.2.1
(1)	(2)	(3a)	(3b)	(4)	(5)	(6)	(7a)	(7b)	(8)	(9)	(10)	(11)		(12)	(13)
0418	FLARES, SURFACE	1	1.1G		1		LQ0	E0		PP		LO01	HA01, HA02, HA03, HA04, HA05, HA06	3	
0419	FLARES, SURFACE	1	1.2G		1		LQ0	E0		PP		LO01	HA01, HA03, HA04, HA05, HA06	3	
0420	FLARES, AERIAL	1	1.1G		1		LQ0	E0		PP		LO01	HA01, HA02, HA03, HA04, HA05, HA06	3	
0421	FLARES, AERIAL	1	1.2G		1		LQ0	E0		PP		LO01	HA01, HA03, HA04, HA05, HA06	3	
0424	PROJECTILES, inert with tracer	1	1.3G		1		LQ0	E0		PP		LO01	HA01, HA03, HA04, HA05, HA06	3	
0425	PROJECTILES, inert with tracer	1	1.4G		1.4		LQ0	E0		PP		LO01	HA01, HA03, HA04, HA05, HA06	1	
0426	PROJECTILES with burster or expelling charge	1	1.2F		1		LQ0	E0		PP		LO01	HA01, HA02, HA03, HA04, HA05, HA06	3	
0427	PROJECTILES with burster or expelling charge	1	1.4F		1.4		LQ0	E0		PP		LO01	HA01, HA02, HA03, HA04, HA05, HA06	1	

UN No. or ID No.	Name and description	Class	Classi-fication Code	Packing group	Labels	Special provi-sions	Limited and excepted quantities		Carriage permitted	Equipment required	Venti-lation	Provisions concerning loading, unloading and carriage	Number of blue cones/ lights	Remarks
	3.1.2	2.2	2.2	2.1.1.3	5.2.2	3.3	3.4.6	3.5.1.2	3.2.1	8.1.5	7.1.6	7.1.6	7.1.5	3.2.1
(1)	(2)	(3a)	(3b)	(4)	(5)	(6)	(7a)	(7b)	(8)	(9)	(10)	(11)	(12)	(13)
0428	ARTICLES, PYROTECHNIC for technical purposes	1	1.1G		1		LQ0	E0		PP		LO01 HA01, HA02, HA03, HA04, HA05, HA06	3	
0429	ARTICLES, PYROTECHNIC for technical purposes	1	1.2G		1		LQ0	E0		PP		LO01 HA01, HA03, HA04, HA05, HA06	3	
0430	ARTICLES, PYROTECHNIC for technical purposes	1	1.3G		1		LQ0	E0		PP		LO01 HA01, HA03, HA04, HA05, HA06	3	
0431	ARTICLES, PYROTECHNIC for technical purposes	1	1.4G		1.4		LQ0	E0		PP		LO01 HA01, HA03, HA04, HA05, HA06	1	
0432	ARTICLES, PYROTECHNIC for technical purposes	1	1.4S		1.4		LQ0	E0		PP		LO01 HA01, HA03, HA04, HA05, HA06	0	
0433	POWDER CAKE (POWDER PASTE), WETTED with not less than 17% alcohol, by mass	1	1.1C		1	266	LQ0	E0		PP		LO01 HA01, HA02, HA03, HA04, HA05, HA06	3	
0434	PROJECTILES with burster or expelling charge	1	1.2G		1		LQ0	E0		PP		LO01 HA01, HA03, HA04, HA05, HA06	3	
0435	PROJECTILES with burster or expelling charge	1	1.4G		1.4		LQ0	E0		PP		LO01 HA01, HA03, HA04, HA05, HA06	1	

UN No. or ID No.	Name and description	Class	Classi-fication Code	Packing group	Labels	Special provi-sions	Limited and excepted quantities		Carriage permitted	Equipment required	Venti-lation	Provisions concerning loading, unloading and carriage		Number of blue cones/ lights	Remarks
	3.1.2	2.2	2.2	2.1.1.3	5.2.2	3.3	3.4.6	3.5.1.2	3.2.1	8.1.5	7.1.6	7.1.6		7.1.5	3.2.1
(1)	(2)	(3a)	(3b)	(4)	(5)	(6)	(7a)	(7b)	(8)	(9)	(10)	(11)		(12)	(13)
0436	ROCKETS with expelling charge	1	1.2C		1		LQ0	E0		PP		LO01	HA01, HA03, HA04, HA05, HA06	3	
0437	ROCKETS with expelling charge	1	1.3C		1		LQ0	E0		PP		LO01	HA01, HA03, HA04, HA05, HA06	3	
0438	ROCKETS with expelling charge	1	1.4C		1.4		LQ0	E0		PP		LO01	HA01, HA03, HA04, HA05, HA06	1	
0439	CHARGES, SHAPED, without detonator	1	1.2D		1		LQ0	E0		PP		LO01	HA01, HA03, HA04, HA05, HA06	3	
0440	CHARGES, SHAPED, without detonator	1	1.4D		1.4		LQ0	E0		PP		LO01	HA01, HA03, HA04, HA05, HA06	1	
0441	CHARGES, SHAPED, without detonator	1	1.4S		1.4		LQ0	E0		PP		LO01	HA01, HA03, HA04, HA05, HA06	0	
0442	CHARGES, EXPLOSIVE, COMMERCIAL without detonator	1	1.1D		1		LQ0	E0		PP		LO01	HA01, HA02, HA03, HA04, HA05, HA06	3	
0443	CHARGES, EXPLOSIVE, COMMERCIAL without detonator	1	1.2D		1		LQ0	E0		PP		LO01	HA01, HA03, HA04, HA05, HA06	3	
0444	CHARGES, EXPLOSIVE, COMMERCIAL without detonator	1	1.4D		1.4		LQ0	E0		PP		LO01	HA01, HA03, HA04, HA05, HA06	1	

UN No. or ID No.	Name and description	Class	Classi-fication Code	Packing group	Labels	Special provi-sions	Limited and excepted quantities		Carriage permitted	Equipment required	Venti-lation	Provisions concerning loading, unloading and carriage		Number of blue cones/ lights	Remarks
	3.1.2	2.2	2.2	2.1.1.3	5.2.2	3.3	3.4.6	3.5.1.2	3.2.1	8.1.5	7.1.6	7.1.6		7.1.5	3.2.1
(1)	(2)	(3a)	(3b)	(4)	(5)	(6)	(7a)	(7b)	(8)	(9)	(10)	(11)		(12)	(13)
0445	CHARGES, EXPLOSIVE, COMMERCIAL without detonator	1	1.4S		1.4		LQ0	E0		PP		LO01	HA01, HA03, HA04, HA05, HA06	0	
0446	CASES, COMBUSTIBLE, EMPTY, WITHOUT PRIMER	1	1.4C		1.4		LQ0	E0		PP		LO01	HA01, HA03, HA04, HA05, HA06	1	
0447	CASES, COMBUSTIBLE, EMPTY, WITHOUT PRIMER	1	1.3C		1		LQ0	E0		PP		LO01	HA01, HA03, HA04, HA05, HA06	3	
0448	5-MERCAPTOTETRAZOL-1-ACETIC ACID	1	1.4C		1.4		LQ0	E0		PP		LO01	HA01, HA03, HA04, HA05, HA06	1	
0449	TORPEDOES, LIQUID FUELLED with or without bursting charge	1	1.1J		1		LQ0	E0		PP		LO01	HA01, HA02, HA03, HA04, HA05, HA06	3	
0450	TORPEDOES, LIQUID FUELLED with inert head	1	1.3J		1		LQ0	E0		PP		LO01	HA01, HA03, HA04, HA05, HA06	3	
0451	TORPEDOES with bursting charge	1	1.1D		1		LQ0	E0		PP		LO01	HA01, HA02, HA03, HA04, HA05, HA06	3	
0452	GRENADES, PRACTICE, hand or rifle	1	1.4G		1.4		LQ0	E0		PP		LO01	HA01, HA03, HA04, HA05, HA06	1	

UN No. or ID No.	Name and description	Class	Classi-fication Code	Packing group	Labels	Special provi-sions	Limited and excepted quantities		Carriage permitted	Equipment required	Venti-lation	Provisions concerning loading, unloading and carriage		Number of blue cones/ lights	Remarks
	3.1.2	2.2	2.2	2.1.1.3	5.2.2	3.3	3.4.6	3.5.1.2	3.2.1	8.1.5	7.1.6	7.1.6		7.1.5	3.2.1
(1)	(2)	(3a)	(3b)	(4)	(5)	(6)	(7a)	(7b)	(8)	(9)	(10)	(11)		(12)	(13)
0453	ROCKETS, LINE-THROWING	1	1.4G		1.4		LQ0	E0		PP		LO01	HA01, HA03, HA04, HA05, HA06	1	
0454	IGNITERS	1	1.4S		1.4		LQ0	E0		PP		LO01	HA01, HA03, HA04, HA05, HA06	0	
0455	DETONATORS, NON-ELECTRIC for blasting	1	1.4S		1.4		LQ0	E0		PP		LO01	HA01, HA03, HA04, HA05, HA06	0	
0456	DETONATORS, ELECTRIC for blasting	1	1.4S		1.4		LQ0	E0		PP		LO01	HA01, HA03, HA04, HA05, HA06	0	
0457	CHARGES, BURSTING, PLASTICS BONDED	1	1.1D		1		LQ0	E0		PP		LO01	HA01, HA02, HA03, HA04, HA05, HA06	3	
0458	CHARGES, BURSTING, PLASTICS BONDED	1	1.2D		1		LQ0	E0		PP		LO01	HA01, HA03, HA04, HA05, HA06	3	
0459	CHARGES, BURSTING, PLASTICS BONDED	1	1.4D		1.4		LQ0	E0		PP		LO01	HA01, HA03, HA04, HA05, HA06	1	
0460	CHARGES, BURSTING, PLASTICS BONDED	1	1.4S		1.4		LQ0	E0		PP		LO01	HA01, HA03, HA04, HA05, HA06	0	

UN No. or ID No.	Name and description	Class	Classi-fication Code	Packing group	Labels	Special provi-sions	Limited and excepted quantities		Carriage permitted	Equipment required	Venti-lation	Provisions concerning loading, unloading and carriage			Number of blue cones/ lights	Remarks
	3.1.2	2.2	2.2	2.1.1.3	5.2.2	3.3	3.4.6	3.5.1.2	3.2.1	8.1.5	7.1.6	7.1.6			7.1.5	3.2.1
(1)	(2)	(3a)	(3b)	(4)	(5)	(6)	(7a)	(7b)	(8)	(9)	(10)	(11)			(12)	(13)
0461	COMPONENTS, EXPLOSIVE TRAIN, N.O.S.	1	1.1B		1	178 274	LQ0	E0		PP		LO01	HA01, HA02, HA03, HA04, HA05, HA06		3	
0462	ARTICLES, EXPLOSIVE, N.O.S.	1	1.1C		1	178 274	LQ0	E0		PP		LO01	HA01, HA02, HA03, HA04, HA05, HA06		3	
0463	ARTICLES, EXPLOSIVE, N.O.S.	1	1.1D		1	178 274	LQ0	E0		PP		LO01	HA01, HA02, HA03, HA04, HA05, HA06		3	
0464	ARTICLES, EXPLOSIVE, N.O.S.	1	1.1E		1	178 274	LQ0	E0		PP		LO01	HA01, HA02, HA03, HA04, HA05, HA06		3	
0465	ARTICLES, EXPLOSIVE, N.O.S.	1	1.1F		1	178 274	LQ0	E0		PP		LO01	HA01, HA02, HA03, HA04, HA05, HA06		3	
0466	ARTICLES, EXPLOSIVE, N.O.S.	1	1.2C		1	178 274	LQ0	E0		PP		LO01	HA01, HA03, HA04, HA05, HA06		3	
0467	ARTICLES, EXPLOSIVE, N.O.S.	1	1.2D		1	178 274	LQ0	E0		PP		LO01	HA01, HA03, HA04, HA05, HA06		3	
0468	ARTICLES, EXPLOSIVE, N.O.S.	1	1.2E		1	178 274	LQ0	E0		PP		LO01	HA01, HA03, HA04, HA05, HA06		3	

UN No. or ID No.	Name and description	Class	Classi-fication Code	Packing group	Labels	Special provi-sions	Limited and excepted quantities		Carriage permitted	Equipment required	Venti-lation	Provisions concerning loading, unloading and carriage		Number of blue cones/ lights	Remarks
	3.1.2	2.2	2.2	2.1.1.3	5.2.2	3.3	3.4.6	3.5.1.2	3.2.1	8.1.5	7.1.6	7.1.6		7.1.5	3.2.1
(1)	(2)	(3a)	(3b)	(4)	(5)	(6)	(7a)	(7b)	(8)	(9)	(10)	(11)		(12)	(13)
0469	ARTICLES, EXPLOSIVE, N.O.S.	1	1.2F		1	178 274	LQ0	E0		PP		LO01	HA01, HA02, HA03, HA04, HA05, HA06	3	
0470	ARTICLES, EXPLOSIVE, N.O.S.	1	1.3C		1	178 274	LQ0	E0		PP		LO01	HA01, HA03, HA04, HA05, HA06	3	
0471	ARTICLES, EXPLOSIVE, N.O.S.	1	1.4E		1.4	178 274	LQ0	E0		PP		LO01	HA01, HA03, HA04, HA05, HA06	1	
0472	ARTICLES, EXPLOSIVE, N.O.S.	1	1.4F		1.4	178 274	LQ0	E0		PP		LO01	HA01, HA02, HA03, HA04, HA05, HA06	1	
0473	SUBSTANCES, EXPLOSIVE, N.O.S.	1	1.1A		1	178 274	LQ0	E0		PP		LO01	HA01, HA02, HA03, HA04, HA05, HA06	3	
0474	SUBSTANCES, EXPLOSIVE, N.O.S.	1	1.1C		1	178 274	LQ0	E0		PP		LO01	HA01, HA02, HA03, HA04, HA05, HA06	3	
0475	SUBSTANCES, EXPLOSIVE, N.O.S.	1	1.1D		1	178 274	LQ0	E0		PP		LO01	HA01, HA02, HA03, HA04, HA05, HA06	3	
0476	SUBSTANCES, EXPLOSIVE, N.O.S.	1	1.1G		1	178 274	LQ0	E0		PP		LO01	HA01, HA02, HA03, HA04, HA05, HA06	3	

UN No. or ID No.	Name and description	Class	Classi-fication Code	Packing group	Labels	Special provi-sions	Limited and excepted quantities		Carriage permitted	Equipment required	Venti-lation	Provisions concerning loading, unloading and carriage		Number of blue cones/ lights	Remarks
	3.1.2	2.2	2.2	2.1.1.3	5.2.2	3.3	3.4.6	3.5.1.2	3.2.1	8.1.5	7.1.6	7.1.6		7.1.5	3.2.1
(1)	(2)	(3a)	(3b)	(4)	(5)	(6)	(7a)	(7b)	(8)	(9)	(10)	(11)		(12)	(13)
0477	SUBSTANCES, EXPLOSIVE, N.O.S.	1	1.3C		1	178 274	LQ0	E0		PP		LO01	HA01, HA03, HA04, HA05, HA06	3	
0478	SUBSTANCES, EXPLOSIVE, N.O.S.	1	1.3G		1	178 274	LQ0	E0		PP		LO01	HA01, HA03, HA04, HA05, HA06	3	
0479	SUBSTANCES, EXPLOSIVE, N.O.S.	1	1.4C		1.4	178 274	LQ0	E0		PP		LO01	HA01, HA03, HA04, HA05, HA06	1	
0480	SUBSTANCES, EXPLOSIVE, N.O.S.	1	1.4D		1.4	178 274	LQ0	E0		PP		LO01	HA01, HA03, HA04, HA05, HA06	1	
0481	SUBSTANCES, EXPLOSIVE, N.O.S.	1	1.4S		1.4	178 274	LQ0	E0		PP		LO01	HA01, HA03, HA04, HA05, HA06	0	
0482	SUBSTANCES, EXPLOSIVE, VERY INSENSITIVE (SUBSTANCES, EVI), N.O.S.	1	1.5D		1.5	178 274	LQ0	E0		PP		LO01	HA01, HA03, HA04, HA05, HA06	3	
0483	CYCLOTRIMETHYLENETRINITRAMINE (CYCLONITE; HEXOGEN; RDX), DESENSITIZED	1	1.1D		1		LQ0	E0		PP		LO01	HA01, HA02, HA03, HA04, HA05, HA06	3	
0484	CYCLOTETRAMETHYLENE-TETRANITRAMINE (HMX; OCTOGEN), DESENSITIZED	1	1.1D		1		LQ0	E0		PP		LO01	HA01, HA02, HA03, HA04, HA05, HA06	3	

UN No. or ID No.	Name and description	Class	Classi-fication Code	Packing group	Labels	Special provi-sions	Limited and excepted quantities		Carriage permitted	Equipment required	Venti-lation	Provisions concerning loading, unloading and carriage		Number of blue cones/ lights	Remarks
	3.1.2	2.2	2.2	2.1.1.3	5.2.2	3.3	3.4.6	3.5.1.2	3.2.1	8.1.5	7.1.6	7.1.6		7.1.5	3.2.1
(1)	(2)	(3a)	(3b)	(4)	(5)	(6)	(7a)	(7b)	(8)	(9)	(10)	(11)		(12)	(13)
0485	SUBSTANCES, EXPLOSIVE, N.O.S.	1	1.4G		1.4	178 274	LQ0	E0		PP		LO01	HA01, HA03, HA04, HA05, HA06	1	
0486	ARTICLES, EXPLOSIVE, EXTREMELY INSENSITIVE (ARTICLES, EEI)	1	1.6N		1.6		LQ0	E0		PP		LO01	HA01, HA03, HA04, HA05, HA06	3	
0487	SIGNALS, SMOKE	1	1.3G		1		LQ0	E0		PP		LO01	HA01, HA03, HA04, HA05, HA06	3	
0488	AMMUNITION, PRACTICE	1	1.3G		1		LQ0	E0		PP		LO01	HA01, HA03, HA04, HA05, HA06	3	
0489	DINITROGLYCOLURIL (DINGU)	1	1.1D		1		LQ0	E0		PP		LO01	HA01, HA02, HA03, HA04, HA05, HA06	3	
0490	NITROTRIAZOLONE (NTO)	1	1.1D		1		LQ0	E0		PP		LO01	HA01, HA02, HA03, HA04, HA05, HA06	3	
0491	CHARGES, PROPELLING	1	1.4C		1.4		LQ0	E0		PP		LO01	HA01, HA03, HA04, HA05, HA06	1	
0492	SIGNALS, RAILWAY TRACK, EXPLOSIVE	1	1.3G		1		LQ0	E0		PP		LO01	HA01, HA03, HA04, HA05, HA06	3	

UN No. or ID No.	Name and description 3.1.2	Class 2.2	Classi-fication Code 2.2	Packing group 2.1.1.3	Labels 5.2.2	Special provi-sions 3.3	Limited and excepted quantities 3.4.6	3.5.1.2	Carriage permitted 3.2.1	Equipment required 8.1.5	Venti-lation 7.1.6	Provisions concerning loading, unloading and carriage 7.1.6		Number of blue cones/ lights 7.1.5	Remarks 3.2.1
(1)	(2)	(3a)	(3b)	(4)	(5)	(6)	(7a)	(7b)	(8)	(9)	(10)	(11)		(12)	(13)
0493	SIGNALS, RAILWAY TRACK, EXPLOSIVE	1	1.4G		1.4		LQ0	E0		PP		LO01	HA01, HA03, HA04, HA05, HA06	1	
0494	JET PERFORATING GUNS, CHARGED, oil well, without detonator	1	1.4D		1.4		LQ0	E0		PP		LO01	HA01, HA03, HA04, HA05, HA06	1	
0495	PROPELLANT, LIQUID	1	1.3C		1	224	LQ0	E0		PP		LO01	HA01, HA03, HA04, HA05, HA06	3	
0496	OCTONAL	1	1.1D		1		LQ0	E0		PP		LO01	HA01, HA02, HA03, HA04, HA05, HA06	3	
0497	PROPELLANT, LIQUID	1	1.1C		1	224	LQ0	E0		PP		LO01	HA01, HA02, HA03, HA04, HA05, HA06	3	
0498	PROPELLANT, SOLID	1	1.1C		1		LQ0	E0		PP		LO01	HA01, HA02, HA03, HA04, HA05, HA06	3	
0499	PROPELLANT, SOLID	1	1.3C		1		LQ0	E0		PP		LO01	HA01, HA03, HA04, HA05, HA06	3	
0500	DETONATOR ASSEMBLIES, NON-ELECTRIC for blasting	1	1.4S		1.4		LQ0	E0		PP		LO01	HA01, HA03, HA04, HA05, HA06	0	

UN No. or ID No.	Name and description	Class	Classi-fication Code	Packing group	Labels	Special provi-sions	Limited and excepted quantities		Carriage permitted	Equipment required	Venti-lation	Provisions concerning loading, unloading and carriage			Number of blue cones/ lights	Remarks
	3.1.2	2.2	2.2	2.1.1.3	5.2.2	3.3	3.4.6	3.5.1.2	3.2.1	8.1.5	7.1.6	7.1.6			7.1.5	3.2.1
(1)	(2)	(3a)	(3b)	(4)	(5)	(6)	(7a)	(7b)	(8)	(9)	(10)	(11)			(12)	(13)
0501	PROPELLANT, SOLID	1	1.4C		1.4		LQ0	E0		PP		LO01	HA01, HA03, HA04, HA05, HA06		1	
0502	ROCKETS with inert head	1	1.2C		1		LQ0	E0		PP		LO01	HA01, HA03, HA04, HA05, HA06		3	
0503	AIR BAG INFLATORS or AIR BAG MODULES or SEAT-BELT PRETENSIONERS	1	1.4G		1.4	235 289	LQ0	E0		PP		LO01	HA01, HA03, HA04, HA05, HA06		1	
0504	1H-TETRAZOLE	1	1.1D		1		LQ0	E0		PP		LO01	HA01, HA02, HA03, HA04, HA05, HA06		3	
0505	SIGNALS, DISTRESS, ship	1	1.4G		1.4		LQ0	E0		PP		LO01	HA01, HA03, HA04, HA05, HA06		1	
0506	SIGNALS, DISTRESS, ship	1	1.4S		1.4		LQ0	E0		PP		LO01	HA01, HA03, HA04, HA05, HA06		0	
0507	SIGNALS, SMOKE	1	1.4S		1.4		LQ0	E0		PP		LO01	HA01, HA03, HA04, HA05, HA06		0	
0508	1-HYDROXY-BENZOTRIAZOLE, ANHYDROUS, dry or wetted with less than 20% water, by mass	1	1.3C		1		LQ0	E0		PP		LO01	HA01, HA03, HA04, HA05, HA06		3	
1001	ACETYLENE, DISSOLVED	2	4F		2.1		LQ0	E0		PP, EX, A	VE01				1	
1002	AIR, COMPRESSED	2	1A		2.2	292	LQ1	E1		PP					0	
1003	AIR, REFRIGERATED LIQUID	2	3O		2.2+5.1		LQ0	E0		PP					0	
1005	AMMONIA, ANHYDROUS	2	2TC		2.3+8	23	LQ0	E0	**T**	PP, EP, TOX, A	VE02				2	

UN No. or ID No.	Name and description	Class	Classi-fication Code	Packing group	Labels	Special provi-sions	Limited and excepted quantities		Carriage permitted	Equipment required	Venti-lation	Provisions concerning loading, unloading and carriage	Number of blue cones/ lights	Remarks
	3.1.2	2.2	2.2	2.1.1.3	5.2.2	3.3	3.4.6	3.5.1.2	3.2.1	8.1.5	7.1.6	7.1.6	7.1.5	3.2.1
(1)	(2)	(3a)	(3b)	(4)	(5)	(6)	(7a)	(7b)	(8)	(9)	(10)	(11)	(12)	(13)
1006	ARGON, COMPRESSED	2	1A		2.2		LQ1	E1		PP			0	
1008	BORON TRIFLUORIDE	2	2TC		2.3+8		LQ0	E0		PP, EP, TOX, A	VE02		2	
1009	BROMOTRIFLUOROMETHANE (REFRIGERANT GAS R 13B1)	2	2A		2.2		LQ1	E1		PP			0	
1010	BUTADIENES, STABILIZED or BUTADIENES AND HYDROCARBON MIXTURE, STABILIZED, having a vapour pressure at 70 °C not exceeding 1.1 MPa (11 bar) and a density at 50 °C not lower than 0.525 kg/l	2	2F		2.1	618	LQ0	E0	T	PP, EX, A	VE01		1	
1011	BUTANE	2	2F		2.1		LQ0	E0	T	PP, EX, A	VE01		1	
1012	BUTYLENES MIXTURE or 1-BUTYLENE or CIS-2-BUTYLENE or TRANS-2-BUTYLENE	2	2F		2.1		LQ0	E0	T	PP, EX, A	VE01		1	
1013	CARBON DIOXIDE	2	2A		2.2	584 653	LQ1	E1		PP			0	
1016	CARBON MONOXIDE, COMPRESSED	2	1TF		2.3+2.1		LQ0	E0		PP, EP, EX, TOX, A	VE01, VE02		2	
1017	CHLORINE	2	2TOC		2.3+5.1+8		LQ0	E0		PP, EP, TOX, A	VE02		2	
1018	CHLORODIFLUOROMETHANE (REFRIGERANT GAS R 22)	2	2A		2.2		LQ1	E1		PP			0	
1020	CHLOROPENTAFLUOROETHANE (REFRIGERANT GAS R 115)	2	2A		2.2		LQ1	E1	T	PP			0	
1021	1-CHLORO- 1,2,2,2-TETRAFLUOROETHANE (REFRIGERANT GAS R 124)	2	2A		2.2		LQ1	E1		PP			0	
1022	CHLOROTRIFLUOROMETHANE (REFRIGERANT GAS R 13)	2	2A		2.2		LQ1	E1		PP			0	
1023	COAL GAS, COMPRESSED	2	1TF		2.3+2.1		LQ0	E0		PP, EP, EX, TOX, A	VE01, VE02		2	
1026	CYANOGEN	2	2TF		2.3+2.1		LQ0	E0		PP, EP, EX, TOX, A	VE01, VE02		2	
1027	CYCLOPROPANE	2	2F		2.1		LQ0	E0		PP, EX, A	VE01		1	
1028	DICHLORODIFLUOROMETHANE (REFRIGERANT GAS R 12)	2	2A		2.2		LQ1	E1		PP			0	
1029	DICHLOROFLUOROMETHANE (REFRIGERANT GAS R 21)	2	2A		2.2		LQ1	E1		PP			0	
1030	1,1-DIFLUOROETHANE (REFRIGERANT GAS R 152a)	2	2F		2.1		LQ0	E0	T	PP, EX, A	VE01		1	
1032	DIMETHYLAMINE, ANHYDROUS	2	2F		2.1		LQ0	E0		PP, EX, A	VE01		1	
1033	DIMETHYL ETHER	2	2F		2.1		LQ0	E0	T	PP, EX, A	VE01		1	
1035	ETHANE	2	2F		2.1		LQ0	E0		PP, EX, A	VE01		1	
1036	ETHYLAMINE	2	2F		2.1		LQ0	E0		PP, EX, A	VE01		1	
1037	ETHYL CHLORIDE	2	2F		2.1		LQ0	E0		PP, EX, A	VE01		1	
1038	ETHYLENE, REFRIGERATED LIQUID	2	3F		2.1		LQ0	E0		PP, EX, A	VE01		1	
1039	ETHYL METHYL ETHER	2	2F		2.1		LQ0	E0		PP, EX, A	VE01		1	

UN No. or ID No.	Name and description	Class	Classi-fication Code	Packing group	Labels	Special provi-sions	Limited and excepted quantities		Carriage permitted	Equipment required	Venti-lation	Provisions concerning loading, unloading and carriage	Number of blue cones/ lights	Remarks
	3.1.2	2.2	2.2	2.1.1.3	5.2.2	3.3	3.4.6	3.5.1.2	3.2.1	8.1.5	7.1.6	7.1.6	7.1.5	3.2.1
(1)	(2)	(3a)	(3b)	(4)	(5)	(6)	(7a)	(7b)	(8)	(9)	(10)	(11)	(12)	(13)
1040	ETHYLENE OXIDE	2	2TF		2.3+2.1		LQ0	E0		PP, EP, EX, TOX, A	VE01, VE02		2	
1040	ETHYLENE OXIDE WITH NITROGEN up to a total pressure of 1 MPa (10 bar) at 50 °C	2	2TF		2.3+2.1		LQ0	E0	T	PP, EP, EX, TOX, A	VE01, VE02		2	
1041	ETHYLENE OXIDE AND CARBON DIOXIDE MIXTURE with more than 9% but not more than 87% ethylene oxide	2	2F		2.1		LQ0	E0		PP, EX, A	VE01		1	
1043	FERTILIZER AMMONIATING SOLUTION with free ammonia	2	4A		2.2		LQ1	E0		PP			0	
1044	FIRE EXTINGUISHERS with compressed or liquefied gas	2	6A		2.2	225 594	LQ0	E0		PP			0	
1045	FLUORINE, COMPRESSED	2	1TOC		2.3+5.1+8		LQ0	E0		PP, EP, TOX, A	VE02		2	
1046	HELIUM, COMPRESSED	2	1A		2.2		LQ1	E1		PP			0	
1048	HYDROGEN BROMIDE, ANHYDROUS	2	2TC		2.3+8		LQ0	E0		PP, EP, TOX, A	VE02		2	
1049	HYDROGEN, COMPRESSED	2	1F		2.1		LQ0	E0		PP, EX, A	VE01		1	
1050	HYDROGEN CHLORIDE, ANHYDROUS	2	2TC		2.3+8		LQ0	E0		PP, EP, TOX, A	VE02		2	
1051	HYDROGEN CYANIDE, STABILIZED containing less than 3% water	6.1	TF1	I	6.1+3	603 802	LQ0	E5		PP, EP, EX, TOX, A	VE01, VE02		2	
1052	HYDROGEN FLUORIDE, ANHYDROUS	8	CT1	I	8+6.1	802	LQ0	E0		PP, EP, TOX, A	VE02		2	
1053	HYDROGEN SULPHIDE	2	2TF		2.3+2.1		LQ0	E0		PP, EP, EX, TOX, A	VE01, VE02		2	
1055	ISOBUTYLENE	2	2F		2.1		LQ0	E0	T	PP, EX, A	VE01		1	
1056	KRYPTON, COMPRESSED	2	1A		2.2		LQ1	E1		PP			0	
1057	LIGHTERS or LIGHTER REFILLS containing flammable gas	2	6F		2.1	201 654	LQ0	E0		PP, EX, A	VE01		1	
1058	LIQUEFIED GASES, non-flammable, charged with nitrogen, carbon dioxide or air	2	2A		2.2		LQ1	E1		PP			0	
1060	METHYLACETYLENE AND PROPADIENE MIXTURE, STABILIZED such as mixture P1 or mixture P2	2	2F		2.1	581	LQ0	E0		PP, EX, A	VE01		1	
1061	METHYLAMINE, ANHYDROUS	2	2F		2.1		LQ0	E0		PP, EX, A	VE01		1	
1062	METHYL BROMIDE with not more than 2% chloropicrin	2	2T		2.3	23	LQ0	E0		PP, EP, TOX, A	VE02		2	
1063	METHYL CHLORIDE (REFRIGERANT GAS R 40)	2	2F		2.1		LQ0	E0	T	PP, EX, A	VE01		1	
1064	METHYL MERCAPTAN	2	2TF		2.3+2.1		LQ0	E0		PP, EP, EX, TOX, A	VE01, VE02		2	
1065	NEON, COMPRESSED	2	1A		2.2		LQ1	E1		PP			0	
1066	NITROGEN, COMPRESSED	2	1A		2.2		LQ1	E1		PP			0	
1067	DINITROGEN TETROXIDE (NITROGEN DIOXIDE)	2	2TOC		2.3+5.1+8		LQ0	E0		PP, EP, TOX, A	VE02		2	

UN No. or ID No.	Name and description	Class	Classi-fication Code	Packing group	Labels	Special provi-sions	Limited and excepted quantities		Carriage permitted	Equipment required	Venti-lation	Provisions concerning loading, unloading and carriage	Number of blue cones/ lights	Remarks
	3.1.2	2.2	2.2	2.1.1.3	5.2.2	3.3	3.4.6	3.5.1.2	3.2.1	8.1.5	7.1.6	7.1.6	7.1.5	3.2.1
(1)	(2)	(3a)	(3b)	(4)	(5)	(6)	(7a)	(7b)	(8)	(9)	(10)	(11)	(12)	(13)
1069	NITROSYL CHLORIDE	2	2TC		2.3+8		LQ0	E0		PP, EP, TOX, A	VE02		2	
1070	NITROUS OXIDE	2	2O		2.2+5.1	584	LQ0	E0		PP			0	
1071	OIL GAS, COMPRESSED	2	1TF		2.3+2.1		LQ0	E0		PP, EP, EX, TOX, A	VE01, VE02		2	
1072	OXYGEN, COMPRESSED	2	1O		2.2+5.1		LQ0	E0		PP			0	
1073	OXYGEN, REFRIGERATED LIQUID	2	3O		2.2+5.1		LQ0	E0		PP			0	
1075	PETROLEUM GASES, LIQUEFIED	2	2F		2.1	274 583 639	LQ0	E0		PP, EX, A	VE01		1	
1076	PHOSGENE	2	2TC		2.3+8		LQ0	E0		PP, EP, TOX, A	VE02		2	
1077	PROPYLENE	2	2F		2.1		LQ0	E0	T	PP, EX, A	VE01		1	
1078	REFRIGERANT GAS, N.O.S., such as mixture F1, mixture F2 or mixture F3	2	2A		2.2	274 582	LQ1	E1		PP			0	
1079	SULPHUR DIOXIDE	2	2TC		2.3+8		LQ0	E0		PP, EP, TOX, A	VE02		2	
1080	SULPHUR HEXAFLUORIDE	2	2A		2.2		LQ1	E1		PP			0	
1081	TETRAFLUOROETHYLENE, STABILIZED	2	2F		2.1		LQ0	E0		PP, EX, A	VE01		1	
1082	TRIFLUOROCHLOROETHYLENE, STABILIZED	2	2TF		2.3+2.1		LQ0	E0		PP, EP, EX, TOX, A	VE01, VE02		2	
1083	TRIMETHYLAMINE, ANHYDROUS	2	2F		2.1		LQ0	E0	T	PP, EX, A	VE01		1	
1085	VINYL BROMIDE, STABILIZED	2	2F		2.1		LQ0	E0		PP, EX, A	VE01		1	
1086	VINYL CHLORIDE, STABILIZED	2	2F		2.1		LQ0	E0	T	PP, EX, A	VE01		1	
1087	VINYL METHYL ETHER, STABILIZED	2	2F		2.1		LQ0	E0		PP, EX, A	VE01		1	
1088	ACETAL	3	F1	II	3		LQ4	E2	T	PP, EX, A	VE01		1	
1089	ACETALDEHYDE	3	F1	I	3		LQ3	E3	T	PP, EX, A	VE01		1	
1090	ACETONE	3	F1	II	3		LQ4	E2	T	PP, EX, A	VE01		1	
1091	ACETONE OILS	3	F1	II	3		LQ4	E2		PP, EX, A	VE01		1	
1092	ACROLEIN, STABILIZED	6.1	TF1	I	6.1+3	802	LQ0	E5	T	PP, EP, EX, TOX, A	VE01, VE02		2	
1093	ACRYLONITRILE, STABILIZED	3	FT1	I	3+6.1	802	LQ0	E0	T	PP, EP, EX, TOX, A	VE01, VE02		2	
1098	ALLYL ALCOHOL	6.1	TF1	I	6.1+3	802	LQ0	E5	T	PP, EP, EX, TOX, A	VE01, VE02		2	
1099	ALLYL BROMIDE	3	FT1	I	3+6.1	802	LQ0	E0		PP, EP, EX, TOX, A	VE01, VE02		2	
1100	ALLYL CHLORIDE	3	FT1	I	3+6.1	802	LQ0	E0	T	PP, EP, EX, TOX, A	VE01, VE02		2	
1104	AMYL ACETATES	3	F1	III	3		LQ7	E1		PP, EX, A	VE01		0	
1105	PENTANOLS	3	F1	II	3		LQ4	E2		PP, EX, A	VE01		1	
1105	PENTANOLS	3	F1	III	3		LQ7	E1	T	PP, EX, A	VE01		0	
1106	AMYLAMINE	3	FC	II	3+8		LQ4	E2	T	PP, EP, EX, A	VE01		1	
1106	AMYLAMINE	3	FC	III	3+8		LQ7	E1		PP, EP, EX, A	VE01		0	
1107	AMYL CHLORIDE	3	F1	II	3		LQ4	E2	T	PP, EX, A	VE01		1	
1108	1-PENTENE (n-AMYLENE)	3	F1	I	3		LQ3	E3	T	PP, EX, A	VE01		1	

UN No. or ID No.	Name and description	Class	Classi-fication Code	Packing group	Labels	Special provi-sions	Limited and excepted quantities		Carriage permitted	Equipment required	Venti-lation	Provisions concerning loading, unloading and carriage	Number of blue cones/ lights	Remarks
	3.1.2	2.2	2.2	2.1.1.3	5.2.2	3.3	3.4.6	3.5.1.2	3.2.1	8.1.5	7.1.6	7.1.6	7.1.5	3.2.1
(1)	(2)	(3a)	(3b)	(4)	(5)	(6)	(7a)	(7b)	(8)	(9)	(10)	(11)	(12)	(13)
1109	AMYL FORMATES	3	F1	III	3		LQ7	E1		PP, EX, A	VE01		0	
1110	n-AMYL METHYL KETONE	3	F1	III	3		LQ7	E1		PP, EX, A	VE01		0	
1111	AMYL MERCAPTAN	3	F1	II	3		LQ4	E2		PP, EX, A	VE01		1	
1112	AMYL NITRATE	3	F1	III	3		LQ7	E1		PP, EX, A	VE01		0	
1113	AMYL NITRITE	3	F1	II	3		LQ4	E2		PP, EX, A	VE01		1	
1114	BENZENE	3	F1	II	3		LQ4	E2	T	PP, EX, A	VE01		1	
1120	BUTANOLS	3	F1	II	3		LQ4	E2	T	PP, EX, A	VE01		1	
1120	BUTANOLS	3	F1	III	3		LQ7	E1	T	PP, EX, A	VE01		0	
1123	BUTYL ACETATES	3	F1	II	3		LQ4	E2	T	PP, EX, A	VE01		1	
1123	BUTYL ACETATES	3	F1	III	3		LQ7	E1	T	PP, EX, A	VE01		0	
1125	n-BUTYLAMINE	3	FC	II	3+8		LQ4	E2	T	PP, EP, EX, A	VE01		1	
1126	1-BROMOBUTANE	3	F1	II	3		LQ4	E2		PP, EX, A	VE01		1	
1127	CHLOROBUTANES	3	F1	II	3		LQ4	E2	T	PP, EX, A	VE01		1	
1128	n-BUTYL FORMATE	3	F1	II	3		LQ4	E2		PP, EX, A	VE01		1	
1129	BUTYRALDEHYDE	3	F1	II	3		LQ4	E2	T	PP, EX, A	VE01		1	
1130	CAMPHOR OIL	3	F1	III	3		LQ7	E1		PP, EX, A	VE01		0	
1131	CARBON DISULPHIDE	3	FT1	I	3+6.1	802	LQ0	E0	T	PP, EP, EX, TOX, A	VE01, VE02		2	
1133	ADHESIVES containing flammable liquid	3	F1	I	3		LQ3	E3		PP, EX, A	VE01		1	
1133	ADHESIVES containing flammable liquid (vapour pressure at 50 °C more than 110 kPa)	3	F1	II	3	640C	LQ6	E2		PP, EX, A	VE01		1	
1133	ADHESIVES containing flammable liquid (vapour pressure at 50 °C not more than 110 kPa)	3	F1	II	3	640D	LQ6	E2		PP, EX, A	VE01		1	
1133	ADHESIVES containing flammable liquid	3	F1	III	3	640E	LQ7	E1		PP, EX, A	VE01		0	
1133	ADHESIVES containing flammable liquid (having a flash-point below 23 °C and viscous according to 2.2.3.1.4) (boiling point not more than 35° C)	3	F1	III	3	640F	LQ7	E1		PP, EX, A	VE01		0	
1133	ADHESIVES containing flammable liquid (having a flash-point below 23 °C and viscous according to 2.2.3.1.4) (vapour pressure at 50 °C more than 110 kPa , boiling point of more than 35° C)	3	F1	III	3	640G	LQ7	E1		PP, EX, A	VE01		0	
1133	ADHESIVES containing flammable liquid (having a flash-point below 23 °C and viscous according to 2.2.3.1.4) (vapour pressure at 50 °C not more than 110 kPa)	3	F1	III	3	640H	LQ7	E1		PP, EX, A	VE01		0	
1134	CHLOROBENZENE	3	F1	III	3		LQ7	E1	T	PP, EX, A	VE01		0	
1135	ETHYLENE CHLOROHYDRIN	6.1	TF1	I	6.1+3	802	LQ0	E5	T	PP, EP, EX, TOX, A	VE01, VE02		2	
1136	COAL TAR DISTILLATES, FLAMMABLE	3	F1	II	3		LQ4	E2		PP, EX, A	VE01		1	
1136	COAL TAR DISTILLATES, FLAMMABLE	3	F1	III	3		LQ7	E1		PP, EX, A	VE01		0	

UN No. or ID No.	Name and description 3.1.2	Class 2.2	Classi-fication Code 2.2	Packing group 2.1.1.3	Labels 5.2.2	Special provi-sions 3.3	Limited and excepted quantities 3.4.6	3.5.1.2	Carriage permitted 3.2.1	Equipment required 8.1.5	Venti-lation 7.1.6	Provisions concerning loading, unloading and carriage 7.1.6	Number of blue cones/ lights 7.1.5	Remarks 3.2.1
(1)	(2)	(3a)	(3b)	(4)	(5)	(6)	(7a)	(7b)	(8)	(9)	(10)	(11)	(12)	(13)
1139	COATING SOLUTION (includes surface treatments or coatings used for industrial or other purposes such as vehicle under coating, drum or barrel lining)	3	F1	I	3		LQ3	E3		PP, EX, A	VE01		1	
1139	COATING SOLUTION (includes surface treatments or coatings used for industrial or other purposes such as vehicle under coating, drum or barrel lining) (vapour pressure at 50 °C more than 110 kPa)	3	F1	II	3	640C	LQ6	E2		PP, EX, A	VE01		1	
1139	COATING SOLUTION (includes surface treatments or coatings used for industrial or other purposes such as vehicle under coating, drum or barrel lining) (vapour pressure at 50 °C not more than 110 kPa)	3	F1	II	3	640D	LQ6	E2		PP, EX, A	VE01		1	
1139	COATING SOLUTION (includes surface treatments or coatings used for industrial or other purposes such as vehicle under coating, drum or barrel lining)	3	F1	III	3	640E	LQ7	E1		PP, EX, A	VE01		0	
1139	COATING SOLUTION (includes surface treatments or coatings used for industrial or other purposes such as vehicle under coating, drum or barrel lining) (having a flash-point below 23 °C and viscous according to 2.2.3.1.4) (boiling point not more than 35° C)	3	F1	III	3	640F	LQ7	E1		PP, EX, A	VE01		0	
1139	COATING SOLUTION (includes surface treatments or coatings used for industrial or other purposes such as vehicle under coating, drum or barrel lining) (having a flash-point below 23 °C and viscous according to 2.2.3.1.4) (vapour pressure at 50 °C , boiling point of more than 35° C)	3	F1	III	3	640G	LQ7	E1		PP, EX, A	VE01		0	
1139	COATING SOLUTION (includes surface treatments or coatings used for industrial or other purposes such as vehicle under coating, drum or barrel lining) (having a flash-point below 23 °C and viscous according to 2.2.3.1.4) (vapour pressure at 50 °C not more than 110 kPa)	3	F1	III	3	640H	LQ7	E1		PP, EX, A	VE01		0	
1143	CROTONALDEHYDE or CROTONALDEHYDE, STABILIZED	6.1	TF1	I	6.1+3	324 802	LQ0	E5	T	PP, EP, EX, TOX, A	VE01, VE02		2	
1144	CROTONYLENE	3	F1	I	3		LQ3	E3		PP, EX, A	VE01		1	
1145	CYCLOHEXANE	3	F1	II	3		LQ4	E2	T	PP, EX, A	VE01		1	
1146	CYCLOPENTANE	3	F1	II	3		LQ4	E2	T	PP, EX, A	VE01		1	
1147	DECAHYDRO-NAPHTHALENE	3	F1	III	3		LQ7	E1		PP, EX, A	VE01		0	
1148	DIACETONE ALCOHOL	3	F1	II	3		LQ4	E2		PP, EX, A	VE01		1	
1148	DIACETONE ALCOHOL	3	F1	III	3		LQ7	E1		PP, EX, A	VE01		0	

UN No. or ID No.	Name and description	Class	Classi-fication Code	Packing group	Labels	Special provi-sions	Limited and excepted quantities		Carriage permitted	Equipment required	Venti-lation	Provisions concerning loading, unloading and carriage	Number of blue cones/ lights	Remarks
	3.1.2	2.2	2.2	2.1.1.3	5.2.2	3.3	3.4.6	3.5.1.2	3.2.1	8.1.5	7.1.6	7.1.6	7.1.5	3.2.1
(1)	(2)	(3a)	(3b)	(4)	(5)	(6)	(7a)	(7b)	(8)	(9)	(10)	(11)	(12)	(13)
1149	DIBUTYL ETHERS	3	F1	III	3		LQ7	E1		PP, EX, A	VE01		0	
1150	1,2-DICHLOROETHYLENE	3	F1	II	3		LQ4	E2	T	PP, EX, A	VE01		1	
1152	DICHLOROPENTANES	3	F1	III	3		LQ7	E1		PP, EX, A	VE01		0	
1153	ETHYLENE GLYCOL DIETHYL ETHER	3	F1	II	3		LQ4	E2	T	PP, EX, A	VE01		1	
1153	ETHYLENE GLYCOL DIETHYL ETHER	3	F1	III	3		LQ7	E1	T	PP, EX, A	VE01		0	
1154	DIETHYLAMINE	3	FC	II	3+8		LQ4	E2	T	PP, EP, EX, A	VE01		1	
1155	DIETHYL ETHER (ETHYL ETHER)	3	F1	I	3		LQ3	E3	T	PP, EX, A	VE01		1	
1156	DIETHYL KETONE	3	F1	II	3		LQ4	E2		PP, EX, A	VE01		1	
1157	DIISOBUTYL KETONE	3	F1	III	3		LQ7	E1	T	PP, EX, A	VE01		0	
1158	DIISOPROPYLAMINE	3	FC	II	3+8		LQ4	E2		PP, EP, EX, A	VE01		1	
1159	DIISOPROPYL ETHER	3	F1	II	3		LQ4	E2	T	PP, EX, A	VE01		1	
1160	DIMETHYLAMINE AQUEOUS SOLUTION	3	FC	II	3+8		LQ4	E2	T	PP, EP, EX, A	VE01		1	
1161	DIMETHYL CARBONATE	3	F1	II	3		LQ4	E2		PP, EX, A	VE01		1	
1162	DIMETHYLDICHLOROSILANE	3	FC	II	3+8		LQ4	E2		PP, EP, EX, A	VE01		1	
1163	DIMETHYLHYDRAZINE, UNSYMMETRICAL	6.1	TFC	I	6.1+3+8	802	LQ0	E5	T	PP, EP, EX, TOX, A	VE01, VE02		2	
1164	DIMETHYL SULPHIDE	3	F1	II	3		LQ4	E2		PP, EX, A	VE01		1	
1165	DIOXANE	3	F1	II	3		LQ4	E2	T	PP, EX, A	VE01		1	
1166	DIOXOLANE	3	F1	II	3		LQ4	E2		PP, EX, A	VE01		1	
1167	DIVINYL ETHER, STABILIZED	3	F1	I	3		LQ3	E3	T	PP, EX, A	VE01		1	
1169	EXTRACTS, AROMATIC, LIQUID	3	F1	I	3		LQ3	E3		PP, EX, A	VE01		1	
1169	EXTRACTS, AROMATIC, LIQUID (vapour pressure at 50 °C more than 110 kPa)	3	F1	II	3	601 640C	LQ6	E2		PP, EX, A	VE01		1	
1169	EXTRACTS, AROMATIC, LIQUID (vapour pressure at 50 °C not more than 110 kPa)	3	F1	II	3	601 640D	LQ6	E2		PP, EX, A	VE01		1	
1169	EXTRACTS, AROMATIC, LIQUID	3	F1	III	3	601 640E	LQ7	E1		PP, EX, A	VE01		0	
1169	EXTRACTS, AROMATIC, LIQUID (having a flash-point below 23 °C and viscous according to 2.2.3.1.4) (boiling point not more than 35° C)	3	F1	III	3	601 640F	LQ7	E1		PP, EX, A	VE01		0	
1169	EXTRACTS, AROMATIC, LIQUID (having a flash-point below 23 °C and viscous according to 2.2.3.1.4) (vapour pressure at 50 °C more than 110 kPa , boiling point of more than 35° C)	3	F1	III	3	601 640G	LQ7	E1		PP, EX, A	VE01		0	
1169	EXTRACTS, AROMATIC, LIQUID (having a flash-point below 23 °C and viscous according to 2.2.3.1.4) (vapour pressure at 50 °C not more than 110 kPa)	3	F1	III	3	601 640H	LQ7	E1		PP, EX, A	VE01		0	
1170	ETHANOL (ETHYL ALCOHOL) or ETHANOL SOLUTION (ETHYL ALCOHOL SOLUTION)	3	F1	II	3	144 601	LQ4	E2	T	PP, EX, A	VE01		1	
1170	ETHANOL SOLUTION (ETHYL ALCOHOL SOLUTION)	3	F1	III	3	144 601	LQ7	E1	T	PP, EX, A	VE01		0	
1171	ETHYLENE GLYCOL MONOETHYL ETHER	3	F1	III	3		LQ7	E1	T	PP, EX, A	VE01		0	

UN No. or ID No.	Name and description 3.1.2	Class 2.2	Classi-fication Code 2.2	Packing group 2.1.1.3	Labels 5.2.2	Special provi-sions 3.3	Limited and excepted quantities 3.4.6	 3.5.1.2	Carriage permitted 3.2.1	Equipment required 8.1.5	Venti-lation 7.1.6	Provisions concerning loading, unloading and carriage 7.1.6	Number of blue cones/ lights 7.1.5	Remarks 3.2.1
(1)	(2)	(3a)	(3b)	(4)	(5)	(6)	(7a)	(7b)	(8)	(9)	(10)	(11)	(12)	(13)
1172	ETHYLENE GLYCOL MONOETHYL ETHER ACETATE	3	F1	III	3		LQ7	E1	T	PP, EX, A	VE01		0	
1173	ETHYL ACETATE	3	F1	II	3		LQ4	E2	T	PP, EX, A	VE01		1	
1175	ETHYLBENZENE	3	F1	II	3		LQ4	E2	T	PP, EX, A	VE01		1	
1176	ETHYL BORATE	3	F1	II	3		LQ4	E2		PP, EX, A	VE01		1	
1177	2-ETHYLBUTYL ACETATE	3	F1	III	3		LQ7	E1	T	PP, EX, A	VE01		0	
1178	2-ETHYLBUTYRALDEHYDE	3	F1	II	3		LQ4	E2		PP, EX, A	VE01		1	
1179	ETHYL BUTYL ETHER	3	F1	II	3		LQ4	E2		PP, EX, A	VE01		1	
1180	ETHYL BUTYRATE	3	F1	III	3		LQ7	E1		PP, EX, A	VE01		0	
1181	ETHYL CHLOROACETATE	6.1	TF1	II	6.1+3	802	LQ17	E4		PP, EP, EX, TOX, A	VE01, VE02		2	
1182	ETHYL CHLOROFORMATE	6.1	TFC	I	6.1+3+8	802	LQ0	E5		PP, EP, EX, TOX, A	VE01, VE02		2	
1183	ETHYLDICHLOROSILANE	4.3	WFC	I	4.3+3+8		LQ0	E0		PP, EP, EX, A	VE01	HA08	1	
1184	ETHYLENE DICHLORIDE	3	FT1	II	3+6.1	802	LQ0	E2	T	PP, EP, EX, TOX, A	VE01, VE02		2	
1185	ETHYLENEIMINE, STABILIZED	6.1	TF1	I	6.1+3	802	LQ0	E5		PP, EP, EX, TOX, A	VE01, VE02		2	
1188	ETHYLENE GLYCOL MONOMETHYL	3	F1	III	3		LQ7	E1	T	PP, EX, A	VE01		0	
1189	ETHYLENE GLYCOL MONOMETHYL ETHER ACETATE	3	F1	III	3		LQ7	E1		PP, EX, A	VE01		0	
1190	ETHYL FORMATE	3	F1	II	3		LQ4	E2		PP, EX, A	VE01		1	
1191	OCTYL ALDEHYDES	3	F1	III	3		LQ7	E1	T	PP, EX, A	VE01		0	
1192	ETHYL LACTATE	3	F1	III	3		LQ7	E1		PP, EX, A	VE01		0	
1193	ETHYL METHYL KETONE (METHYL ETHYL KETONE)	3	F1	II	3		LQ4	E2	T	PP, EX, A	VE01		1	
1194	ETHYL NITRITE SOLUTION	3	FT1	I	3+6.1	802	LQ0	E0		PP, EP, EX, TOX, A	VE01, VE02		2	
1195	ETHYL PROPIONATE	3	F1	II	3		LQ4	E2		PP, EX, A	VE01		1	
1196	ETHYLTRICHLOROSILANE	3	FC	II	3+8		LQ4	E2		PP, EP, EX, A	VE01		1	
1197	EXTRACTS, FLAVOURING, LIQUID	3	F1	I	3		LQ3	E3		PP, EX, A	VE01		1	
1197	EXTRACTS, FLAVOURING, LIQUID (vapour pressure at 50 °C more than 110 kPa)	3	F1	II	3	601 640C	LQ6	E2		PP, EX, A	VE01		1	
1197	EXTRACTS, FLAVOURING, LIQUID (vapour pressure at 50 °C not more than 110 kPa)	3	F1	II	3	601 640D	LQ6	E2		PP, EX, A	VE01		1	
1197	EXTRACTS, FLAVOURING, LIQUID	3	F1	III	3	601 640E	LQ7	E1		PP, EX, A	VE01		0	
1197	EXTRACTS, FLAVOURING, LIQUID (having a flash-point below 23 °C and viscous according to 2.2.3.1.4) (boiling point not more than 35° C)	3	F1	III	3	601 640F	LQ7	E1		PP, EX, A	VE01		0	
1197	EXTRACTS, FLAVOURING, LIQUID (having a flash-point below 23 °C and viscous according to 2.2.3.1.4) (vapour pressure at 50 °C more than 110 kPa , boiling point of more than 35° C)	3	F1	III	3	601 640G	LQ7	E1		PP, EX, A	VE01		0	

UN No. or ID No.	Name and description	Class	Classi-fication Code	Packing group	Labels	Special provi-sions	Limited and excepted quantities		Carriage permitted	Equipment required	Venti-lation	Provisions concerning loading, unloading and carriage	Number of blue cones/ lights	Remarks
	3.1.2	2.2	2.2	2.1.1.3	5.2.2	3.3	3.4.6	3.5.1.2	3.2.1	8.1.5	7.1.6	7.1.6	7.1.5	3.2.1
(1)	(2)	(3a)	(3b)	(4)	(5)	(6)	(7a)	(7b)	(8)	(9)	(10)	(11)	(12)	(13)
1197	EXTRACTS, FLAVOURING, LIQUID (having a flash-point below 23 °C and viscous according to 2.2.3.1.4) (vapour pressure at 50 °C not more than 110 kPa)	3	F1	III	3	601 640H	LQ7	E1		PP, EX, A	VE01		0	
1198	FORMALDEHYDE SOLUTION, FLAMMABLE	3	FC	III	3+8		LQ7	E1	T	PP, EP, EX, A	VE01		0	
1199	FURALDEHYDES	6.1	TF1	II	6.1+3	802	LQ0	E4	T	PP, EP, EX, TOX, A	VE01, VE02		2	
1201	FUSEL OIL	3	F1	II	3		LQ4	E2		PP, EX, A	VE01		1	
1201	FUSEL OIL	3	F1	III	3		LQ7	E1		PP, EX, A	VE01		0	
1202	GAS OIL or DIESEL FUEL or HEATING OIL, LIGHT (flash-point not more than 60 °C)	3	F1	III	3	640K	LQ7	E1	T	PP, EX, A	VE01		0	
1202	DIESEL FUEL complying with standard EN 590:2004 or GAS OIL or HEATING OIL, LIGHT with a flash-point as specified in EN 590:2004	3	F1	III	3	640L	LQ7	E1	T	PP, EX, A	VE01		0	
1202	GAS OIL or DIESEL FUEL or HEATING OIL, LIGHT (flash-point more than 60 °C and not more than 100 °C)	3	F1	III	3	640M	LQ7	E1	T	PP, EX, A	VE01		0	
1203	MOTOR SPIRIT or GASOLINE or PETROL	3	F1	II	3	243 534	LQ4	E2	T	PP, EX, A	VE01		1	
1204	NITROGLYCERIN SOLUTION IN ALCOHOL with not more than 1% nitroglycerin	3	D	II	3	601	LQ0	E0		PP, EX, A	VE01		1	
1206	HEPTANES	3	F1	II	3		LQ4	E2	T	PP, EX, A	VE01		1	
1207	HEXALDEHYDE	3	F1	III	3		LQ7	E1		PP, EX, A	VE01		0	
1208	HEXANES	3	F1	II	3		LQ4	E2	T	PP, EX, A	VE01		1	
1210	PRINTING INK, flammable or PRINTING INK RELATED MATERIAL (including printing ink thinning or reducing compound), flammable	3	F1	I	3	163	LQ3	E3		PP, EX, A	VE01		1	
1210	PRINTING INK, flammable or PRINTING INK RELATED MATERIAL (including printing ink thinning or reducing compound), flammable (vapour pressure at 50 °C more than 110 kPa)	3	F1	II	3	163 640C	LQ6	E2		PP, EX, A	VE01		1	
1210	PRINTING INK, flammable or PRINTING INK RELATED MATERIAL (including printing ink thinning or reducing compound), flammable (vapour pressure at 50 °C not more than 110 kPa)	3	F1	II	3	163 640D	LQ6	E2		PP, EX, A	VE01		1	
1210	PRINTING INK, flammable or PRINTING INK RELATED MATERIAL (including printing ink thinning or reducing compound), flammable	3	F1	III	3	163 640E	LQ7	E1		PP, EX, A	VE01		0	
1210	PRINTING INK, flammable or PRINTING INK RELATED MATERIAL (including printing ink thinning or reducing compound), flammable (having a flash-point below 23 °C and viscous according to 2.2.3.1.4) (boiling point not more than 35° C)	3	F1	III	3	163 640F	LQ7	E1		PP, EX, A	VE01		0	

UN No. or ID No.	Name and description	Class	Classi-fication Code	Packing group	Labels	Special provi-sions	Limited and excepted quantities		Carriage permitted	Equipment required	Venti-lation	Provisions concerning loading, unloading and carriage	Number of blue cones/ lights	Remarks
	3.1.2	2.2	2.2	2.1.1.3	5.2.2	3.3	3.4.6	3.5.1.2	3.2.1	8.1.5	7.1.6	7.1.6	7.1.5	3.2.1
(1)	(2)	(3a)	(3b)	(4)	(5)	(6)	(7a)	(7b)	(8)	(9)	(10)	(11)	(12)	(13)
1210	PRINTING INK, flammable or PRINTING INK RELATED MATERIAL (including printing ink thinning or reducing compound), flammable (having a flash-point below 23 °C and viscous according to 2.2.3.1.4) (vapour pressure at 50 °C more than 110 kPa , boiling point of more than 35° C)	3	F1	III	3	163 640G	LQ7	E1		PP, EX, A	VE01		0	
1210	PRINTING INK, flammable or PRINTING INK RELATED MATERIAL (including printing ink thinning or reducing compound), flammable (having a flash-point below 23 °C and viscous according to 2.2.3.1.4) (vapour pressure at 50 °C not more than 110 kPa)	3	F1	III	3	163 640H	LQ7	E1		PP, EX, A	VE01		0	
1212	ISOBUTANOL (ISOBUTYL ALCOHOL)	3	F1	III	3		LQ7	E1	T	PP, EX, A	VE01		0	
1213	ISOBUTYL ACETATE	3	F1	II	3		LQ4	E2	T	PP, EX, A	VE01		1	
1214	ISOBUTYLAMINE	3	FC	II	3+8		LQ4	E2	T	PP, EP, EX, A	VE01		1	
1216	ISOOCTENES	3	F1	II	3		LQ4	E2	T	PP, EX, A	VE01		1	
1218	ISOPRENE, STABILIZED	3	F1	I	3		LQ3	E3	T	PP, EX, A	VE01		1	
1219	ISOPROPANOL (ISOPROPYL ALCOHOL)	3	F1	II	3	601	LQ4	E2	T	PP, EX, A	VE01		1	
1220	ISOPROPYL ACETATE	3	F1	II	3		LQ4	E2	T	PP, EX, A	VE01		1	
1221	ISOPROPYLAMINE	3	FC	I	3+8		LQ3	E0	T	PP, EP, EX, A	VE01		1	
1222	ISOPROPYL NITRATE	3	F1	II	3		LQ4	E2		PP, EX, A	VE01		1	
1223	KEROSENE	3	F1	III	3		LQ7	E1	T	PP, EX, A	VE01		0	
1224	KETONES, LIQUID, N.O.S. (vapour pressure at 50 °C more than 110 kPa)	3	F1	II	3	274 640C	LQ4	E2	T	PP, EX, A	VE01		1	
1224	KETONES, LIQUID, N.O.S. (vapour pressure at 50 °C not more than 110 kPa)	3	F1	II	3	274 640D	LQ4	E2	T	PP, EX, A	VE01		1	
1224	KETONES, LIQUID, N.O.S.	3	F1	III	3	274	LQ7	E1	T	PP, EX, A	VE01		0	
1228	MERCAPTANS, LIQUID, FLAMMABLE, TOXIC, N.O.S. or MERCAPTAN MIXTURE, LIQUID, FLAMMABLE, TOXIC, N.O.S.	3	FT1	II	3+6.1	274 802	LQ0	E2		PP, EP, EX, TOX, A	VE01, VE02		2	
1228	MERCAPTANS, LIQUID, FLAMMABLE, TOXIC, N.O.S. or MERCAPTAN MIXTURE, LIQUID, FLAMMABLE, TOXIC, N.O.S.	3	FT1	III	3+6.1	274 802	LQ7	E1		PP, EP, EX, TOX, A	VE01, VE02		0	
1229	MESITYL OXIDE	3	F1	III	3		LQ7	E1	T	PP, EX, A	VE01		0	
1230	METHANOL	3	FT1	II	3+6.1	279 802	LQ0	E2	T	PP, EP, EX, TOX, A	VE01, VE02		2	
1231	METHYL ACETATE	3	F1	II	3		LQ4	E2	T	PP, EX, A	VE01		1	
1233	METHYLAMYL ACETATE	3	F1	III	3		LQ7	E1		PP, EX, A	VE01		0	
1234	METHYLAL	3	F1	II	3		LQ4	E2		PP, EX, A	VE01		1	
1235	METHYLAMINE, AQUEOUS SOLUTION	3	FC	II	3+8		LQ4	E2	T	PP, EP, EX, A	VE01		1	
1237	METHYL BUTYRATE	3	F1	II	3		LQ4	E2		PP, EX, A	VE01		1	
1238	METHYL CHLOROFORMATE	6.1	TFC	I	6.1+3+8	802	LQ0	E5		PP, EP, EX, TOX, A	VE01, VE02		2	
1239	METHYL CHLOROMETHYL ETHER	6.1	TF1	I	6.1+3	802	LQ0	E5		PP, EP, EX, TOX, A	VE01, VE02		2	

UN No. or ID No.	Name and description	Class	Classi-fication Code	Packing group	Labels	Special provi-sions	Limited and excepted quantities		Carriage permitted	Equipment required	Venti-lation	Provisions concerning loading, unloading and carriage	Number of blue cones/ lights	Remarks
	3.1.2	2.2	2.2	2.1.1.3	5.2.2	3.3	3.4.6	3.5.1.2	3.2.1	8.1.5	7.1.6	7.1.6	7.1.5	3.2.1
(1)	(2)	(3a)	(3b)	(4)	(5)	(6)	(7a)	(7b)	(8)	(9)	(10)	(11)	(12)	(13)
1242	METHYLDICHLOROSILANE	4.3	WFC	I	4.3+3+8		LQ0	E0		PP, EP, EX, A	VE01	HA08	1	
1243	METHYL FORMATE	3	F1	I	3		LQ3	E3	T	PP, EX, A	VE01		1	
1244	METHYLHYDRAZINE	6.1	TFC	I	6.1+3+8	802	LQ0	E5	T	PP, EP, EX, TOX, A	VE01, VE02		2	
1245	METHYL ISOBUTYL KETONE	3	F1	II	3		LQ4	E2	T	PP, EX, A	VE01		1	
1246	METHYL ISOPROPENYL KETONE, STABILIZED	3	F1	II	3		LQ4	E2		PP, EX, A	VE01		1	
1247	METHYL METHACRYLATE MONOMER, STABILIZED	3	F1	II	3		LQ4	E2	T	PP, EX, A	VE01		1	
1248	METHYL PROPIONATE	3	F1	II	3		LQ4	E2		PP, EX, A	VE01		1	
1249	METHYL PROPYL KETONE	3	F1	II	3		LQ4	E2		PP, EX, A	VE01		1	
1250	METHYLTRICHLOROSILANE	3	FC	II	3+8		LQ4	E2		PP, EP, EX, A	VE01		1	
1251	METHYL VINYL KETONE, STABILIZED	6.1	TFC	I	6.1+3+8	802	LQ0	E5		PP, EP, EX, TOX, A	VE01, VE02		2	
1259	NICKEL CARBONYL	6.1	TF1	I	6.1+3	802	LQ0	E5		PP, EP, EX, TOX, A	VE01, VE02		2	
1261	NITROMETHANE	3	F1	II	3		LQ4	E2		PP, EX, A	VE01		1	
1262	OCTANES	3	F1	II	3		LQ4	E2	T	PP, EX, A	VE01		1	
1263	PAINT (including paint, lacquer, enamel, stain, shellac, varnish, polish, liquid filler and liquid lacquer base) or PAINT RELATED MATERIAL (including paint thinning and reducing compound)	3	F1	I	3	163 650	LQ3	E3		PP, EX, A	VE01		1	
1263	PAINT (including paint, lacquer, enamel, stain, shellac, varnish, polish, liquid filler and liquid lacquer base) or PAINT RELATED MATERIAL (including paint thinning and reducing compound) (vapour pressure at 50 °C more than 110 kPa)	3	F1	II	3	163 640C 650	LQ6	E2		PP, EX, A	VE01		1	
1263	PAINT (including paint, lacquer, enamel, stain, shellac, varnish, polish, liquid filler and liquid lacquer base) or PAINT RELATED MATERIAL (including paint thinning and reducing compound) (vapour pressure at 50 °C not more than 110 kPa)	3	F1	II	3	163 640D 650	LQ6	E2		PP, EX, A	VE01		1	
1263	PAINT (including paint, lacquer, enamel, stain, shellac, varnish, polish, liquid filler and liquid lacquer base) or PAINT RELATED MATERIAL (including paint thinning and reducing compound)	3	F1	III	3	163 640E 650	LQ7	E1		PP, EX, A	VE01		0	

UN No. or ID No.	Name and description 3.1.2	Class 2.2	Classi-fication Code 2.2	Packing group 2.1.1.3	Labels 5.2.2	Special provi-sions 3.3	Limited and excepted quantities 3.4.6	3.5.1.2	Carriage permitted 3.2.1	Equipment required 8.1.5	Venti-lation 7.1.6	Provisions concerning loading, unloading and carriage 7.1.6	Number of blue cones/ lights 7.1.5	Remarks 3.2.1
(1)	(2)	(3a)	(3b)	(4)	(5)	(6)	(7a)	(7b)	(8)	(9)	(10)	(11)	(12)	(13)
1263	PAINT (including paint, lacquer, enamel, stain, shellac, varnish, polish, liquid filler and liquid lacquer base) or PAINT RELATED MATERIAL (including paint thinning and reducing compound) (having a flash-point below 23 °C and viscous according to 2.2.3.1 (boiling point not more than 35° C).	3	F1	III	3	163 640G 650	LQ7	E1		PP, EX, A	VE01		0	
1263	PAINT (including paint, lacquer, enamel, stain, shellac, varnish, polish, liquid filler and liquid lacquer base) or PAINT RELATED MATERIAL (including paint thinning and reducing compound) (having a flash-point below 23 °C and viscous according to 2.2.3.1) (vapour pressure at 50 °C more than 110 kPa , boiling point of more than 35° C).	3	F1	III	3	163 640G 650	LQ7	E1		PP, EX, A	VE01		0	
1263	PAINT (including paint, lacquer, enamel, stain, shellac, varnish, polish, liquid filler and liquid lacquer base) or PAINT RELATED MATERIAL (including paint thinning and reducing compound) (having a flash-point below 23 °C and viscous according to 2.2.3.1) (vapour pressure at 50 °C not more than 110 kPa).	3	F1	III	3	163 640H 650	LQ7	E1		PP, EX, A	VE01		0	
1264	PARALDEHYDE	3	F1	III	3		LQ7	E1	T	PP, EX, A	VE01		0	
1265	PENTANES, liquid	3	F1	I	3		LQ3	E3	T	PP, EX, A	VE01		1	
1265	PENTANES, liquid	3	F1	II	3		LQ4	E2	T	PP, EX, A	VE01		1	
1266	PERFUMERY PRODUCTS with flammable solvents	3	F1	I	3		LQ3	E3		PP, EX, A	VE01		1	
1266	PERFUMERY PRODUCTS with flammable solvents (vapour pressure at 50 °C more than 110 kPa)	3	F1	II	3	640C	LQ6	E2		PP, EX, A	VE01		1	
1266	PERFUMERY PRODUCTS with flammable solvents (vapour pressure at 50 °C not more than 110 kPa)	3	F1	II	3	640D	LQ6	E2		PP, EX, A	VE01		1	
1266	PERFUMERY PRODUCTS with flammable solvents	3	F1	III	3	640E	LQ7	E1		PP, EX, A	VE01		0	
1266	PERFUMERY PRODUCTS with flammable solvents (having a flash-point below 23 °C and viscous according to 2.2.3.1.4) (boiling point not more than 35° C)	3	F1	III	3	640F	LQ7	E1		PP, EX, A	VE01		0	
1266	PERFUMERY PRODUCTS with flammable solvents (having a flash-point below 23 °C and viscous according to 2.2.3.1.4) (vapour pressure at 50 °C more than 110 kPa , boiling point of more than 35° C)	3	F1	III	3	640G	LQ7	E1		PP, EX, A	VE01		0	

UN No. or ID No.	Name and description	Class	Classi-fication Code	Packing group	Labels	Special provi-sions	Limited and excepted quantities		Carriage permitted	Equipment required	Venti-lation	Provisions concerning loading, unloading and carriage	Number of blue cones/ lights	Remarks
	3.1.2	2.2	2.2	2.1.1.3	5.2.2	3.3	3.4.6	3.5.1.2	3.2.1	8.1.5	7.1.6	7.1.6	7.1.5	3.2.1
(1)	(2)	(3a)	(3b)	(4)	(5)	(6)	(7a)	(7b)	(8)	(9)	(10)	(11)	(12)	(13)
1266	PERFUMERY PRODUCTS with flammable solvents (having a flash-point below 23 °C and viscous according to 2.2.3.1.4) (vapour pressure at 50 °C not more than 110 kPa)	3	F1	III	3	640H	LQ7	E1		PP, EX, A	VE01		0	
1267	PETROLEUM CRUDE OIL	3	F1	I	3	649	LQ3	E3	T	PP, EX, A	VE01		1	
1267	PETROLEUM CRUDE OIL (vapour pressure at 50 °C more than 110 kPa)	3	F1	II	3	640C 649	LQ4	E2	T	PP, EX, A	VE01		1	
1267	PETROLEUM CRUDE OIL (vapour pressure at 50 °C not more than 110 kPa)	3	F1	II	3	640D 649	LQ4	E2	T	PP, EX, A	VE01		1	
1267	PETROLEUM CRUDE OIL	3	F1	III	3		LQ7	E1	T	PP, EX, A	VE01		0	
1268	PETROLEUM DISTILLATES, N.O.S. or PETROLEUM PRODUCTS, N.O.S.	3	F1	I	3	649	LQ3	E3	T	PP, EX, A	VE01		1	
1268	PETROLEUM DISTILLATES, N.O.S. or PETROLEUM PRODUCTS, N.O.S. (vapour pressure at 50 °C more than 110 kPa)	3	F1	II	3	640C 649	LQ4	E2	T	PP, EX, A	VE01		1	
1268	PETROLEUM DISTILLATES, N.O.S. or PETROLEUM PRODUCTS, N.O.S. (vapour pressure at 50 °C not more than 110 kPa)	3	F1	II	3	640D 649	LQ4	E2	T	PP, EX, A	VE01		1	
1268	PETROLEUM DISTILLATES, N.O.S. or PETROLEUM PRODUCTS, N.O.S.	3	F1	III	3		LQ7	E1	T	PP, EX, A	VE01		0	
1272	PINE OIL	3	F1	III	3		LQ7	E1		PP, EX, A	VE01		0	
1274	n-PROPANOL (PROPYL ALCOHOL, NORMAL)	3	F1	II	3		LQ4	E2	T	PP, EX, A	VE01		1	
1274	n-PROPANOL (PROPYL ALCOHOL, NORMAL)	3	F1	III	3		LQ7	E1	T	PP, EX, A	VE01		0	
1275	PROPIONALDEHYDE	3	F1	II	3		LQ4	E2	T	PP, EX, A	VE01		1	
1276	n-PROPYL ACETATE	3	F1	II	3		LQ4	E2	T	PP, EX, A	VE01		1	
1277	PROPYLAMINE	3	FC	II	3+8		LQ4	E2	T	PP, EP, EX, A	VE01		1	
1278	1-CHLOROPROPANE	3	F1	II	3		LQ4	E2	T	PP, EX, A	VE01		1	
1279	1,2-DICHLOROPROPANE	3	F1	II	3		LQ4	E2	T	PP, EX, A	VE01		1	
1280	PROPYLENE OXIDE	3	F1	I	3		LQ3	E3	T	PP, EX, A	VE01		1	
1281	PROPYL FORMATES	3	F1	II	3		LQ4	E2		PP, EX, A	VE01		1	
1282	PYRIDINE	3	F1	II	3		LQ4	E2	T	PP, EX, A	VE01		1	
1286	ROSIN OIL	3	F1	I	3		LQ3	E3		PP, EX, A	VE01		1	
1286	ROSIN OIL (vapour pressure at 50 °C more than 110 kPa)	3	F1	II	3	640C	LQ6	E2		PP, EX, A	VE01		1	
1286	ROSIN OIL (vapour pressure at 50 °C not more than 110 kPa)	3	F1	II	3	640D	LQ6	E2		PP, EX, A	VE01		1	
1286	ROSIN OIL	3	F1	III	3	640E	LQ7	E1		PP, EX, A	VE01		0	
1286	ROSIN OIL (having a flash-point below 23 °C and viscous according to 2.2.3.1.4) (boiling point not more than 35° C)	3	F1	III	3	640F	LQ7	E1		PP, EX, A	VE01		0	
1286	ROSIN OIL (having a flash-point below 23 °C and viscous according to 2.2.3.1.4) (vapour pressure at 50 °C more than 110 kPa , boiling point of more than 35° C)	3	F1	III	3	640G	LQ7	E1		PP, EX, A	VE01		0	

UN No. or ID No.	Name and description	Class	Classi-fication Code	Packing group	Labels	Special provi-sions	Limited and excepted quantities		Carriage permitted	Equipment required	Venti-lation	Provisions concerning loading, unloading and carriage	Number of blue cones/ lights	Remarks
	3.1.2	2.2	2.2	2.1.1.3	5.2.2	3.3	3.4.6	3.5.1.2	3.2.1	8.1.5	7.1.6	7.1.6	7.1.5	3.2.1
(1)	(2)	(3a)	(3b)	(4)	(5)	(6)	(7a)	(7b)	(8)	(9)	(10)	(11)	(12)	(13)
1286	ROSIN OIL (having a flash-point below 23 °C and viscous according to 2.2.3.1.4) (vapour pressure at 50 °C not more than 110 kPa)	3	F1	III	3	640H	LQ7	E1		PP, EX, A	VE01		0	
1287	RUBBER SOLUTION	3	F1	I	3		LQ3	E3		PP, EX, A	VE01		1	
1287	RUBBER SOLUTION (vapour pressure at 50 °C more than 110 kPa)	3	F1	II	3	640C	LQ6	E2		PP, EX, A	VE01		1	
1287	RUBBER SOLUTION (vapour pressure at 50 °C not more than 110 kPa)	3	F1	II	3	640D	LQ6	E2		PP, EX, A	VE01		1	
1287	RUBBER SOLUTION	3	F1	III	3	640E	LQ7	E1		PP, EX, A	VE01		0	
1287	RUBBER SOLUTION (having a flash-point below 23 °C and viscous according to 2.2.3.1.4) (boiling point not more than 35° C)	3	F1	III	3	640F	LQ7	E1		PP, EX, A	VE01		0	
1287	RUBBER SOLUTION (having a flash-point below 23 °C and viscous according to 2.2.3.1.4) (vapour pressure at 50 °C more than 110 kPa , boiling point of more than 35° C)	3	F1	III	3	640G	LQ7	E1		PP, EX, A	VE01		0	
1287	RUBBER SOLUTION (having a flash-point below 23 °C and viscous according to 2.2.3.1.4) (vapour pressure at 50 °C not more than 110 kPa)	3	F1	III	3	640H	LQ7	E1		PP, EX, A	VE01		0	
1288	SHALE OIL	3	F1	II	3		LQ4	E2		PP, EX, A	VE01		1	
1288	SHALE OIL	3	F1	III	3		LQ7	E1		PP, EX, A	VE01		0	
1289	SODIUM METHYLATE SOLUTION in alcohol	3	FC	II	3+8		LQ4	E2		PP, EP, EX, A	VE01		1	
1289	SODIUM METHYLATE SOLUTION in alcohol	3	FC	III	3+8		LQ7	E1	T	PP, EP, EX, A	VE01		0	
1292	TETRAETHYL SILICATE	3	F1	III	3		LQ7	E1		PP, EX, A	VE01		0	
1293	TINCTURES, MEDICINAL	3	F1	II	3	601	LQ4	E2		PP, EX, A	VE01		1	
1293	TINCTURES, MEDICINAL	3	F1	III	3	601	LQ7	E1		PP, EX, A	VE01		0	
1294	TOLUENE	3	F1	II	3		LQ4	E2	T	PP, EX, A	VE01		1	
1295	TRICHLOROSILANE	4.3	WFC	I	4.3+3+8		LQ0	E0		PP, EP, EX, A	VE01	HA08	1	
1296	TRIETHYLAMINE	3	FC	II	3+8		LQ4	E2	T	PP, EP, EX, A	VE01		1	
1297	TRIMETHYLAMINE, AQUEOUS SOLUTION, not more than 50% trimethylamine, by mass	3	FC	I	3+8		LQ3	E0		PP, EP, EX, A	VE01		1	
1297	TRIMETHYLAMINE, AQUEOUS SOLUTION, not more than 50% trimethylamine, by mass	3	FC	II	3+8		LQ4	E2		PP, EP, EX, A	VE01		1	
1297	TRIMETHYLAMINE, AQUEOUS SOLUTION, not more than 50% trimethylamine, by mass	3	FC	III	3+8		LQ7	E1		PP, EP, EX, A	VE01		0	
1298	TRIMETHYLCHLOROSILANE	3	FC	II	3+8		LQ4	E2		PP, EP, EX, A	VE01		1	
1299	TURPENTINE	3	F1	III	3		LQ7	E1		PP, EX, A	VE01		0	
1300	TURPENTINE SUBSTITUTE	3	F1	II	3		LQ4	E2		PP, EX, A	VE01		1	
1300	TURPENTINE SUBSTITUTE	3	F1	III	3		LQ7	E1	T	PP, EX, A	VE01		0	
1301	VINYL ACETATE, STABILIZED	3	F1	II	3		LQ4	E2	T	PP, EX, A	VE01		1	
1302	VINYL ETHYL ETHER, STABILIZED	3	F1	I	3		LQ3	E3		PP, EX, A	VE01		1	
1303	VINYLIDENE CHLORIDE, STABILIZED	3	F1	I	3		LQ3	E3		PP, EX, A	VE01		1	
1304	VINYL ISOBUTYL ETHER, STABILIZED	3	F1	II	3		LQ4	E2		PP, EX, A	VE01		1	

UN No. or ID No.	Name and description	Class	Classi-fication Code	Packing group	Labels	Special provi-sions	Limited and excepted quantities		Carriage permitted	Equipment required	Venti-lation	Provisions concerning loading, unloading and carriage	Number of blue cones/ lights	Remarks
	3.1.2	2.2	2.2	2.1.1.3	5.2.2	3.3	3.4.6	3.5.1.2	3.2.1	8.1.5	7.1.6	7.1.6	7.1.5	3.2.1
(1)	(2)	(3a)	(3b)	(4)	(5)	(6)	(7a)	(7b)	(8)	(9)	(10)	(11)	(12)	(13)
1305	VINYLTRICHLOROSILANE, STABILIZED	3	FC	II	3+8		LQ4	E2		PP, EP, EX, A	VE01		1	
1306	WOOD PRESERVATIVES, LIQUID (vapour pressure at 50 °C more than 110 kPa)	3	F1	II	3	640C	LQ6	E2		PP, EX, A	VE01		1	
1306	WOOD PRESERVATIVES, LIQUID (vapour pressure at 50 °C not more than 110 kPa)	3	F1	II	3	640D	LQ6	E2		PP, EX, A	VE01		1	
1306	WOOD PRESERVATIVES, LIQUID	3	F1	III	3	640E	LQ7	E1		PP, EX, A	VE01		0	
1306	WOOD PRESERVATIVES, LIQUID (having a flash-point below 23 °C and viscous according to 2.2.3.1.4) (boiling point not more than 35° C)	3	F1	III	3	640F	LQ7	E1		PP, EX, A	VE01		0	
1306	WOOD PRESERVATIVES, LIQUID (having a flash-point below 23 °C and viscous according to 2.2.3.1.4) (vapour pressure at 50 °C more than 110 kPa , boiling point of more than 35° C)	3	F1	III	3	640G	LQ7	E1		PP, EX, A	VE01		0	
1306	WOOD PRESERVATIVES, LIQUID (having a flash-point below 23 °C and viscous according to 2.2.3.1.4) (vapour pressure at 50 °C not more than 110 kPa)	3	F1	III	3	640H	LQ7	E1		PP, EX, A	VE01		0	
1307	XYLENES	3	F1	II	3		LQ4	E2	T	PP, EX, A	VE01		1	
1307	XYLENES	3	F1	III	3		LQ7	E1	T	PP, EX, A	VE01		0	
1308	ZIRCONIUM SUSPENDED IN A FLAMMABLE LIQUID	3	F1	I	3		LQ3	E3		PP, EX, A	VE01		1	
1308	ZIRCONIUM SUSPENDED IN A FLAMMABLE LIQUID (vapour pressure at 50 °C more than 110 kPa)	3	F1	II	3	640C	LQ4	E2		PP, EX, A	VE01		1	
1308	ZIRCONIUM SUSPENDED IN A FLAMMABLE LIQUID (vapour pressure at 50 °C not more than 110 kPa)	3	F1	II	3	640D	LQ4	E2		PP, EX, A	VE01		1	
1308	ZIRCONIUM SUSPENDED IN A FLAMMABLE LIQUID	3	F1	III	3		LQ7	E1		PP, EX, A	VE01		0	
1309	ALUMINIUM POWDER, COATED	4.1	F3	II	4.1		LQ8	E2		PP			1	
1309	ALUMINIUM POWDER, COATED	4.1	F3	III	4.1		LQ9	E1		PP			0	
1310	AMMONIUM PICRATE, WETTED with not less than 10% water, by mass	4.1	D	I	4.1		LQ0	E0		PP			1	
1312	BORNEOL	4.1	F1	III	4.1		LQ9	E1		PP			0	
1313	CALCIUM RESINATE	4.1	F3	III	4.1		LQ9	E1		PP			0	
1314	CALCIUM RESINATE, FUSED	4.1	F3	III	4.1		LQ9	E1		PP			0	
1318	COBALT RESINATE, PRECIPITATED	4.1	F3	III	4.1		LQ9	E1		PP			0	
1320	DINITROPHENOL, WETTED with not less than 15% water, by mass	4.1	DT	I	4.1+6.1	802	LQ0	E0		PP			2	
1321	DINITROPHENOLATES, WETTED with not less than 15% water, by mass	4.1	DT	I	4.1+6.1	802	LQ0	E0		PP			2	
1322	DINITRORESORCINOL, WETTED with not less than 15% water, by mass	4.1	D	I	4.1		LQ0	E0		PP			1	
1323	FERROCERIUM	4.1	F3	II	4.1	249	LQ8	E2		PP			1	

UN No. or ID No.	Name and description	Class	Classi-fication Code	Packing group	Labels	Special provi-sions	Limited and excepted quantities		Carriage permitted	Equipment required	Venti-lation	Provisions concerning loading, unloading and carriage			Number of blue cones/ lights	Remarks
	3.1.2	2.2	2.2	2.1.1.3	5.2.2	3.3	3.4.6	3.5.1.2	3.2.1	8.1.5	7.1.6	7.1.6			7.1.5	3.2.1
(1)	(2)	(3a)	(3b)	(4)	(5)	(6)	(7a)	(7b)	(8)	(9)	(10)	(11)			(12)	(13)
1324	FILMS, NITROCELLULOSE BASE, gelatin coated, except scrap	4.1	F1	III	4.1		LQ9	E1		PP					0	
1325	FLAMMABLE SOLID, ORGANIC, N.O.S.	4.1	F1	II	4.1	274	LQ8	E2		PP					1	
1325	FLAMMABLE SOLID, ORGANIC, N.O.S.	4.1	F1	III	4.1	274	LQ9	E1		PP					0	
1326	HAFNIUM POWDER, WETTED with not less than 25% water	4.1	F3	II	4.1	586	LQ8	E2		PP					1	
1327	Hay, Straw or Bhusa	4.1	F1	NOT SUBJECT TO ADN												
1328	HEXAMETHYLENETETRAMINE	4.1	F1	III	4.1		LQ9	E1		PP					0	
1330	MANGANESE RESINATE	4.1	F3	III	4.1		LQ9	E1		PP					0	
1331	MATCHES, 'STRIKE ANYWHERE'	4.1	F1	III	4.1	293	LQ9	E1		PP					0	
1332	METALDEHYDE	4.1	F1	III	4.1		LQ9	E1		PP					0	
1333	CERIUM, slabs, ingots or rods	4.1	F3	II	4.1		LQ8	E2		PP					1	
1334	NAPHTHALENE, CRUDE or NAPHTHALENE, REFINED	4.1	F1	III	4.1	501	LQ9	E1	**B**	PP		CO01			0	
1336	NITROGUANIDINE (PICRITE), WETTED with not less than 20% water, by mass	4.1	D	I	4.1		LQ0	E0		PP					1	
1337	NITROSTARCH, WETTED with not less than 20% water, by mass	4.1	D	I	4.1		LQ0	E0		PP					1	
1338	PHOSPHORUS, AMORPHOUS	4.1	F3	III	4.1		LQ9	E1		PP					0	
1339	PHOSPHORUS HEPTASULPHIDE, free from yellow and white phosphorus	4.1	F3	II	4.1	602	LQ8	E2		PP					1	
1340	PHOSPHORUS PENTASULPHIDE, free from yellow and white phosphorus	4.3	WF2	II	4.3+4.1	602	LQ11	E2		PP, EX, A	VE01		HA08		1	
1341	PHOSPHORUS SESQUISULPHIDE, free from yellow and white phosphorus	4.1	F3	II	4.1	602	LQ8	E2		PP					1	
1343	PHOSPHORUS TRISULPHIDE, free from yellow and white phosphorus	4.1	F3	II	4.1	602	LQ8	E2		PP					1	
1344	TRINITROPHENOL (PICRIC ACID), WETTED with not less than 30% water, by mass	4.1	D	I	4.1		LQ0	E0		PP					1	
1345	RUBBER SCRAP or RUBBER SHODDY, powdered or granulated	4.1	F1	II	4.1		LQ8	E2		PP					1	
1346	SILICON POWDER, AMORPHOUS	4.1	F3	III	4.1	32	LQ9	E1		PP					0	
1347	SILVER PICRATE, WETTED with not less than 30% water, by mass	4.1	D	I	4.1		LQ0	E0		PP					1	
1348	SODIUM DINITRO-o-CRESOLATE, WETTED with not less than 15% water, by mass	4.1	DT	I	4.1+6.1	802	LQ0	E0		PP					2	
1349	SODIUM PICRAMATE, WETTED with not less than 20% water, by mass	4.1	D	I	4.1		LQ0	E0		PP					1	
1350	SULPHUR	4.1	F3	III	4.1	242	LQ9	E1	**B**	PP					0	
1352	TITANIUM POWDER, WETTED with not less than 25% water	4.1	F3	II	4.1	586	LQ8	E2		PP					1	
1353	FIBRES or FABRICS IMPREGNATED WITH WEAKLY NITRATED NITROCELLULOSE, N.O.S.	4.1	F1	III	4.1	274 502	LQ9	E1		PP					0	

UN No. or ID No.	Name and description	Class	Classi-fication Code	Packing group	Labels	Special provi-sions	Limited and excepted quantities		Carriage permitted	Equipment required	Venti-lation	Provisions concerning loading, unloading and carriage	Number of blue cones/ lights	Remarks
	3.1.2	2.2	2.2	2.1.1.3	5.2.2	3.3	3.4.6	3.5.1.2	3.2.1	8.1.5	7.1.6	7.1.6	7.1.5	3.2.1
(1)	(2)	(3a)	(3b)	(4)	(5)	(6)	(7a)	(7b)	(8)	(9)	(10)	(11)	(12)	(13)
1354	TRINITROBENZENE, WETTED with not less than 30% water, by mass	4.1	D	I	4.1		LQ0	E0		PP			1	
1355	TRINITROBENZOIC ACID, WETTED with not less than 30% water, by mass	4.1	D	I	4.1		LQ0	E0		PP			1	
1356	TRINITROTOLUENE (TNT), WETTED with not less than 30% water, by mass	4.1	D	I	4.1		LQ0	E0		PP			1	
1357	UREA NITRATE, WETTED with not less than 20% water, by mass	4.1	D	I	4.1	227	LQ0	E0		PP			1	
1358	ZIRCONIUM POWDER, WETTED with not less than 25% water	4.1	F3	II	4.1	586	LQ8	E2		PP			1	
1360	CALCIUM PHOSPHIDE	4.3	WT2	I	4.3+6.1	802	LQ0	E0		PP, EP, EX, TOX, A	VE01, VE02	HA08	2	
1361	CARBON, animal or vegetable origin	4.2	S2	II	4.2		LQ0	E2		PP			0	
1361	CARBON, animal or vegetable origin	4.2	S2	III	4.2		LQ0	E1		PP			0	
1362	CARBON, ACTIVATED	4.2	S2	III	4.2	646	LQ0	E1		PP			0	
1363	COPRA	4.2	S2	III	4.2		LQ0	E1	**B**	PP		IN01, IN02	0	IN01 and IN02 apply only when this substance is carried in bulk or without packaging
1364	COTTON WASTE, OILY	4.2	S2	III	4.2		LQ0	E1	**B**	PP			0	
1365	COTTON, WET	4.2	S2	III	4.2		LQ0	E1	**B**	PP			0	
1369	p-NITROSODIMETHYLANILINE	4.2	S2	II	4.2		LQ0	E2		PP			0	
1372	Fibres, animal or fibres, vegetable burnt, wet or damp	4.2	S2	**NOT SUBJECT TO ADN**										
1373	FIBRES or FABRICS, ANIMAL or VEGETABLE or SYNTHETIC, N.O.S. with oil	4.2	S2	III	4.2	274	LQ0	E1	**B**	PP			0	
1374	FISH MEAL (FISH SCRAP), UNSTABILIZED	4.2	S2	II	4.2	300	LQ0	E2		PP			0	
1376	IRON OXIDE, SPENT or IRON SPONGE, SPENT obtained from coal gas purification	4.2	S4	III	4.2	592	LQ0	E1	**B**	PP			0	
1378	METAL CATALYST, WETTED with a visible excess of liquid	4.2	S4	II	4.2	274	LQ0	E2		PP			0	
1379	PAPER, UNSATURATED OIL TREATED, incompletely dried (including carbon paper)	4.2	S2	III	4.2		LQ0	E1	**B**	PP			0	
1380	PENTABORANE	4.2	ST3	I	4.2+6.1	802	LQ0	E0		PP, EP, TOX, A	VE02		2	
1381	PHOSPHORUS, WHITE or YELLOW, UNDER WATER or IN SOLUTION	4.2	ST3	I	4.2+6.1	503 802	LQ0	E0		PP, EP, TOX, A	VE02		2	
1381	PHOSPHORUS, WHITE or YELLOW, DRY	4.2	ST4	I	4.2+6.1	503 802	LQ0	E0		PP, EP			2	
1382	POTASSIUM SULPHIDE, ANHYDROUS or POTASSIUM SULPHIDE with less than 30% water of crystallization	4.2	S4	II	4.2	504	LQ0	E2		PP			0	
1383	PYROPHORIC METAL, N.O.S. or PYROPHORIC ALLOY, N.O.S.	4.2	S4	I	4.2	274	LQ0	E0		PP			0	

UN No. or ID No.	Name and description	Class	Classi-fication Code	Packing group	Labels	Special provi-sions	Limited and excepted quantities		Carriage permitted	Equipment required	Venti-lation	Provisions concerning loading, unloading and carriage			Number of blue cones/ lights	Remarks
	3.1.2	2.2	2.2	2.1.1.3	5.2.2	3.3	3.4.6	3.5.1.2	3.2.1	8.1.5	7.1.6	7.1.6			7.1.5	3.2.1
(1)	(2)	(3a)	(3b)	(4)	(5)	(6)	(7a)	(7b)	(8)	(9)	(10)	(11)			(12)	(13)
1384	SODIUM DITHIONITE (SODIUM HYDROSULPHITE)	4.2	S4	II	4.2		LQ0	E2		PP					0	
1385	SODIUM SULPHIDE, ANHYDROUS or SODIUM SULPHIDE with less than 30% water of crystallization	4.2	S4	II	4.2	504	LQ0	E2		PP					0	
1386	SEED CAKE with more than 1.5% oil and not more than 11% moisture	4.2	S2	III	4.2	800	LQ0	E1	B	PP				IN01, IN02	0	IN01 and IN02 apply only when this substance is carried in bulk or without packaging
1387	Wool waste, wet	4.2	S2	NOT SUBJECT TO ADN												
1389	ALKALI METAL AMALGAM, LIQUID	4.3	W1	I	4.3	182 274	LQ0	E0		PP, EX, A	VE01		HA08		0	
1390	ALKALI METAL AMIDES	4.3	W2	II	4.3	182 274 505	LQ11	E2		PP, EX, A	VE01		HA08		0	
1391	ALKALI METAL DISPERSION or ALKALINE EARTH METAL DISPERSION having a flash-point above 60 °C	4.3	W1	I	4.3	182 183 274 506	LQ0	E0		PP, EX, A	VE01		HA08		0	
1391	ALKALI METAL DISPERSION or ALKALINE EARTH METAL DISPERSION having a flash-point of not more than 60 °C	4.3	WF1	I	4.3 3	182 183 274 506	LQ0	E0		PP, EX, A	VE01		HA08		0	
1392	ALKALINE EARTH METAL AMALGAM, LIQUID	4.3	W1	I	4.3	183 274 506	LQ0	E0		PP, EX, A	VE01		HA08		0	
1393	ALKALINE EARTH METAL ALLOY, N.O.S.	4.3	W2	II	4.3	183 274 506	LQ11	E2		PP, EX, A	VE01		HA08		0	
1394	ALUMINIUM CARBIDE	4.3	W2	II	4.3		LQ11	E2		PP, EX, A	VE01		HA08		0	
1395	ALUMINIUM FERROSILICON POWDER	4.3	WT2	II	4.3+6.1	802	LQ11	E2		PP, EP, EX, TOX, A	VE01, VE02		HA08		2	
1396	ALUMINIUM POWDER, UNCOATED	4.3	W2	II	4.3		LQ12	E2		PP, EX, A	VE01		HA08		0	
1396	ALUMINIUM POWDER, UNCOATED	4.3	W2	III	4.3		LQ12	E1		PP, EX, A	VE01		HA08		0	
1397	ALUMINIUM PHOSPHIDE	4.3	WT2	I	4.3+6.1	507 802	LQ0	E0		PP, EP, EX, TOX, A	VE01, VE02		HA08		2	
1398	ALUMINIUM SILICON POWDER, UNCOATED	4.3	W2	III	4.3	37	LQ12	E1	B	PP, EX, A	VE01, VE03	LO03	HA07, HA08	IN01, IN03	0	VE03, LO03, HA07, IN01 and IN03 apply only when this substance is carried in bulk or without packaging
1400	BARIUM	4.3	W2	II	4.3		LQ11	E2		PP, EX, A	VE01		HA08		0	
1401	CALCIUM	4.3	W2	II	4.3		LQ11	E2		PP, EX, A	VE01		HA08		0	
1402	CALCIUM CARBIDE	4.3	W2	I	4.3		LQ0	E0		PP, EX, A	VE01		HA08		0	

UN No. or ID No.	Name and description	Class	Classi-fication Code	Packing group	Labels	Special provi-sions	Limited and excepted quantities		Carriage permitted	Equipment required	Venti-lation	Provisions concerning loading, unloading and carriage	Number of blue cones/ lights	Remarks
	3.1.2	2.2	2.2	2.1.1.3	5.2.2	3.3	3.4.6	3.5.1.2	3.2.1	8.1.5	7.1.6	7.1.6	7.1.5	3.2.1
(1)	(2)	(3a)	(3b)	(4)	(5)	(6)	(7a)	(7b)	(8)	(9)	(10)	(11)	(12)	(13)
1402	CALCIUM CARBIDE	4.3	W2	II	4.3		LQ11	E2		PP, EX, A	VE01	HA08	0	
1403	CALCIUM CYANAMIDE with more than 0.1% calcium carbide	4.3	W2	III	4.3	38	LQ12	E1		PP, EX, A	VE01	HA08	0	
1404	CALCIUM HYDRIDE	4.3	W2	I	4.3		LQ0	E0		PP, EX, A	VE01	HA08	0	
1405	CALCIUM SILICIDE	4.3	W2	II	4.3		LQ11	E2		PP, EX, A	VE01	HA08	0	
1405	CALCIUM SILICIDE	4.3	W2	III	4.3		LQ12	E1		PP, EX, A	VE01	HA08	0	
1407	CAESIUM	4.3	W2	I	4.3		LQ0	E0		PP, EX, A	VE01	HA08	0	
1408	FERROSILICON with 30% or more but less than 90% silicon	4.3	WT2	III	4.3+6.1	39 802	LQ12	E1	**B**	PP, EP, EX, TOX, A	VE01, VE02, VE03	LO03 HA07, HA08 IN01, IN02, IN03	0	VE03, LO03, HA07, IN01, IN02 and IN03 apply only when this substance is carried in bulk or without packaging
1409	METAL HYDRIDES, WATER-REACTIVE, N.O.S.	4.3	W2	I	4.3	274 508	LQ0	E0		PP, EX, A	VE01	HA08	0	
1409	METAL HYDRIDES, WATER-REACTIVE, N.O.S.	4.3	W2	II	4.3	274 508	LQ11	E2		PP, EX, A	VE01	HA08	0	
1410	LITHIUM ALUMINIUM HYDRIDE	4.3	W2	I	4.3		LQ0	E0		PP, EX, A	VE01	HA08	0	
1411	LITHIUM ALUMINIUM HYDRIDE, ETHEREAL	4.3	WF1	I	4.3+3		LQ0	E0		PP, EX, A	VE01	HA08	1	
1413	LITHIUM BOROHYDRIDE	4.3	W2	I	4.3		LQ0	E0		PP, EX, A	VE01	HA08	0	
1414	LITHIUM HYDRIDE	4.3	W2	I	4.3		LQ0	E0		PP, EX, A	VE01	HA08	0	
1415	LITHIUM	4.3	W2	I	4.3		LQ0	E0		PP, EX, A	VE01	HA08	0	
1417	LITHIUM SILICON	4.3	W2	II	4.3		LQ11	E2		PP, EX, A	VE01	HA08	0	
1418	MAGNESIUM POWDER or MAGNESIUM ALLOYS POWDER	4.3	WS	I	4.3+4.2		LQ0	E0		PP, EX, A	VE01	HA08	0	
1418	MAGNESIUM POWDER or MAGNESIUM ALLOYS POWDER	4.3	WS	II	4.3+4.2		LQ11	E2		PP, EX, A	VE01	HA08	0	
1418	MAGNESIUM POWDER or MAGNESIUM ALLOYS POWDER	4.3	WS	III	4.3+4.2		LQ12	E1		PP, EX, A	VE01	HA08	0	
1419	MAGNESIUM ALUMINIUM PHOSPHIDE	4.3	WT2	I	4.3+6.1	802	LQ0	E0		PP, EP, EX, TOX, A	VE01, VE02	HA08	2	
1420	POTASSIUM METAL ALLOYS, LIQUID	4.3	W1	I	4.3		LQ0	E0		PP, EX, A	VE01	HA08	0	
1421	ALKALI METAL ALLOY, LIQUID, N.O.S.	4.3	W1	I	4.3	182 274	LQ0	E0		PP, EX, A	VE01	HA08	0	
1422	POTASSIUM SODIUM ALLOYS, LIQUID	4.3	W1	I	4.3		LQ0	E0		PP, EX, A	VE01	HA08	0	
1423	RUBIDIUM	4.3	W2	I	4.3		LQ0	E0		PP, EX, A	VE01	HA08	0	
1426	SODIUM BOROHYDRIDE	4.3	W2	I	4.3		LQ0	E0		PP, EX, A	VE01	HA08	0	
1427	SODIUM HYDRIDE	4.3	W2	I	4.3		LQ0	E0		PP, EX, A	VE01	HA08	0	
1428	SODIUM	4.3	W2	I	4.3		LQ0	E0		PP, EX, A	VE01	HA08	0	
1431	SODIUM METHYLATE	4.2	SC4	II	4.2+8		LQ0	E2		PP			0	
1432	SODIUM PHOSPHIDE	4.3	WT2	I	4.3+6.1	802	LQ0	E0		PP, EP, EX, TOX, A	VE01, VE02	HA08	2	
1433	STANNIC PHOSPHIDES	4.3	WT2	I	4.3+6.1	802	LQ0	E0		PP, EP, EX, TOX, A	VE01, VE02	HA08	2	

UN No. or ID No.	Name and description 3.1.2	Class 2.2	Classi-fication Code 2.2	Packing group 2.1.1.3	Labels 5.2.2	Special provi-sions 3.3	Limited and excepted quantities 3.4.6	 3.5.1.2	Carriage permitted 3.2.1	Equipment required 8.1.5	Venti-lation 7.1.6	Provisions concerning loading, unloading and carriage 7.1.6			Number of blue cones/ lights 7.1.5	Remarks 3.2.1
(1)	(2)	(3a)	(3b)	(4)	(5)	(6)	(7a)	(7b)	(8)	(9)	(10)	(11)			(12)	(13)
1435	ZINC ASHES	4.3	W2	III	4.3		LQ12	E1	**B**	PP, EX, A	VE01, VE03	LO03	HA07, HA08	IN01, IN03	0	VE03, LO03, HA07, IN01 and IN03 apply only when this substance is carried in bulk or without packaging
1436	ZINC POWDER or ZINC DUST	4.3	WS	I	4.3+4.2		LQ0	E0		PP, EX, A	VE01		HA08		0	
1436	ZINC POWDER or ZINC DUST	4.3	WS	II	4.3+4.2		LQ11	E2		PP, EX, A	VE01		HA08		0	
1436	ZINC POWDER or ZINC DUST	4.3	WS	III	4.3+4.2		LQ12	E1		PP, EX, A	VE01		HA08		0	
1437	ZIRCONIUM HYDRIDE	4.1	F3	II	4.1		LQ8	E2		PP					1	
1438	ALUMINIUM NITRATE	5.1	O2	III	5.1		LQ12	E1	**B**	PP		CO02 LO04			0	CO02 and LO04 apply only when this substance is carried in bulk or without packaging
1439	AMMONIUM DICHROMATE	5.1	O2	II	5.1		LQ11	E2		PP					0	
1442	AMMONIUM PERCHLORATE	5.1	O2	II	5.1	152	LQ11	E2		PP					0	
1444	AMMONIUM PERSULPHATE	5.1	O2	III	5.1		LQ12	E1		PP					0	
1445	BARIUM CHLORATE, SOLID	5.1	OT2	II	5.1+6.1	802	LQ11	E2		PP					2	
1446	BARIUM NITRATE	5.1	OT2	II	5.1+6.1	802	LQ11	E2		PP					2	
1447	BARIUM PERCHLORATE, SOLID	5.1	OT2	II	5.1+6.1	802	LQ11	E2		PP					2	
1448	BARIUM PERMANGANATE	5.1	OT2	II	5.1+6.1	802	LQ11	E2		PP					2	
1449	BARIUM PEROXIDE	5.1	OT2	II	5.1+6.1	802	LQ11	E2		PP					2	
1450	BROMATES, INORGANIC, N.O.S.	5.1	O2	II	5.1	274 604	LQ11	E2		PP					0	
1451	CAESIUM NITRATE	5.1	O2	III	5.1		LQ12	E1	**B**	PP		CO02 LO04			0	CO02 and LO04 apply only when this substance is carried in bulk or without packaging
1452	CALCIUM CHLORATE	5.1	O2	II	5.1		LQ11	E2		PP					0	
1453	CALCIUM CHLORITE	5.1	O2	II	5.1		LQ11	E2		PP					0	
1454	CALCIUM NITRATE	5.1	O2	III	5.1	208	LQ12	E1	**B**	PP		CO02 LO04			0	CO02 and LO04 apply only when this substance is carried in bulk or without packaging
1455	CALCIUM PERCHLORATE	5.1	O2	II	5.1		LQ11	E2		PP					0	
1456	CALCIUM PERMANGANATE	5.1	O2	II	5.1		LQ11	E2		PP					0	
1457	CALCIUM PEROXIDE	5.1	O2	II	5.1		LQ11	E2		PP					0	
1458	CHLORATE AND BORATE MIXTURE	5.1	O2	II	5.1		LQ11	E2		PP					0	
1458	CHLORATE AND BORATE MIXTURE	5.1	O2	III	5.1		LQ12	E1		PP					0	
1459	CHLORATE AND MAGNESIUM CHLORIDE MIXTURE, SOLID	5.1	O2	II	5.1		LQ11	E2		PP					0	
1459	CHLORATE AND MAGNESIUM CHLORIDE MIXTURE, SOLID	5.1	O2	III	5.1		LQ12	E1		PP					0	

UN No. or ID No.	Name and description	Class	Classi-fication Code	Packing group	Labels	Special provi-sions	Limited and excepted quantities		Carriage permitted	Equipment required	Venti-lation	Provisions concerning loading, unloading and carriage			Number of blue cones/ lights	Remarks
	3.1.2	2.2	2.2	2.1.1.3	5.2.2	3.3	3.4.6	3.5.1.2	3.2.1	8.1.5	7.1.6	7.1.6			7.1.5	3.2.1
(1)	(2)	(3a)	(3b)	(4)	(5)	(6)	(7a)	(7b)	(8)	(9)	(10)	(11)			(12)	(13)
1461	CHLORATES, INORGANIC, N.O.S.	5.1	O2	II	5.1	274 605	LQ11	E2		PP					0	
1462	CHLORITES, INORGANIC, N.O.S.	5.1	O2	II	5.1	274 509 606	LQ11	E2		PP					0	
1463	CHROMIUM TRIOXIDE, ANHYDROUS	5.1	OTC	II	5.1+6.1+8	510	LQ11	E2		PP					0	
1465	DIDYMIUM NITRATE	5.1	O2	III	5.1		LQ12	E1	**B**	PP		CO02 LO04			0	CO02 and LO04 apply only when this substance is carried in bulk or without packaging
1466	FERRIC NITRATE	5.1	O2	III	5.1		LQ12	E1	**B**	PP		CO02 LO04			0	CO02 and LO04 apply only when this substance is carried in bulk or without packaging
1467	GUANIDINE NITRATE	5.1	O2	III	5.1		LQ12	E1	**B**	PP		CO02 LO04			0	CO02 and LO04 apply only when this substance is carried in bulk or without packaging
1469	LEAD NITRATE	5.1	OT2	II	5.1+6.1	802	LQ11	E2		PP					2	
1470	LEAD PERCHLORATE, SOLID	5.1	OT2	II	5.1+6.1	802	LQ11	E2		PP					2	
1471	LITHIUM HYPOCHLORITE, DRY or LITHIUM HYPOCHLORITE MIXTURE	5.1	O2	II	5.1		LQ11	E2		PP					0	
1472	LITHIUM PEROXIDE	5.1	O2	II	5.1		LQ11	E2		PP					0	
1473	MAGNESIUM BROMATE	5.1	O2	II	5.1		LQ11	E2		PP					0	
1474	MAGNESIUM NITRATE	5.1	O2	III	5.1	332	LQ12	E1	**B**	PP		CO02 LO04			0	CO02 and LO04 apply only when this substance is carried in bulk or without packaging
1475	MAGNESIUM PERCHLORATE	5.1	O2	II	5.1		LQ11	E2		PP					0	
1476	MAGNESIUM PEROXIDE	5.1	O2	II	5.1		LQ11	E2		PP					0	
1477	NITRATES, INORGANIC, N.O.S.	5.1	O2	II	5.1	274 511	LQ11	E2		PP					0	
1477	NITRATES, INORGANIC, N.O.S.	5.1	O2	III	5.1	274 511	LQ12	E1	**B**	PP		CO02 LO04			0	CO02 and LO04 apply only when this substance is carried in bulk or without packaging
1479	OXIDIZING SOLID, N.O.S.	5.1	O2	I	5.1	274	LQ0	E0		PP					0	
1479	OXIDIZING SOLID, N.O.S.	5.1	O2	II	5.1	274	LQ11	E2		PP					0	
1479	OXIDIZING SOLID, N.O.S.	5.1	O2	III	5.1	274	LQ12	E1		PP					0	
1481	PERCHLORATES, INORGANIC, N.O.S.	5.1	O2	II	5.1	274	LQ11	E2		PP					0	
1481	PERCHLORATES, INORGANIC, N.O.S.	5.1	O2	III	5.1	274	LQ12	E1		PP					0	

UN No. or ID No.	Name and description	Class	Classi-fication Code	Packing group	Labels	Special provi-sions	Limited and excepted quantities		Carriage permitted	Equipment required	Venti-lation	Provisions concerning loading, unloading and carriage	Number of blue cones/ lights	Remarks
	3.1.2	2.2	2.2	2.1.1.3	5.2.2	3.3	3.4.6	3.5.1.2	3.2.1	8.1.5	7.1.6	7.1.6	7.1.5	3.2.1
(1)	(2)	(3a)	(3b)	(4)	(5)	(6)	(7a)	(7b)	(8)	(9)	(10)	(11)	(12)	(13)
1482	PERMANGANATES, INORGANIC, N.O.S.	5.1	O2	II	5.1	274 608	LQ11	E2		PP			0	
1482	PERMANGANATES, INORGANIC, N.O.S.	5.1	O2	III	5.1	274 608	LQ12	E1		PP			0	
1483	PEROXIDES, INORGANIC, N.O.S.	5.1	O2	II	5.1	274	LQ11	E2		PP			0	
1483	PEROXIDES, INORGANIC, N.O.S.	5.1	O2	III	5.1	274	LQ12	E1		PP			0	
1484	POTASSIUM BROMATE	5.1	O2	II	5.1		LQ11	E2		PP			0	
1485	POTASSIUM CHLORATE	5.1	O2	II	5.1		LQ11	E2		PP			0	
1486	POTASSIUM NITRATE	5.1	O2	III	5.1		LQ12	E1	**B**	PP		CO02 LO04	0	CO02 and LO04 apply only when this substance is carried in bulk or without packaging
1487	POTASSIUM NITRATE AND SODIUM NITRITE MIXTURE	5.1	O2	II	5.1	607	LQ11	E2		PP			0	
1488	POTASSIUM NITRITE	5.1	O2	II	5.1		LQ11	E2		PP			0	
1489	POTASSIUM PERCHLORATE	5.1	O2	II	5.1		LQ11	E2		PP			0	
1490	POTASSIUM PERMANGANATE	5.1	O2	II	5.1		LQ11	E2		PP			0	
1491	POTASSIUM PEROXIDE	5.1	O2	I	5.1		LQ0	E0		PP			0	
1492	POTASSIUM PERSULPHATE	5.1	O2	III	5.1		LQ12	E1		PP			0	
1493	SILVER NITRATE	5.1	O2	II	5.1		LQ11	E2		PP			0	
1494	SODIUM BROMATE	5.1	O2	II	5.1		LQ11	E2		PP			0	
1495	SODIUM CHLORATE	5.1	O2	II	5.1		LQ11	E2		PP			0	
1496	SODIUM CHLORITE	5.1	O2	II	5.1		LQ11	E2		PP			0	
1498	SODIUM NITRATE	5.1	O2	III	5.1		LQ12	E1	**B**	PP		CO02 LO04	0	CO02 and LO04 apply only when this substance is carried in bulk or without packaging
1499	SODIUM NITRATE AND POTASSIUM NITRATE MIXTURE	5.1	O2	III	5.1		LQ12	E1	**B**	PP		CO02 LO04	0	CO02 and LO04 apply only when this substance is carried in bulk or without packaging
1500	SODIUM NITRITE	5.1	OT2	III	5.1+6.1	802	LQ12	E1		PP			0	
1502	SODIUM PERCHLORATE	5.1	O2	II	5.1		LQ11	E2		PP			0	
1503	SODIUM PERMANGANATE	5.1	O2	II	5.1		LQ11	E2		PP			0	
1504	SODIUM PEROXIDE	5.1	O2	I	5.1		LQ0	E0		PP			0	
1505	SODIUM PERSULPHATE	5.1	O2	III	5.1		LQ12	E1		PP			0	
1506	STRONTIUM CHLORATE	5.1	O2	II	5.1		LQ11	E2		PP			0	
1507	STRONTIUM NITRATE	5.1	O2	III	5.1		LQ12	E1	**B**	PP		CO02 LO04	0	CO02 and LO04 apply only when this substance is carried in bulk or without packaging
1508	STRONTIUM PERCHLORATE	5.1	O2	II	5.1		LQ11	E2		PP			0	

UN No. or ID No.	Name and description	Class	Classi-fication Code	Packing group	Labels	Special provi-sions	Limited and excepted quantities		Carriage permitted	Equipment required	Venti-lation	Provisions concerning loading, unloading and carriage	Number of blue cones/ lights	Remarks
	3.1.2	2.2	2.2	2.1.1.3	5.2.2	3.3	3.4.6	3.5.1.2	3.2.1	8.1.5	7.1.6	7.1.6	7.1.5	3.2.1
(1)	(2)	(3a)	(3b)	(4)	(5)	(6)	(7a)	(7b)	(8)	(9)	(10)	(11)	(12)	(13)
1509	STRONTIUM PEROXIDE	5.1	O2	II	5.1		LQ11	E2		PP			0	
1510	TETRANITROMETHANE	5.1	OT1	I	5.1+6.1	609 802	LQ0	E0		PP, EP, TOX, A	VE02		2	
1511	UREA HYDROGEN PEROXIDE	5.1	OC2	III	5.1+8		LQ12	E1		PP			0	
1512	ZINC AMMONIUM NITRITE	5.1	O2	II	5.1		LQ11	E2		PP			0	
1513	ZINC CHLORATE	5.1	O2	II	5.1		LQ11	E2		PP			0	
1514	ZINC NITRATE	5.1	O2	II	5.1		LQ11	E2		PP			0	
1515	ZINC PERMANGANATE	5.1	O2	II	5.1		LQ11	E2		PP			0	
1516	ZINC PEROXIDE	5.1	O2	II	5.1		LQ11	E2		PP			0	
1517	ZIRCONIUM PICRAMATE, WETTED with not less than 20% water, by mass	4.1	D	I	4.1		LQ0	E0		PP			1	
1541	ACETONE CYANOHYDRIN, STABILIZED	6.1	T1	I	6.1	802	LQ0	E5	T	PP, EP, TOX, A	VE02		2	
1544	ALKALOIDS, SOLID, N.O.S. or ALKALOID SALTS, SOLID, N.O.S.	6.1	T2	I	6.1	43 274 802	LQ0	E5		PP, EP			2	
1544	ALKALOIDS, SOLID, N.O.S. or ALKALOID SALTS, SOLID, N.O.S.	6.1	T2	II	6.1	43 274 802	LQ18	E4		PP, EP			2	
1544	ALKALOIDS, SOLID, N.O.S. or ALKALOID SALTS, SOLID, N.O.S.	6.1	T2	III	6.1	43 274 802	LQ9	E1		PP, EP			0	
1545	ALLYL ISOTHIOCYANATE, STABILIZED	6.1	TF1	II	6.1+3	802	LQ17	E4	T	PP, EP, EX, TOX, A	VE01, VE02		2	
1546	AMMONIUM ARSENATE	6.1	T5	II	6.1	802	LQ18	E4		PP, EP			2	
1547	ANILINE	6.1	T1	II	6.1	279 802	LQ17	E4	T	PP, EP, TOX, A	VE02		2	
1548	ANILINE HYDROCHLORIDE	6.1	T2	III	6.1	802	LQ9	E1		PP, EP			0	
1549	ANTIMONY COMPOUND, INORGANIC, SOLID, N.O.S.	6.1	T5	III	6.1	45 274 512 802	LQ9	E1		PP, EP			0	
1550	ANTIMONY LACTATE	6.1	T5	III	6.1	802	LQ9	E1		PP, EP			0	
1551	ANTIMONY POTASSIUM TARTRATE	6.1	T5	III	6.1	802	LQ9	E1		PP, EP			0	
1553	ARSENIC ACID, LIQUID	6.1	T4	I	6.1	802	LQ0	E5		PP, EP, TOX, A	VE02		2	
1554	ARSENIC ACID, SOLID	6.1	T5	II	6.1	802	LQ18	E4		PP, EP			2	
1555	ARSENIC BROMIDE	6.1	T5	II	6.1	802	LQ18	E4		PP, EP			2	
1556	ARSENIC COMPOUND, LIQUID, N.O.S., inorganic, including: Arsenates, n.o.s., Arsenites, n.o.s.; and Arsenic sulphides, n.o.s.	6.1	T4	I	6.1	43 274 802	LQ0	E5		PP, EP, TOX, A	VE02		2	
1556	ARSENIC COMPOUND, LIQUID, N.O.S., inorganic, including: Arsenates, n.o.s., Arsenites, n.o.s.; and Arsenic sulphides, n.o.s.	6.1	T4	II	6.1	43 274 802	LQ17	E4		PP, EP, TOX, A	VE02		2	

UN No. or ID No.	Name and description	Class	Classi-fication Code	Packing group	Labels	Special provi-sions	Limited and excepted quantities		Carriage permitted	Equipment required	Venti-lation	Provisions concerning loading, unloading and carriage	Number of blue cones/ lights	Remarks
	3.1.2	2.2	2.2	2.1.1.3	5.2.2	3.3	3.4.6	3.5.1.2	3.2.1	8.1.5	7.1.6	7.1.6	7.1.5	3.2.1
(1)	(2)	(3a)	(3b)	(4)	(5)	(6)	(7a)	(7b)	(8)	(9)	(10)	(11)	(12)	(13)
1556	ARSENIC COMPOUND, LIQUID, N.O.S., inorganic, including: Arsenates, n.o.s., Arsenites, n.o.s.; and Arsenic sulphides, n.o.s.	6.1	T4	III	6.1	43 274 802	LQ7	E1		PP, EP, TOX, A	VE02		0	
1557	ARSENIC COMPOUND, SOLID, N.O.S., inorganic, including: Arsenates, n.o.s.; Arsenites, n.o.s.; and Arsenic sulphides, n.o.s.	6.1	T5	I	6.1	43 274 802	LQ0	E5		PP, EP			2	
1557	ARSENIC COMPOUND, SOLID, N.O.S., inorganic, including: Arsenates, n.o.s.; Arsenites, n.o.s.; and Arsenic sulphides, n.o.s.	6.1	T5	II	6.1	43 274 802	LQ18	E4		PP, EP			2	
1557	ARSENIC COMPOUND, SOLID, N.O.S., inorganic, including: Arsenates, n.o.s.; Arsenites, n.o.s.; and Arsenic sulphides, n.o.s.	6.1	T5	III	6.1	43 274 802	LQ9	E1		PP, EP			0	
1558	ARSENIC	6.1	T5	II	6.1	802	LQ18	E4		PP, EP			2	
1559	ARSENIC PENTOXIDE	6.1	T5	II	6.1	802	LQ18	E4		PP, EP			2	
1560	ARSENIC TRICHLORIDE	6.1	T4	I	6.1	802	LQ0	E5		PP, EP, TOX, A	VE02		2	
1561	ARSENIC TRIOXIDE	6.1	T5	II	6.1	802	LQ18	E4		PP, EP			2	
1562	ARSENICAL DUST	6.1	T5	II	6.1	802	LQ18	E4		PP, EP			2	
1564	BARIUM COMPOUND, N.O.S.	6.1	T5	II	6.1	177 274 513 587 802	LQ18	E4		PP, EP			2	
1564	BARIUM COMPOUND, N.O.S.	6.1	T5	III	6.1	177 274 513 587 802	LQ9	E1		PP, EP			0	
1565	BARIUM CYANIDE	6.1	T5	I	6.1	802	LQ0	E5		PP, EP			2	
1566	BERYLLIUM COMPOUND, N.O.S.	6.1	T5	II	6.1	274 514 802	LQ18	E4		PP, EP			2	
1566	BERYLLIUM COMPOUND, N.O.S.	6.1	T5	III	6.1	274 514 802	LQ9	E1		PP, EP			0	
1567	BERYLLIUM POWDER	6.1	TF3	II	6.1+4.1	802	LQ18	E4		PP, EP			2	
1569	BROMOACETONE	6.1	TF1	II	6.1+3	802	LQ17	E4		PP, EP, EX, TOX, A	VE01, VE02		2	
1570	BRUCINE	6.1	T2	I	6.1	43 802	LQ0	E5		PP, EP			2	
1571	BARIUM AZIDE, WETTED with not less than 50% water, by mass	4.1	DT	I	4.1+6.1	568 802	LQ0	E0		PP			2	

UN No. or ID No.	Name and description	Class	Classi-fication Code	Packing group	Labels	Special provi-sions	Limited and excepted quantities		Carriage permitted	Equipment required	Venti-lation	Provisions concerning loading, unloading and carriage	Number of blue cones/ lights	Remarks
	3.1.2	2.2	2.2	2.1.1.3	5.2.2	3.3	3.4.6	3.5.1.2	3.2.1	8.1.5	7.1.6	7.1.6	7.1.5	3.2.1
(1)	(2)	(3a)	(3b)	(4)	(5)	(6)	(7a)	(7b)	(8)	(9)	(10)	(11)	(12)	(13)
1572	CACODYLIC ACID	6.1	T5	II	6.1	802	LQ18	E4		PP, EP			2	
1573	CALCIUM ARSENATE	6.1	T5	II	6.1	802	LQ18	E4		PP, EP			2	
1574	CALCIUM ARSENATE AND CALCIUM ARSENITE MIXTURE, SOLID	6.1	T5	II	6.1	802	LQ18	E4		PP, EP			2	
1575	CALCIUM CYANIDE	6.1	T5	I	6.1	802	LQ0	E5		PP, EP			2	
1577	CHLORODINITROBENZENES, LIQUID	6.1	T1	II	6.1	279 802	LQ17	E4		PP, EP, TOX, A	VE02		2	
1578	CHLORONITROBENZENES, SOLID	6.1	T2	II	6.1	279 802	LQ18	E4	T	PP, EP, TOX, A	VE02		2	
1579	4-CHLORO-o-TOLUIDINE HYDROCHLORIDE, SOLID	6.1	T2	III	6.1	802	LQ9	E1		PP, EP			0	
1580	CHLOROPICRIN	6.1	T1	I	6.1	802	LQ0	E5		PP, EP, TOX, A	VE02		2	
1581	CHLOROPICRIN AND METHYL BROMIDE MIXTURE with more than 2% chloropicrin	2	2T		2.3		LQ0	E0		PP, EP, TOX, A	VE02		2	
1582	CHLOROPICRIN AND METHYL CHLORIDE MIXTURE	2	2T		2.3		LQ0	E0		PP, EP, TOX, A	VE02		2	
1583	CHLOROPICRIN MIXTURE, N.O.S.	6.1	T1	I	6.1	274 315 515 802	LQ0	E5		PP, EP, TOX, A	VE02		2	
1583	CHLOROPICRIN MIXTURE, N.O.S.	6.1	T1	II	6.1	274 515 802	LQ17	E4		PP, EP, TOX, A	VE02		2	
1583	CHLOROPICRIN MIXTURE, N.O.S.	6.1	T1	III	6.1	274 515 802	LQ7	E1		PP, EP, TOX, A	VE02		0	
1585	COPPER ACETOARSENITE	6.1	T5	II	6.1	802	LQ18	E4		PP, EP			2	
1586	COPPER ARSENITE	6.1	T5	II	6.1	802	LQ18	E4		PP, EP			2	
1587	COPPER CYANIDE	6.1	T5	II	6.1	802	LQ18	E4		PP, EP			2	
1588	CYANIDES, INORGANIC, SOLID, N.O.S.	6.1	T5	I	6.1	47 47 802	LQ0	E5		PP, EP			2	
1588	CYANIDES, INORGANIC, SOLID, N.O.S.	6.1	T5	II	6.1	47 274 802	LQ18	E4		PP, EP			2	
1588	CYANIDES, INORGANIC, SOLID, N.O.S.	6.1	T5	III	6.1	47 274 802	LQ9	E1		PP, EP			0	
1589	CYANOGEN CHLORIDE, STABILIZED	2	2TC		2.3+8		LQ0	E0		PP, EP, TOX, A	VE02		2	
1590	DICHLOROANILINES, LIQUID	6.1	T1	II	6.1	279 802	LQ17	E4		PP, EP, TOX, A	VE02		2	
1591	o-DICHLOROBENZENE	6.1	T1	III	6.1	279 802	LQ7	E1	T	PP, EP, TOX, A	VE02		0	

UN No. or ID No.	Name and description	Class	Classi-fication Code	Packing group	Labels	Special provi-sions	Limited and excepted quantities		Carriage permitted	Equipment required	Venti-lation	Provisions concerning loading, unloading and carriage	Number of blue cones/ lights	Remarks
	3.1.2	2.2	2.2	2.1.1.3	5.2.2	3.3	3.4.6	3.5.1.2	3.2.1	8.1.5	7.1.6	7.1.6	7.1.5	3.2.1
(1)	(2)	(3a)	(3b)	(4)	(5)	(6)	(7a)	(7b)	(8)	(9)	(10)	(11)	(12)	(13)
1593	DICHLOROMETHANE	6.1	T1	III	6.1	516 802	LQ7	E1	**T**	PP, EP, TOX, A	VE02		0	
1594	DIETHYL SULPHATE	6.1	T1	II	6.1	802	LQ17	E4	**T**	PP, EP, TOX, A	VE02		2	
1595	DIMETHYL SULPHATE	6.1	TC1	I	6.1+8	802	LQ0	E5	**T**	PP, EP, TOX, A	VE02		2	
1596	DINITROANILINES	6.1	T2	II	6.1	802	LQ18	E4		PP, EP			2	
1597	DINITROBENZENES, LIQUID	6.1	T1	II	6.1	802	LQ17	E4		PP, EP, TOX, A	VE02		2	
1597	DINITROBENZENES, LIQUID	6.1	T1	III	6.1	802	LQ7	E1		PP, EP, TOX, A	VE02		0	
1598	DINITRO-o-CRESOL	6.1	T2	II	6.1	43 802	LQ18	E4		PP, EP			2	
1599	DINITROPHENOL SOLUTION	6.1	T1	II	6.1	802	LQ17	E4		PP, EP, A			2	
1599	DINITROPHENOL SOLUTION	6.1	T1	III	6.1	802	LQ7	E1		PP, EP, A			0	
1600	DINITROTOLUENES, MOLTEN	6.1	T1	II	6.1	802	LQ0	E0		PP, EP, TOX, A	VE02		2	
1601	DISINFECTANT, SOLID, TOXIC, N.O.S.	6.1	T2	I	6.1	274 802	LQ0	E5		PP, EP			2	
1601	DISINFECTANT, SOLID, TOXIC, N.O.S.	6.1	T2	II	6.1	274 802	LQ18	E4		PP, EP			2	
1601	DISINFECTANT, SOLID, TOXIC, N.O.S.	6.1	T2	III	6.1	274 802	LQ9	E1		PP, EP			0	
1602	DYE, LIQUID, TOXIC, N.O.S. or DYE INTERMEDIATE, LIQUID, TOXIC, N.O.S.	6.1	T1	I	6.1	274 802	LQ0	E5		PP, EP, TOX, A	VE02		2	
1602	DYE, LIQUID, TOXIC, N.O.S. or DYE INTERMEDIATE, LIQUID, TOXIC, N.O.S.	6.1	T1	II	6.1	274 802	LQ17	E4		PP, EP, TOX, A	VE02		2	
1602	DYE, LIQUID, TOXIC, N.O.S. or DYE INTERMEDIATE, LIQUID, TOXIC, N.O.S.	6.1	T1	III	6.1	274 802	LQ7	E1		PP, EP, TOX, A	VE02		0	
1603	ETHYL BROMOACETATE	6.1	TF1	II	6.1+3	802	LQ17	E4		PP, EP, EX, TOX, A	VE01, VE02		2	
1604	ETHYLENEDIAMINE	8	CF1	II	8+3		LQ22	E2	**T**	PP, EP, EX, A	VE01		1	
1605	ETHYLENE DIBROMIDE	6.1	T1	I	6.1	802	LQ0	E5	**T**	PP, EP, TOX, A	VE02		2	
1606	FERRIC ARSENATE	6.1	T5	II	6.1	802	LQ18	E4		PP, EP			2	
1607	FERRIC ARSENITE	6.1	T5	II	6.1	802	LQ18	E4		PP, EP			2	
1608	FERROUS ARSENATE	6.1	T5	II	6.1	802	LQ18	E4		PP, EP			2	
1611	HEXAETHYL TETRAPHOSPHATE	6.1	T1	II	6.1	802	LQ17	E4		PP, EP, TOX, A	VE02		2	
1612	HEXAETHYL TETRAPHOSPHATE AND COMPRESSED GAS MIXTURE	2	1T		2.3		LQ0	E0		PP, EP, TOX, A	VE02		2	
1613	HYDROCYANIC ACID, AQUEOUS SOLUTION (HYDROGEN CYANIDE, AQUEOUS SOLUTION) with not more than 20% hydrogen cyanide	6.1	TF1	I	6.1+3	48 802	LQ0	E5		PP, EP, EX, TOX, A	VE01, VE02		2	

UN No. or ID No.	Name and description	Class	Classi-fication Code	Packing group	Labels	Special provi-sions	Limited and excepted quantities		Carriage permitted	Equipment required	Venti-lation	Provisions concerning loading, unloading and carriage			Number of blue cones/ lights	Remarks
	3.1.2	2.2	2.2	2.1.1.3	5.2.2	3.3	3.4.6	3.5.1.2	3.2.1	8.1.5	7.1.6	7.1.6			7.1.5	3.2.1
(1)	(2)	(3a)	(3b)	(4)	(5)	(6)	(7a)	(7b)	(8)	(9)	(10)	(11)			(12)	(13)
1614	HYDROGEN CYANIDE, STABILIZED, containing less than 3% water and absorbed in a porous inert material	6.1	TF1	I	6.1+3	603 802	LQ0	E5		PP, EP, EX, TOX, A	VE01, VE02				2	
1616	LEAD ACETATE	6.1	T5	III	6.1	802	LQ9	E1		PP, EP					0	
1617	LEAD ARSENATES	6.1	T5	II	6.1	802	LQ18	E4		PP, EP					2	
1618	LEAD ARSENITES	6.1	T5	II	6.1	802	LQ18	E4		PP, EP					2	
1620	LEAD CYANIDE	6.1	T5	II	6.1	802	LQ18	E4		PP, EP					2	
1621	LONDON PURPLE	6.1	T5	II	6.1	43 802	LQ18	E4		PP, EP					2	
1622	MAGNESIUM ARSENATE	6.1	T5	II	6.1	802	LQ18	E4		PP, EP					2	
1623	MERCURIC ARSENATE	6.1	T5	II	6.1	802	LQ18	E4		PP, EP					2	
1624	MERCURIC CHLORIDE	6.1	T5	II	6.1	802	LQ18	E4		PP, EP					2	
1625	MERCURIC NITRATE	6.1	T5	II	6.1	802	LQ18	E4		PP, EP					2	
1626	MERCURIC POTASSIUM CYANIDE	6.1	T5	I	6.1	802	LQ0	E5		PP, EP					2	
1627	MERCUROUS NITRATE	6.1	T5	II	6.1	802	LQ18	E4		PP, EP					2	
1629	MERCURY ACETATE	6.1	T5	II	6.1	802	LQ18	E4		PP, EP					2	
1630	MERCURY AMMONIUM CHLORIDE	6.1	T5	II	6.1	802	LQ18	E4		PP, EP					2	
1631	MERCURY BENZOATE	6.1	T5	II	6.1	802	LQ18	E4		PP, EP					2	
1634	MERCURY BROMIDES	6.1	T5	II	6.1	802	LQ18	E4		PP, EP					2	
1636	MERCURY CYANIDE	6.1	T5	II	6.1	802	LQ18	E4		PP, EP					2	
1637	MERCURY GLUCONATE	6.1	T5	II	6.1	802	LQ18	E4		PP, EP					2	
1638	MERCURY IODIDE	6.1	T5	II	6.1	802	LQ18	E4		PP, EP					2	
1639	MERCURY NUCLEATE	6.1	T5	II	6.1	802	LQ18	E4		PP, EP					2	
1640	MERCURY OLEATE	6.1	T5	II	6.1	802	LQ18	E4		PP, EP					2	
1641	MERCURY OXIDE	6.1	T5	II	6.1	802	LQ18	E4		PP, EP					2	
1642	MERCURY OXYCYANIDE, DESENSITIZED	6.1	T5	II	6.1	802	LQ18	E4		PP, EP					2	
1643	MERCURY POTASSIUM IODIDE	6.1	T5	II	6.1	802	LQ18	E4		PP, EP					2	
1644	MERCURY SALICYLATE	6.1	T5	II	6.1	802	LQ18	E4		PP, EP					2	
1645	MERCURY SULPHATE	6.1	T5	II	6.1	802	LQ18	E4		PP, EP					2	
1646	MERCURY THIOCYANATE	6.1	T5	II	6.1	802	LQ18	E4		PP, EP					2	
1647	METHYL BROMIDE AND ETHYLENE DIBROMIDE MIXTURE, LIQUID	6.1	T1	I	6.1	802	LQ0	E5		PP, EP, TOX, A	VE02				2	
1648	ACETONITRILE	3	F1	II	3		LQ4	E2	T	PP, EX, A	VE01				1	
1649	MOTOR FUEL ANTI-KNOCK MIXTURE having a flash-point above 60 °C	6.1	T3	I	6.1	802	LQ0	E5		PP, EP, TOX, A	VE02				2	
1649	MOTOR FUEL ANTI-KNOCK MIXTURE having a flash-point of not more than 60 °C	6.1	TF1	I	6.1 3	802	LQ0	E5		PP, EP, EX, TOX, A	VE01, VE02				2	
1650	beta-NAPHTHYLAMINE, SOLID	6.1	T2	II	6.1	802	LQ18	E4		PP, EP					2	
1651	NAPHTHYLTHIOUREA	6.1	T2	II	6.1	43 802	LQ18	E4		PP, EP					2	
1652	NAPHTHYLUREA	6.1	T2	II	6.1	802	LQ18	E4		PP, EP					2	
1653	NICKEL CYANIDE	6.1	T5	II	6.1	802	LQ18	E4		PP, EP					2	
1654	NICOTINE	6.1	T1	II	6.1	802	LQ17	E4		PP, EP, TOX, A	VE02				2	

UN No. or ID No.	Name and description	Class	Classi-fication Code	Packing group	Labels	Special provi-sions	Limited and excepted quantities		Carriage permitted	Equipment required	Venti-lation	Provisions concerning loading, unloading and carriage	Number of blue cones/ lights	Remarks
	3.1.2	2.2	2.2	2.1.1.3	5.2.2	3.3	3.4.6	3.5.1.2	3.2.1	8.1.5	7.1.6	7.1.6	7.1.5	3.2.1
(1)	(2)	(3a)	(3b)	(4)	(5)	(6)	(7a)	(7b)	(8)	(9)	(10)	(11)	(12)	(13)
1655	NICOTINE COMPOUND, SOLID, N.O.S. or NICOTINE PREPARATION, SOLID, N.O.S.	6.1	T2	I	6.1	43 274 802	LQ0	E5		PP, EP			2	
1655	NICOTINE COMPOUND, SOLID, N.O.S. or NICOTINE PREPARATION, SOLID, N.O.S.	6.1	T2	II	6.1	43 274 802	LQ18	E4		PP, EP			2	
1655	NICOTINE COMPOUND, SOLID, N.O.S. or NICOTINE PREPARATION, SOLID, N.O.S.	6.1	T2	III	6.1	43 274 802	LQ9	E1		PP, EP			0	
1656	NICOTINE HYDROCHLORIDE, LIQUID or SOLUTION	6.1	T1	II	6.1	43 802	LQ17	E4		PP, EP, TOX, A	VE02		2	
1656	NICOTINE HYDROCHLORIDE, LIQUID or SOLUTION	6.1	T1	III	6.1	43 802	LQ7	E1		PP, EP, TOX, A	VE02		0	
1657	NICOTINE SALICYLATE	6.1	T2	II	6.1	802	LQ18	E4		PP, EP			2	
1658	NICOTINE SULPHATE, SOLUTION	6.1	T1	II	6.1	802	LQ17	E4		PP, EP, TOX, A	VE02		2	
1658	NICOTINE SULPHATE, SOLUTION	6.1	T1	III	6.1	802	LQ7	E1		PP, EP, TOX, A	VE02		0	
1659	NICOTINE TARTRATE	6.1	T2	II	6.1	802	LQ18	E4		PP, EP			2	
1660	NITRIC OXIDE, COMPRESSED	2	1TOC		2.3+5.1+8		LQ0	E0		PP, EP, TOX, A	VE02		2	
1661	NITROANILINES (o-, m-, p-)	6.1	T2	II	6.1	279 802	LQ18	E4		PP, EP			2	
1662	NITROBENZENE	6.1	T1	II	6.1	279 802	LQ17	E4	**T**	PP, EP, TOX, A	VE02		2	
1663	NITROPHENOLS (o-, m-, p-)	6.1	T2	III	6.1	279 802	LQ9	E1	**T**	PP, EP			0	
1664	NITROTOLUENES, LIQUID	6.1	T1	II	6.1	802	LQ17	E4	**T**	PP, EP, TOX, A	VE02		2	
1665	NITROXYLENES, LIQUID	6.1	T1	II	6.1	802	LQ17	E4		PP, EP, TOX, A	VE02		2	
1669	PENTACHLOROETHANE	6.1	T1	II	6.1	802	LQ17	E4		PP, EP, TOX, A	VE02		2	
1670	PERCHLOROMETHYL MERCAPTAN	6.1	T1	I	6.1	802	LQ0	E5		PP, EP, TOX, A	VE02		2	
1671	PHENOL, SOLID	6.1	T2	II	6.1	279 802	LQ18	E4		PP, EP			2	
1672	PHENYLCARBYLAMINE CHLORIDE	6.1	T1	I	6.1	802	LQ0	E5		PP, EP, TOX, A	VE02		2	
1673	PHENYLENEDIAMINES (o-, m-, p-)	6.1	T2	III	6.1	279 802	LQ9	E1		PP, EP			0	
1674	PHENYLMERCURIC ACETATE	6.1	T3	II	6.1	43 802	LQ18	E4		PP, EP, TOX, A	VE02		2	
1677	POTASSIUM ARSENATE	6.1	T5	II	6.1	802	LQ18	E4		PP, EP			2	
1678	POTASSIUM ARSENITE	6.1	T5	II	6.1	802	LQ18	E4		PP, EP			2	
1679	POTASSIUM CUPROCYANIDE	6.1	T5	II	6.1	802	LQ18	E4		PP, EP			2	

UN No. or ID No.	Name and description	Class	Classi-fication Code	Packing group	Labels	Special provi-sions	Limited and excepted quantities		Carriage permitted	Equipment required	Venti-lation	Provisions concerning loading, unloading and carriage	Number of blue cones/ lights	Remarks
	3.1.2	2.2	2.2	2.1.1.3	5.2.2	3.3	3.4.6	3.5.1.2	3.2.1	8.1.5	7.1.6	7.1.6	7.1.5	3.2.1
(1)	(2)	(3a)	(3b)	(4)	(5)	(6)	(7a)	(7b)	(8)	(9)	(10)	(11)	(12)	(13)
1680	POTASSIUM CYANIDE, SOLID	6.1	T5	I	6.1	802	LQ0	E5		PP, EP			2	
1683	SILVER ARSENITE	6.1	T5	II	6.1	802	LQ18	E4		PP, EP			2	
1684	SILVER CYANIDE	6.1	T5	II	6.1	802	LQ18	E4		PP, EP			2	
1685	SODIUM ARSENATE	6.1	T5	II	6.1	802	LQ18	E4		PP, EP			2	
1686	SODIUM ARSENITE, AQUEOUS SOLUTION	6.1	T4	II	6.1	43 802	LQ17	E4		PP, EP			2	
1686	SODIUM ARSENITE, AQUEOUS SOLUTION	6.1	T4	III	6.1	43 802	LQ7	E1		PP, EP			0	
1687	SODIUM AZIDE	6.1	T5	II	6.1	802	LQ18	E4		PP, EP			2	
1688	SODIUM CACODYLATE	6.1	T5	II	6.1	802	LQ18	E4		PP, EP			2	
1689	SODIUM CYANIDE, SOLID	6.1	T5	I	6.1	802	LQ0	E5		PP, EP			2	
1690	SODIUM FLUORIDE, SOLID	6.1	T5	III	6.1	802	LQ9	E1	**B**	PP, EP			0	
1691	STRONTIUM ARSENITE	6.1	T5	II	6.1	802	LQ18	E4		PP, EP			2	
1692	STRYCHNINE or STRYCHNINE SALTS	6.1	T2	I	6.1	802	LQ0	E5		PP, EP			2	
1693	TEAR GAS SUBSTANCE, LIQUID, N.O.S.	6.1	T1	I	6.1	274 802	LQ0	E5		PP, EP, TOX, A	VE02		2	
1693	TEAR GAS SUBSTANCE, LIQUID, N.O.S.	6.1	T1	II	6.1	274 802	LQ17	E4		PP, EP, TOX, A	VE02		2	
1694	BROMOBENZYL CYANIDES, LIQUID	6.1	T1	I	6.1	138 302	LQ0	E5		PP, EP, TOX, A	VE02		2	
1695	CHLOROACETONE, STABILIZED	6.1	TFC	I	6.1+3+8	802	LQ0	E5		PP, EP, EX, TOX, A	VE01, VE02		2	
1697	CHLOROACETOPHENONE, SOLID	6.1	T2	II	6.1	802	LQ18	E4		PP, EP, TOX, A	VE02		2	
1698	DIPHENYLAMINE CHLOROARSINE	6.1	T3	I	6.1	802	LQ0	E5		PP, EP, TOX, A	VE02		2	
1699	DIPHENYLCHLOROARSINE, LIQUID	6.1	T3	I	6.1	802	LQ0	E5		PP, EP, TOX, A	VE02		2	
1700	TEAR GAS CANDLES	6.1	TF3	II	6.1+4.1	802	LQ18	E0		PP, EP			2	
1701	XYLYL BROMIDE, LIQUID	6.1	T1	II	6.1	802	LQ17	E4		PP, EP, TOX, A	VE02		2	
1702	1,1,2,2-TETRACHLOROETHANE	6.1	T1	II	6.1	802	LQ17	E4		PP, EP, TOX, A	VE02		2	
1704	TETRAETHYL DITHIOPYROPHOSPHATE	6.1	T2	II	6.1	43 802	LQ18	E4		PP, EP			2	
1707	THALLIUM COMPOUND, N.O.S.	6.1	T5	II	6.1	43 274 802	LQ18	E4		PP, EP			2	
1708	TOLUIDINES, LIQUID	6.1	T1	II	6.1	279 802	LQ17	E4	**T**	PP, EP, TOX, A	VE02		2	
1709	2,4-TOLUYLENEDIAMINE, SOLID	6.1	T2	III	6.1	802	LQ9	E1		PP, EP			0	
1710	TRICHLOROETHYLENE	6.1	T1	III	6.1	802	LQ7	E1	**T**	PP, EP, TOX, A	VE02		0	
1711	XYLIDINES, LIQUID	6.1	T1	II	6.1	802	LQ17	E4		PP, EP, TOX, A	VE02		2	

UN No. or ID No.	Name and description	Class	Classi-fication Code	Packing group	Labels	Special provi-sions	Limited and excepted quantities		Carriage permitted	Equipment required	Venti-lation	Provisions concerning loading, unloading and carriage	Number of blue cones/ lights	Remarks
	3.1.2	2.2	2.2	2.1.1.3	5.2.2	3.3	3.4.6	3.5.1.2	3.2.1	8.1.5	7.1.6	7.1.6	7.1.5	3.2.1
(1)	(2)	(3a)	(3b)	(4)	(5)	(6)	(7a)	(7b)	(8)	(9)	(10)	(11)	(12)	(13)
1712	ZINC ARSENATE, ZINC ARSENITE or ZINC ARSENATE AND ZINC ARSENITE MIXTURE	6.1	T5	II	6.1	802	LQ18	E4		PP, EP			2	
1713	ZINC CYANIDE	6.1	T5	I	6.1	802	LQ0	E5		PP, EP			2	
1714	ZINC PHOSPHIDE	4.3	WT2	I	4.3+6.1	802	LQ0	E0		PP, EP, EX, TOX, A	VE01, VE02	HA08	2	
1715	ACETIC ANHYDRIDE	8	CF1	II	8+3		LQ22	E2	T	PP, EP, EX, A	VE01		1	
1716	ACETYL BROMIDE	8	C3	II	8		LQ22	E2		PP, EP			0	
1717	ACETYL CHLORIDE	3	FC	II	3+8		LQ4	E2	T	PP, EP, EX, A	VE01		1	
1718	BUTYL ACID PHOSPHATE	8	C3	III	8		LQ7	E1	T	PP, EP			0	
1719	CAUSTIC ALKALI LIQUID, N.O.S.	8	C5	II	8	274	LQ22	E2	T	PP, EP			0	
1719	CAUSTIC ALKALI LIQUID, N.O.S.	8	C5	III	8	274	LQ7	E1	T	PP, EP			0	
1722	ALLYL CHLOROFORMATE	6.1	TFC	I	6.1+3+8	802	LQ0	E5		PP, EP, EX, TOX, A	VE01, VE02		2	
1723	ALLYL IODIDE	3	FC	II	3+8		LQ4	E2		PP, EP, EX, A	VE01		1	
1724	ALLYLTRICHLOROSILANE, STABILIZED	8	CF1	II	8+3		LQ22	E2		PP, EP, EX, A	VE01		1	
1725	ALUMINIUM BROMIDE, ANHYDROUS	8	C2	II	8	588	LQ23	E2		PP, EP			0	
1726	ALUMINIUM CHLORIDE, ANHYDROUS	8	C2	II	8	588	LQ23	E2		PP, EP			0	
1727	AMMONIUM HYDROGENDIFLUORIDE, SOLID	8	C2	II	8		LQ23	E2		PP, EP			0	
1728	AMYLTRICHLOROSILANE	8	C3	II	8		LQ22	E2		PP, EP			0	
1729	ANISOYL CHLORIDE	8	C4	II	8		LQ23	E2		PP, EP			0	
1730	ANTIMONY PENTACHLORIDE, LIQUID	8	C1	II	8		LQ22	E2		PP, EP			0	
1731	ANTIMONY PENTACHLORIDE SOLUTION	8	C1	II	8		LQ22	E2		PP, EP			0	
1731	ANTIMONY PENTACHLORIDE SOLUTION	8	C1	III	8		LQ7	E1		PP, EP			0	
1732	ANTIMONY PENTAFLUORIDE	8	CT1	II	8+6.1	802	LQ22	E2		PP, EP, TOX, A	VE02		2	
1733	ANTIMONY TRICHLORIDE	8	C2	II	8		LQ23	E2		PP, EP			0	
1736	BENZOYL CHLORIDE	8	C3	II	8		LQ22	E2		PP, EP			0	
1737	BENZYL BROMIDE	6.1	TC1	II	6.1+8	802	LQ17	E4		PP, EP, TOX, A	VE02		2	
1738	BENZYL CHLORIDE	6.1	TC1	II	6.1+8	802	LQ17	E4	T	PP, EP, TOX, A	VE02		2	
1739	BENZYL CHLOROFORMATE	8	C9	I	8		LQ0	E0		PP, EP			0	
1740	HYDROGENDIFLUORIDES, SOLID, N.O.S.	8	C2	II	8	274 517	LQ23	E2		PP, EP			0	
1740	HYDROGENDIFLUORIDES, SOLID, N.O.S.	8	C2	III	8	274 517	LQ24	E1		PP, EP			0	
1741	BORON TRICHLORIDE	2	2TC		2.3+8		LQ0	E0		PP, EP, TOX, A	VE02		2	
1742	BORON TRIFLUORIDE ACETIC ACID COMPLEX, LIQUID	8	C3	II	8		LQ22	E2	T	PP, EP			0	
1743	BORON TRIFLUORIDE PROPIONIC ACID COMPLEX, LIQUID	8	C3	II	8		LQ22	E2		PP, EP			0	
1744	BROMINE or BROMINE SOLUTION	8	CT1	I	8+6.1	802	LQ0	E0		PP, EP, TOX, A	VE02		2	

UN No. or ID No.	Name and description	Class	Classi-fication Code	Packing group	Labels	Special provi-sions	Limited and excepted quantities		Carriage permitted	Equipment required	Venti-lation	Provisions concerning loading, unloading and carriage	Number of blue cones/ lights	Remarks
	3.1.2	2.2	2.2	2.1.1.3	5.2.2	3.3	3.4.6	3.5.1.2	3.2.1	8.1.5	7.1.6	7.1.6	7.1.5	3.2.1
(1)	(2)	(3a)	(3b)	(4)	(5)	(6)	(7a)	(7b)	(8)	(9)	(10)	(11)	(12)	(13)
1745	BROMINE PENTAFLUORIDE	5.1	OTC	I	5.1+6.1+8	802	LQ0	E0		PP, EP, TOX, A	VE02		2	
1746	BROMINE TRIFLUORIDE	5.1	OTC	I	5.1+6.1+8	802	LQ0	E0		PP, EP, TOX, A	VE02		2	
1747	BUTYLTRICHLOROSILANE	8	CF1	II	8+3		LQ22	E2		PP, EP, EX, A	VE01		1	
1748	CALCIUM HYPOCHLORITE, DRY or CALCIUM HYPOCHLORITE MIXTURE, DRY with more than 39% available chlorine (8.8% available oxygen)	5.1	O2	II	5.1	313 314 589	LQ11	E2		PP			0	
1748	CALCIUM HYPOCHLORITE, DRY or CALCIUM HYPOCHLORITE MIXTURE, DRY with more than 39% available chlorine (8.8% available oxygen)	5.1	O2	III	5.1	316 589	LQ12	E1		PP			0	
1749	CHLORINE TRIFLUORIDE	2	2TOC		2.3+5.1+8		LQ0	E0		PP, EP, TOX, A	VE02		2	
1750	CHLOROACETIC ACID SOLUTION	6.1	TC1	II	6.1+8	802	LQ17	E4	T	PP, EP, TOX, A	VE02		2	
1751	CHLOROACETIC ACID, SOLID	6.1	TC2	II	6.1+8	802	LQ18	E4		PP, EP			2	
1752	CHLOROACETYL CHLORIDE	6.1	TC1	I	6.1+8	802	LQ0	E5		PP, EP, TOX, A	VE02		2	
1753	CHLOROPHENYLTRICHLOROSILANE	8	C3	II	8		LQ22	E2		PP, EP			0	
1754	CHLOROSULPHONIC ACID (with or without sulphur trioxide)	8	C1	I	8		LQ0	E0		PP, EP			0	
1755	CHROMIC ACID SOLUTION	8	C1	II	8	518	LQ22	E2		PP, EP			0	
1755	CHROMIC ACID SOLUTION	8	C1	III	8	518	LQ7	E1		PP, EP			0	
1756	CHROMIC FLUORIDE, SOLID	8	C2	II	8		LQ23	E2		PP, EP			0	
1757	CHROMIC FLUORIDE SOLUTION	8	C1	II	8		LQ22	E2		PP, EP			0	
1757	CHROMIC FLUORIDE SOLUTION	8	C1	III	8		LQ7	E1		PP, EP			0	
1758	CHROMIUM OXYCHLORIDE	8	C1	I	8		LQ0	E0		PP, EP			0	
1759	CORROSIVE SOLID, N.O.S.	8	C10	I	8	274	LQ0	E0		PP, EP			0	
1759	CORROSIVE SOLID, N.O.S.	8	C10	II	8	274	LQ23	E2		PP, EP			0	
1759	CORROSIVE SOLID, N.O.S.	8	C10	III	8	274	LQ24	E1		PP, EP			0	
1760	CORROSIVE LIQUID, N.O.S.	8	C9	I	8	274	LQ0	E0	T	PP, EP			0	
1760	CORROSIVE LIQUID, N.O.S.	8	C9	II	8	274	LQ22	E2	T	PP, EP			0	
1760	CORROSIVE LIQUID, N.O.S.	8	C9	III	8	274	LQ7	E1	T	PP, EP			0	
1761	CUPRIETHYLENEDIAMINE SOLUTION	8	CT1	II	8+6.1	802	LQ22	E2		PP, EP, A			2	
1761	CUPRIETHYLENEDIAMINE SOLUTION	8	CT1	III	8+6.1	802	LQ7	E1		PP, EP, A			0	
1762	CYCLOHEXENYLTRICHLOROSILANE	8	C3	II	8		LQ22	E2		PP, EP			0	
1763	CYCLOHEXYLTRICHLOROSILANE	8	C3	II	8		LQ22	E2		PP, EP			0	
1764	DICHLOROACETIC ACID	8	C3	II	8		LQ22	E2	T	PP, EP			0	
1765	DICHLOROACETYL CHLORIDE	8	C3	II	8		LQ22	E2		PP, EP			0	
1766	DICHLOROPHENYLTRICHLOROSILANE	8	C3	II	8		LQ22	E2		PP, EP			0	
1767	DIETHYLDICHLOROSILANE	8	CF1	II	8+3		LQ22	E2		PP, EP, EX, A	VE01		1	
1768	DIFLUOROPHOSPHORIC ACID, ANHYDROUS	8	C1	II	8		LQ22	E2		PP, EP			0	
1769	DIPHENYLDICHLOROSILANE	8	C3	II	8		LQ22	E2		PP, EP			0	

UN No. or ID No.	Name and description 3.1.2	Class 2.2	Classi-fication Code 2.2	Packing group 2.1.1.3	Labels 5.2.2	Special provi-sions 3.3	Limited and excepted quantities 3.4.6	3.5.1.2	Carriage permitted 3.2.1	Equipment required 8.1.5	Venti-lation 7.1.6	Provisions concerning loading, unloading and carriage 7.1.6	Number of blue cones/ lights 7.1.5	Remarks 3.2.1
(1)	(2)	(3a)	(3b)	(4)	(5)	(6)	(7a)	(7b)	(8)	(9)	(10)	(11)	(12)	(13)
1770	DIPHENYLMETHYL BROMIDE	8	C10	II	8		LQ23	E2		PP, EP			0	
1771	DODECYLTRICHLOROSILANE	8	C3	II	8		LQ22	E2		PP, EP			0	
1773	FERRIC CHLORIDE, ANHYDROUS	8	C2	III	8	590	LQ24	E1		PP, EP			0	
1774	FIRE EXTINGUISHER CHARGES, corrosive liquid	8	C11	II	8		LQ22	E0		PP, EP			0	
1775	FLUOROBORIC ACID	8	C1	II	8		LQ22	E2		PP, EP			0	
1776	FLUOROPHOSPHORIC ACID, ANHYDROUS	8	C1	II	8		LQ22	E2		PP, EP			0	
1777	FLUOROSULPHONIC ACID	8	C1	I	8		LQ0	E0		PP, EP			0	
1778	FLUOROSILICIC ACID	8	C1	II	8		LQ22	E2	T	PP, EP			0	
1779	FORMIC ACID with more than 85% acid by mass	8	CF1	II	8+3		LQ22	E2	T	PP, EP, EX, A			0	
1780	FUMARYL CHLORIDE	8	C3	II	8		LQ22	E2	T	PP, EP			0	
1781	HEXADECYLTRICHLOROSILANE	8	C3	II	8		LQ22	E2		PP, EP			0	
1782	HEXAFLUOROPHOSPHORIC ACID	8	C1	II	8		LQ22	E2		PP, EP			0	
1783	HEXAMETHYLENEDIAMINE SOLUTION	8	C7	II	8		LQ22	E2	T	PP, EP			0	
1783	HEXAMETHYLENEDIAMINE SOLUTION	8	C7	III	8		LQ7	E1	T	PP, EP			0	
1784	HEXYLTRICHLOROSILANE	8	C3	II	8		LQ22	E2		PP, EP			0	
1786	HYDROFLUORIC ACID AND SULPHURIC ACID MIXTURE	8	CT1	I	8+6.1	802	LQ0	E0		PP, EP, TOX, A	VE02		2	
1787	HYDRIODIC ACID	8	C1	II	8		LQ22	E2		PP, EP			0	
1787	HYDRIODIC ACID	8	C1	III	8		LQ7	E1		PP, EP			0	
1788	HYDROBROMIC ACID	8	C1	II	8	519	LQ22	E2		PP, EP			0	
1788	HYDROBROMIC ACID	8	C1	III	8	519	LQ7	E1		PP, EP			0	
1789	HYDROCHLORIC ACID	8	C1	II	8	520	LQ22	E2	T	PP, EP			0	
1789	HYDROCHLORIC ACID	8	C1	III	8	520	LQ7	E1	T	PP, EP			0	
1790	HYDROFLUORIC ACID with more than 85% hydrofluoric acid	8	CT1	I	8+6.1	640I 802	LQ0	E0		PP, EP, TOX, A	VE02		2	
1790	HYDROFLUORIC ACID with more than 60% but not more than 85% hydrofluoric acid	8	CT1	I	8+6.1	640J 802	LQ0	E0		PP, EP, TOX, A	VE02		2	
1790	HYDROFLUORIC ACID with not more than 60% hydrofluoric acid	8	CT1	II	8+6.1	802	LQ22	E2		PP, EP, TOX, A	VE02		2	
1791	HYPOCHLORITE SOLUTION	8	C9	II	8	521	LQ22	E2		PP, EP			0	
1791	HYPOCHLORITE SOLUTION	8	C9	III	8	521	LQ7	E1		PP, EP			0	
1792	IODINE MONOCHLORIDE	8	C1	II	8		LQ22	E2		PP, EP			0	
1793	ISOPROPYL ACID PHOSPHATE	8	C3	III	8		LQ7	E1		PP, EP			0	
1794	LEAD SULPHATE with more than 3% free acid	8	C2	II	8	591	LQ23	E2		PP, EP			0	
1796	NITRATING ACID MIXTURE with more than 50% nitric acid	8	CO1	I	8+5.1		LQ0	E0		PP, EP			0	
1796	NITRATING ACID MIXTURE with not more than 50% nitric acid	8	C1	II	8		LQ22	E2		PP, EP			0	
1798	NITROHYDROCHLORIC ACID	8	COT	**CARRIAGE PROHIBITED**										
1799	NONYLTRICHLOROSILANE	8	C3	II	8		LQ22	E2		PP, EP			0	
1800	OCTADECYLTRICHLORO-SILANE	8	C3	II	8		LQ22	E2		PP, EP			0	
1801	OCTYLTRICHLOROSILANE	8	C3	II	8		LQ22	E2		PP, EP			0	

UN No. or ID No.	Name and description	Class	Classi-fication Code	Packing group	Labels	Special provi-sions	Limited and excepted quantities		Carriage permitted	Equipment required	Venti-lation	Provisions concerning loading, unloading and carriage	Number of blue cones/ lights	Remarks
	3.1.2	2.2	2.2	2.1.1.3	5.2.2	3.3	3.4.6	3.5.1.2	3.2.1	8.1.5	7.1.6	7.1.6	7.1.5	3.2.1
(1)	(2)	(3a)	(3b)	(4)	(5)	(6)	(7a)	(7b)	(8)	(9)	(10)	(11)	(12)	(13)
1802	PERCHLORIC ACID with not more than 50% acid, by mass	8	CO1	II	8+5.1	522	LQ22	E2		PP, EP			0	
1803	PHENOLSULPHONIC ACID, LIQUID	8	C3	II	8		LQ22	E2		PP, EP			0	
1804	PHENYLTRICHLOROSILANE	8	C3	II	8		LQ22	E2		PP, EP			0	
1805	PHOSPHORIC ACID, SOLUTION	8	C1	III	8		LQ7	E1	**T**	PP, EP			0	
1806	PHOSPHORUS PENTACHLORIDE	8	C2	II	8		LQ23	E2		PP, EP			0	
1807	PHOSPHORUS PENTOXIDE	8	C2	II	8		LQ23	E2		PP, EP			0	
1808	PHOSPHORUS TRIBROMIDE	8	C1	II	8		LQ22	E2		PP, EP			0	
1809	PHOSPHORUS TRICHLORIDE	6.1	TC3	I	6.1+8	802	LQ0	E5		PP, EP, TOX, A	VE02		2	
1810	PHOSPHORUS OXYCHLORIDE	8	C1	II	8		LQ22	E2		PP, EP			0	
1811	POTASSIUM HYDROGENDIFLUORIDE, SOLID	8	CT2	II	8+6.1	802	LQ23	E2		PP, EP			2	
1812	POTASSIUM FLUORIDE, SOLID	6.1	T5	III	6.1	802	LQ9	E1	**B**	PP, EP			0	
1813	POTASSIUM HYDROXIDE, SOLID	8	C6	II	8		LQ23	E2		PP, EP			0	
1814	POTASSIUM HYDROXIDE SOLUTION	8	C5	II	8		LQ22	E2	**T**	PP, EP			0	
1814	POTASSIUM HYDROXIDE SOLUTION	8	C5	III	8		LQ7	E1	**T**	PP, EP			0	
1815	PROPIONYL CHLORIDE	3	FC	II	3+8		LQ4	E2		PP, EP, EX, A	VE01		1	
1816	PROPYLTRICHLOROSILANE	8	CF1	II	8+3		LQ22	E2		PP, EP, EX, A	VE01		1	
1817	PYROSULPHURYL CHLORIDE	8	C1	II	8		LQ22	E2		PP, EP			0	
1818	SILICON TETRACHLORIDE	8	C1	II	8		LQ0	E2		PP, EP			0	
1819	SODIUM ALUMINATE SOLUTION	8	C5	II	8		LQ22	E2		PP, EP			0	
1819	SODIUM ALUMINATE SOLUTION	8	C5	III	8		LQ7	E1		PP, EP			0	
1823	SODIUM HYDROXIDE, SOLID	8	C6	II	8		LQ23	E2	**T**	PP, EP			0	
1824	SODIUM HYDROXIDE SOLUTION	8	C5	II	8		LQ22	E2	**T**	PP, EP			0	
1824	SODIUM HYDROXIDE SOLUTION	8	C5	III	8		LQ7	E1	**T**	PP, EP			0	
1825	SODIUM MONOXIDE	8	C6	II	8		LQ23	E2		PP, EP			0	
1826	NITRATING ACID MIXTURE, SPENT, with more than 50% nitric acid	8	CO1	I	8+5.1	113	LQ0	E0		PP, EP			0	
1826	NITRATING ACID MIXTURE, SPENT, with not more than 50% nitric acid	8	C1	II	8	113	LQ22	E2		PP, EP			0	
1827	STANNIC CHLORIDE, ANHYDROUS	8	C1	II	8		LQ22	E2		PP, EP			0	
1828	SULPHUR CHLORIDES	8	C1	I	8		LQ0	E0		PP, EP			0	
1829	SULPHUR TRIOXIDE, STABILIZED	8	C1	I	8	623	LQ0	E0		PP, EP			0	
1830	SULPHURIC ACID with more than 51% acid	8	C1	II	8		LQ22	E2	**T**	PP, EP			0	
1831	SULPHURIC ACID, FUMING	8	CT1	I	8+6.1	802	LQ0	E0	**T**	PP, EP, TOX, A	VE02		2	
1832	SULPHURIC ACID, SPENT	8	C1	II	8	113	LQ22	E2	**T**	PP, EP			0	
1833	SULPHUROUS ACID	8	C1	II	8		LQ22	E2		PP, EP			0	
1834	SULPHURYL CHLORIDE	8	C1	I	8		LQ0	E0		PP, EP			0	
1835	TETRAMETHYLAMMONIUM HYDROXIDE, SOLUTION	8	C7	II	8		LQ22	E2		PP, EP			0	
1835	TETRAMETHYLAMMONIUM HYDROXIDE SOLUTION	8	C7	III	8		LQ7	E1		PP, EP			0	
1836	THIONYL CHLORIDE	8	C1	I	8		LQ0	E0		PP, EP			0	
1837	THIOPHOSPHORYL CHLORIDE	8	C1	II	8		LQ22	E2		PP, EP			0	

UN No. or ID No.	Name and description	Class	Classi-fication Code	Packing group	Labels	Special provi-sions	Limited and excepted quantities		Carriage permitted	Equipment required	Venti-lation	Provisions concerning loading, unloading and carriage	Number of blue cones/ lights	Remarks
	3.1.2	2.2	2.2	2.1.1.3	5.2.2	3.3	3.4.6	3.5.1.2	3.2.1	8.1.5	7.1.6	7.1.6	7.1.5	3.2.1
(1)	(2)	(3a)	(3b)	(4)	(5)	(6)	(7a)	(7b)	(8)	(9)	(10)	(11)	(12)	(13)
1838	TITANIUM TETRACHLORIDE	8	C1	II	8		LQ22	E2		PP, EP			0	
1839	TRICHLOROACETIC ACID	8	C4	II	8		LQ23	E2		PP, EP			0	
1840	ZINC CHLORIDE SOLUTION	8	C1	III	8		LQ7	E1		PP, EP			0	
1841	ACETALDEHYDE AMMONIA	9	M11	III	9		LQ27	E1		PP			0	
1843	AMMONIUM DINITRO-o-CRESOLATE, SOLID	6.1	T2	II	6.1	802	LQ18	E4		PP, EP			2	
1845	Carbon dioxide, solid (Dry ice)	9	M11	NOT SUBJECT TO ADN										
1846	CARBON TETRACHLORIDE	6.1	T1	II	6.1	802	LQ17	E4	T	PP, EP, TOX, A	VE02		2	
1847	POTASSIUM SULPHIDE, HYDRATED with not less than 30% water of crystallization	8	C6	II	8	523	LQ23	E2		PP, EP			0	
1848	PROPIONIC ACID with not less than 10% and less than 90% acid by mass	8	C3	III	8		LQ7	E1	T	PP, EP			0	
1849	SODIUM SULPHIDE, HYDRATED with not less than 30% water	8	C6	II	8	523	LQ23	E2		PP, EP			0	
1851	MEDICINE, LIQUID, TOXIC, N.O.S.	6.1	T1	II	6.1	221 274 601 802	LQ17	E4		PP, EP, TOX, A	VE02		2	
1851	MEDICINE, LIQUID, TOXIC, N.O.S.	6.1	T1	III	6.1	221 274 601 802	LQ7	E1		PP, EP, TOX, A	VE02		0	
1854	BARIUM ALLOYS, PYROPHORIC	4.2	S4	I	4.2		LQ0	E0		PP			0	
1855	CALCIUM, PYROPHORIC or CALCIUM ALLOYS, PYROPHORIC	4.2	S4	I	4.2		LQ0	E0		PP			0	
1856	Rags, oily	4.2	S2	NOT SUBJECT TO ADN										
1857	Textile waste, wet	4.2	S2	NOT SUBJECT TO ADN										
1858	HEXAFLUOROPROPYLENE (REFRIGERANT GAS R 1216)	2	2A		2.2		LQ1	E1		PP			0	
1859	SILICON TETRAFLUORIDE	2	2TC		2.3+8		LQ0	E0		PP, EP, TOX, A	VE02		2	
1860	VINYL FLUORIDE, STABILIZED	2	2F		2.1		LQ0	E0		PP, EX, A	VE01		1	
1862	ETHYL CROTONATE	3	F1	II	3		LQ4	E2		PP, EX, A	VE01		1	
1863	FUEL, AVIATION, TURBINE ENGINE	3	F1	I	3		LQ3	E3	T	PP, EX, A	VE01		1	
1863	FUEL, AVIATION, TURBINE ENGINE (vapour pressure at 50 °C more than 110 kPa)	3	F1	II	3	640C	LQ4	E2	T	PP, EX, A	VE01		1	
1863	FUEL, AVIATION, TURBINE ENGINE (vapour pressure at 50 °C not more than 110 kPa)	3	F1	II	3	640D	LQ4	E2	T	PP, EX, A	VE01		1	
1863	FUEL, AVIATION, TURBINE ENGINE	3	F1	III	3		LQ7	E1	T	PP, EX, A	VE01		0	
1865	n-PROPYL NITRATE	3	F1	II	3		LQ4	E2		PP, EX, A	VE01		1	
1866	RESIN SOLUTION, flammable	3	F1	I	3		LQ3	E3		PP, EX, A	VE01		1	

UN No. or ID No.	Name and description	Class	Classi-fication Code	Packing group	Labels	Special provi-sions	Limited and excepted quantities		Carriage permitted	Equipment required	Venti-lation	Provisions concerning loading, unloading and carriage			Number of blue cones/ lights	Remarks
	3.1.2	2.2	2.2	2.1.1.3	5.2.2	3.3	3.4.6	3.5.1.2	3.2.1	8.1.5	7.1.6	7.1.6			7.1.5	3.2.1
(1)	(2)	(3a)	(3b)	(4)	(5)	(6)	(7a)	(7b)	(8)	(9)	(10)	(11)			(12)	(13)
1866	RESIN SOLUTION, flammable (vapour pressure at 50 °C more than 110 kPa)	3	F1	II	3	640C	LQ6	E2		PP, EX, A	VE01				1	
1866	RESIN SOLUTION, flammable (vapour pressure at 50 °C not more than 110 kPa)	3	F1	II	3	640D	LQ6	E2		PP, EX, A	VE01				1	
1866	RESIN SOLUTION, flammable	3	F1	III	3	640E	LQ7	E1		PP, EX, A	VE01				0	
1866	RESIN SOLUTION, flammable (having a flash-point below 23 °C and viscous according to 2.2.3.1.4) (boiling point not more than 35° C)	3	F1	III	3	640F	LQ7	E1		PP, EX, A	VE01				0	
1866	RESIN SOLUTION, flammable (having a flash-point below 23 °C and viscous according to 2.2.3.1.4) (vapour pressure at50 °C more than 110 kPa , boiling point of more than 35° C)	3	F1	III	3	640G	LQ7	E1		PP, EX, A	VE01				0	
1866	RESIN SOLUTION, flammable (having a flash-point below 23 °C and viscous according to 2.2.3.1.4) (vapour pressure at 50 °C not more than 110 kPa)	3	F1	III	3	640H	LQ7	E1		PP, EX, A	VE01				0	
1868	DECABORANE	4.1	FT2	II	4.1+6.1	802	LQ0	E2		PP					2	
1869	MAGNESIUM or MAGNESIUM ALLOYS with more than 50% magnesium in pellets, turnings or ribbons	4.1	F3	III	4.1	59	LQ9	E1		PP					0	
1870	POTASSIUM BOROHYDRIDE	4.3	W2	I	4.3		LQ0	E0		PP, EX, A	VE01		HA08		0	
1871	TITANIUM HYDRIDE	4.1	F3	II	4.1		LQ8	E2		PP					1	
1872	LEAD DIOXIDE	5.1	OT2	III	5.1+6.1	802	LQ12	E1		PP					0	
1873	PERCHLORIC ACID with more than 50% but not more than 72% acid, by mass	5.1	OC1	I	5.1+8	60	LQ0	E0		PP, EP					0	
1884	BARIUM OXIDE	6.1	T5	III	6.1	802	LQ9	E1		PP, EP					0	
1885	BENZIDINE	6.1	T2	II	6.1	802	LQ18	E4		PP, EP					2	
1886	BENZYLIDENE CHLORIDE	6.1	T1	II	6.1	802	LQ17	E4		PP, EP, TOX, A	VE02				2	
1887	BROMOCHLOROMETHANE	6.1	T1	III	6.1	802	LQ7	E1		PP, EP, TOX, A	VE02				0	
1888	CHLOROFORM	6.1	T1	III	6.1	802	LQ7	E1	**T**	PP, EP, TOX, A	VE02				0	
1889	CYANOGEN BROMIDE	6.1	TC2	I	6.1+8	802	LQ0	E5		PP, EP					2	
1891	ETHYL BROMIDE	6.1	T1	II	6.1	802	LQ17	E4		PP, EP, TOX, A	VE02				2	
1892	ETHYLDICHLOROARSINE	6.1	T3	I	6.1	802	LQ0	E5		PP, EP, TOX, A	VE02				2	
1894	PHENYLMERCURIC HYDROXIDE	6.1	T3	II	6.1	802	LQ18	E4		PP, EP, TOX, A	VE02				2	
1895	PHENYLMERCURIC NITRATE	6.1	T3	II	6.1	802	LQ18	E4		PP, EP, TOX, A	VE02				2	
1897	TETRACHLOROETHYLENE	6.1	T1	III	6.1	802	LQ7	E1	**T**	PP, EP, TOX, A	VE02				0	
1898	ACETYL IODIDE	8	C3	II	8		LQ22	E2		PP, EP					0	
1902	DIISOOCTYL ACID PHOSPHATE	8	C3	III	8		LQ7	E1		PP, EP					0	

UN No. or ID No.	Name and description 3.1.2	Class 2.2	Classi-fication Code 2.2	Packing group 2.1.1.3	Labels 5.2.2	Special provi-sions 3.3	Limited and excepted quantities 3.4.6	 3.5.1.2	Carriage permitted 3.2.1	Equipment required 8.1.5	Venti-lation 7.1.6	Provisions concerning loading, unloading and carriage 7.1.6	Number of blue cones/ lights 7.1.5	Remarks 3.2.1
(1)	(2)	(3a)	(3b)	(4)	(5)	(6)	(7a)	(7b)	(8)	(9)	(10)	(11)	(12)	(13)
1903	DISINFECTANT, LIQUID, CORROSIVE, N.O.S.	8	C9	I	8	274	LQ0	E0		PP, EP			0	
1903	DISINFECTANT, LIQUID, CORROSIVE, N.O.S.	8	C9	II	8	274	LQ22	E2		PP, EP			0	
1903	DISINFECTANT, LIQUID, CORROSIVE, N.O.S.	8	C9	III	8	274	LQ7	E1		PP, EP			0	
1905	SELENIC ACID	8	C2	I	8		LQ0	E0		PP, EP			0	
1906	SLUDGE ACID	8	C1	II	8		LQ22	E2		PP, EP			0	
1907	SODA LIME with more than 4% sodium hydroxide	8	C6	III	8	62	LQ24	E1		PP, EP			0	
1908	CHLORITE SOLUTION	8	C9	II	8	521	LQ22	E2		PP, EP			0	
1908	CHLORITE SOLUTION	8	C9	III	8	521	LQ7	E1		PP, EP			0	
1910	Calcium oxide	8	C6	NOT SUBJECT TO ADN										
1911	DIBORANE	2	2TF		2.3+2.1		LQ0	E0		PP, EP, EX, TOX, A	VE01, VE02		2	
1912	METHYL CHLORIDE AND METHYLENE CHLORIDE MIXTURE	2	2F		2.1	228	LQ0	E0	**T**	PP, EX, A	VE01		1	
1913	NEON, REFRIGERATED LIQUID	2	3A		2.2	593	LQ1	E1		PP			0	
1914	BUTYL PROPIONATES	3	F1	III	3		LQ7	E1		PP, EX, A	VE01		0	
1915	CYCLOHEXANONE	3	F1	III	3		LQ7	E1	**T**	PP, EX, A	VE01		0	
1916	2,2'-DICHLORODIETHYL ETHER	6.1	TF1	II	6.1+3	802	LQ17	E4		PP, EP, EX, TOX, A	VE01, VE02		2	
1917	ETHYL ACRYLATE, STABILIZED	3	F1	II	3		LQ4	E2	**T**	PP, EX, A	VE01		1	
1918	ISOPROPYLBENZENE	3	F1	III	3		LQ7	E1	**T**	PP, EX, A	VE01		0	
1919	METHYL ACRYLATE, STABILIZED	3	F1	II	3		LQ4	E2	**T**	PP, EX, A	VE01		1	
1920	NONANES	3	F1	III	3		LQ7	E1	**T**	PP, EX, A	VE01		0	
1921	PROPYLENEIMINE, STABILIZED	3	FT1	I	3+6.1	802	LQ0	E0		PP, EP, EX, TOX, A	VE01, VE02		2	
1922	PYRROLIDINE	3	FC	II	3+8		LQ4	E2	**T**	PP, EP, EX, A	VE01		1	
1923	CALCIUM DITHIONITE (CALCIUM HYDROSULPHITE)	4.2	S4	II	4.2		LQ0	E2		PP			0	
1928	METHYL MAGNESIUM BROMIDE IN ETHYL ETHER	4.3	WF1	I	4.3+3		LQ0	E0		PP, EX, A	VE01	HA08	1	
1929	POTASSIUM DITHIONITE (POTASSIUM HYDROSULPHITE)	4.2	S4	II	4.2		LQ0	E2		PP			0	
1931	ZINC DITHIONITE (ZINC HYDROSULPHITE)	9	M11	III	9		LQ27	E1		PP			0	
1932	ZIRCONIUM SCRAP	4.2	S4	III	4.2	524 592	LQ0	E1		PP			0	
1935	CYANIDE SOLUTION, N.O.S.	6.1	T4	I	6.1	274 525 802	LQ0	E5		PP, EP, TOX, A	VE02		2	
1935	CYANIDE SOLUTION, N.O.S.	6.1	T4	II	6.1	274 525 802	LQ17	E4		PP, EP, TOX, A	VE02		2	

UN No. or ID No.	Name and description	Class	Classi-fication Code	Packing group	Labels	Special provi-sions	Limited and excepted quantities		Carriage permitted	Equipment required	Venti-lation	Provisions concerning loading, unloading and carriage			Number of blue cones/ lights	Remarks
	3.1.2	2.2	2.2	2.1.1.3	5.2.2	3.3	3.4.6	3.5.1.2	3.2.1	8.1.5	7.1.6	7.1.6			7.1.5	3.2.1
(1)	(2)	(3a)	(3b)	(4)	(5)	(6)	(7a)	(7b)	(8)	(9)	(10)	(11)			(12)	(13)
1935	CYANIDE SOLUTION, N.O.S.	6.1	T4	III	6.1	274 525 802	LQ7	E1		PP, EP, TOX, A	VE02				0	
1938	BROMOACETIC ACID, SOLUTION	8	C3	II	8		LQ22	E2		PP, EP					0	
1938	BROMOACETIC ACID SOLUTION	8	C3	III	8		LQ7	E1		PP, EP					0	
1939	PHOSPHORUS OXYBROMIDE	8	C2	II	8		LQ23	E2		PP, EP					0	
1940	THIOGLYCOLIC ACID	8	C3	II	8		LQ22	E2		PP, EP					0	
1941	DIBROMODIFLUOROMETHANE	9	M11	III	9		LQ28	E1		PP					0	
1942	AMMONIUM NITRATE with not more than 0.2% total combustible material, including any organic substance calculated as carbon, to the exclusion of any other added substance	5.1	O2	III	5.1	306 611	LQ12	E1	**B**	PP		ST01, CO02	HA09		0	CO02 and HA09 apply only when this substance is carried in bulk or without packaging
1944	MATCHES, SAFETY (book, card or strike on box)	4.1	F1	III	4.1	293	LQ9	E1		PP					0	
1945	MATCHES, WAX 'VESTA'	4.1	F1	III	4.1	293	LQ9	E1		PP					0	
1950	AEROSOLS, asphyxiant	2	5A		2.2	190 327 625	LQ2	E0		PP	VE04				0	
1950	AEROSOLS, corrosive	2	5C		2.2+8	190 327 625	LQ2	E0		PP, EP	VE04				0	
1950	AEROSOLS, corrosive, oxidizing	2	5CO		2.2+5.1+8	190 327 625	LQ2	E0		PP, EP	VE04				0	
1950	AEROSOLS, flammable	2	5F		2.1	190 327 625	LQ2	E0		PP, EX, A	VE01, VE04				1	
1950	AEROSOLS, flammable, corrosive	2	5FC		2.1+8	190 327 625	LQ2	E0		PP, EX, A	VE01, VE04				1	
1950	AEROSOLS, oxidizing	2	5O		2.2+5.1	190 327 625	LQ2	E0		PP	VE04				0	
1950	AEROSOLS, toxic	2	5T		2.2+6.1	190 327 625	LQ1	E0		PP, EP, TOX, A	VE02, VE04				2	
1950	AEROSOLS, toxic, corrosive	2	5TC		2.2+6.1+8	190 327 625	LQ1	E0		PP, EP, TOX, A	VE02, VE04				2	
1950	AEROSOLS, toxic, flammable	2	5TF		2.1+6.1	190 327 625	LQ1	E0		PP, EP, EX, TOX, A	VE01, VE02, VE04				2	
1950	AEROSOLS, toxic, flammable, corrosive	2	5TFC		2.1+6.1+8	190 327 625	LQ1	E0		PP, EP, EX, TOX, A	VE01, VE02				2	

UN No. or ID No.	Name and description	Class	Classi-fication Code	Packing group	Labels	Special provi-sions	Limited and excepted quantities		Carriage permitted	Equipment required	Venti-lation	Provisions concerning loading, unloading and carriage	Number of blue cones/ lights	Remarks
	3.1.2	2.2	2.2	2.1.1.3	5.2.2	3.3	3.4.6	3.5.1.2	3.2.1	8.1.5	7.1.6	7.1.6	7.1.5	3.2.1
(1)	(2)	(3a)	(3b)	(4)	(5)	(6)	(7a)	(7b)	(8)	(9)	(10)	(11)	(12)	(13)
1950	AEROSOLS, toxic, oxidizing	2	5TO		2.2+5.1+6.1	190 327 625	LQ1	E0		PP, EP, TOX, A	VE02, VE04		2	
1950	AEROSOLS, toxic, oxidizing, corrosive	2	5TOC		2.2+5.1+6.1+8	190 327 625	LQ1	E0		PP, EP, TOX, A	VE02, VE04		2	
1951	ARGON, REFRIGERATED LIQUID	2	3A		2.2	593	LQ1	E1		PP			0	
1952	ETHYLENE OXIDE AND CARBON DIOXIDE MIXTURE with not more than 9% ethylene oxide	2	2A		2.2		LQ1	E1		PP			0	
1953	COMPRESSED GAS, TOXIC, FLAMMABLE, N.O.S.	2	1TF		2.3+2.1	274	LQ0	E0		PP, EP, EX, TOX, A	VE01, VE02		2	
1954	COMPRESSED GAS, FLAMMABLE, N.O.S.	2	1F		2.1	274	LQ0	E0		PP, EX, A	VE01		1	
1955	COMPRESSED GAS, TOXIC, N.O.S.	2	1T		2.3	274	LQ0	E0		PP, EP, TOX, A	VE02		2	
1956	COMPRESSED GAS, N.O.S.	2	1A		2.2	274 292 567	LQ1	E1		PP			0	
1957	DEUTERIUM, COMPRESSED	2	1F		2.1		LQ0	E0		PP, EX, A	VE01		1	
1958	1,2-DICHLORO-1,1,2,2-TETRAFLUOROETHANE (REFRIGERANT GAS R 114)	2	2A		2.2		LQ1	E1		PP			0	
1959	1,1-DIFLUOROETHYLENE (REFRIGERANT GAS R 1132a)	2	2F		2.1		LQ0	E0		PP, EX, A	VE01		1	
1961	ETHANE, REFRIGERATED LIQUID	2	3F		2.1		LQ0	E0		PP, EX, A	VE01		1	
1962	ETHYLENE	2	2F		2.1		LQ0	E0		PP, EX, A	VE01		1	
1963	HELIUM, REFRIGERATED LIQUID	2	3A		2.2	593	LQ1	E1		PP			0	
1964	HYDROCARBON GAS MIXTURE, COMPRESSED, N.O.S.	2	1F		2.1	274	LQ0	E0		PP, EX, A	VE01		1	
1965	HYDROCARBON GAS MIXTURE, LIQUEFIED, N.O.S. such as mixtures A, A01, A02, A0, A1, B1, B2, B or C	2	2F		2.1	274 583	LQ0	E0	T	PP, EX, A	VE01		1	
1966	HYDROGEN, REFRIGERATED LIQUID	2	3F		2.1		LQ0	E0		PP, EX, A	VE01		1	
1967	INSECTICIDE GAS, TOXIC, N.O.S.	2	2T		2.3	274	LQ0	E0		PP, EP, TOX, A	VE02		2	
1968	INSECTICIDE GAS, N.O.S.	2	2A		2.2	274	LQ1	E1		PP			0	
1969	ISOBUTANE	2	2F		2.1		LQ0	E0	T	PP, EX, A	VE01		1	
1970	KRYPTON, REFRIGERATED LIQUID	2	3A		2.2	593	LQ1	E1		PP			0	
1971	METHANE, COMPRESSED or NATURAL GAS, COMPRESSED with high methane content	2	1F		2.1		LQ0	E0		PP, EX, A	VE01		1	
1972	METHANE, REFRIGERATED LIQUID or NATURAL GAS, REFRIGERATED LIQUID with high methane content	2	3F		2.1		LQ0	E0		PP, EX, A	VE01		1	

UN No. or ID No.	Name and description	Class	Classi-fication Code	Packing group	Labels	Special provi-sions	Limited and excepted quantities		Carriage permitted	Equipment required	Venti-lation	Provisions concerning loading, unloading and carriage	Number of blue cones/ lights	Remarks
	3.1.2	2.2	2.2	2.1.1.3	5.2.2	3.3	3.4.6	3.5.1.2	3.2.1	8.1.5	7.1.6	7.1.6	7.1.5	3.2.1
(1)	(2)	(3a)	(3b)	(4)	(5)	(6)	(7a)	(7b)	(8)	(9)	(10)	(11)	(12)	(13)
1973	CHLORODIFLUOROMETHANE AND CHLOROPENTAFLUOROETHANE MIXTURE with fixed boiling point, with approximately 49% chlorodifluoromethane (REFRIGERANT GAS R 502)	2	2A		2.2		LQ1	E1		PP			0	
1974	CHLORODIFLUOROBROMOMETHANE (REFRIGERANT GAS R 12B1)	2	2A		2.2		LQ1	E1		PP			0	
1975	NITRIC OXIDE AND DINITROGEN TETROXIDE MIXTURE (NITRIC OXIDE AND NITROGEN DIOXIDE MIXTURE)	2	2TOC		2.3+5.1+8		LQ0	E0		PP, EP, TOX, A	VE02		2	
1976	OCTAFLUOROCYCLOBUTANE (REFRIGERANT GAS RC 318)	2	2A		2.2		LQ1	E1		PP			0	
1977	NITROGEN, REFRIGERATED LIQUID	2	3A		2.2	593	LQ1	E1		PP			0	
1978	PROPANE	2	2F		2.1		LQ0	E0	T	PP, EX, A	VE01		1	
1982	TETRAFLUOROMETHANE (REFRIGERANT GAS R 14)	2	2A		2.2		LQ1	E1		PP			0	
1983	1-CHLORO-2,2,2-TRIFLUOROETHANE (REFRIGERANT GAS R 133a)	2	2A		2.2		LQ1	E1		PP			0	
1984	TRIFLUOROMETHANE (REFRIGERANT GAS R 23)	2	2A		2.2		LQ1	E1		PP			0	
1986	ALCOHOLS, FLAMMABLE, TOXIC, N.O.S.	3	FT1	I	3+6.1	274 802	LQ0	E0	T	PP, EP, EX, TOX, A	VE01, VE02		2	
1986	ALCOHOLS, FLAMMABLE, TOXIC, N.O.S.	3	FT1	II	3+6.1	274 802	LQ0	E2	T	PP, EP, EX, TOX, A	VE01, VE02		2	
1986	ALCOHOLS, FLAMMABLE, TOXIC, N.O.S.	3	FT1	III	3+6.1	274 802	LQ7	E1	T	PP, EP, EX, TOX, A	VE01, VE02		0	
1987	ALCOHOLS, N.O.S. (vapour pressure at 50 °C more than 110 kPa)	3	F1	II	3	274 601 640C	LQ4	E2	T	PP, EX, A	VE01		1	
1987	ALCOHOLS, N.O.S. (vapour pressure at 50 °C not more than 110 kPa)	3	F1	II	3	274 601 640D	LQ4	E2	T	PP, EX, A	VE01		1	
1987	ALCOHOLS, N.O.S.	3	F1	III	3	274 601	LQ7	E1	T	PP, EX, A	VE01		0	
1988	ALDEHYDES, FLAMMABLE, TOXIC, N.O.S.	3	FT1	I	3+6.1	274 802	LQ0	E0		PP, EP, EX, TOX, A	VE01, VE02		2	
1988	ALDEHYDES, FLAMMABLE, TOXIC, N.O.S.	3	FT1	II	3+6.1	274 802	LQ0	E2		PP, EP, EX, TOX, A	VE01, VE02		2	
1988	ALDEHYDES, FLAMMABLE, TOXIC, N.O.S.	3	FT1	III	3+6.1	274 802	LQ7	E1		PP, EP, EX, TOX, A	VE01, VE02		0	
1989	ALDEHYDES, N.O.S.	3	F1	I	3	274	LQ3	E3		PP, EX, A	VE01		1	
1989	ALDEHYDES, N.O.S. (vapour pressure at 50 °C more than 110 kPa)	3	F1	II	3	274 640C	LQ4	E2	T	PP, EX, A	VE01		1	
1989	ALDEHYDES, N.O.S. (vapour pressure at 50 °C not more than 110 kPa)	3	F1	II	3	274 640D	LQ4	E2	T	PP, EX, A	VE01		1	
1989	ALDEHYDES, N.O.S.	3	F1	III	3	274	LQ7	E1	T	PP, EX, A	VE01		0	

UN No. or ID No.	Name and description 3.1.2	Class 2.2	Classi-fication Code 2.2	Packing group 2.1.1.3	Labels 5.2.2	Special provi-sions 3.3	Limited and excepted quantities 3.4.6	3.5.1.2	Carriage permitted 3.2.1	Equipment required 8.1.5	Venti-lation 7.1.6	Provisions concerning loading, unloading and carriage 7.1.6	Number of blue cones/ lights 7.1.5	Remarks 3.2.1
(1)	(2)	(3a)	(3b)	(4)	(5)	(6)	(7a)	(7b)	(8)	(9)	(10)	(11)	(12)	(13)
1990	BENZALDEHYDE	9	M11	III	9		LQ28	E1		PP			0	
1991	CHLOROPRENE, STABILIZED	3	FT1	I	3+6.1	802	LQ0	E0	T	PP, EP, EX, TOX, A	VE01, VE02		2	
1992	FLAMMABLE LIQUID, TOXIC, N.O.S.	3	FT1	I	3+6.1	274 802	LQ0	E0	T	PP, EP, EX, TOX, A	VE01, VE02		2	
1992	FLAMMABLE LIQUID, TOXIC, N.O.S.	3	FT1	II	3+6.1	274 802	LQ0	E2	T	PP, EP, EX, TOX, A	VE01, VE02		2	
1992	FLAMMABLE LIQUID, TOXIC, N.O.S.	3	FT1	III	3+6.1	274 802	LQ7	E1	T	PP, EP, EX, TOX, A	VE01, VE02		0	
1993	FLAMMABLE LIQUID, N.O.S.	3	F1	I	3	274	LQ3	E3	T	PP, EX, A	VE01		1	
1993	FLAMMABLE LIQUID, N.O.S. (vapour pressure at 50 °C more than 110 kPa)	3	F1	II	3	274 601 640C	LQ4	E2	T	PP, EX, A	VE01		1	
1993	FLAMMABLE LIQUID, N.O.S. (vapour pressure at 50 °C not more than 110 kPa)	3	F1	II	3	274 601 640D	LQ4	E2	T	PP, EX, A	VE01		1	
1993	FLAMMABLE LIQUID, N.O.S.	3	F1	III	3	274 601 640E	LQ7	E1	T	PP, EX, A	VE01		0	
1993	FLAMMABLE LIQUID, N.O.S. (having a flash-point below 23 °C and viscous according to 2.2.3.1.4) (boiling point not more than 35° C)	3	F1	III	3	274 601 640F	LQ7	E1	T	PP, EX, A	VE01		0	
1993	FLAMMABLE LIQUID, N.O.S. (having a flash-point below 23 °C and viscous according to 2.2.3.1.4) (vapour pressure at 50 °C more than 110 kPa , boiling point of more than 35° C)	3	F1	III	3	274 601 640G	LQ7	E1	T	PP, EX, A	VE01		0	
1993	FLAMMABLE LIQUID, N.O.S. (having a flash-point below 23 °C and viscous according to 2.2.3.1.4) (vapour pressure at 50 °C not more than 110 kPa)	3	F1	III	3	274 601 640H	LQ7	E1	T	PP, EX, A	VE01		0	
1994	IRON PENTACARBONYL	6.1	TF1	I	6.1+3	802	LQ0	E5		PP, EP, EX, TOX, A	VE01, VE02		2	
1999	TARS, LIQUID, including road asphalt and oils, bitumen and cut backs (vapour pressure at 50 °C more than 110 kPa)	3	F1	II	3	640C	LQ6	E2		PP, EX, A	VE01		1	
1999	TARS, LIQUID, including road asphalt and oils, bitumen and cut backs (vapour pressure at 50 °C not more than 110 kPa)	3	F1	II	3	640D	LQ6	E2		PP, EX, A	VE01		1	
1999	TARS, LIQUID, including road asphalt and oils, bitumen and cut backs	3	F1	III	3	640E	LQ7	E1	T	PP, EX, A	VE01		0	
1999	TARS, LIQUID, including road asphalt and oils, bitumen and cut backs (having a flash-point below 23 °C and viscous according to 2.2.3.1.4) (boiling point not more than 35° C)	3	F1	III	3	640F	LQ7	E1		PP, EX, A	VE01		0	

UN No. or ID No.	Name and description	Class	Classi-fication Code	Packing group	Labels	Special provi-sions	Limited and excepted quantities		Carriage permitted	Equipment required	Venti-lation	Provisions concerning loading, unloading and carriage	Number of blue cones/ lights	Remarks
	3.1.2	2.2	2.2	2.1.1.3	5.2.2	3.3	3.4.6	3.5.1.2	3.2.1	8.1.5	7.1.6	7.1.6	7.1.5	3.2.1
(1)	(2)	(3a)	(3b)	(4)	(5)	(6)	(7a)	(7b)	(8)	(9)	(10)	(11)	(12)	(13)
1999	TARS, LIQUID, including road asphalt and oils, bitumen and cut backs (having a flash-point below 23 °C and viscous according to 2.2.3.1.4) (vapour pressure at 50 °C more than 110 kPa , boiling point of more than 35° C)	3	F1	III	3	640G	LQ7	E1		PP, EX, A	VE01		0	
1999	TARS, LIQUID, including road asphalt and oils, bitumen and cut backs (having a flash-point below 23 °C and viscous according to 2.2.3.1.4) (vapour pressure at 50 °C not more than 110 kPa)	3	F1	III	3	640H	LQ7	E1		PP, EX, A	VE01		0	
2000	CELLULOID in block, rods, rolls, sheets, tubes, etc., except scrap	4.1	F1	III	4.1	502	LQ9	E1		PP			0	
2001	COBALT NAPHTHENATES, POWDER	4.1	F3	III	4.1		LQ9	E1		PP			0	
2002	CELLULOID, SCRAP	4.2	S2	III	4.2	526 592	LQ0	E1		PP			0	
2004	MAGNESIUM DIAMIDE	4.2	S4	II	4.2		LQ0	E2		PP			0	
2006	PLASTICS, NITROCELLULOSE-BASED, SELF-HEATING, N.O.S.	4.2	S2	III	4.2	274 528	LQ0	E1		PP			0	
2008	ZIRCONIUM POWDER, DRY	4.2	S4	I	4.2	524 540	LQ0	E0		PP			0	
2008	ZIRCONIUM POWDER, DRY	4.2	S4	II	4.2	524 540	LQ0	E2		PP			0	
2008	ZIRCONIUM POWDER, DRY	4.2	S4	III	4.2	524 540	LQ0	E1		PP			0	
2009	ZIRCONIUM, DRY, finished sheets, strip or coiled wire	4.2	S4	III	4.2	524 592	LQ0	E1		PP			0	
2010	MAGNESIUM HYDRIDE	4.3	W2	I	4.3		LQ0	E0		PP, EX, A	VE01	HA08	0	
2011	MAGNESIUM PHOSPHIDE	4.3	WT2	I	4.3+6.1	802	LQ0	E0		PP, EP, EX, TOX, A	VE01, VE02	HA08	2	
2012	POTASSIUM PHOSPHIDE	4.3	WT2	I	4.3+6.1	802	LQ0	E0		PP, EP, EX, TOX, A	VE01, VE02	HA08	2	
2013	STRONTIUM PHOSPHIDE	4.3	WT2	I	4.3+6.1	802	LQ0	E0		PP, EP, EX, TOX, A	VE01, VE02	HA08	2	
2014	HYDROGEN PEROXIDE, AQUEOUS SOLUTION with not less than 20% but not more than 60% hydrogen peroxide (stabilized as necessary)	5.1	OC1	II	5.1+8		LQ10	E2	**T**	PP, EP			0	
2015	HYDROGEN PEROXIDE, AQUEOUS SOLUTION, STABILIZED with more than 70% hydrogen peroxide	5.1	OC1	I	5.1+8	640N	LQ0	E0		PP, EP			0	
2015	HYDROGEN PEROXIDE, AQUEOUS SOLUTION, STABILIZED with more than 60% hydrogen peroxide and not more than 70% hydrogen peroxide	5.1	OC1	I	5.1+8	640O	LQ0	E0		PP, EP			0	
2016	AMMUNITION, TOXIC, NON-EXPLOSIVE without burster or expelling charge, non-fuzed	6.1	T2	II	6.1	802	LQ0	E0		PP, EP			2	

UN No. or ID No.	Name and description	Class	Classi-fication Code	Packing group	Labels	Special provi-sions	Limited and excepted quantities		Carriage permitted	Equipment required	Venti-lation	Provisions concerning loading, unloading and carriage	Number of blue cones/ lights	Remarks
	3.1.2	2.2	2.2	2.1.1.3	5.2.2	3.3	3.4.6	3.5.1.2	3.2.1	8.1.5	7.1.6	7.1.6	7.1.5	3.2.1
(1)	(2)	(3a)	(3b)	(4)	(5)	(6)	(7a)	(7b)	(8)	(9)	(10)	(11)	(12)	(13)
2017	AMMUNITION, TEAR-PRODUCING, NON-EXPLOSIVE without burster or expelling charge, non-fuzed	6.1	TC2	II	6.1+8	802	LQ0	E0		PP, EP			2	
2018	CHLOROANILINES, SOLID	6.1	T2	II	6.1	802	LQ18	E4		PP, EP			2	
2019	CHLOROANILINES, LIQUID	6.1	T1	II	6.1	802	LQ17	E4		PP, EP, TOX, A	VE02		2	
2020	CHLOROPHENOLS, SOLID	6.1	T2	III	6.1	205 802	LQ9	E1		PP, EP			0	
2021	CHLOROPHENOLS, LIQUID	6.1	T1	III	6.1	802	LQ7	E1	T	PP, EP, TOX, A	VE02		0	
2022	CRESYLIC ACID	6.1	TC1	II	6.1+8	802	LQ17	E4	T	PP, EP, TOX, A	VE02		2	
2023	EPICHLOROHYDRIN	6.1	TF1	II	6.1+3	279 802	LQ17	E4	T	PP, EP, EX, TOX, A	VE01, VE02		2	
2024	MERCURY COMPOUND, LIQUID, N.O.S.	6.1	T4	I	6.1	43 274 802	LQ0	E5		PP, EP, TOX, A	VE02		2	
2024	MERCURY COMPOUND, LIQUID, N.O.S.	6.1	T4	II	6.1	43 274 802	LQ17	E4		PP, EP, TOX, A	VE02		2	
2024	MERCURY COMPOUND, LIQUID, N.O.S.	6.1	T4	III	6.1	43 274 802	LQ7	E1		PP, EP, TOX, A	VE02		0	
2025	MERCURY COMPOUND, SOLID, N.O.S.	6.1	T5	I	6.1	43 274 529 585 802	LQ0	E5		PP, EP			2	
2025	MERCURY COMPOUND, SOLID, N.O.S.	6.1	T5	II	6.1	43 274 29 585 802	LQ18	E4		PP, EP			2	
2025	MERCURY COMPOUND, SOLID, N.O.S.	6.1	T5	III	6.1	43 274 529 585 802	LQ9	E1		PP, EP			0	
2026	PHENYLMERCURIC COMPOUND, N.O.S.	6.1	T3	I	6.1	43 274 802	LQ0	E5		PP, EP, TOX, A	VE02		2	
2026	PHENYLMERCURIC COMPOUND, N.O.S.	6.1	T3	II	6.1	43 274 802	LQ18	E4		PP, EP, TOX, A	VE02		2	

UN No. or ID No.	Name and description	Class	Classi-fication Code	Packing group	Labels	Special provi-sions	Limited and excepted quantities		Carriage permitted	Equipment required	Venti-lation	Provisions concerning loading, unloading and carriage	Number of blue cones/ lights	Remarks
	3.1.2	2.2	2.2	2.1.1.3	5.2.2	3.3	3.4.6	3.5.1.2	3.2.1	8.1.5	7.1.6	7.1.6	7.1.5	3.2.1
(1)	(2)	(3a)	(3b)	(4)	(5)	(6)	(7a)	(7b)	(8)	(9)	(10)	(11)	(12)	(13)
2026	PHENYLMERCURIC COMPOUND, N.O.S.	6.1	T3	III	6.1	43 274 802	LQ9	E1		PP, EP, TOX, A	VE02		0	
2027	SODIUM ARSENITE, SOLID	6.1	T5	II	6.1	43 802	LQ18	E4		PP, EP			2	
2028	BOMBS, SMOKE, NON-EXPLOSIVE with corrosive liquid, without initiating device	8	C11	II	8		LQ0	E0		PP, EP			0	
2029	HYDRAZINE, ANHYDROUS	8	CFT	I	8+3+6.1	802	LQ0	E0		PP, EP, EX, TOX, A	VE01, VE02		2	
2030	HYDRAZINE AQUEOUS SOLUTION, with more than 37% hydrazine by mass, having a flash-point above 60 °C	8	CT1	I	8+6.1	530 802	LQ0	E0		PP, EP, TOX, A	VE02		2	
2030	HYDRAZINE AQUEOUS SOLUTION, with more than 37% hydrazine by mass having a flash-point of not more than 60 °C	8	CFT	I	8 3 6.1	530 802	LQ0	E0		PP, EP, EX, TOX, A	VE01, VE02		2	
2030	HYDRAZINE AQUEOUS SOLUTION, with more than 37% hydrazine by mass	8	CT1	II	8+6.1	530 802	LQ22	E2		PP, EP, TOX, A	VE02		2	
2030	HYDRAZINE AQUEOUS SOLUTION, with more than 37% hydrazine by mass	8	CT1	III	8+6.1	530 802	LQ7	E1		PP, EP, TOX, A	VE02		0	
2031	NITRIC ACID, other than red fuming, with more than 70% nitric acid	8	CO1	I	8+5.1		LQ0	E0	**T**	PP, EP			0	
2031	NITRIC ACID, other than red fuming, with at least 65%, but not more than 70% nitric acid	8	CO1	II	8+5.1		LQ22	E2	**T**	PP, EP			0	
2031	NITRIC ACID, other than red fuming, with less than 65% nitric acid	8	C1	II	8		LQ22	E2	T	PP,EP			0	
2032	NITRIC ACID, RED FUMING	8	COT	I	8+5.1+6.1	802	LQ0	E0	**T**	PP, EP, TOX, A	VE02		2	
2033	POTASSIUM MONOXIDE	8	C6	II	8		LQ23	E2		PP, EP			0	
2034	HYDROGEN AND METHANE MIXTURE, COMPRESSED	2	1F		2.1		LQ0	E0		PP, EX, A	VE01		1	
2035	1,1,1-TRIFLUOROETHANE (REFRIGERANT GAS R 143a)	2	2F		2.1		LQ0	E0		PP, EX, A	VE01		1	
2036	XENON	2	2A		2.2		LQ1	E1		PP			0	
2037	RECEPTACLES, SMALL, CONTAINING GAS (GAS CARTRIDGES) without a release device, non-refillable	2	5A		2.2	191 303	LQ2	E0		PP			0	
2037	RECEPTACLES, SMALL, CONTAINING GAS (GAS CARTRIDGES) without a release device, non-refillable	2	5F		2.1	191 303	LQ2	E0		PP, EX, A	VE01		1	
2037	RECEPTACLES, SMALL, CONTAINING GAS (GAS CARTRIDGES) without a release device, non-refillable	2	5O		2.2+5.1	191 303	LQ2	E0		PP			0	
2037	RECEPTACLES, SMALL, CONTAINING GAS (GAS CARTRIDGES) without a release device, non-refillable	2	5T		2.3	303	LQ1	E0		PP, EP, TOX, A	VE02		2	

UN No. or ID No.	Name and description	Class	Classi-fication Code	Packing group	Labels	Special provi-sions	Limited and excepted quantities		Carriage permitted	Equipment required	Venti-lation	Provisions concerning loading, unloading and carriage	Number of blue cones/ lights	Remarks
	3.1.2	2.2	2.2	2.1.1.3	5.2.2	3.3	3.4.6	3.5.1.2	3.2.1	8.1.5	7.1.6	7.1.6	7.1.5	3.2.1
(1)	(2)	(3a)	(3b)	(4)	(5)	(6)	(7a)	(7b)	(8)	(9)	(10)	(11)	(12)	(13)
2037	RECEPTACLES, SMALL, CONTAINING GAS (GAS CARTRIDGES) without a release device, non-refillable	2	5TC		2.3+8	303	LQ1	E0		PP, EP, TOX, A	VE02		2	
2037	RECEPTACLES, SMALL, CONTAINING GAS (GAS CARTRIDGES) without a release device, non-refillable	2	5TF		2.3+2.1	303	LQ1	E0		PP, EP, EX, TOX, A	VE01, VE02		2	
2037	RECEPTACLES, SMALL, CONTAINING GAS (GAS CARTRIDGES) without a release device, non-refillable	2	5TFC		2.3+2.1+8	303	LQ1	E0		PP, EP, EX, TOX, A	VE01, VE02		2	
2037	RECEPTACLES, SMALL, CONTAINING GAS (GAS CARTRIDGES) without a release device, non-refillable	2	5TO		2.3+5.1	303	LQ1	E0		PP, EP, TOX, A	VE02		2	
2037	RECEPTACLES, SMALL, CONTAINING GAS (GAS CARTRIDGES) without a release device, non-refillable	2	5TOC		2.3+5.1+8	303	LQ1	E0		PP, EP, TOX, A	VE02		2	
2038	DINITROTOLUENES, LIQUID	6.1	T1	II	6.1	802	LQ17	E4		PP, EP, TOX, A	VE02		2	
2044	2,2-DIMETHYLPROPANE	2	2F		2.1		LQ0	E0		PP, EX, A	VE01		1	
2045	ISOBUTYRALDEHYDE (ISOBUTYL ALDEHYDE)	3	F1	II	3		LQ4	E2	T	PP, EX, A	VE01		1	
2046	CYMENES	3	F1	III	3		LQ7	E1	T	PP, EX, A	VE01		0	
2047	DICHLOROPROPENES	3	F1	II	3		LQ4	E2	T	PP, EX, A	VE01		1	
2047	DICHLOROPROPENES	3	F1	III	3		LQ7	E1	T	PP, EX, A	VE01		0	
2048	DICYCLOPENTADIENE	3	F1	III	3		LQ7	E1	T	PP, EX, A	VE01		0	
2049	DIETHYLBENZENE	3	F1	III	3		LQ7	E1		PP, EX, A	VE01		0	
2050	DIISOBUTYLENE, ISOMERIC COMPOUNDS	3	F1	II	3		LQ4	E2	T	PP, EX, A	VE01		1	
2051	2-DIMETHYLAMINOETHANOL	8	CF1	II	8+3		LQ22	E2	T	PP, EP, EX, A	VE01		1	
2052	DIPENTENE	3	F1	III	3		LQ7	E1		PP, EX, A	VE01		0	
2053	METHYL ISOBUTYL CARBINOL	3	F1	III	3		LQ7	E1	T	PP, EX, A	VE01		0	
2054	MORPHOLINE	8	CF1	I	8+3		LQ0	E0	T	PP, EP, EX, A	VE01		1	
2055	STYRENE MONOMER, STABILIZED	3	F1	III	3		LQ7	E1	T	PP, EX, A	VE01		0	
2056	TETRAHYDROFURAN	3	F1	II	3		LQ4	E2	T	PP, EX, A	VE01		1	
2057	TRIPROPYLENE	3	F1	II	3		LQ4	E2	T	PP, EX, A	VE01		1	
2057	TRIPROPYLENE	3	F1	III	3		LQ7	E1	T	PP, EX, A	VE01		0	
2058	VALERALDEHYDE	3	F1	II	3		LQ4	E2		PP, EX, A	VE01		1	
2059	NITROCELLULOSE SOLUTION, FLAMMABLE with not more than 12.6% nitrogen, by dry mass, and not more than 55% nitrocellulose	3	D	I	3	198 531	LQ3	E0		PP, EX, A	VE01		1	
2059	NITROCELLULOSE SOLUTION, FLAMMABLE with not more than 12.6% nitrogen, by dry mass, and not more than 55% nitrocellulose (vapour pressure at 50 °C more than 110 kPa)	3	D	II	3	198 531 640C	LQ4	E0		PP, EX, A	VE01		1	

UN No. or ID No.	Name and description	Class	Classi-fication Code	Packing group	Labels	Special provi-sions	Limited and excepted quantities		Carriage permitted	Equipment required	Venti-lation	Provisions concerning loading, unloading and carriage			Number of blue cones/ lights	Remarks
	3.1.2	2.2	2.2	2.1.1.3	5.2.2	3.3	3.4.6	3.5.1.2	3.2.1	8.1.5	7.1.6	7.1.6			7.1.5	3.2.1
(1)	(2)	(3a)	(3b)	(4)	(5)	(6)	(7a)	(7b)	(8)	(9)	(10)	(11)			(12)	(13)
2059	NITROCELLULOSE SOLUTION, FLAMMABLE with not more than 12.6% nitrogen, by dry mass, and not more than 55% nitrocellulose (vapour pressure at 50 °C not more than 110 kPa)	3	D	II	3	198 531 640D	LQ4	E0		PP, EX, A	VE01				1	
2059	NITROCELLULOSE SOLUTION, FLAMMABLE with not more than 12.6% nitrogen, by dry mass, and not more than 55% nitrocellulose	3	D	III	3	198 531	LQ7	E0		PP, EX, A	VE01				0	
2067	AMMONIUM NITRATE BASED FERTILIZER	5.1	O2	III	5.1	186 306 307	LQ12	E1	**B**	PP		CO02, ST01, LO04	HA09		0	CO02, LO04 and HA09 apply only when this substance is carried in bulk or without packaging
2071	Ammonium nitrate based fertilizers, uniform mixtures of the nitrogen/phosphate, nitrogen/potash or nitrogen/phosphate/potash type, containing not more than 70% ammonium nitrate and not more than 0.4% total combustible/organic material calculated as carbon or with not more than 45% ammonium nitrate and unrestricted combustible material	9	M11			186 193			**B**	PP		CO02, ST02	HA09		0	Dangerous only in bulk or without packaging. CO02, ST02 and HA09 apply only when this substance is carried in bulk or without packaging
2073	AMMONIA SOLUTION, relative density less than 0.880 at 15 °C in water, with more than 35% but not more than 50% ammonia	2	4A		2.2	532	LQ1	E1		PP					0	
2074	ACRYLAMIDE, SOLID	6.1	T2	III	6.1	802	LQ9	E1	**T**	PP, EP					0	
2075	CHLORAL, ANHYDROUS, STABILIZED	6.1	T1	II	6.1	802	LQ17	E4		PP, EP, TOX, A	VE02				2	
2076	CRESOLS, LIQUID	6.1	TC1	II	6.1+8	802	LQ17	E4		PP, EP, TOX, A	VE02				2	
2077	alpha-NAPHTHYLAMINE	6.1	T2	III	6.1	802	LQ9	E1		PP, EP					0	
2078	TOLUENE DIISOCYANATE	6.1	T1	II	6.1	279 802	LQ17	E4	**T***	PP, EP, TOX, A	VE02				2	* only for 2,4-TOLUENE DIISOCYANATE
2079	DIETHYLENETRIAMINE	8	C7	II	8		LQ22	E2	**T**	PP, EP					0	
2186	HYDROGEN CHLORIDE, REFRIGERATED LIQUID	2	3TC	**CARRIAGE PROHIBITED**												
2187	CARBON DIOXIDE, REFRIGERATED LIQUID	2	3A		2.2	593	LQ1	E1		PP					0	
2188	ARSINE	2	2TF		2.3+2.1		LQ0	E0		PP, EP, EX, TOX, A	VE01, VE02				2	
2189	DICHLOROSILANE	2	2TFC		2.3+2.1+8		LQ0	E0		PP, EP, EX, TOX, A	VE01, VE02				2	
2190	OXYGEN DIFLUORIDE, COMPRESSED	2	1TOC		2.3+5.1+8		LQ0	E0		PP, EP, TOX, A	VE02				2	

UN No. or ID No.	Name and description	Class	Classi-fication Code	Packing group	Labels	Special provi-sions	Limited and excepted quantities		Carriage permitted	Equipment required	Venti-lation	Provisions concerning loading, unloading and carriage			Number of blue cones/ lights	Remarks
	3.1.2	2.2	2.2	2.1.1.3	5.2.2	3.3	3.4.6	3.5.1.2	3.2.1	8.1.5	7.1.6	7.1.6			7.1.5	3.2.1
(1)	(2)	(3a)	(3b)	(4)	(5)	(6)	(7a)	(7b)	(8)	(9)	(10)	(11)			(12)	(13)
2191	SULPHURYL FLUORIDE	2	2T		2.3		LQ0	E0		PP, EP, TOX, A	VE02				2	
2192	GERMANE	2	2TF		2.3+2.1	632	LQ0	E0		PP, EP, EX, TOX, A	VE01, VE02				2	
2193	HEXAFLUOROETHANE (REFRIGERANT GAS R 116)	2	2A		2.2		LQ1	E1		PP					0	
2194	SELENIUM HEXAFLUORIDE	2	2TC		2.3+8		LQ0	E0		PP, EP, TOX, A	VE02				2	
2195	TELLURIUM HEXAFLUORIDE	2	2TC		2.3+8		LQ0	E0		PP, EP, TOX, A	VE02				2	
2196	TUNGSTEN HEXAFLUORIDE	2	2TC		2.3+8		LQ0	E0		PP, EP, TOX, A	VE02				2	
2197	HYDROGEN IODIDE, ANHYDROUS	2	2TC		2.3+8		LQ0	E0		PP, EP, TOX, A	VE02				2	
2198	PHOSPHORUS PENTAFLUORIDE	2	2TC		2.3+8		LQ0	E0		PP, EP, TOX, A	VE02				2	
2199	PHOSPHINE	2	2TF		2.3+2.1	632	LQ0	E0		PP, EP, EX, TOX, A	VE01, VE02				2	
2200	PROPADIENE, STABILIZED	2	2F		2.1		LQ0	E0		PP, EX, A	VE01				1	
2201	NITROUS OXIDE, REFRIGERATED LIQUID	2	3O		2.2+5.1		LQ0	E0		PP					0	
2202	HYDROGEN SELENIDE, ANHYDROUS	2	2TF		2.3+2.1		LQ0	E0		PP, EP, EX, TOX, A	VE01, VE02				2	
2203	SILANE	2	2F		2.1	632	LQ0	E0		PP, EX, A	VE01				1	
2204	CARBONYL SULPHIDE	2	2TF		2.3+2.1		LQ0	E0		PP, EP, EX, TOX, A	VE01, VE02				2	
2205	ADIPONITRILE	6.1	T1	III	6.1	802	LQ7	E1	**T**	PP, EP, TOX, A	VE02				0	
2206	ISOCYANATES, TOXIC, N.O.S. or ISOCYANATE SOLUTION, TOXIC, N.O.S.	6.1	T1	II	6.1	274 551 802	LQ17	E4	**T**	PP, EP, TOX, A	VE02				2	
2206	ISOCYANATES, TOXIC, N.O.S. or ISOCYANATE SOLUTION, TOXIC, N.O.S.	6.1	T1	III	6.1	274 551 802	LQ7	E1		PP, EP, TOX, A	VE02				0	
2208	CALCIUM HYPOCHLORITE MIXTURE, DRY with more than 10% but not more than 39% available chlorine	5.1	O2	III	5.1	313 314	LQ12	E1		PP					0	
2209	FORMALDEHYDE SOLUTION with not less than 25% formaldehyde	8	C9	III	8	533	LQ7	E1	**T**	PP, EP					0	
2210	MANEB or MANEB PREPARATION with not less than 60% maneb	4.2	SW	III	4.2+4.3	273	LQ0	E1	**B**	PP, EX, A	VE01, VE03			IN01, IN03	0	VE03, IN01 and IN03 apply only when this substance is carried in bulk or without packaging

UN No. or ID No.	Name and description	Class	Classi-fication Code	Packing group	Labels	Special provi-sions	Limited and excepted quantities		Carriage permitted	Equipment required	Venti-lation	Provisions concerning loading, unloading and carriage	Number of blue cones/ lights	Remarks
	3.1.2	2.2	2.2	2.1.1.3	5.2.2	3.3	3.4.6	3.5.1.2	3.2.1	8.1.5	7.1.6	7.1.6	7.1.5	3.2.1
(1)	(2)	(3a)	(3b)	(4)	(5)	(6)	(7a)	(7b)	(8)	(9)	(10)	(11)	(12)	(13)
2211	POLYMERIC BEADS, EXPANDABLE, evolving flammable vapour	9	M3	III	none	207 633	LQ27	E1	**B**	PP, EX, EP, A	VE01, VE03	IN01	0	VE03 and IN01 apply only when this substance is carried in bulk or without packaging
2212	BLUE ASBESTOS (crocidolite) or BROWN ASBESTOS (amosite, mysorite)	9	M1	II	9	168 802	LQ25	E2		PP			0	
2213	PARAFORMALDEHYDE	4.1	F1	III	4.1		LQ9	E1		PP			0	
2214	PHTHALIC ANHYDRIDE with more than 0.05% of maleic anhydride	8	C4	III	8	169	LQ24	E1		PP, EP			0	
2215	MALEIC ANHYDRIDE, MOLTEN	8	C3	III	8		LQ0	E0	**T**	PP, EP			0	
2215	MALEIC ANHYDRIDE	8	C4	III	8		LQ24	E1		PP, EP			0	
2216	FISH MEAL, STABILISED or FISH SCRAP, STABILISED	9	M11						**B**	PP			0	
2217	SEED CAKE with not more than 1.5% oil and not more than 11% moisture	4.2	S2	III	4.2	142 800	LQ0	E1	**B**	PP		IN01	0	IN01 applies only when this substance is carried in bulk or without packaging
2218	ACRYLIC ACID, STABILIZED	8	CF1	II	8+3		LQ22	E2	**T**	PP, EP, EX, A	VE01		1	
2219	ALLYL GLYCIDYL ETHER	3	F1	III	3		LQ7	E1		PP, EX, A	VE01		0	
2222	ANISOLE	3	F1	III	3		LQ7	E1		PP, EX, A	VE01		0	
2224	BENZONITRILE	6.1	T1	II	6.1	802	LQ17	E4		PP, EP, TOX, A	VE02		2	
2225	BENZENESULPHONYL CHLORIDE	8	C3	III	8		LQ7	E1		PP, EP			0	
2226	BENZOTRICHLORIDE	8	C9	II	8		LQ22	E2		PP, EP			0	
2227	n-BUTYL METHACRYLATE, STABILIZED	3	F1	III	3		LQ7	E1	**T**	PP, EX, A	VE01		0	
2232	2-CHLOROETHANAL	6.1	T1	I	6.1	802	LQ0	E5		PP, EP, TOX, A	VE02		2	
2233	CHLOROANISIDINES	6.1	T2	III	6.1	802	LQ9	E1		PP, EP			0	
2234	CHLOROBENZOTRIFLUORIDES	3	F1	III	3		LQ7	E1		PP, EX, A	VE01		0	
2235	CHLOROBENZYL CHLORIDES, LIQUID	6.1	T1	III	6.1	802	LQ7	E1		PP, EP			0	
2236	3-CHLORO-4-METHYLPHENYL ISOCYANATE, LIQUID	6.1	T1	II	6.1	802	LQ17	E4		PP, EP			2	
2237	CHLORONITROANILINES	6.1	T2	III	6.1	802	LQ9	E1		PP, EP			0	
2238	CHLOROTOLUENES	3	F1	III	3		LQ7	E1	**T**	PP, EX, A	VE01		0	
2239	CHLOROTOLUIDINES, SOLID	6.1	T2	III	6.1	802	LQ9	E1		PP, EP			0	
2240	CHROMOSULPHURIC ACID	8	C1	I	8		LQ0	E0		PP, EP			0	
2241	CYCLOHEPTANE	3	F1	II	3		LQ4	E2	**T**	PP, EX, A	VE01		1	
2242	CYCLOHEPTENE	3	F1	II	3		LQ4	E2		PP, EX, A	VE01		1	
2243	CYCLOHEXYL ACETATE	3	F1	III	3		LQ7	E1		PP, EX, A	VE01		0	
2244	CYCLOPENTANOL	3	F1	III	3		LQ7	E1		PP, EX, A	VE01		0	
2245	CYCLOPENTANONE	3	F1	III	3		LQ7	E1		PP, EX, A	VE01		0	
2246	CYCLOPENTENE	3	F1	II	3		LQ4	E2		PP, EX, A	VE01		1	
2247	n-DECANE	3	F1	III	3		LQ7	E1	**T**	PP, EX, A	VE01		0	
2248	DI-n-BUTYLAMINE	8	CF1	II	8+3		LQ22	E2	**T**	PP, EP, EX, A	VE01		1	

UN No. or ID No.	Name and description	Class	Classi-fication Code	Packing group	Labels	Special provi-sions	Limited and excepted quantities		Carriage permitted	Equipment required	Venti-lation	Provisions concerning loading, unloading and carriage	Number of blue cones/ lights	Remarks
	3.1.2	2.2	2.2	2.1.1.3	5.2.2	3.3	3.4.6	3.5.1.2	3.2.1	8.1.5	7.1.6	7.1.6	7.1.5	3.2.1
(1)	(2)	(3a)	(3b)	(4)	(5)	(6)	(7a)	(7b)	(8)	(9)	(10)	(11)	(12)	(13)
2249	DICHLORODIMETHYL ETHER, SYMMETRICAL	6.1	TF1	CARRIAGE PROHIBITED										
2250	DICHLOROPHENYL ISOCYANATES	6.1	T2	II	6.1	802	LQ17	E4		PP, EP			2	
2251	BICYCLO[2.2.1]HEPTA-2,5-DIENE, STABILIZED (2,5-NORBORNADIENE, STABILIZED)	3	F1	II	3		LQ4	E2		PP, EX, A	VE01		1	
2252	1,2-DIMETHOXYETHANE	3	F1	II	3		LQ4	E2		PP, EX, A	VE01		1	
2253	N,N-DIMETHYLANILINE	6.1	T1	II	6.1	802	LQ17	E4		PP, EP, TOX, A	VE02		2	
2254	MATCHES, FUSEE	4.1	F1	III	4.1	293	LQ9	E1		PP			0	
2256	CYCLOHEXENE	3	F1	II	3		LQ4	E2		PP, EX, A	VE01		1	
2257	POTASSIUM	4.3	W2	I	4.3		LQ0	E0		PP, EX, A	VE01	HA08	0	
2258	1,2-PROPYLENEDIAMINE	8	CF1	II	8+3		LQ22	E2		PP, EP, EX, A	VE01		1	
2259	TRIETHYLENETETRAMINE	8	C7	II	8		LQ22	E2	T	PP, EP			0	
2260	TRIPROPYLAMINE	3	FC	III	3+8		LQ7	E1		PP, EP, EX, A	VE01		0	
2261	XYLENOLS, SOLID	6.1	T2	II	6.1	802	LQ18	E4		PP, EP			2	
2262	DIMETHYLCARBAMOYL CHLORIDE	8	C3	II	8		LQ22	E2		PP, EP			0	
2263	DIMETHYLCYCLOHEXANES	3	F1	II	3		LQ4	E2	T	PP, EX, A	VE01		1	
2264	N,N-DIMETHYLCYCLOHEXYLAMINE	8	CF1	II	8+3		LQ22	E2	T	PP, EP, EX, A	VE01		1	
2265	N,N-DIMETHYLFORMAMIDE	3	F1	III	3		LQ7	E1	T	PP, EX, A	VE01		0	
2266	DIMETHYL-N-PROPYLAMINE	3	FC	II	3+8		LQ4	E2	T	PP, EP, EX, A	VE01		1	
2267	DIMETHYL THIOPHOSPHORYL CHLORIDE	6.1	TC1	II	6.1+8	802	LQ17	E4		PP, EP, TOX, A	VE02		2	
2269	3,3'-IMINODIPROPYLAMINE	8	C7	III	8		LQ7	E1		PP, EP			0	
2270	ETHYLAMINE, AQUEOUS SOLUTION with not less than 50% but not more than 70% ethylamine	3	FC	II	3+8		LQ4	E2		PP, EP, EX, A	VE01		1	
2271	ETHYL AMYL KETONE	3	F1	III	3		LQ7	E1		PP, EX, A	VE01		0	
2272	N-ETHYLANILINE	6.1	T1	III	6.1	802	LQ7	E1		PP, EP, TOX, A	VE02		0	
2273	2-ETHYLANILINE	6.1	T1	III	6.1	802	LQ7	E1		PP, EP, TOX, A	VE02		0	
2274	N-ETHYL-N-BENZYLANILINE	6.1	T1	III	6.1	802	LQ7	E1		PP, EP, TOX, A	VE02		0	
2275	2-ETHYLBUTANOL	3	F1	III	3		LQ7	E1		PP, EX, A	VE01		0	
2276	2-ETHYLHEXYLAMINE	3	FC	III	3+8		LQ7	E1	T	PP, EP, EX, A	VE01		0	
2277	ETHYL METHACRYLATE, STABILIZED	3	F1	II	3		LQ4	E2		PP, EX, A	VE01		1	
2278	n-HEPTENE	3	F1	II	3		LQ4	E2	T	PP, EX, A	VE01		1	
2279	HEXACHLOROBUTADIENE	6.1	T1	III	6.1	802	LQ7	E1		PP, EP, TOX, A	VE02		0	
2280	HEXAMETHYLENEDIAMINE, SOLID	8	C8	III	8		LQ24	E1	T	PP, EP			0	
2281	HEXAMETHYLENE DIISOCYANATE	6.1	T1	II	6.1	802	LQ17	E4		PP, EP, TOX, A	VE02		2	
2282	HEXANOLS	3	F1	III	3		LQ7	E1	T	PP, EX, A	VE01		0	
2283	ISOBUTYL METHACRYLATE, STABILIZED	3	F1	III	3		LQ7	E1		PP, EX, A	VE01		0	

UN No. or ID No.	Name and description	Class	Classi-fication Code	Packing group	Labels	Special provi-sions	Limited and excepted quantities		Carriage permitted	Equipment required	Venti-lation	Provisions concerning loading, unloading and carriage	Number of blue cones/ lights	Remarks
	3.1.2	2.2	2.2	2.1.1.3	5.2.2	3.3	3.4.6	3.5.1.2	3.2.1	8.1.5	7.1.6	7.1.6	7.1.5	3.2.1
(1)	(2)	(3a)	(3b)	(4)	(5)	(6)	(7a)	(7b)	(8)	(9)	(10)	(11)	(12)	(13)
2284	ISOBUTYRONITRILE	3	FT1	II	3+6.1	802	LQ0	E2		PP, EP, EX, TOX, A	VE01, VE02		2	
2285	ISOCYANATOBENZOTRIFLUORIDES	6.1	TF1	II	6.1+3	802	LQ17	E4		PP, EP, EX, TOX, A	VE01, VE02		2	
2286	PENTAMETHYLHEPTANE	3	F1	III	3		LQ7	E1	T	PP, EX, A	VE01		0	
2287	ISOHEPTENES	3	F1	II	3		LQ4	E2		PP, EX, A	VE01		1	
2288	ISOHEXENES	3	F1	II	3		LQ4	E2		PP, EX, A	VE01		1	
2289	ISOPHORONEDIAMINE	8	C7	III	8		LQ7	E1	T	PP, EP			0	
2290	ISOPHORONE DIISOCYANATE	6.1	T1	III	6.1	802	LQ7	E1		PP, EP, TOX, A	VE02		0	
2291	LEAD COMPOUND, SOLUBLE, N.O.S.	6.1	T5	III	6.1	199 274 535 802	LQ9	E1		PP, EP			0	
2293	4-METHOXY-4-METHYLPENTAN-2-ONE	3	F1	III	3		LQ7	E1		PP, EX, A	VE01		0	
2294	N-METHYLANILINE	6.1	T1	III	6.1	802	LQ7	E1		PP, EP, TOX, A	VE02		0	
2295	METHYL CHLOROACETATE	6.1	TF1	I	6.1+3	802	LQ0	E5		PP, EP, EX, TOX, A	VE01, VE02		2	
2296	METHYLCYCLOHEXANE	3	F1	II	3		LQ4	E2		PP, EX, A	VE01		1	
2297	METHYLCYCLOHEXANONE	3	F1	III	3		LQ7	E1		PP, EX, A	VE01		0	
2298	METHYLCYCLOPENTANE	3	F1	II	3		LQ4	E2		PP, EX, A	VE01		1	
2299	METHYL DICHLOROACETATE	6.1	T1	III	6.1	802	LQ7	E1		PP, EP, TOX, A	VE02		0	
2300	2-METHYL-5-ETHYLPYRIDINE	6.1	T1	III	6.1	802	LQ7	E1		PP, EP, TOX, A	VE02		0	
2301	2-METHYLFURAN	3	F1	II	3		LQ4	E2		PP, EX, A	VE01		1	
2302	5-METHYLHEXAN-2-ONE	3	F1	III	3		LQ7	E1	T	PP, EX, A	VE01		0	
2303	ISOPROPENYLBENZENE	3	F1	III	3		LQ7	E1	T	PP, EX, A	VE01		0	
2304	NAPHTHALENE, MOLTEN	4.1	F2	III	4.1	536	LQ0	E0		PP			0	
2305	NITROBENZENESULPHONIC ACID	8	C4	II	8		LQ23	E2		PP, EP			0	
2306	NITROBENZOTRIFLUORIDES, LIQUID	6.1	T1	II	6.1	802	LQ17	E4		PP, EP, TOX, A	VE02		2	
2307	3-NITRO-4-CHLOROBENZOTRIFLUORIDE	6.1	T1	II	6.1	802	LQ17	E4		PP, EP, TOX, A	VE02		2	
2308	NITROSYLSULPHURIC ACID, LIQUID	8	C1	II	8		LQ22	E2		PP, EP			0	
2309	OCTADIENE	3	F1	II	3		LQ4	E2	T	PP, EX, A	VE01		1	
2310	PENTANE-2,4-DIONE	3	FT1	III	3+6.1	802	LQ7	E1		PP, EP, EX, TOX, A	VE01, VE02		0	
2311	PHENETIDINES	6.1	T1	III	6.1	279 802	LQ7	E1	T	PP, EP, TOX, A	VE02		0	
2312	PHENOL, MOLTEN	6.1	T1	II	6.1	802	LQ0	E0	T	PP, EP, TOX, A	VE02		2	
2313	PICOLINES	3	F1	III	3		LQ7	E1		PP, EX, A	VE01		0	
2315	POLYCHLORINATED BIPHENYLS, LIQUID	9	M2	II	9	305 802	LQ26	E2		PP, EP			0	

UN No. or ID No.	Name and description	Class	Classi-fication Code	Packing group	Labels	Special provi-sions	Limited and excepted quantities		Carriage permitted	Equipment required	Venti-lation	Provisions concerning loading, unloading and carriage	Number of blue cones/ lights	Remarks
	3.1.2	2.2	2.2	2.1.1.3	5.2.2	3.3	3.4.6	3.5.1.2	3.2.1	8.1.5	7.1.6	7.1.6	7.1.5	3.2.1
(1)	(2)	(3a)	(3b)	(4)	(5)	(6)	(7a)	(7b)	(8)	(9)	(10)	(11)	(12)	(13)
2316	SODIUM CUPROCYANIDE, SOLID	6.1	T5	I	6.1	802	LQ0	E5		PP, EP			2	
2317	SODIUM CUPROCYANIDE SOLUTION	6.1	T4	I	6.1	802	LQ0	E5		PP, EP			2	
2318	SODIUM HYDROSULPHIDE with less than 25% water of crystallization	4.2	S4	II	4.2	504	LQ0	E2		PP			0	
2319	TERPENE HYDROCARBONS, N.O.S.	3	F1	III	3		LQ7	E1		PP, EX, A	VE01		0	
2320	TETRAETHYLENEPENTAMINE	8	C7	III	8		LQ7	E1	T	PP, EP			0	
2321	TRICHLOROBENZENES, LIQUID	6.1	T1	III	6.1	802	LQ7	E1	T	PP, EP, TOX, A	VE02		0	
2322	TRICHLOROBUTENE	6.1	T1	II	6.1	802	LQ17	E4		PP, EP, TOX, A	VE02		2	
2323	TRIETHYL PHOSPHITE	3	F1	III	3		LQ7	E1	T	PP, EX, A	VE01		0	
2324	TRIISOBUTYLENE	3	F1	III	3		LQ7	E1	T	PP, EX, A	VE01		0	
2325	1,3,5-TRIMETHYLBENZENE	3	F1	III	3		LQ7	E1	T	PP, EX, A	VE01		0	
2326	TRIMETHYLCYCLOHEXYLAMINE	8	C7	III	8		LQ7	E1		PP, EP			0	
2327	TRIMETHYLHEXAMETHYLENEDIAMINES	8	C7	III	8		LQ7	E1		PP, EP			0	
2328	TRIMETHYLHEXAMETHYLENE DIISOCYANATE	6.1	T1	III	6.1	802	LQ7	E1		PP, EP, TOX, A	VE02		0	
2329	TRIMETHYL PHOSPHITE	3	F1	III	3		LQ7	E1		PP, EX, A	VE01		0	
2330	UNDECANE	3	F1	III	3		LQ7	E1		PP, EX, A	VE01		0	
2331	ZINC CHLORIDE, ANHYDROUS	8	C2	III	8		LQ24	E1		PP, EP			0	
2332	ACETALDEHYDE OXIME	3	F1	III	3		LQ7	E1		PP, EX, A	VE01		0	
2333	ALLYL ACETATE	3	FT1	II	3+6.1	802	LQ0	E2	T	PP, EP, EX, TOX, A	VE01, VE02		2	
2334	ALLYLAMINE	6.1	TF1	I	6.1+3	802	LQ0	E5		PP, EP, EX, TOX, A	VE01, VE02		2	
2335	ALLYL ETHYL ETHER	3	FT1	II	3+6.1	802	LQ0	E2		PP, EP, EX, TOX, A	VE01, VE02		2	
2336	ALLYL FORMATE	3	FT1	I	3+6.1	802	LQ0	E0		PP, EP, EX, TOX, A	VE01, VE02		2	
2337	PHENYL MERCAPTAN	6.1	TF1	I	6.1+3	802	LQ0	E5		PP, EP, EX, TOX, A	VE01, VE02		2	
2338	BENZOTRIFLUORIDE	3	F1	II	3		LQ4	E2		PP, EX, A	VE01		1	
2339	2-BROMOBUTANE	3	F1	II	3		LQ4	E2		PP, EX, A	VE01		1	
2340	2-BROMOETHYL ETHYL ETHER	3	F1	II	3		LQ4	E2		PP, EX, A	VE01		1	
2341	1-BROMO-3-METHYLBUTANE	3	F1	III	3		LQ7	E1		PP, EX, A	VE01		0	
2342	BROMOMETHYLPROPANES	3	F1	II	3		LQ4	E2		PP, EX, A	VE01		1	
2343	2-BROMOPENTANE	3	F1	II	3		LQ4	E2		PP, EX, A	VE01		1	
2344	BROMOPROPANES	3	F1	II	3		LQ4	E2		PP, EX, A	VE01		1	
2344	BROMOPROPANES	3	F1	III	3		LQ7	E1		PP, EX, A	VE01		0	
2345	3-BROMOPROPYNE	3	F1	II	3		LQ4	E2		PP, EX, A	VE01		1	
2346	BUTANEDIONE	3	F1	II	3		LQ4	E2		PP, EX, A	VE01		1	
2347	BUTYL MERCAPTAN	3	F1	II	3		LQ4	E2		PP, EX, A	VE01		1	
2348	BUTYL ACRYLATES, STABILIZED	3	F1	III	3		LQ7	E1	T	PP, EX, A	VE01		0	
2350	BUTYL METHYL ETHER	3	F1	II	3		LQ4	E2	T	PP, EX, A	VE01		1	
2351	BUTYL NITRITES	3	F1	II	3		LQ4	E2		PP, EX, A	VE01		1	
2351	BUTYL NITRITES	3	F1	III	3		LQ7	E1		PP, EX, A	VE01		0	

UN No. or ID No.	Name and description	Class	Classi-fication Code	Packing group	Labels	Special provi-sions	Limited and excepted quantities		Carriage permitted	Equipment required	Venti-lation	Provisions concerning loading, unloading and carriage	Number of blue cones/ lights	Remarks
	3.1.2	2.2	2.2	2.1.1.3	5.2.2	3.3	3.4.6	3.5.1.2	3.2.1	8.1.5	7.1.6	7.1.6	7.1.5	3.2.1
(1)	(2)	(3a)	(3b)	(4)	(5)	(6)	(7a)	(7b)	(8)	(9)	(10)	(11)	(12)	(13)
2352	BUTYL VINYL ETHER, STABILIZED	3	F1	II	3		LQ4	E2		PP, EX, A	VE01		1	
2353	BUTYRYL CHLORIDE	3	FC	II	3+8		LQ4	E2		PP, EP, EX, A	VE01		1	
2354	CHLOROMETHYL ETHYL ETHER	3	FT1	II	3+6.1	802	LQ0	E2		PP, EP, EX, TOX, A	VE01, VE02		2	
2356	2-CHLOROPROPANE	3	F1	I	3		LQ3	E3	T	PP, EX, A	VE01		1	
2357	CYCLOHEXYLAMINE	8	CF1	II	8+3		LQ22	E2	T	PP, EP, EX, A	VE01		1	
2358	CYCLOOCTATETRAENE	3	F1	II	3		LQ4	E2		PP, EX, A	VE01		1	
2359	DIALLYLAMINE	3	FTC	II	3+6.1+8	802	LQ0	E2		PP, EP, EX, TOX, A	VE01, VE02		2	
2360	DIALLYL ETHER	3	FT1	II	3+6.1	802	LQ0	E2		PP, EP, EX, TOX, A	VE01, VE02		2	
2361	DIISOBUTYLAMINE	3	FC	III	3+8		LQ7	E1		PP, EP, EX, A	VE01		0	
2362	1,1-DICHLOROETHANE	3	F1	II	3		LQ4	E2	T	PP, EX, A	VE01		1	
2363	ETHYL MERCAPTAN	3	F1	I	3		LQ3	E3		PP, EX, A	VE01		1	
2364	n-PROPYLBENZENE	3	F1	III	3		LQ7	E1		PP, EX, A	VE01		0	
2366	DIETHYL CARBONATE	3	F1	III	3		LQ7	E1		PP, EX, A	VE01		0	
2367	alpha-METHYLVALERALDEHYDE	3	F1	II	3		LQ4	E2		PP, EX, A	VE01		1	
2368	alpha-PINENE	3	F1	III	3		LQ7	E1		PP, EX, A	VE01		0	
2370	1-HEXENE	3	F1	II	3		LQ4	E2	T	PP, EX, A	VE01		1	
2371	ISOPENTENES	3	F1	I	3		LQ3	E3		PP, EX, A	VE01		1	
2372	1,2-DI-(DIMETHYLAMINO) ETHANE	3	F1	II	3		LQ4	E2		PP, EX, A	VE01		1	
2373	DIETHOXYMETHANE	3	F1	II	3		LQ4	E2		PP, EX, A	VE01		1	
2374	3,3-DIETHOXYPROPENE	3	F1	II	3		LQ4	E2		PP, EX, A	VE01		1	
2375	DIETHYL SULPHIDE	3	F1	II	3		LQ4	E2		PP, EX, A	VE01		1	
2376	2,3-DIHYDROPYRAN	3	F1	II	3		LQ4	E2		PP, EX, A	VE01		1	
2377	1,1-DIMETHOXYETHANE	3	F1	II	3		LQ4	E2		PP, EX, A	VE01		1	
2378	2-DIMETHYLAMINOACETONITRILE	3	FT1	II	3+6.1	802	LQ0	E2		PP, EP, EX, TOX, A	VE01, VE02		2	
2379	1,3-DIMETHYLBUTYLAMINE	3	FC	II	3+8		LQ4	E2		PP, EP, EX, A	VE01		1	
2380	DIMETHYLDIETHOXYSILANE	3	F1	II	3		LQ4	E2		PP, EX, A	VE01		1	
2381	DIMETHYL DISULPHIDE	3	F1	II	3		LQ4	E2		PP, EX, A	VE01		1	
2382	DIMETHYLHYDRAZINE, SYMMETRICAL	6.1	TF1	I	6.1+3	802	LQ0	E5	T	PP, EP, EX, TOX, A	VE01, VE02		2	
2383	DIPROPYLAMINE	3	FC	II	3+8		LQ4	E2	T	PP, EP, EX, A	VE01		1	
2384	DI-n-PROPYL ETHER	3	F1	II	3		LQ4	E2		PP, EX, A	VE01		1	
2385	ETHYL ISOBUTYRATE	3	F1	II	3		LQ4	E2		PP, EX, A	VE01		1	
2386	1-ETHYLPIPERIDINE	3	FC	II	3+8		LQ4	E2		PP, EP, EX, A	VE01		1	
2387	FLUOROBENZENE	3	F1	II	3		LQ4	E2		PP, EX, A	VE01		1	
2388	FLUOROTOLUENES	3	F1	II	3		LQ4	E2		PP, EX, A	VE01		1	
2389	FURAN	3	F1	I	3		LQ3	E3		PP, EX, A	VE01		1	
2390	2-IODOBUTANE	3	F1	II	3		LQ4	E2		PP, EX, A	VE01		1	
2391	IODOMETHYLPROPANES	3	F1	II	3		LQ4	E2		PP, EX, A	VE01		1	
2392	IODOPROPANES	3	F1	III	3		LQ7	E1		PP, EX, A	VE01		0	
2393	ISOBUTYL FORMATE	3	F1	II	3		LQ4	E2		PP, EX, A	VE01		1	
2394	ISOBUTYL PROPIONATE	3	F1	III	3		LQ7	E1		PP, EX, A	VE01		0	
2395	ISOBUTYRYL CHLORIDE	3	FC	II	3+8		LQ4	E2		PP, EP, EX, A	VE01		1	

UN No. or ID No.	Name and description	Class	Classi-fication Code	Packing group	Labels	Special provi-sions	Limited and excepted quantities		Carriage permitted	Equipment required	Venti-lation	Provisions concerning loading, unloading and carriage	Number of blue cones/ lights	Remarks
	3.1.2	2.2	2.2	2.1.1.3	5.2.2	3.3	3.4.6	3.5.1.2	3.2.1	8.1.5	7.1.6	7.1.6	7.1.5	3.2.1
(1)	(2)	(3a)	(3b)	(4)	(5)	(6)	(7a)	(7b)	(8)	(9)	(10)	(11)	(12)	(13)
2396	METHACRYLALDEHYDE, STABILIZED	3	FT1	II	3+6.1	802	LQ0	E2		PP, EP, EX, TOX, A	VE01, VE02		2	
2397	3-METHYLBUTAN-2-ONE	3	F1	II	3		LQ4	E2	**T**	PP, EX, A	VE01		1	
2398	METHYL tert-BUTYL ETHER	3	F1	II	3		LQ4	E2	**T**	PP, EX, A	VE01		1	
2399	1-METHYLPIPERIDINE	3	FC	II	3+8		LQ4	E2		PP, EP, EX, A	VE01		1	
2400	METHYL ISOVALERATE	3	F1	II	3		LQ4	E2		PP, EX, A	VE01		1	
2401	PIPERIDINE	8	CF1	I	8+3		LQ0	E0		PP, EP, EX, A	VE01		1	
2402	PROPANETHIOLS	3	F1	II	3		LQ4	E2		PP, EX, A	VE01		1	
2403	ISOPROPENYL ACETATE	3	F1	II	3		LQ4	E2		PP, EX, A	VE01		1	
2404	PROPIONITRILE	3	FT1	II	3+6.1	802	LQ0	E2	**T**	PP, EP, EX, TOX, A	VE01, VE02		2	
2405	ISOPROPYL BUTYRATE	3	F1	III	3		LQ7	E1		PP, EX, A	VE01		0	
2406	ISOPROPYL ISOBUTYRATE	3	F1	II	3		LQ4	E2		PP, EX, A	VE01		1	
2407	ISOPROPYL CHLOROFORMATE	6.1	TFC	I	6.1+3+8	802	LQ0	E5		PP, EP, EX, TOX, A	VE01, VE02		2	
2409	ISOPROPYL PROPIONATE	3	F1	II	3		LQ4	E2		PP, EX, A	VE01		1	
2410	1,2,3,6-TETRAHYDROPYRIDINE	3	F1	II	3		LQ4	E2		PP, EX, A	VE01		1	
2411	BUTYRONITRILE	3	FT1	II	3+6.1	802	LQ0	E2		PP, EP, EX, TOX, A	VE01, VE02		2	
2412	TETRAHYDROTHIOPHENE	3	F1	II	3		LQ4	E2		PP, EX, A	VE01		1	
2413	TETRAPROPYL ORTHOTITANATE	3	F1	III	3		LQ7	E1		PP, EX, A	VE01		0	
2414	THIOPHENE	3	F1	II	3		LQ4	E2	**T**	PP, EX, A	VE01		1	
2416	TRIMETHYL BORATE	3	F1	II	3		LQ4	E2		PP, EX, A	VE01		1	
2417	CARBONYL FLUORIDE	2	2TC		2.3+8		LQ0	E0		PP, EP, TOX, A	VE02		2	
2418	SULPHUR TETRAFLUORIDE	2	2TC		2.3+8		LQ0	E0		PP, EP, TOX, A	VE02		2	
2419	BROMOTRIFLUOROETHYLENE	2	2F		2.1		LQ0	E0		PP, EX, A	VE01		1	
2420	HEXAFLUOROACETONE	2	2TC		2.3+8		LQ0	E0		PP, EP, TOX, A	VE02		2	
2421	NITROGEN TRIOXIDE	2	2TOC	**CARRIAGE PROHIBITED**										
2422	OCTAFLUOROBUT-2-ENE (REFRIGERANT GAS R 1318)	2	2A		2.2		LQ1	E1		PP			0	
2424	OCTAFLUOROPROPANE (REFRIGERANT GAS R 218)	2	2A		2.2		LQ1	E1		PP			0	
2426	AMMONIUM NITRATE, LIQUID, hot concentrated solution, in a concentration of more than 80% but not more than 93%	5.1	O1		5.1	252 644	LQ0	E0		PP			0	
2427	POTASSIUM CHLORATE, AQUEOUS SOLUTION	5.1	O1	II	5.1		LQ10	E2		PP			0	
2427	POTASSIUM CHLORATE, AQUEOUS SOLUTION	5.1	O1	III	5.1		LQ13	E1		PP			0	
2428	SODIUM CHLORATE, AQUEOUS SOLUTION	5.1	O1	II	5.1		LQ10	E2		PP			0	
2428	SODIUM CHLORATE, AQUEOUS SOLUTION	5.1	O1	III	5.1		LQ13	E1		PP			0	

UN No. or ID No.	Name and description	Class	Classi-fication Code	Packing group	Labels	Special provi-sions	Limited and excepted quantities		Carriage permitted	Equipment required	Venti-lation	Provisions concerning loading, unloading and carriage	Number of blue cones/ lights	Remarks
	3.1.2	2.2	2.2	2.1.1.3	5.2.2	3.3	3.4.6	3.5.1.2	3.2.1	8.1.5	7.1.6	7.1.6	7.1.5	3.2.1
(1)	(2)	(3a)	(3b)	(4)	(5)	(6)	(7a)	(7b)	(8)	(9)	(10)	(11)	(12)	(13)
2429	CALCIUM CHLORATE, AQUEOUS SOLUTION	5.1	O1	II	5.1		LQ10	E2		PP			0	
2429	CALCIUM CHLORATE, AQUEOUS SOLUTION	5.1	O1	III	5.1		LQ13	E1		PP			0	
2430	ALKYLPHENOLS, SOLID, N.O.S. (including C_2-C_{12} homologues)	8	C4	I	8	274	LQ0	E0		PP, EP			0	
2430	ALKYLPHENOLS, SOLID, N.O.S. (including C_2-C_{12} homologues)	8	C4	II	8	274	LQ23	E2	T	PP, EP			0	
2430	ALKYLPHENOLS, SOLID, N.O.S. (including C_2-C_{12} homologues)	8	C4	III	8	274	LQ24	E1		PP, EP			0	
2431	ANISIDINES	6.1	T1	III	6.1	802	LQ7	E1		PP, EP, TOX, A	VE02		0	
2432	N,N-DIETHYLANILINE	6.1	T1	III	6.1	279 802	LQ7	E1	T	PP, EP, TOX, A	VE02		0	
2433	CHLORONITROTOLUENES, LIQUID	6.1	T1	III	6.1	802	LQ7	E1		PP, EP, TOX, A	VE02		0	
2434	DIBENZYLDICHLOROSILANE	8	C3	II	8		LQ22	E2		PP, EP			0	
2435	ETHYLPHENYLDICHLOROSILANE	8	C3	II	8		LQ22	E2		PP, EP			0	
2436	THIOACETIC ACID	3	F1	II	3		LQ4	E2		PP, EX, A	VE01		1	
2437	METHYLPHENYLDICHLOROSILANE	8	C3	II	8		LQ22	E2		PP, EP			0	
2438	TRIMETHYLACETYL CHLORIDE	6.1	TFC	I	6.1+3+8	802	LQ0	E5		PP, EP, EX, TOX, A	VE01, VE02		2	
2439	SODIUM HYDROGENDIFLUORIDE	8	C2	II	8		LQ23	E2		PP, EP			0	
2440	STANNIC CHLORIDE PENTAHYDRATE	8	C2	III	8		LQ24	E1		PP, EP			0	
2441	TITANIUM TRICHLORIDE, PYROPHORIC or TITANIUM TRICHLORIDE MIXTURE, PYROPHORIC	4.2	SC4	I	4.2+8	537	LQ0	E0		PP			0	
2442	TRICHLOROACETYL CHLORIDE	8	C3	II	8		LQ22	E2		PP, EP			0	
2443	VANADIUM OXYTRICHLORIDE	8	C1	II	8		LQ22	E2		PP, EP			0	
2444	VANADIUM TETRACHLORIDE	8	C1	I	8		LQ0	E0		PP, EP			0	
2446	NITROCRESOLS, SOLID	6.1	T2	III	6.1	802	LQ9	E1		PP, EP			0	
2447	PHOSPHORUS, WHITE, MOLTEN	4.2	ST3	I	4.2+6.1	802	LQ0	E0		PP, EP, TOX, A	VE02		2	
2448	SULPHUR, MOLTEN	4.1	F3	III	4.1	538	LQ0	E0	T	PP			0	
2451	NITROGEN TRIFLUORIDE	2	2O		2.2+5.1		LQ0	E0		PP			0	
2452	ETHYLACETYLENE, STABILIZED	2	2F		2.1		LQ0	E0		PP, EX, A	VE01		1	
2453	ETHYL FLUORIDE (REFRIGERANT GAS R 161)	2	2F		2.1		LQ0	E0		PP, EX, A	VE01		1	
2454	METHYL FLUORIDE (REFRIGERANT GAS R 41)	2	2F		2.1		LQ0	E0		PP, EX, A	VE01		1	
2455	METHYL NITRITE	2	2A	CARRIAGE PROHIBITED										
2456	2-CHLOROPROPENE	3	F1	I	3		LQ3	E3		PP, EX, A	VE01		1	
2457	2,3-DIMETHYLBUTANE	3	F1	II	3		LQ4	E2		PP, EX, A	VE01		1	
2458	HEXADIENES	3	F1	II	3		LQ4	E2	T	PP, EX, A	VE01		1	
2459	2-METHYL-1-BUTENE	3	F1	I	3		LQ3	E3		PP, EX, A	VE01		1	
2460	2-METHYL-2-BUTENE	3	F1	II	3		LQ4	E2		PP, EX, A	VE01		1	

UN No. or ID No.	Name and description 3.1.2	Class 2.2	Classi-fication Code 2.2	Packing group 2.1.1.3	Labels 5.2.2	Special provi-sions 3.3	Limited and excepted quantities 3.4.6	 3.5.1.2	Carriage permitted 3.2.1	Equipment required 8.1.5	Venti-lation 7.1.6	Provisions concerning loading, unloading and carriage 7.1.6	Number of blue cones/ lights 7.1.5	Remarks 3.2.1
(1)	(2)	(3a)	(3b)	(4)	(5)	(6)	(7a)	(7b)	(8)	(9)	(10)	(11)	(12)	(13)
2461	METHYLPENTADIENE	3	F1	II	3		LQ4	E2		PP, EX, A	VE01		1	
2463	ALUMINIUM HYDRIDE	4.3	W2	I	4.3		LQ0	E0		PP, EX, A	VE01	HA08	0	
2464	BERYLLIUM NITRATE	5.1	OT2	II	5.1+6.1	802	LQ11	E2		PP			2	
2465	DICHLOROISOCYANURIC ACID, DRY or DICHLOROISOCYANURIC ACID SALTS	5.1	O2	II	5.1	135	LQ11	E2		PP			0	
2466	POTASSIUM SUPEROXIDE	5.1	O2	I	5.1		LQ0	E0		PP			0	
2468	TRICHLOROISOCYANURIC ACID, DRY	5.1	O2	II	5.1		LQ11	E2		PP			0	
2469	ZINC BROMATE	5.1	O2	III	5.1		LQ12	E1		PP			0	
2470	PHENYLACETONITRILE, LIQUID	6.1	T1	III	6.1	802	LQ7	E1		PP, EP, TOX, A	VE02		0	
2471	OSMIUM TETROXIDE	6.1	T5	I	6.1	802	LQ0	E5		PP, EP			2	
2473	SODIUM ARSANILATE	6.1	T3	III	6.1	802	LQ9	E1		PP, EP, TOX, A	VE02		0	
2474	THIOPHOSGENE	6.1	T1	II	6.1	279 802	LQ17	E4		PP, EP, TOX, A	VE02		2	
2475	VANADIUM TRICHLORIDE	8	C2	III	8		LQ24	E1		PP, EP			0	
2477	METHYL ISOTHIOCYANATE	6.1	TF1	I	6.1+3	802	LQ0	E5	**T**	PP, EP, EX, TOX, A	VE01, VE02		2	
2478	ISOCYANATES, FLAMMABLE, TOXIC, N.O.S. or ISOCYANATE SOLUTION, FLAMMABLE, TOXIC, N.O.S.	3	FT1	II	3+6.1	274 539 802	LQ0	E2		PP, EP, EX, TOX, A	VE01, VE02		2	
2478	ISOCYANATES, FLAMMABLE, TOXIC, N.O.S. or ISOCYANATE SOLUTION, FLAMMABLE, TOXIC, N.O.S.	3	FT1	III	3+6.1	274 802	LQ7	E1		PP, EP, EX, TOX, A	VE01, VE02		0	
2480	METHYL ISOCYANATE	6.1	TF1	I	6.1+3	802	LQ0	E5		PP, EP, EX, TOX, A	VE01, VE02		2	
2481	ETHYL ISOCYANATE	3	FT1	I	3+6.1	802	LQ0	E0		PP, EP, EX, TOX, A	VE01, VE02		2	
2482	n-PROPYL ISOCYANATE	6.1	TF1	I	6.1+3	802	LQ0	E5		PP, EP, EX, TOX, A	VE01, VE02		2	
2483	ISOPROPYL ISOCYANATE	3	FT1	I	3+6.1	802	LQ0	E0		PP, EP, EX, TOX, A	VE01, VE02		2	
2484	tert-BUTYL ISOCYANATE	6.1	TF1	I	6.1+3	802	LQ0	E5		PP, EP, EX, TOX, A	VE01, VE02		2	
2485	n-BUTYL ISOCYANATE	6.1	TF1	I	6.1+3	802	LQ0	E5	**T**	PP, EP, EX, TOX, A	VE01, VE02		2	
2486	ISOBUTYL ISOCYANATE	3	FT1	II	3+6.1	802	LQ0	E2	**T**	PP, EP, EX, TOX, A	VE01, VE02		2	
2487	PHENYL ISOCYANATE	6.1	TF1	I	6.1+3	802	LQ0	E5	**T**	PP, EP, EX, TOX, A	VE01, VE02		2	
2488	CYCLOHEXYL ISOCYANATE	6.1	TF1	I	6.1+3	802	LQ0	E5		PP, EP, EX, TOX, A	VE01, VE02		2	
2490	DICHLOROISOPROPYL ETHER	6.1	T1	II	6.1	802	LQ17	E4	**T**	PP, EP, TOX, A	VE02		2	
2491	ETHANOLAMINE or ETHANOLAMINE SOLUTION	8	C7	III	8		LQ7	E1	**T**	PP, EP			0	

UN No. or ID No.	Name and description	Class	Classi-fication Code	Packing group	Labels	Special provi-sions	Limited and excepted quantities		Carriage permitted	Equipment required	Venti-lation	Provisions concerning loading, unloading and carriage	Number of blue cones/ lights	Remarks
	3.1.2	2.2	2.2	2.1.1.3	5.2.2	3.3	3.4.6	3.5.1.2	3.2.1	8.1.5	7.1.6	7.1.6	7.1.5	3.2.1
(1)	(2)	(3a)	(3b)	(4)	(5)	(6)	(7a)	(7b)	(8)	(9)	(10)	(11)	(12)	(13)
2493	HEXAMETHYLENEIMINE	3	FC	II	3+8		LQ4	E2	**T**	PP, EP, EX, A	VE01		1	
2495	IODINE PENTAFLUORIDE	5.1	OTC	I	5.1+6.1+8	802	LQ0	E0		PP, EP, TOX, A	VE02		2	
2496	PROPIONIC ANHYDRIDE	8	C3	III	8		LQ7	E1	**T**	PP, EP			0	
2498	1,2,3,6-TETRAHYDROBENZALDEHYDE	3	F1	III	3		LQ7	E1		PP, EX, A	VE01		0	
2501	TRIS-(1-AZIRIDINYL) PHOSPHINE OXIDE SOLUTION	6.1	T1	II	6.1	802	LQ17	E4		PP, EP, TOX, A	VE02		2	
2501	TRIS-(1-AZIRIDINYL) PHOSPHINE OXIDE SOLUTION	6.1	T1	III	6.1	802	LQ7	E1		PP, EP, TOX, A	VE02		0	
2502	VALERYL CHLORIDE	8	CF1	II	8+3		LQ22	E2		PP, EP, EX, A	VE01		1	
2503	ZIRCONIUM TETRACHLORIDE	8	C2	III	8		LQ24	E1		PP, EP			0	
2504	TETRABROMOETHANE	6.1	T1	III	6.1	802	LQ7	E1		PP, EP, TOX, A	VE02		0	
2505	AMMONIUM FLUORIDE	6.1	T5	III	6.1	802	LQ9	E1	**B**	PP, EP			0	
2506	AMMONIUM HYDROGEN SULPHATE	8	C2	II	8		LQ23	E2	**B**	PP, EP		CO03	0	CO03 applies only when this substance is carried in bulk or without packaging
2507	CHLOROPLATINIC ACID, SOLID	8	C2	III	8		LQ24	E1		PP, EP			0	
2508	MOLYBDENUM PENTACHLORIDE	8	C2	III	8		LQ24	E1		PP, EP			0	
2509	POTASSIUM HYDROGEN SULPHATE	8	C2	II	8		LQ23	E2	**B**	PP, EP		CO03	0	CO03 applies only when this substance is carried in bulk or without packaging
2511	2-CHLOROPROPIONIC ACID	8	C3	III	8		LQ7	E1		PP, EP			0	
2512	AMINOPHENOLS (o-, m-, p-)	6.1	T2	III	6.1	279 802	LQ9	E1		PP, EP			0	
2513	BROMOACETYL BROMIDE	8	C3	II	8		LQ22	E2		PP, EP			0	
2514	BROMOBENZENE	3	F1	III	3		LQ7	E1		PP, EX, A	VE01		0	
2515	BROMOFORM	6.1	T1	III	6.1	802	LQ7	E1		PP, EP, TOX, A	VE02		0	
2516	CARBON TETRABROMIDE	6.1	T2	III	6.1	802	LQ9	E1		PP, EP			0	
2517	1-CHLORO-1,1-DIFLUOROETHANE (REFRIGERANT GAS R 142b)	2	2F		2.1		LQ0	E0		PP, EX, A	VE01		1	
2518	1,5,9-CYCLODODECATRIENE	6.1	T1	III	6.1	802	LQ7	E1	**T**	PP, EP, TOX, A	VE02		0	
2520	CYCLOOCTADIENES	3	F1	III	3		LQ7	E1		PP, EX, A	VE01		0	
2521	DIKETENE, STABILIZED	6.1	TF1	I	6.1+3	802	LQ0	E5		PP, EP, EX, TOX, A	VE01, VE02		2	
2522	2-DIMETHYLAMINOETHYL METHACRYLATE	6.1	T1	II	6.1	802	LQ17	E4		PP, EP, TOX, A	VE02		2	
2524	ETHYL ORTHOFORMATE	3	F1	III	3		LQ7	E1		PP, EX, A	VE01		0	
2525	ETHYL OXALATE	6.1	T1	III	6.1	802	LQ7	E1		PP, EP, TOX, A	VE02		0	
2526	FURFURYLAMINE	3	FC	III	3+8		LQ7	E1		PP, EP, EX, A	VE01		0	
2527	ISOBUTYL ACRYLATE, STABILIZED	3	F1	III	3		LQ7	E1	**T**	PP, EX, A	VE01		0	

UN No. or ID No.	Name and description	Class	Classi-fication Code	Packing group	Labels	Special provi-sions	Limited and excepted quantities		Carriage permitted	Equipment required	Venti-lation	Provisions concerning loading, unloading and carriage	Number of blue cones/ lights	Remarks
	3.1.2	2.2	2.2	2.1.1.3	5.2.2	3.3	3.4.6	3.5.1.2	3.2.1	8.1.5	7.1.6	7.1.6	7.1.5	3.2.1
(1)	(2)	(3a)	(3b)	(4)	(5)	(6)	(7a)	(7b)	(8)	(9)	(10)	(11)	(12)	(13)
2528	ISOBUTYL ISOBUTYRATE	3	F1	III	3		LQ7	E1	T	PP, EX, A	VE01		0	
2529	ISOBUTYRIC ACID	3	FC	III	3+8		LQ7	E1		PP, EP, EX, A	VE01		0	
2531	METHACRYLIC ACID, STABILIZED	8	C3	II	8		LQ22	E2	T	PP, EP			0	
2533	METHYL TRICHLOROACETATE	6.1	T1	III	6.1	802	LQ7	E1		PP, EP, TOX, A	VE02		0	
2534	METHYLCHLOROSILANE	2	2TFC		2.3+2.1+8		LQ0	E0		PP, EP, EX, TOX, A	VE01, VE02		2	
2535	4-METHYLMORPHOLINE (N-METHYLMORPHOLINE)	3	FC	II	3+8		LQ4	E2		PP, EP, EX, A	VE01		1	
2536	METHYL-TETRAHYDROFURAN	3	F1	II	3		LQ4	E2		PP, EX, A	VE01		1	
2538	NITRONAPHTHALENE	4.1	F1	III	4.1		LQ9	E1		PP			0	
2541	TERPINOLENE	3	F1	III	3		LQ7	E1		PP, EX, A	VE01		0	
2542	TRIBUTYLAMINE	6.1	T1	II	6.1	802	LQ17	E4		PP, EP, TOX, A	VE02		2	
2545	HAFNIUM POWDER, DRY	4.2	S4	I	4.2	540	LQ0	E0		PP			0	
2545	HAFNIUM POWDER, DRY	4.2	S4	II	4.2	540	LQ0	E2		PP			0	
2545	HAFNIUM POWDER, DRY	4.2	S4	III	4.2	540	LQ0	E1		PP			0	
2546	TITANIUM POWDER, DRY	4.2	S4	I	4.2	540	LQ0	E0		PP			0	
2546	TITANIUM POWDER, DRY	4.2	S4	II	4.2	540	LQ0	E2		PP			0	
2546	TITANIUM POWDER, DRY	4.2	S4	III	4.2	540	LQ0	E1		PP			0	
2547	SODIUM SUPEROXIDE	5.1	O2	I	5.1		LQ0	E0		PP			0	
2548	CHLORINE PENTAFLUORIDE	2	2TOC		2.3+5.1+8		LQ0	E0		PP, EP, TOX, A	VE02		2	
2552	HEXAFLUOROACETONE HYDRATE, LIQUID	6.1	T1	II	6.1	802	LQ17	E4		PP, EP, TOX, A	VE02		2	
2554	METHYLALLYL CHLORIDE	3	F1	II	3		LQ4	E2		PP, EX, A	VE01		1	
2555	NITROCELLULOSE WITH WATER (not less than 25% water, by mass)	4.1	D	II	4.1	541	LQ0	E0		PP			0	
2556	NITROCELLULOSE WITH ALCOHOL (not less than 25% alcohol, by mass, and not more than 12.6% nitrogen, by dry mass)	4.1	D	II	4.1	541	LQ0	E0		PP			0	
2557	NITROCELLULOSE, with not more than 12.6% nitrogen, by dry mass, MIXTURE WITH or WITHOUT PLASTICIZER, WITH or WITHOUT PIGMENT	4.1	D	II	4.1	241 541	LQ0	E0		PP			0	
2558	EPIBROMOHYDRIN	6.1	TF1	I	6.1+3	802	LQ0	E5		PP, EP, EX, TOX, A	VE01, VE02		2	
2560	2-METHYLPENTAN-2-OL	3	F1	III	3		LQ7	E1		PP, EX, A	VE01		0	
2561	3-METHYL-1-BUTENE	3	F1	I	3		LQ3	E3		PP, EX, A	VE01		1	
2564	TRICHLOROACETIC ACID SOLUTION	8	C3	II	8		LQ22	E2	T	PP, EP			0	
2564	TRICHLOROACETIC ACID SOLUTION	8	C3	III	8		LQ7	E1	T	PP, EP			0	
2565	DICYCLOHEXYLAMINE	8	C7	III	8		LQ7	E1		PP, EP			0	
2567	SODIUM PENTACHLOROPHENATE	6.1	T2	II	6.1	802	LQ18	E4		PP, EP			2	
2570	CADMIUM COMPOUND	6.1	T5	I	6.1	274 596 802	LQ0	E5		PP, EP			2	

UN No. or ID No.	Name and description	Class	Classi-fication Code	Packing group	Labels	Special provi-sions	Limited and excepted quantities		Carriage permitted	Equipment required	Venti-lation	Provisions concerning loading, unloading and carriage	Number of blue cones/ lights	Remarks
	3.1.2	2.2	2.2	2.1.1.3	5.2.2	3.3	3.4.6	3.5.1.2	3.2.1	8.1.5	7.1.6	7.1.6	7.1.5	3.2.1
(1)	(2)	(3a)	(3b)	(4)	(5)	(6)	(7a)	(7b)	(8)	(9)	(10)	(11)	(12)	(13)
2570	CADMIUM COMPOUND	6.1	T5	II	6.1	274 596 802	LQ18	E4		PP, EP			2	
2570	CADMIUM COMPOUND	6.1	T5	III	6.1	274 596 802	LQ9	E1		PP, EP			0	
2571	ALKYLSULPHURIC ACIDS	8	C3	II	8		LQ22	E2		PP, EP			0	
2572	PHENYLHYDRAZINE	6.1	T1	II	6.1	802	LQ17	E4		PP, EP, TOX, A	VE02		2	
2573	THALLIUM CHLORATE	5.1	OT2	II	5.1+6.1	802	LQ11	E2		PP			2	
2574	TRICRESYL PHOSPHATE with more than 3% ortho isomer	6.1	T1	II	6.1	802	LQ17	E4	T	PP, EP, TOX, A	VE02		2	
2576	PHOSPHORUS OXYBROMIDE, MOLTEN	8	C1	II	8		LQ0	E0		PP, EP			0	
2577	PHENYLACETYL CHLORIDE	8	C3	II	8		LQ22	E2		PP, EP			0	
2578	PHOSPHORUS TRIOXIDE	8	C2	III	8		LQ24	E1		PP, EP			0	
2579	PIPERAZINE	8	C8	III	8		LQ24	E1	T	PP, EP			0	
2580	ALUMINIUM BROMIDE SOLUTION	8	C1	III	8		LQ7	E1		PP, EP			0	
2581	ALUMINIUM CHLORIDE SOLUTION	8	C1	III	8		LQ7	E1		PP, EP			0	
2582	FERRIC CHLORIDE SOLUTION	8	C1	III	8		LQ7	E1		PP, EP			0	
2583	ALKYLSULPHONIC ACIDS, SOLID or ARYLSULPHONIC ACIDS, SOLID with more than 5% free sulphuric acid	8	C2	II	8	274	LQ23	E2		PP, EP			0	
2584	ALKYLSULPHONIC ACIDS, LIQUID or ARYLSULPHONIC ACIDS, LIQUID with more than 5% free sulphuric acid	8	C1	II	8	274	LQ22	E2		PP, EP			0	
2585	ALKYLSULPHONIC ACIDS, SOLID or ARYLSULPHONIC ACIDS, SOLID with not more than 5% free sulphuric acid	8	C4	III	8	274	LQ24	E1		PP, EP			0	
2586	ALKYLSULPHONIC ACIDS, LIQUID or ARYLSULPHONIC ACIDS, LIQUID with not more than 5% free sulphuric acid	8	C3	III	8	274	LQ7	E1	T	PP, EP			0	
2587	BENZOQUINONE	6.1	T2	II	6.1	802	LQ18	E4		PP, EP			2	
2588	PESTICIDE, SOLID, TOXIC, N.O.S.	6.1	T7	I	6.1	61 274 648 802	LQ0	E5		PP, EP			2	
2588	PESTICIDE, SOLID, TOXIC, N.O.S.	6.1	T7	II	6.1	61 274 648 802	LQ18	E4		PP, EP			2	
2588	PESTICIDE, SOLID, TOXIC, N.O.S.	6.1	T7	III	6.1	61 274 648 802	LQ9	E1		PP, EP			0	
2589	VINYL CHLOROACETATE	6.1	TF1	II	6.1+3	802	LQ17	E4		PP, EP, EX, TOX, A	VE01, VE02		2	

UN No. or ID No.	Name and description	Class	Classi-fication Code	Packing group	Labels	Special provi-sions	Limited and excepted quantities		Carriage permitted	Equipment required	Venti-lation	Provisions concerning loading, unloading and carriage	Number of blue cones/ lights	Remarks
	3.1.2	2.2	2.2	2.1.1.3	5.2.2	3.3	3.4.6	3.5.1.2	3.2.1	8.1.5	7.1.6	7.1.6	7.1.5	3.2.1
(1)	(2)	(3a)	(3b)	(4)	(5)	(6)	(7a)	(7b)	(8)	(9)	(10)	(11)	(12)	(13)
2590	WHITE ASBESTOS (chrysotile, actinolite, anthophyllite, tremolite)	9	M1	III	9	168 542 802	LQ27	E1		PP			0	
2591	XENON, REFRIGERATED LIQUID	2	3A		2.2	593	LQ1	E1		PP			0	
2599	CHLOROTRIFLUORO-METHANE AND TRIFLUOROMETHANE AZEOTROPIC MIXTURE with approximately 60% chlorotrifluoromethane (REFRIGERANT GAS R 503)	2	2A		2.2		LQ1	E1		PP			0	
2601	CYCLOBUTANE	2	2F		2.1		LQ0	E0		PP, EX, A	VE01		1	
2602	DICHLORODIFLUOROMETHANE AND 1,1-DIFLUOROETHANE AZEOTROPIC MIXTURE with approximately 74% dichlorodifluoromethane (REFRIGERANT GAS R 500)	2	2A		2.2		LQ1	E1		PP			0	
2603	CYCLOHEPTATRIENE	3	FT1	II	3+6.1	802	LQ0	E2		PP, EP, EX, TOX, A	VE01, VE02		2	
2604	BORON TRIFLUORIDE DIETHYL ETHERATE	8	CF1	I	8+3		LQ0	E0		PP, EP, EX, A	VE01		1	
2605	METHOXYMETHYL ISOCYANATE	3	FT1	I	3+6.1	802	LQ0	E0		PP, EP, EX, TOX, A	VE01, VE02		2	
2606	METHYL ORTHOSILICATE	6.1	TF1	I	6.1+3	802	LQ0	E5		PP, EP, EX, TOX, A	VE01, VE02		2	
2607	ACROLEIN DIMER, STABILIZED	3	F1	III	3		LQ7	E1		PP, EX, A	VE01		0	
2608	NITROPROPANES	3	F1	III	3		LQ7	E1	T	PP, EX, A	VE01		0	
2609	TRIALLYL BORATE	6.1	T1	III	6.1	802	LQ7	E1		PP, EP, TOX, A	VE02		0	
2610	TRIALLYLAMINE	3	FC	III	3+8		LQ7	E1		PP, EP, EX, A	VE01		0	
2611	PROPYLENE CHLOROHYDRIN	6.1	TF1	II	6.1+3	802	LQ17	E4		PP, EP, EX, TOX, A	VE01, VE02		2	
2612	METHYL PROPYL ETHER	3	F1	II	3		LQ4	E2		PP, EX, A	VE01		1	
2614	METHALLYL ALCOHOL	3	F1	III	3		LQ7	E1		PP, EX, A	VE01		0	
2615	ETHYL PROPYL ETHER	3	F1	II	3		LQ4	E2	T	PP, EX, A	VE01		1	
2616	TRIISOPROPYL BORATE	3	F1	II	3		LQ4	E2		PP, EX, A	VE01		1	
2616	TRIISOPROPYL BORATE	3	F1	III	3		LQ7	E1		PP, EX, A	VE01		0	
2617	METHYLCYCLOHEXANOLS, flammable	3	F1	III	3		LQ7	E1		PP, EX, A	VE01		0	
2618	VINYLTOLUENES, STABILIZED	3	F1	III	3		LQ7	E1	T	PP, EX, A	VE01		0	
2619	BENZYLDIMETHYLAMINE	8	CF1	II	8+3		LQ22	E2		PP, EP, EX, A	VE01		1	
2620	AMYL BUTYRATES	3	F1	III	3		LQ7	E1		PP, EX, A	VE01		0	
2621	ACETYL METHYL CARBINOL	3	F1	III	3		LQ7	E1		PP, EX, A	VE01		0	
2622	GLYCIDALDEHYDE	3	FT1	II	3+6.1	802	LQ0	E2		PP, EP, EX, TOX, A	VE01, VE02		2	
2623	FIRELIGHTERS, SOLID with flammable liquid	4.1	F1	III	4.1		LQ9	E1		PP			0	
2624	MAGNESIUM SILICIDE	4.3	W2	II	4.3		LQ11	E2		PP, EX, A	VE01	HA08	0	
2626	CHLORIC ACID, AQUEOUS SOLUTION with not more than 10% chloric acid	5.1	O1	II	5.1	613	LQ10	E2		PP			0	

UN No. or ID No.	Name and description	Class	Classi-fication Code	Packing group	Labels	Special provi-sions	Limited and excepted quantities		Carriage permitted	Equipment required	Venti-lation	Provisions concerning loading, unloading and carriage	Number of blue cones/ lights	Remarks
	3.1.2	2.2	2.2	2.1.1.3	5.2.2	3.3	3.4.6	3.5.1.2	3.2.1	8.1.5	7.1.6	7.1.6	7.1.5	3.2.1
(1)	(2)	(3a)	(3b)	(4)	(5)	(6)	(7a)	(7b)	(8)	(9)	(10)	(11)	(12)	(13)
2627	NITRITES, INORGANIC, N.O.S.	5.1	O2	II	5.1	103 274	LQ11	E2		PP			0	
2628	POTASSIUM FLUOROACETATE	6.1	T2	I	6.1	802	LQ0	E5		PP, EP			2	
2629	SODIUM FLUOROACETATE	6.1	T2	I	6.1	802	LQ0	E5		PP, EP			2	
2630	SELENATES or SELENITES	6.1	T5	I	6.1	274 802	LQ0	E5		PP, EP			2	
2642	FLUOROACETIC ACID	6.1	T2	I	6.1	802	LQ0	E5		PP, EP			2	
2643	METHYL BROMOACETATE	6.1	T1	II	6.1	802	LQ17	E4		PP, EP, TOX, A	VE02		2	
2644	METHYL IODIDE	6.1	T1	I	6.1	802	LQ0	E5		PP, EP, TOX, A	VE02		2	
2645	PHENACYL BROMIDE	6.1	T2	II	6.1	802	LQ18	E4		PP, EP			2	
2646	HEXACHLOROCYCLOPENTADIENE	6.1	T1	I	6.1	802	LQ0	E5		PP, EP, TOX, A	VE02		2	
2647	MALONONITRILE	6.1	T2	II	6.1	802	LQ18	E4		PP, EP			2	
2648	1,2-DIBROMOBUTAN-3-ONE	6.1	T1	II	6.1	802	LQ17	E4		PP, EP, TOX, A	VE02		2	
2649	1,3-DICHLOROACETONE	6.1	T2	II	6.1	802	LQ18	E4		PP, EP			2	
2650	1,1-DICHLORO-1-NITROETHANE	6.1	T1	II	6.1	802	LQ17	E4		PP, EP, TOX, A	VE02		2	
2651	4,4'-DIAMINODIPHENYLMETHANE	6.1	T2	III	6.1	802	LQ9	E1	T	PP, EP			0	
2653	BENZYL IODIDE	6.1	T1	II	6.1	802	LQ17	E4		PP, EP, TOX, A	VE02		2	
2655	POTASSIUM FLUOROSILICATE	6.1	T5	III	6.1	802	LQ9	E1		PP, EP			0	
2656	QUINOLINE	6.1	T1	III	6.1	802	LQ7	E1		PP, EP, TOX, A	VE02		0	
2657	SELENIUM DISULPHIDE	6.1	T5	II	6.1	802	LQ18	E4		PP, EP			2	
2659	SODIUM CHLOROACETATE	6.1	T2	III	6.1	802	LQ9	E1		PP, EP			0	
2660	NITROTOLUIDINES (MONO)	6.1	T2	III	6.1	802	LQ9	E1		PP, EP			0	
2661	HEXACHLOROACETONE	6.1	T1	III	6.1	802	LQ7	E1		PP, EP, TOX, A	VE02		0	
2664	DIBROMOMETHANE	6.1	T1	III	6.1	802	LQ7	E1		PP, EP, TOX, A	VE02		0	
2667	BUTYLTOLUENES	6.1	T1	III	6.1	802	LQ7	E1		PP, EP, TOX, A	VE02		0	
2668	CHLOROACETONITRILE	6.1	TF1	II	6.1+3	802	LQ17	E4		PP, EP, EX, TOX, A	VE01, VE02		2	
2669	CHLOROCRESOLS, SOLUTION	6.1	T1	II	6.1	802	LQ17	E4		PP, EP, TOX, A	VE02		2	
2669	CHLOROCRESOLS, SOLUTION	6.1	T1	III	6.1	802	LQ7	E1		PP, EP, TOX, A	VE02		0	
2670	CYANURIC CHLORIDE	8	C4	II	8		LQ23	E2		PP, EP			0	
2671	AMINOPYRIDINES (o-, m-, p-)	6.1	T2	II	6.1	802	LQ18	E4		PP, EP			2	

UN No. or ID No.	Name and description	Class	Classi-fication Code	Packing group	Labels	Special provi-sions	Limited and excepted quantities		Carriage permitted	Equipment required	Venti-lation	Provisions concerning loading, unloading and carriage	Number of blue cones/ lights	Remarks
	3.1.2	2.2	2.2	2.1.1.3	5.2.2	3.3	3.4.6	3.5.1.2	3.2.1	8.1.5	7.1.6	7.1.6	7.1.5	3.2.1
(1)	(2)	(3a)	(3b)	(4)	(5)	(6)	(7a)	(7b)	(8)	(9)	(10)	(11)	(12)	(13)
2672	AMMONIA SOLUTION, relative density between 0.880 and 0.957 at 15 °C in water, with more than 10% but not more than 35% ammonia	8	C5	III	8	543	LQ7	E1	T	PP, EP			0	
2673	2-AMINO-4-CHLOROPHENOL	6.1	T2	II	6.1	802	LQ18	E4		PP, EP			2	
2674	SODIUM FLUOROSILICATE	6.1	T5	III	6.1	802	LQ9	E1		PP, EP			0	
2676	STIBINE	2	2TF		2.3+2.1		LQ0	E0		PP, EP, EX, TOX, A	VE01, VE02		2	
2677	RUBIDIUM HYDROXIDE SOLUTION	8	C5	II	8		LQ22	E2		PP, EP			0	
2677	RUBIDIUM HYDROXIDE SOLUTION	8	C5	III	8		LQ7	E1		PP, EP			0	
2678	RUBIDIUM HYDROXIDE	8	C6	II	8		LQ23	E2		PP, EP			0	
2679	LITHIUM HYDROXIDE SOLUTION	8	C5	II	8		LQ22	E2		PP, EP			0	
2679	LITHIUM HYDROXIDE SOLUTION	8	C5	III	8		LQ7	E1		PP, EP			0	
2680	LITHIUM HYDROXIDE	8	C6	II	8		LQ23	E2		PP, EP			0	
2681	CAESIUM HYDROXIDE SOLUTION	8	C5	II	8		LQ22	E2		PP, EP			0	
2681	CAESIUM HYDROXIDE SOLUTION	8	C5	III	8		LQ7	E1		PP, EP			0	
2682	CAESIUM HYDROXIDE	8	C6	II	8		LQ23	E2		PP, EP			0	
2683	AMMONIUM SULPHIDE SOLUTION	8	CFT	II	8+3+6.1	802	LQ22	E2	T	PP, EP, EX, TOX, A	VE01, VE02		2	
2684	3-DIETHYLAMINOPROPYLAMINE	3	FC	III	3+8		LQ7	E1		PP, EP, EX, A	VE01		0	
2685	N,N-DIETHYLETHYLENEDIAMINE	8	CF1	II	8+3		LQ22	E2		PP, EP, EX, A	VE01		1	
2686	2-DIETHYLAMINOETHANOL	8	CF1	II	8+3		LQ22	E2		PP, EP, EX, A	VE01		1	
2687	DICYCLOHEXYLAMMONIUM NITRITE	4.1	F3	III	4.1		LQ9	E1		PP			0	
2688	1-BROMO-3-CHLOROPROPANE	6.1	T1	III	6.1	802	LQ7	E1		PP, EP, TOX, A	VE02		0	
2689	GLYCEROL alpha-MONOCHLOROHYDRIN	6.1	T1	III	6.1	802	LQ7	E1		PP, EP, TOX, A	VE02		0	
2690	N,n-BUTYLIMIDAZOLE	6.1	T1	II	6.1	802	LQ17	E4		PP, EP, TOX, A	VE02		2	
2691	PHOSPHORUS PENTABROMIDE	8	C2	II	8		LQ23	E2		PP, EP			0	
2692	BORON TRIBROMIDE	8	C1	I	8		LQ0	E0		PP, EP			0	
2693	BISULPHITES, AQUEOUS SOLUTION, N.O.S.	8	C1	III	8	274	LQ7	E1	T	PP, EP			0	
2698	TETRAHYDROPHTHALIC ANHYDRIDES with more than 0.05% of maleic anhydride	8	C4	III	8	169	LQ24	E1		PP, EP			0	
2699	TRIFLUOROACETIC ACID	8	C3	I	8		LQ0	E0		PP, EP			0	
2705	1-PENTOL	8	C9	II	8		LQ22	E2		PP, EP			0	
2707	DIMETHYLDIOXANES	3	F1	II	3		LQ4	E2		PP, EX, A	VE01		1	
2707	DIMETHYLDIOXANES	3	F1	III	3		LQ7	E1		PP, EX, A	VE01		0	
2709	BUTYLBENZENES	3	F1	III	3		LQ7	E1	T	PP, EX, A	VE01		0	
2710	DIPROPYL KETONE	3	F1	III	3		LQ7	E1		PP, EX, A	VE01		0	
2713	ACRIDINE	6.1	T2	III	6.1	802	LQ9	E1		PP, EP			0	
2714	ZINC RESINATE	4.1	F3	III	4.1		LQ9	E1		PP			0	
2715	ALUMINIUM RESINATE	4.1	F3	III	4.1		LQ9	E1		PP			0	
2716	1,4-BUTYNEDIOL	6.1	T2	III	6.1	802	LQ9	E1		PP, EP			0	
2717	CAMPHOR, synthetic	4.1	F1	III	4.1		LQ9	E1		PP			0	

UN No. or ID No.	Name and description	Class	Classi-fication Code	Packing group	Labels	Special provi-sions	Limited and excepted quantities		Carriage permitted	Equipment required	Venti-lation	Provisions concerning loading, unloading and carriage			Number of blue cones/ lights	Remarks
	3.1.2	2.2	2.2	2.1.1.3	5.2.2	3.3	3.4.6	3.5.1.2	3.2.1	8.1.5	7.1.6	7.1.6			7.1.5	3.2.1
(1)	(2)	(3a)	(3b)	(4)	(5)	(6)	(7a)	(7b)	(8)	(9)	(10)	(11)			(12)	(13)
2719	BARIUM BROMATE	5.1	OT2	II	5.1+6.1	802	LQ11	E2		PP					2	
2720	CHROMIUM NITRATE	5.1	O2	III	5.1		LQ12	E1	**B**	PP		CO02, LO04			0	CO02 and LO04 apply only when this substance is carried in bulk or without packaging
2721	COPPER CHLORATE	5.1	O2	II	5.1		LQ11	E2		PP					0	
2722	LITHIUM NITRATE	5.1	O2	III	5.1		LQ12	E1	**B**	PP		CO02, LO04			0	CO02 and LO04 apply only when this substance is carried in bulk or without packaging
2723	MAGNESIUM CHLORATE	5.1	O2	II	5.1		LQ11	E2		PP					0	
2724	MANGANESE NITRATE	5.1	O2	III	5.1		LQ12	E1	**B**	PP		CO02, LO04			0	CO02 and LO04 apply only when this substance is carried in bulk or without packaging
2725	NICKEL NITRATE	5.1	O2	III	5.1		LQ12	E1	**B**	PP		CO02, LO04			0	CO02 and LO04 apply only when this substance is carried in bulk or without packaging
2726	NICKEL NITRITE	5.1	O2	III	5.1		LQ12	E1		PP					0	
2727	THALLIUM NITRATE	6.1	TO2	II	6.1+5.1	802	LQ18	E4		PP, EP					2	
2728	ZIRCONIUM NITRATE	5.1	O2	III	5.1		LQ12	E1	**B**	PP		CO02, LO04			0	CO02 and LO04 apply only when this substance is carried in bulk or without packaging
2729	HEXACHLOROBENZENE	6.1	T2	III	6.1	802	LQ9	E1		PP, EP					0	
2730	NITROANISOLES, LIQUID	6.1	T1	III	6.1	279 802	LQ7	E1		PP, EP, TOX, A	VE02				0	
2732	NITROBROMOBENZENES, LIQUID	6.1	T1	III	6.1	802	LQ7	E1		PP, EP, TOX, A	VE02				0	
2733	AMINES, FLAMMABLE, CORROSIVE, N.O.S. or POLYAMINES, FLAMMABLE, CORROSIVE, N.O.S.	3	FC	I	3+8	274 544	LQ3	E0		PP, EP, EX, A	VE01				1	
2733	AMINES, FLAMMABLE, CORROSIVE, N.O.S. or POLYAMINES, FLAMMABLE, CORROSIVE, N.O.S.	3	FC	II	3+8	274 544	LQ4	E2	**T**	PP, EP, EX, A	VE01				1	
2733	AMINES, FLAMMABLE, CORROSIVE, N.O.S. or POLYAMINES, FLAMMABLE, CORROSIVE, N.O.S.	3	FC	III	3+8	274 544	LQ7	E1		PP, EP, EX, A	VE01				0	

UN No. or ID No.	Name and description	Class	Classi-fication Code	Packing group	Labels	Special provi-sions	Limited and excepted quantities		Carriage permitted	Equipment required	Venti-lation	Provisions concerning loading, unloading and carriage	Number of blue cones/ lights	Remarks
	3.1.2	2.2	2.2	2.1.1.3	5.2.2	3.3	3.4.6	3.5.1.2	3.2.1	8.1.5	7.1.6	7.1.6	7.1.5	3.2.1
(1)	(2)	(3a)	(3b)	(4)	(5)	(6)	(7a)	(7b)	(8)	(9)	(10)	(11)	(12)	(13)
2734	AMINES, LIQUID, CORROSIVE, FLAMMABLE, N.O.S. or POLYAMINES, LIQUID, CORROSIVE, FLAMMABLE, N.O.S.	8	CF1	I	8+3	274	LQ0	E0		PP, EP, EX, A	VE01		1	
2734	AMINES, LIQUID, CORROSIVE, FLAMMABLE, N.O.S. or POLYAMINES, LIQUID, CORROSIVE, FLAMMABLE, N.O.S.	8	CF1	II	8+3	274	LQ22	E2		PP, EP, EX, A	VE01		1	
2735	AMINES, LIQUID, CORROSIVE, N.O.S. or POLYAMINES, LIQUID, CORROSIVE, N.O.S.	8	C7	I	8	274	LQ0	E0	T	PP, EP			0	
2735	AMINES, LIQUID, CORROSIVE, N.O.S. or POLYAMINES, LIQUID, CORROSIVE, N.O.S.	8	C7	II	8	274	LQ22	E2	T	PP, EP			0	
2735	AMINES, LIQUID, CORROSIVE, N.O.S. or POLYAMINES, LIQUID, CORROSIVE, N.O.S.	8	C7	III	8	274	LQ7	E1	T	PP, EP			0	
2738	N-BUTYLANILINE	6.1	T1	II	6.1	802	LQ17	E4		PP, EP, TOX, A	VE02		2	
2739	BUTYRIC ANHYDRIDE	8	C3	III	8		LQ7	E1		PP, EP			0	
2740	n-PROPYL CHLOROFORMATE	6.1	TFC	I	6.1+3+8	802	LQ0	E5		PP, EP, EX, TOX, A	VE01, VE02		2	
2741	BARIUM HYPOCHLORITE with more than 22% available chlorine	5.1	OT2	II	5.1+6.1	802	LQ11	E2		PP			2	
2742	CHLOROFORMATES, TOXIC, CORROSIVE, FLAMMABLE, N.O.S.	6.1	TFC	II	6.1+3+8	274 561 802	LQ17	E4		PP, EP, EX, TOX, A	VE01, VE02		2	
2743	n-BUTYL CHLOROFORMATE	6.1	TFC	II	6.1+3+8	802	LQ17	E4		PP, EP, EX, TOX, A	VE01, VE02		2	
2744	CYCLOBUTYL CHLOROFORMATE	6.1	TFC	II	6.1+3+8	802	LQ17	E4		PP, EP, EX, TOX, A	VE01, VE02		2	
2745	CHLOROMETHYL CHLOROFORMATE	6.1	TC1	II	6.1+8	802	LQ17	E4		PP, EP, TOX, A	VE02		2	
2746	PHENYL CHLOROFORMATE	6.1	TC1	II	6.1+8	802	LQ17	E4		PP, EP, TOX, A	VE02		2	
2747	tert-BUTYLCYCLOHEXYL CHLOROFORMATE	6.1	T1	III	6.1	802	LQ7	E1		PP, EP, TOX, A	VE02		0	
2748	2-ETHYLHEXYL CHLOROFORMATE	6.1	TC1	II	6.1+8	802	LQ17	E4		PP, EP, TOX, A	VE02		2	
2749	TETRAMETHYLSILANE	3	F1	I	3		LQ3	E3		PP, EX, A	VE01		1	
2750	1,3-DICHLOROPROPANOL-2	6.1	T1	II	6.1	802	LQ17	E4		PP, EP, TOX, A	VE02		2	
2751	DIETHYLTHIOPHOSPHORYL CHLORIDE	8	C3	II	8		LQ22	E2		PP, EP			0	
2752	1,2-EPOXY-3-ETHOXYPROPANE	3	F1	III	3		LQ7	E1		PP, EX, A	VE01		0	
2753	N-ETHYLBENZYLTOLUIDINES, LIQUID	6.1	T1	III	6.1	802	LQ7	E1		PP, EP, TOX, A	VE02		0	
2754	N-ETHYLTOLUIDINES	6.1	T1	II	6.1	802	LQ17	E4	T	PP, EP, TOX, A	VE02		2	

UN No. or ID No.	Name and description	Class	Classi-fication Code	Packing group	Labels	Special provi-sions	Limited and excepted quantities		Carriage permitted	Equipment required	Venti-lation	Provisions concerning loading, unloading and carriage			Number of blue cones/ lights	Remarks
	3.1.2	2.2	2.2	2.1.1.3	5.2.2	3.3	3.4.6	3.5.1.2	3.2.1	8.1.5	7.1.6	7.1.6			7.1.5	3.2.1
(1)	(2)	(3a)	(3b)	(4)	(5)	(6)	(7a)	(7b)	(8)	(9)	(10)	(11)			(12)	(13)
2757	CARBAMATE PESTICIDE, SOLID, TOXIC	6.1	T7	I	6.1	61 274 648 802	LQ0	E5		PP, EP					2	
2757	CARBAMATE PESTICIDE, SOLID, TOXIC	6.1	T7	II	6.1	61 274 648 802	LQ18	E4		PP, EP					2	
2757	CARBAMATE PESTICIDE, SOLID, TOXIC	6.1	T7	III	6.1	61 274 648 802	LQ9	E1		PP, EP					0	
2758	CARBAMATE PESTICIDE, LIQUID, FLAMMABLE, TOXIC, flash-point less than 23 °C	3	FT2	I	3+6.1	61 274 802	LQ3	E0		PP, EP, EX, TOX, A	VE01, VE02				2	
2758	CARBAMATE PESTICIDE, LIQUID, FLAMMABLE, TOXIC, flash-point less than 23 °C	3	FT2	II	3+6.1	61 274 802	LQ4	E2		PP, EP, EX, TOX, A	VE01, VE02				2	
2759	ARSENICAL PESTICIDE, SOLID, TOXIC	6.1	T7	I	6.1	61 274 648 802	LQ0	E5		PP, EP					2	
2759	ARSENICAL PESTICIDE, SOLID, TOXIC	6.1	T7	II	6.1	61 274 648 802	LQ18	E4		PP, EP					2	
2759	ARSENICAL PESTICIDE, SOLID, TOXIC	6.1	T7	III	6.1	61 274 648 802	LQ9	E1		PP, EP					0	
2760	ARSENICAL PESTICIDE, LIQUID, FLAMMABLE, TOXIC, flash-point less than 23 °C	3	FT2	I	3+6.1	61 274 802	LQ3	E0		PP, EP, EX, TOX, A	VE01, VE02				2	
2760	ARSENICAL PESTICIDE, LIQUID, FLAMMABLE, TOXIC, flash-point less than 23 °C	3	FT2	II	3+6.1	61 274 802	LQ4	E2		PP, EP, EX, TOX, A	VE01, VE02				2	
2761	ORGANOCHLORINE PESTICIDE, SOLID, TOXIC	6.1	T7	I	6.1	61 274 648 802	LQ0	E5		PP, EP					2	
2761	ORGANOCHLORINE PESTICIDE, SOLID, TOXIC	6.1	T7	II	6.1	61 274 648 802	LQ18	E4		PP, EP					2	

UN No. or ID No.	Name and description	Class	Classi-fication Code	Packing group	Labels	Special provi-sions	Limited and excepted quantities		Carriage permitted	Equipment required	Venti-lation	Provisions concerning loading, unloading and carriage			Number of blue cones/ lights	Remarks
	3.1.2	2.2	2.2	2.1.1.3	5.2.2	3.3	3.4.6	3.5.1.2	3.2.1	8.1.5	7.1.6	7.1.6			7.1.5	3.2.1
(1)	(2)	(3a)	(3b)	(4)	(5)	(6)	(7a)	(7b)	(8)	(9)	(10)	(11)			(12)	(13)
2761	ORGANOCHLORINE PESTICIDE, SOLID, TOXIC	6.1	T7	III	6.1	61 274 648 802	LQ9	E1		PP, EP					0	
2762	ORGANOCHLORINE PESTICIDE, LIQUID, FLAMMABLE, TOXIC, flash-point less than 23 °C	3	FT2	I	3+6.1	61 274 802	LQ3	E0		PP, EP, EX, TOX, A	VE01, VE02				2	
2762	ORGANOCHLORINE PESTICIDE, LIQUID, FLAMMABLE, TOXIC, flash-point less than 23 °C	3	FT2	II	3+6.1	61 274 802	LQ4	E2		PP, EP, EX, TOX, A	VE01, VE02				2	
2763	TRIAZINE PESTICIDE, SOLID, TOXIC	6.1	T7	I	6.1	61 274 648 802	LQ0	E5		PP, EP					2	
2763	TRIAZINE PESTICIDE, SOLID, TOXIC	6.1	T7	II	6.1	61 274 648 802	LQ18	E4		PP, EP					2	
2763	TRIAZINE PESTICIDE, SOLID, TOXIC	6.1	T7	III	6.1	61 274 648 802	LQ9	E1		PP, EP					0	
2764	TRIAZINE PESTICIDE, LIQUID, FLAMMABLE, TOXIC, flash-point less than 23 °C	3	FT2	I	3+6.1	61 274 802	LQ3	E0		PP, EP, EX, TOX, A	VE01, VE02				2	
2764	TRIAZINE PESTICIDE, LIQUID, FLAMMABLE, TOXIC, flash-point less than 23 °C	3	FT2	II	3+6.1	61 274 802	LQ4	E2		PP, EP, EX, TOX, A	VE01, VE02				2	
2771	THIOCARBAMATE PESTICIDE, SOLID, TOXIC	6.1	T7	I	6.1	61 274 648 802	LQ0	E5		PP, EP					2	
2771	THIOCARBAMATE PESTICIDE, SOLID, TOXIC	6.1	T7	II	6.1	61 274 648 802	LQ18	E4		PP, EP					2	
2771	THIOCARBAMATE PESTICIDE, SOLID, TOXIC	6.1	T7	III	6.1	61 274 648 802	LQ9	E1		PP, EP					0	
2772	THIOCARBAMATE PESTICIDE, LIQUID, FLAMMABLE, TOXIC, flash-point less than 23 °C	3	FT2	I	3+6.1	61 274 802	LQ3	E0		PP, EP, EX, TOX, A	VE01, VE02				2	
2772	THIOCARBAMATE PESTICIDE, LIQUID, FLAMMABLE, TOXIC, flash-point less than 23 °C	3	FT2	II	3+6.1	61 274 802	LQ4	E2		PP, EP, EX, TOX, A	VE01, VE02				2	

UN No. or ID No.	Name and description	Class	Classi-fication Code	Packing group	Labels	Special provi-sions	Limited and excepted quantities		Carriage permitted	Equipment required	Venti-lation	Provisions concerning loading, unloading and carriage	Number of blue cones/ lights	Remarks
	3.1.2	2.2	2.2	2.1.1.3	5.2.2	3.3	3.4.6	3.5.1.2	3.2.1	8.1.5	7.1.6	7.1.6	7.1.5	3.2.1
(1)	(2)	(3a)	(3b)	(4)	(5)	(6)	(7a)	(7b)	(8)	(9)	(10)	(11)	(12)	(13)
2775	COPPER BASED PESTICIDE, SOLID, TOXIC	6.1	T7	I	6.1	61 274 648 802	LQ0	E5		PP, EP			2	
2775	COPPER BASED PESTICIDE, SOLID, TOXIC	6.1	T7	II	6.1	61 274 648 802	LQ18	E4		PP, EP			2	
2775	COPPER BASED PESTICIDE, SOLID, TOXIC	6.1	T7	III	6.1	61 274 648 802	LQ9	E1		PP, EP			0	
2776	COPPER BASED PESTICIDE, LIQUID, FLAMMABLE, TOXIC, flash-point less than 23 °C	3	FT2	I	3+6.1	61 274 802	LQ3	E0		PP, EP, EX, TOX, A	VE01, VE02		2	
2776	COPPER BASED PESTICIDE, LIQUID, FLAMMABLE, TOXIC, flash-point less than 23 °C	3	FT2	II	3+6.1	61 274 802	LQ4	E2		PP, EP, EX, TOX, A	VE01, VE02		2	
2777	MERCURY BASED PESTICIDE, SOLID, TOXIC	6.1	T7	I	6.1	61 274 648 802	LQ0	E5		PP, EP			2	
2777	MERCURY BASED PESTICIDE, SOLID, TOXIC	6.1	T7	II	6.1	61 274 648 802	LQ18	E4		PP, EP			2	
2777	MERCURY BASED PESTICIDE, SOLID, TOXIC	6.1	T7	III	6.1	61 274 648 802	LQ9	E1		PP, EP			0	
2778	MERCURY BASED PESTICIDE, LIQUID, FLAMMABLE, TOXIC, flash-point less than 23 °C	3	FT2	I	3+6.1	61 274 802	LQ3	E0		PP, EP, EX, TOX, A	VE01, VE02		2	
2778	MERCURY BASED PESTICIDE, LIQUID, FLAMMABLE, TOXIC, flash-point less than 23 °C	3	FT2	II	3+6.1	61 274 802	LQ4	E2		PP, EP, EX, TOX, A	VE01, VE02		2	
2779	SUBSTITUTED NITROPHENOL PESTICIDE, SOLID, TOXIC	6.1	T7	I	6.1	61 274 648 802	LQ0	E5		PP, EP			2	
2779	SUBSTITUTED NITROPHENOL PESTICIDE, SOLID, TOXIC	6.1	T7	II	6.1	61 274 648 802	LQ18	E4		PP, EP			2	

UN No. or ID No.	Name and description	Class	Classi-fication Code	Packing group	Labels	Special provi-sions	Limited and excepted quantities		Carriage permitted	Equipment required	Venti-lation	Provisions concerning loading, unloading and carriage	Number of blue cones/ lights	Remarks
	3.1.2	2.2	2.2	2.1.1.3	5.2.2	3.3	3.4.6	3.5.1.2	3.2.1	8.1.5	7.1.6	7.1.6	7.1.5	3.2.1
(1)	(2)	(3a)	(3b)	(4)	(5)	(6)	(7a)	(7b)	(8)	(9)	(10)	(11)	(12)	(13)
2779	SUBSTITUTED NITROPHENOL PESTICIDE, SOLID, TOXIC	6.1	T7	III	6.1	61 274 648 802	LQ9	E1		PP, EP			0	
2780	SUBSTITUTED NITROPHENOL PESTICIDE, LIQUID, FLAMMABLE, TOXIC, flash-point less than 23 °C	3	FT2	I	3+6.1	61 274 802	LQ3	E0		PP, EP, EX, TOX, A	VE01, VE02		2	
2780	SUBSTITUTED NITROPHENOL PESTICIDE, LIQUID, FLAMMABLE, TOXIC, flash-point less than 23 °C	3	FT2	II	3+6.1	61 274 802	LQ4	E2		PP, EP, EX, TOX, A	VE01, VE02		2	
2781	BIPYRIDILIUM PESTICIDE, SOLID, TOXIC	6.1	T7	I	6.1	61 274 648 802	LQ0	E5		PP, EP			2	
2781	BIPYRIDILIUM PESTICIDE, SOLID, TOXIC	6.1	T7	II	6.1	61 274 648 802	LQ18	E4		PP, EP			2	
2781	BIPYRIDILIUM PESTICIDE, SOLID, TOXIC	6.1	T7	III	6.1	61 274 648 802	LQ9	E1		PP, EP			0	
2782	BIPYRIDILIUM PESTICIDE, LIQUID, FLAMMABLE, TOXIC, flash-point less than 23 °C	3	FT2	I	3+6.1	61 274 802	LQ3	E0		PP, EP, EX, TOX, A	VE01, VE02		2	
2782	BIPYRIDILIUM PESTICIDE, LIQUID, FLAMMABLE, TOXIC, flash-point less than 23 °C	3	FT2	II	3+6.1	61 274 802	LQ4	E2		PP, EP, EX, TOX, A	VE01, VE02		2	
2783	ORGANOPHOSPHORUS PESTICIDE, SOLID, TOXIC	6.1	T7	I	6.1	61 274 648 802	LQ0	E5		PP, EP			2	
2783	ORGANOPHOSPHORUS PESTICIDE, SOLID, TOXIC	6.1	T7	II	6.1	61 274 648 802	LQ18	E4		PP, EP			2	
2783	ORGANOPHOSPHORUS PESTICIDE, SOLID, TOXIC	6.1	T7	III	6.1	61 274 648 802	LQ9	E1		PP, EP			0	
2784	ORGANOPHOSPHORUS PESTICIDE, LIQUID, FLAMMABLE, TOXIC, flash-point less than 23 °C	3	FT2	I	3+6.1	61 274 802	LQ3	E0		PP, EP, EX, TOX, A	VE01, VE02		2	
2784	ORGANOPHOSPHORUS PESTICIDE, LIQUID, FLAMMABLE, TOXIC, flash-point less than 23 °C	3	FT2	II	3+6.1	61 274 802	LQ4	E2		PP, EP, EX, TOX, A	VE01, VE02		2	

UN No. or ID No.	Name and description	Class	Classi-fication Code	Packing group	Labels	Special provi-sions	Limited and excepted quantities		Carriage permitted	Equipment required	Venti-lation	Provisions concerning loading, unloading and carriage	Number of blue cones/ lights	Remarks
	3.1.2	2.2	2.2	2.1.1.3	5.2.2	3.3	3.4.6	3.5.1.2	3.2.1	8.1.5	7.1.6	7.1.6	7.1.5	3.2.1
(1)	(2)	(3a)	(3b)	(4)	(5)	(6)	(7a)	(7b)	(8)	(9)	(10)	(11)	(12)	(13)
2785	4-THIAPENTANAL	6.1	T1	III	6.1	802	LQ7	E1		PP, EP, TOX, A	VE02		0	
2786	ORGANOTIN PESTICIDE, SOLID, TOXIC	6.1	T7	I	6.1	61 274 648 802	LQ0	E5		PP, EP			2	
2786	ORGANOTIN PESTICIDE, SOLID, TOXIC	6.1	T7	II	6.1	61 274 648 802	LQ18	E4		PP, EP			2	
2786	ORGANOTIN PESTICIDE, SOLID, TOXIC	6.1	T7	III	6.1	61 274 648 802	LQ9	E1		PP, EP			0	
2787	ORGANOTIN PESTICIDE, LIQUID, FLAMMABLE, TOXIC, flash-point less than 23 °C	3	FT2	I	3+6.1	61 274 802	LQ3	E0		PP, EP, EX, TOX, A	VE01, VE02		2	
2787	ORGANOTIN PESTICIDE, LIQUID, FLAMMABLE, TOXIC, flash-point less than 23 °C	3	FT2	II	3+6.1	61 274 802	LQ4	E2		PP, EP, EX, TOX, A	VE01, VE02		2	
2788	ORGANOTIN COMPOUND, LIQUID, N.O.S.	6.1	T3	I	6.1	43 274 802	LQ0	E5		PP, EP, TOX, A	VE02		2	
2788	ORGANOTIN COMPOUND, LIQUID, N.O.S.	6.1	T3	II	6.1	43 274 802	LQ17	E4		PP, EP, TOX, A	VE02		2	
2788	ORGANOTIN COMPOUND, LIQUID, N.O.S.	6.1	T3	III	6.1	43 274 802	LQ7	E1		PP, EP, TOX, A	VE02		0	
2789	ACETIC ACID, GLACIAL or ACETIC ACID SOLUTION, more than 80% acid, by mass	8	CF1	II	8+3		LQ22	E2	**T**	PP, EP, EX, A	VE01		1	
2790	ACETIC ACID SOLUTION, not less than 50% but not more than 80% acid, by mass	8	C3	II	8		LQ22	E2	**T**	PP, EP			0	
2790	ACETIC ACID SOLUTION, more than 10% and less than 50% acid, by mass	8	C3	III	8	597 647	LQ7	E1	**T**	PP, EP			0	
2793	FERROUS METAL BORINGS, SHAVINGS, TURNINGS or CUTTINGS in a form liable to self-heating	4.2	S4	III	4.2	592	LQ0	E1	**B**	PP		LO02	0	LO02 applies only when this substance is carried in bulk or without packaging
2794	BATTERIES, WET, FILLED WITH ACID, electric storage	8	C11		8	295 598	LQ0	E0		PP, EP			0	
2795	BATTERIES, WET, FILLED WITH ALKALI, electric storage	8	C11		8	295 598	LQ0	E0		PP, EP			0	
2796	SULPHURIC ACID with not more than 51% acid or BATTERY FLUID, ACID	8	C1	II	8		LQ22	E2	**T**	PP, EP			0	
2797	BATTERY FLUID, ALKALI	8	C5	II	8		LQ22	E2	**T**	PP, EP			0	

UN No. or ID No.	Name and description	Class	Classi-fication Code	Packing group	Labels	Special provi-sions	Limited and excepted quantities		Carriage permitted	Equipment required	Venti-lation	Provisions concerning loading, unloading and carriage	Number of blue cones/ lights	Remarks
	3.1.2	2.2	2.2	2.1.1.3	5.2.2	3.3	3.4.6	3.5.1.2	3.2.1	8.1.5	7.1.6	7.1.6	7.1.5	3.2.1
(1)	(2)	(3a)	(3b)	(4)	(5)	(6)	(7a)	(7b)	(8)	(9)	(10)	(11)	(12)	(13)
2798	PHENYLPHOSPHORUS DICHLORIDE	8	C3	II	8		LQ22	E2		PP, EP			0	
2799	PHENYLPHOSPHORUS THIODICHLORIDE	8	C3	II	8		LQ22	E2		PP, EP			0	
2800	BATTERIES, WET, NON-SPILLABLE, electric storage	8	C11		8	238 295 598	LQ0	E0		PP, EP			0	
2801	DYE, LIQUID, CORROSIVE, N.O.S. or DYE INTERMEDIATE, LIQUID, CORROSIVE, N.O.S.	8	C9	I	8	274	LQ0	E0		PP, EP			0	
2801	DYE, LIQUID, CORROSIVE, N.O.S. or DYE INTERMEDIATE, LIQUID, CORROSIVE, N.O.S.	8	C9	II	8	274	LQ22	E2		PP, EP			0	
2801	DYE, LIQUID, CORROSIVE, N.O.S. or DYE INTERMEDIATE, LIQUID, CORROSIVE, N.O.S.	8	C9	III	8	274	LQ7	E1		PP, EP			0	
2802	COPPER CHLORIDE	8	C2	III	8		LQ24	E1		PP, EP			0	
2803	GALLIUM	8	C10	III	8		LQ24	E0		PP, EP			0	
2805	LITHIUM HYDRIDE, FUSED SOLID	4.3	W2	II	4.3		LQ11	E2		PP, EX, A	VE01	HA08	0	
2806	LITHIUM NITRIDE	4.3	W2	I	4.3		LQ0	E0		PP, EX, A	VE01	HA08	0	
2807	Magnetized material	9	M11	NOT SUBJECT TO ADN										
2809	MERCURY	8	C9	III	8	599	LQ19	E0		PP, EP			0	
2810	TOXIC LIQUID, ORGANIC, N.O.S.	6.1	T1	I	6.1	274 315 614 802	LQ0	E5	T	PP, EP, TOX, A	VE02		2	
2810	TOXIC LIQUID, ORGANIC, N.O.S.	6.1	T1	II	6.1	274 614 802	LQ17	E4	T	PP, EP, TOX, A	VE02		2	
2810	TOXIC LIQUID, ORGANIC, N.O.S.	6.1	T1	III	6.1	274 614 802	LQ7	E1	T	PP, EP, TOX, A	VE02		0	
2811	TOXIC SOLID, ORGANIC, N.O.S.	6.1	T2	I	6.1	274 614 802	LQ0	E5		PP, EP			2	
2811	TOXIC SOLID, ORGANIC, N.O.S.	6.1	T2	II	6.1	274 614 802	LQ18	E4		PP, EP			2	
2811	TOXIC SOLID, ORGANIC, N.O.S.	6.1	T2	III	6.1	274 614 802	LQ9	E1	T	PP, EP			0	
2812	Sodium aluminate, solid	8	C6	NOT SUBJECT TO ADN										
2813	WATER-REACTIVE SOLID, N.O.S.	4.3	W2	I	4.3	274	LQ0	E0		PP, EX, A	VE01	HA08	0	
2813	WATER-REACTIVE SOLID, N.O.S.	4.3	W2	II	4.3	274	LQ11	E2		PP, EX, A	VE01	HA08	0	
2813	WATER-REACTIVE SOLID, N.O.S.	4.3	W2	III	4.3	274	LQ12	E1		PP, EX, A	VE01	HA08	0	
2814	INFECTIOUS SUBSTANCE, AFFECTING HUMANS	6.2	I1		6.2	318 802	LQ0	E0		PP			0	

UN No. or ID No.	Name and description	Class	Classi-fication Code	Packing group	Labels	Special provi-sions	Limited and excepted quantities		Carriage permitted	Equipment required	Venti-lation	Provisions concerning loading, unloading and carriage	Number of blue cones/ lights	Remarks
	3.1.2	2.2	2.2	2.1.1.3	5.2.2	3.3	3.4.6	3.5.1.2	3.2.1	8.1.5	7.1.6	7.1.6	7.1.5	3.2.1
(1)	(2)	(3a)	(3b)	(4)	(5)	(6)	(7a)	(7b)	(8)	(9)	(10)	(11)	(12)	(13)
2814	INFECTIOUS SUBSTANCE, AFFECTING HUMANS, in refrigerated liquid nitrogen	6.2	I1		6.2 +2.2	318 802	LQ0	E0		PP			0	
2814	INFECTIOUS SUBSTANCE, AFFECTING HUMANS (animal material only)	6.2	I1		6.2	318 802	LQ0	E0		PP			0	
2815	N-AMINOETHYL-PIPERAZINE	8	C7	III	8		LQ7	E1	T	PP, EP			0	
2817	AMMONIUM HYDROGENDIFLUORIDE SOLUTION	8	CT1	II	8+6.1	802	LQ22	E2		PP, EP			2	
2817	AMMONIUM HYDROGENDIFLUORIDE SOLUTION	8	CT1	III	8+6.1	802	LQ7	E1		PP, EP			0	
2818	AMMONIUM POLYSULPHIDE SOLUTION	8	CT1	II	8+6.1	802	LQ22	E2		PP, EP			2	
2818	AMMONIUM POLYSULPHIDE SOLUTION	8	CT1	III	8+6.1	802	LQ7	E1		PP, EP			0	
2819	AMYL ACID PHOSPHATE	8	C3	III	8		LQ7	E1		PP, EP			0	
2820	BUTYRIC ACID	8	C3	III	8		LQ7	E1	T	PP, EP			0	
2821	PHENOL SOLUTION	6.1	T1	II	6.1	802	LQ17	E4		PP, EP, TOX, A	VE02		2	
2821	PHENOL SOLUTION	6.1	T1	III	6.1	802	LQ7	E1		PP, EP, TOX, A	VE02		0	
2822	2-CHLOROPYRIDINE	6.1	T1	II	6.1	802	LQ17	E4		PP, EP, TOX, A	VE02		2	
2823	CROTONIC ACID, SOLID	8	C4	III	8		LQ24	E1		PP, EP			0	
2826	ETHYL CHLOROTHIOFORMATE	8	CF1	II	8+3		LQ22	E2		PP, EP, EX, A	VE01		1	
2829	CAPROIC ACID	8	C3	III	8		LQ7	E1	T	PP, EP			0	
2830	LITHIUM FERROSILICON	4.3	W2	II	4.3		LQ11	E2		PP, EX, A	VE01	HA08	0	
2831	1,1,1-TRICHLOROETHANE	6.1	T1	III	6.1	802	LQ7	E1	T	PP, EP, TOX, A	VE02		0	
2834	PHOSPHOROUS ACID	8	C2	III	8		LQ24	E1		PP, EP			0	
2835	SODIUM ALUMINIUM HYDRIDE	4.3	W2	II	4.3		LQ11	E2		PP, EX, A	VE01	HA08	0	
2837	BISULPHATES, AQUEOUS SOLUTION	8	C1	II	8	274	LQ22	E2		PP, EP			0	
2837	BISULPHATES, AQUEOUS SOLUTION	8	C1	III	8	274	LQ7	E1		PP, EP			0	
2838	VINYL BUTYRATE, STABILIZED	3	F1	II	3		LQ4	E2		PP, EX, A	VE01		1	
2839	ALDOL	6.1	T1	II	6.1	802	LQ17	E4		PP, EP, TOX, A	VE02		2	
2840	BUTYRALDOXIME	3	F1	III	3		LQ7	E1		PP, EX, A	VE01		0	
2841	DI-n-AMYLAMINE	3	FT1	III	3+6.1	802	LQ7	E1		PP, EP, EX, TOX, A	VE01, VE02		2	
2842	NITROETHANE	3	F1	III	3		LQ7	E1		PP, EX, A	VE01		0	
2844	CALCIUM MANGANESE SILICON	4.3	W2	III	4.3		LQ12	E1		PP, EX, A	VE01	HA08	0	
2845	PYROPHORIC LIQUID, ORGANIC, N.O.S.	4.2	S1	I	4.2	274	LQ0	E0		PP			0	
2846	PYROPHORIC SOLID, ORGANIC, N.O.S.	4.2	S2	I	4.2	274	LQ0	E0		PP			0	
2849	3-CHLOROPROPANOL-1	6.1	T1	III	6.1	802	LQ7	E1		PP, EP, TOX, A	VE02		0	
2850	PROPYLENE TETRAMER	3	F1	III	3		LQ7	E1	T	PP, EX, A	VE01		0	
2851	BORON TRIFLUORIDE DIHYDRATE	8	C1	II	8		LQ22	E2		PP, EP			0	
2852	DIPICRYL SULPHIDE, WETTED with not less than 10% water, by mass	4.1	D	I	4.1	545	LQ0	E0		PP			1	
2853	MAGNESIUM FLUOROSILICATE	6.1	T5	III	6.1	802	LQ9	E1		PP, EP			0	

UN No. or ID No.	Name and description	Class	Classi-fication Code	Packing group	Labels	Special provi-sions	Limited and excepted quantities		Carriage permitted	Equipment required	Venti-lation	Provisions concerning loading, unloading and carriage	Number of blue cones/ lights	Remarks
	3.1.2	2.2	2.2	2.1.1.3	5.2.2	3.3	3.4.6	3.5.1.2	3.2.1	8.1.5	7.1.6	7.1.6	7.1.5	3.2.1
(1)	(2)	(3a)	(3b)	(4)	(5)	(6)	(7a)	(7b)	(8)	(9)	(10)	(11)	(12)	(13)
2854	AMMONIUM FLUOROSILICATE	6.1	T5	III	6.1	802	LQ9	E1		PP, EP			0	
2855	ZINC FLUOROSILICATE	6.1	T5	III	6.1	802	LQ9	E1		PP, EP			0	
2856	FLUOROSILICATES, N.O.S.	6.1	T5	III	6.1	274 802	LQ9	E1		PP, EP			0	
2857	REFRIGERATING MACHINES containing non-flammable, non-toxic gases or ammonia solutions (UN 2672)	2	6A		2.2	119	LQ0	E0		PP			0	
2858	ZIRCONIUM, DRY, coiled wire, finished metal sheets, strip (thinner than 254 microns but not thinner than 18 microns)	4.1	F3	III	4.1	546	LQ9	E1		PP			0	
2859	AMMONIUM METAVANADATE	6.1	T5	II	6.1	802	LQ18	E4		PP, EP			2	
2861	AMMONIUM POLYVANADATE	6.1	T5	II	6.1	802	LQ18	E4		PP, EP			2	
2862	VANADIUM PENTOXIDE, non-fused form	6.1	T5	III	6.1	600 802	LQ9	E1		PP, EP			0	
2863	SODIUM AMMONIUM VANADATE	6.1	T5	II	6.1	802	LQ18	E4		PP, EP			2	
2864	POTASSIUM METAVANADATE	6.1	T5	II	6.1	802	LQ18	E4		PP, EP			2	
2865	HYDROXYLAMINE SULPHATE	8	C2	III	8		LQ24	E1		PP, EP			0	
2869	TITANIUM TRICHLORIDE MIXTURE	8	C2	II	8		LQ23	E2		PP, EP			0	
2869	TITANIUM TRICHLORIDE MIXTURE	8	C2	III	8		LQ24	E1		PP, EP			0	
2870	ALUMINIUM BOROHYDRIDE	4.2	SW	I	4.2+4.3		LQ0	E0		PP, EX, A	VE01		0	
2870	ALUMINIUM BOROHYDRIDE IN DEVICES	4.2	SW	I	4.2+4.3		LQ0	E0		PP, EX, A	VE01		0	
2871	ANTIMONY POWDER	6.1	T5	III	6.1	802	LQ9	E1		PP, EP			0	
2872	DIBROMOCHLOROPROPANES	6.1	T1	II	6.1	802	LQ17	E4		PP, EP, TOX, A	VE02		2	
2872	DIBROMOCHLOROPROPANES	6.1	T1	III	6.1	802	LQ7	E1		PP, EP, TOX, A	VE02		0	
2873	DIBUTYLAMINOETHANOL	6.1	T1	III	6.1	802	LQ7	E1		PP, EP, TOX, A	VE02		0	
2874	FURFURYL ALCOHOL	6.1	T1	III	6.1	802	LQ7	E1	T	PP, EP, TOX, A	VE02		0	
2875	HEXACHLOROPHENE	6.1	T2	III	6.1	802	LQ9	E1		PP, EP			0	
2876	RESORCINOL	6.1	T2	III	6.1	802	LQ9	E1		PP, EP			0	
2878	TITANIUM SPONGE GRANULES or TITANIUM SPONGE POWDERS	4.1	F3	III	4.1		LQ9	E1		PP			0	
2879	SELENIUM OXYCHLORIDE	8	CT1	I	8+6.1	802	LQ0	E0		PP, EP, TOX, A	VE02		2	
2880	CALCIUM HYPOCHLORITE, HYDRATED, or CALCIUM HYPOCHLORITE, HYDRATED MIXTURE, with not less than 5.5% but not more than 16% water	5.1	O2	II	5.1	313 314 322	LQ11	E2		PP			0	
2880	CALCIUM HYPOCHLORITE, HYDRATED or CALCIUM HYPOCHLORITE HYDRATED MIXTURE, with not less than 5.5% but not more than 16% water	5.1	O2	III	5.1	313 314	LQ12	E1		PP			0	
2881	METAL CATALYST, DRY	4.2	S4	I	4.2	274	LQ0	E0		PP			0	
2881	METAL CATALYST, DRY	4.2	S4	II	4.2	274	LQ0	E2		PP			0	

UN No. or ID No.	Name and description	Class	Classi-fication Code	Packing group	Labels	Special provi-sions	Limited and excepted quantities		Carriage permitted	Equipment required	Venti-lation	Provisions concerning loading, unloading and carriage	Number of blue cones/ lights	Remarks
	3.1.2	2.2	2.2	2.1.1.3	5.2.2	3.3	3.4.6	3.5.1.2	3.2.1	8.1.5	7.1.6	7.1.6	7.1.5	3.2.1
(1)	(2)	(3a)	(3b)	(4)	(5)	(6)	(7a)	(7b)	(8)	(9)	(10)	(11)	(12)	(13)
2881	METAL CATALYST, DRY	4.2	S4	III	4.2	274	LQ0	E1		PP			0	
2900	INFECTIOUS SUBSTANCE, AFFECTING ANIMALS only	6.2	I2		6.2	318 802	LQ0	E0		PP			0	
2900	INFECTIOUS SUBSTANCE, AFFECTING ANIMALS only, in refrigerated liquid nitrogen	6.2	I2		6.2 +2.2	318 802	LQ0	E0		PP			0	
2900	INFECTIOUS SUBSTANCE, AFFECTING ANIMALS only (animal material only)	6.2	I2		6.2	318 802	LQ0	E0		PP			0	
2901	BROMINE CHLORIDE	2	2TOC		2.3+5.1+8		LQ0	E0		PP, EP, TOX, A	VE02		2	
2902	PESTICIDE, LIQUID, TOXIC, N.O.S.	6.1	T6	I	6.1	61 274 648 802	LQ0	E5		PP, EP, TOX, A	VE02		2	
2902	PESTICIDE, LIQUID, TOXIC, N.O.S.	6.1	T6	II	6.1	61 274 648 802	LQ17	E4		PP, EP, TOX, A	VE02		2	
2902	PESTICIDE, LIQUID, TOXIC, N.O.S.	6.1	T6	III	6.1	61 274 648 802	LQ7	E1		PP, EP, TOX, A	VE02		0	
2903	PESTICIDE, LIQUID, TOXIC, FLAMMABLE, N.O.S., flash-point not less than 23 °C	6.1	TF2	I	6.1+3	61 274 802	LQ0	E5		PP, EP, EX, TOX, A	VE01, VE02		2	
2903	PESTICIDE, LIQUID, TOXIC, FLAMMABLE, N.O.S., flash-point not less than 23 °C	6.1	TF2	II	6.1+3	61 274 802	LQ17	E4		PP, EP, EX, TOX, A	VE01, VE02		2	
2903	PESTICIDE, LIQUID, TOXIC, FLAMMABLE, N.O.S., flash-point not less than 23 °C	6.1	TF2	III	6.1+3	61 274 802	LQ7	E1		PP, EP, EX, TOX, A	VE01, VE02		0	
2904	CHLOROPHENOLATES, LIQUID or PHENOLATES, LIQUID	8	C9	III	8		LQ7	E1	**T** *	PP, EP			0	* applies only to phenolates but not to chlorophenolates
2905	CHLOROPHENOLATES, SOLID or PHENOLATES, SOLID	8	C10	III	8		LQ24	E1		PP, EP			0	
2907	ISOSORBIDE DINITRATE MIXTURE with not less than 60% lactose, mannose, starch or calcium hydrogen phosphate	4.1	D	II	4.1	127	LQ8	E0		PP			0	
2908	RADIOACTIVE MATERIAL, EXCEPTED PACKAGE - EMPTY PACKAGING	7				290	LQ0	E0		PP			0	
2909	RADIOACTIVE MATERIAL, EXCEPTED PACKAGE - ARTICLES MANUFACTURED FROM NATURAL URANIUM or DEPLETED URANIUM or NATURAL THORIUM	7				290	LQ0	E0		PP			0	

UN No. or ID No.	Name and description	Class	Classi-fication Code	Packing group	Labels	Special provi-sions	Limited and excepted quantities		Carriage permitted	Equipment required	Venti-lation	Provisions concerning loading, unloading and carriage	Number of blue cones/ lights	Remarks
	3.1.2	2.2	2.2	2.1.1.3	5.2.2	3.3	3.4.6	3.5.1.2	3.2.1	8.1.5	7.1.6	7.1.6	7.1.5	3.2.1
(1)	(2)	(3a)	(3b)	(4)	(5)	(6)	(7a)	(7b)	(8)	(9)	(10)	(11)	(12)	(13)
2910	RADIOACTIVE MATERIAL, EXCEPTED PACKAGE - LIMITED QUANTITY OF MATERIAL	7				290	LQ0	E0		PP			0	
2911	RADIOACTIVE MATERIAL, EXCEPTED PACKAGE - INSTRUMENTS or ARTICLES	7				290	LQ0	E0		PP			0	
2912	RADIOACTIVE MATERIAL, LOW SPECIFIC ACTIVITY (LSA-I), non fissile or fissile-excepted	7			7X	172 317 325	LQ0	E0	**B**	PP		RA01	2	
2913	RADIOACTIVE MATERIAL, SURFACE CONTAMINATED OBJECTS (SCO-I or SCO-II), non fissile or fissile-excepted	7			7X	172 317 336	LQ0	E0	**B**	PP		RA02, RA03	2	
2915	RADIOACTIVE MATERIAL, TYPE A PACKAGE, non-special form, non fissile or fissile-excepted	7			7X	172 317 325	LQ0	E0		PP			2	
2916	RADIOACTIVE MATERIAL, TYPE B(U) PACKAGE, non fissile or fissile-excepted	7			7X	172 317 337	LQ0	E0		PP			2	
2917	RADIOACTIVE MATERIAL, TYPE B(M) PACKAGE, non fissile or fissile-excepted	7			7X	172 317 337	LQ0	E0		PP			2	
2919	RADIOACTIVE MATERIAL, TRANSPORTED UNDER SPECIAL ARRANGEMENT, non fissile or fissile-excepted	7			7X	172 317	LQ0	E0		PP			2	
2920	CORROSIVE LIQUID, FLAMMABLE, N.O.S.	8	CF1	I	8+3	274	LQ0	E0		PP, EP, EX, A	VE01		1	
2920	CORROSIVE LIQUID, FLAMMABLE, N.O.S.	8	CF1	II	8+3	274	LQ22	E2	**T**	PP, EP, EX, A	VE01		1	
2921	CORROSIVE SOLID, FLAMMABLE, N.O.S.	8	CF2	I	8+4.1	274	LQ0	E0		PP, EP			1	
2921	CORROSIVE SOLID, FLAMMABLE, N.O.S.	8	CF2	II	8+4.1	274	LQ23	E2		PP, EP			1	
2922	CORROSIVE LIQUID, TOXIC, N.O.S.	8	CT1	I	8+6.1	274 802	LQ0	E0	**T**	PP, EP, TOX, A	VE02		2	
2922	CORROSIVE LIQUID, TOXIC, N.O.S.	8	CT1	II	8+6.1	274 802	LQ22	E2	**T**	PP, EP, TOX, A	VE02		2	
2922	CORROSIVE LIQUID, TOXIC, N.O.S.	8	CT1	III	8+6.1	274 802	LQ7	E1	**T**	PP, EP, TOX, A	VE02		0	
2923	CORROSIVE SOLID, TOXIC, N.O.S.	8	CT2	I	8+6.1	274 802	LQ0	E0		PP, EP			2	
2923	CORROSIVE SOLID, TOXIC, N.O.S.	8	CT2	II	8+6.1	274 802	LQ23	E2		PP, EP			2	
2923	CORROSIVE SOLID, TOXIC, N.O.S.	8	CT2	III	8+6.1	274 802	LQ24	E1		PP, EP			0	
2924	FLAMMABLE LIQUID, CORROSIVE, N.O.S.	3	FC	I	3+8	274	LQ3	E0	**T**	PP, EP, EX, A	VE01		1	
2924	FLAMMABLE LIQUID, CORROSIVE, N.O.S.	3	FC	II	3+8	274	LQ4	E2	**T**	PP, EP, EX, A	VE01		1	
2924	FLAMMABLE LIQUID, CORROSIVE, N.O.S.	3	FC	III	3+8	274	LQ7	E1	**T**	PP, EP, EX, A	VE01		0	
2925	FLAMMABLE SOLID, CORROSIVE, ORGANIC, N.O.S.	4.1	FC1	II	4.1+8	274	LQ0	E2		PP			1	

UN No. or ID No.	Name and description	Class	Classi-fication Code	Packing group	Labels	Special provi-sions	Limited and excepted quantities		Carriage permitted	Equipment required	Venti-lation	Provisions concerning loading, unloading and carriage	Number of blue cones/ lights	Remarks
	3.1.2	2.2	2.2	2.1.1.3	5.2.2	3.3	3.4.6	3.5.1.2	3.2.1	8.1.5	7.1.6	7.1.6	7.1.5	3.2.1
(1)	(2)	(3a)	(3b)	(4)	(5)	(6)	(7a)	(7b)	(8)	(9)	(10)	(11)	(12)	(13)
2925	FLAMMABLE SOLID, CORROSIVE, ORGANIC, N.O.S.	4.1	FC1	III	4.1+8	274	LQ0	E1		PP			0	
2926	FLAMMABLE SOLID, TOXIC, ORGANIC, N.O.S.	4.1	FT1	II	4.1+6.1	274 802	LQ0	E2		PP			2	
2926	FLAMMABLE SOLID, TOXIC, ORGANIC, N.O.S.	4.1	FT1	III	4.1+6.1	274 802	LQ0	E1		PP			0	
2927	TOXIC LIQUID, CORROSIVE, ORGANIC, N.O.S.	6.1	TC1	I	6.1+8	274 315 802	LQ0	E5	T	PP, EP, TOX, A	VE02		2	
2927	TOXIC LIQUID, CORROSIVE, ORGANIC, N.O.S.	6.1	TC1	II	6.1+8	274 802	LQ17	E4	T	PP, EP, TOX, A	VE02		2	
2928	TOXIC SOLID, CORROSIVE, ORGANIC, N.O.S.	6.1	TC2	I	6.1+8	274 802	LQ0	E5		PP, EP			2	
2928	TOXIC SOLID, CORROSIVE, ORGANIC, N.O.S.	6.1	TC2	II	6.1+8	274 802	LQ18	E4		PP, EP			2	
2929	TOXIC LIQUID, FLAMMABLE, ORGANIC, N.O.S.	6.1	TF1	I	6.1+3	274 315 802	LQ0	E5	T	PP, EP, EX, TOX, A	VE01, VE02		2	
2929	TOXIC LIQUID, FLAMMABLE, ORGANIC, N.O.S.	6.1	TF1	II	6.1+3	274 802	LQ17	E4	T	PP, EP, EX, TOX, A	VE01, VE02		2	
2930	TOXIC SOLID, FLAMMABLE, ORGANIC, N.O.S.	6.1	TF3	I	6.1+4.1	274 802	LQ0	E5		PP, EP			2	
2930	TOXIC SOLID, FLAMMABLE, ORGANIC, N.O.S.	6.1	TF3	II	6.1+4.1	274 802	LQ18	E4		PP, EP			2	
2931	VANADYL SULPHATE	6.1	T5	II	6.1	802	LQ18	E4		PP, EP			2	
2933	METHYL 2-CHLOROPROPIONATE	3	F1	III	3		LQ7	E1		PP, EX, A	VE01		0	
2934	ISOPROPYL 2-CHLOROPROPIONATE	3	F1	III	3		LQ7	E1		PP, EX, A	VE01		0	
2935	ETHYL 2-CHLOROPROPIONATE	3	F1	III	3		LQ7	E1	T	PP, EX, A	VE01		0	
2936	THIOLACTIC ACID	6.1	T1	II	6.1	802	LQ17	E4		PP, EP, TOX, A	VE02		2	
2937	alpha-METHYLBENZYL ALCOHOL, LIQUID	6.1	T1	III	6.1	802	LQ7	E1		PP, EP, TOX, A	VE02		0	
2940	9-PHOSPHABICYCLONONANES (CYCLOOCTADIENE PHOSPHINES)	4.2	S2	II	4.2		LQ0	E2		PP			0	
2941	FLUOROANILINES	6.1	T1	III	6.1	802	LQ7	E1		PP, EP, TOX, A	VE02		0	
2942	2-TRIFLUOROMETHYLANILINE	6.1	T1	III	6.1	802	LQ7	E1		PP, EP, TOX, A	VE02		0	
2943	TETRAHYDROFURFURYLAMINE	3	F1	III	3		LQ7	E1		PP, EX, A	VE01		0	
2945	N-METHYLBUTYLAMINE	3	FC	II	3+8		LQ4	E2		PP, EP, EX, A	VE01		1	
2946	2-AMINO-5-DIETHYLAMINOPENTANE	6.1	T1	III	6.1	802	LQ7	E1		PP, EP, TOX, A	VE02		0	
2947	ISOPROPYL CHLOROACETATE	3	F1	III	3		LQ7	E1	T	PP, EX, A	VE01		0	
2948	3-TRIFLUOROMETHYLANILINE	6.1	T1	II	6.1	802	LQ17	E4		PP, EP, TOX, A	VE02		2	

UN No. or ID No.	Name and description 3.1.2	Class 2.2	Classi-fication Code 2.2	Packing group 2.1.1.3	Labels 5.2.2	Special provi-sions 3.3	Limited and excepted quantities 3.4.6	3.5.1.2	Carriage permitted 3.2.1	Equipment required 8.1.5	Venti-lation 7.1.6	Provisions concerning loading, unloading and carriage 7.1.6			Number of blue cones/ lights 7.1.5	Remarks 3.2.1
(1)	(2)	(3a)	(3b)	(4)	(5)	(6)	(7a)	(7b)	(8)	(9)	(10)	(11)			(12)	(13)
2949	SODIUM HYDROSULPHIDE , HYDRATED with not less than 25% water of crystallization	8	C6	II	8	523	LQ23	E2		PP, EP					0	
2950	MAGNESIUM GRANULES, COATED, particle size not less than 149 microns	4.3	W2	III	4.3		LQ12	E1		PP, EX, A	VE01		HA08		0	
2956	5-tert-BUTYL-2,4,6-TRINITRO-m-XYLENE (MUSK XYLENE)	4.1	SR1	III	4.1	638	LQ0	E1		PP					0	
2965	BORON TRIFLUORIDE DIMETHYL ETHERATE	4.3	WFC	I	4.3+3+8		LQ0	E0		PP, EP, EX, A	VE01		HA08		1	
2966	THIOGLYCOL	6.1	T1	II	6.1	802	LQ17	E4	**T**	PP, EP, TOX, A	VE02				2	
2967	SULPHAMIC ACID	8	C2	III	8		LQ24	E1		PP, EP					0	
2968	MANEB, STABILIZED or MANEB PREPARATION, STABILIZED against self-heating	4.3	W2	III	4.3	547	LQ12	E1		PP, EX, A	VE01		HA08		0	
2969	CASTOR BEANS or CASTOR MEAL or CASTOR POMACE or CASTOR FLAKE	9	M11	II	9	141	LQ25	E2	**B**	PP					0	
2977	RADIOACTIVE MATERIAL, URANIUM HEXAFLUORIDE, FISSILE	7			7X+7E+8	172	LQ0	E0		PP					2	
2978	RADIOACTIVE MATERIAL, URANIUM HEXAFLUORIDE, non fissile or fissile-excepted	7			7X+8	172 317	LQ0	E0	**B**	PP				RA01	2	
2983	ETHYLENE OXIDE AND PROPYLENE OXIDE MIXTURE, not more than 30% ethylene	3	FT1	I	3+6.1	802	LQ0	E0	**T**	PP, EP, EX, TOX, A	VE01, VE02				2	
2984	HYDROGEN PEROXIDE, AQUEOUS SOLUTION with not less than 8% but less than 20% hydrogen peroxide (stabilized as necessary)	5.1	O1	III	5.1	65	LQ13	E1		PP					0	
2985	CHLOROSILANES, FLAMMABLE, CORROSIVE, N.O.S.	3	FC	II	3+8	274 548	LQ4	E2		PP, EP, EX, A	VE01				1	
2986	CHLOROSILANES, CORROSIVE, FLAMMABLE, N.O.S.	8	CF1	II	8+3	274 548	LQ22	E2		PP, EP, EX, A	VE01				1	
2987	CHLOROSILANES, CORROSIVE, N.O.S.	8	C3	II	8	274 548	LQ22	E2		PP, EP					0	
2988	CHLOROSILANES, WATER-REACTIVE, FLAMMABLE, CORROSIVE, N.O.S.	4.3	WFC	I	4.3+3+8	274 549	LQ0	E0		PP, EP, EX, A	VE01		HA08		1	
2989	LEAD PHOSPHITE, DIBASIC	4.1	F3	II	4.1		LQ8	E2		PP					1	
2989	LEAD PHOSPHITE, DIBASIC	4.1	F3	III	4.1		LQ9	E1		PP					0	
2990	LIFE-SAVING APPLIANCES, SELF-INFLATING	9	M5		9	296 635	LQ0	E0		PP					0	
2991	CARBAMATE PESTICIDE, LIQUID, TOXIC, FLAMMABLE, flash-point not less than 23 °C	6.1	TF2	I	6.1+3	61 274 802	LQ0	E5		PP, EP, EX, TOX, A	VE01, VE02				2	
2991	CARBAMATE PESTICIDE, LIQUID, TOXIC, FLAMMABLE, flash-point not less than 23 °C	6.1	TF2	II	6.1+3	61 274 802	LQ17	E4		PP, EP, EX, TOX, A	VE01, VE02				2	

UN No. or ID No.	Name and description	Class	Classi-fication Code	Packing group	Labels	Special provi-sions	Limited and excepted quantities		Carriage permitted	Equipment required	Venti-lation	Provisions concerning loading, unloading and carriage	Number of blue cones/ lights	Remarks
	3.1.2	2.2	2.2	2.1.1.3	5.2.2	3.3	3.4.6	3.5.1.2	3.2.1	8.1.5	7.1.6	7.1.6	7.1.5	3.2.1
(1)	(2)	(3a)	(3b)	(4)	(5)	(6)	(7a)	(7b)	(8)	(9)	(10)	(11)	(12)	(13)
2991	CARBAMATE PESTICIDE, LIQUID, TOXIC, FLAMMABLE, flash-point not less than 23 °C	6.1	TF2	III	6.1+3	61 274 802	LQ7	E1		PP, EP, EX, TOX, A	VE01, VE02		0	
2992	CARBAMATE PESTICIDE, LIQUID, TOXIC	6.1	T6	I	6.1	61 274 648 802	LQ0	E5		PP, EP, TOX, A	VE02		2	
2992	CARBAMATE PESTICIDE, LIQUID, TOXIC	6.1	T6	II	6.1	61 274 648 802	LQ17	E4		PP, EP, TOX, A	VE02		2	
2992	CARBAMATE PESTICIDE, LIQUID, TOXIC	6.1	T6	III	6.1	61 274 648 802	LQ7	E1		PP, EP, TOX, A	VE02		0	
2993	ARSENICAL PESTICIDE, LIQUID, TOXIC, FLAMMABLE, flash-point not less than 23 °C	6.1	TF2	I	6.1+3	61 274 802	LQ0	E5		PP, EP, EX, TOX, A	VE01, VE02		2	
2993	ARSENICAL PESTICIDE, LIQUID, TOXIC, FLAMMABLE, flash-point not less than 23 °C	6.1	TF2	II	6.1+3	61 274 802	LQ17	E4		PP, EP, EX, TOX, A	VE01, VE02		2	
2993	ARSENICAL PESTICIDE, LIQUID, TOXIC, FLAMMABLE, flash-point not less than 23 °C	6.1	TF2	III	6.1+3	61 274 802	LQ7	E1		PP, EP, EX, TOX, A	VE01, VE02		0	
2994	ARSENICAL PESTICIDE, LIQUID, TOXIC	6.1	T6	I	6.1	61 274 648 802	LQ0	E5		PP, EP, TOX, A	VE02		2	
2994	ARSENICAL PESTICIDE, LIQUID, TOXIC	6.1	T6	II	6.1	61 274 648 802	LQ17	E4		PP, EP, TOX, A	VE02		2	
2994	ARSENICAL PESTICIDE, LIQUID, TOXIC	6.1	T6	III	6.1	61 274 648 802	LQ7	E1		PP, EP, TOX, A	VE02		0	
2995	ORGANOCHLORINE PESTICIDE, LIQUID, TOXIC, FLAMMABLE, flash-point not less than 23 °C	6.1	TF2	I	6.1+3	61 274 802	LQ0	E5		PP, EP, EX, TOX, A	VE01, VE02		2	
2995	ORGANOCHLORINE PESTICIDE, LIQUID, TOXIC, FLAMMABLE, flash-point not less than 23 °C	6.1	TF2	II	6.1+3	61 274 802	LQ17	E4		PP, EP, EX, TOX, A	VE01, VE02		2	
2995	ORGANOCHLORINE PESTICIDE, LIQUID, TOXIC, FLAMMABLE, flash-point not less than 23 °C	6.1	TF2	III	6.1+3	61 274 802	LQ7	E1		PP, EP, EX, TOX, A	VE01, VE02		0	

UN No. or ID No.	Name and description	Class	Classi-fication Code	Packing group	Labels	Special provi-sions	Limited and excepted quantities		Carriage permitted	Equipment required	Venti-lation	Provisions concerning loading, unloading and carriage			Number of blue cones/ lights	Remarks
	3.1.2	2.2	2.2	2.1.1.3	5.2.2	3.3	3.4.6	3.5.1.2	3.2.1	8.1.5	7.1.6	7.1.6			7.1.5	3.2.1
(1)	(2)	(3a)	(3b)	(4)	(5)	(6)	(7a)	(7b)	(8)	(9)	(10)	(11)			(12)	(13)
2996	ORGANOCHLORINE PESTICIDE, LIQUID, TOXIC	6.1	T6	I	6.1	61 274 648 802	LQ0	E5		PP, EP, TOX, A	VE02				2	
2996	ORGANOCHLORINE PESTICIDE, LIQUID, TOXIC	6.1	T6	II	6.1	61 274 648 802	LQ17	E4		PP, EP, TOX, A	VE02				2	
2996	ORGANOCHLORINE PESTICIDE, LIQUID, TOXIC	6.1	T6	III	6.1	61 274 648 802	LQ7	E1		PP, EP, TOX, A	VE02				0	
2997	TRIAZINE PESTICIDE, LIQUID, TOXIC, FLAMMABLE, flash-point not less than 23 °C	6.1	TF2	I	6.1+3	61 274 802	LQ0	E5		PP, EP, EX, TOX, A	VE01, VE02				2	
2997	TRIAZINE PESTICIDE, LIQUID, TOXIC, FLAMMABLE, flash-point not less than 23 °C	6.1	TF2	II	6.1+3	61 274 802	LQ17	E4		PP, EP, EX, TOX, A	VE01, VE02				2	
2997	TRIAZINE PESTICIDE, LIQUID, TOXIC, FLAMMABLE, flash-point not less than 23 °C	6.1	TF2	III	6.1+3	61 274 802	LQ7	E1		PP, EP, EX, TOX, A	VE01, VE02				0	
2998	TRIAZINE PESTICIDE, LIQUID, TOXIC	6.1	T6	I	6.1	61 274 648 802	LQ0	E5		PP, EP, TOX, A	VE02				2	
2998	TRIAZINE PESTICIDE, LIQUID, TOXIC	6.1	T6	II	6.1	61 274 648 802	LQ17	E4		PP, EP, TOX, A	VE02				2	
2998	TRIAZINE PESTICIDE, LIQUID, TOXIC	6.1	T6	III	6.1	61 274 648 802	LQ7	E1		PP, EP, TOX, A	VE02				0	
3005	THIOCARBAMATE PESTICIDE, LIQUID, TOXIC, FLAMMABLE, flash-point not less than 23 °C	6.1	TF2	I	6.1+3	61 274 802	LQ0	E5		PP, EP, EX, TOX, A	VE01, VE02				2	
3005	THIOCARBAMATE PESTICIDE, LIQUID, TOXIC, FLAMMABLE, flash-point not less than 23 °C	6.1	TF2	II	6.1+3	61 274 802	LQ17	E4		PP, EP, EX, TOX, A	VE01, VE02				2	
3005	THIOCARBAMATE PESTICIDE, LIQUID, TOXIC, FLAMMABLE, flash-point not less than 23 °C	6.1	TF2	III	6.1+3	61 274 802	LQ7	E1		PP, EP, EX, TOX, A	VE01, VE02				0	
3006	THIOCARBAMATE PESTICIDE, LIQUID, TOXIC	6.1	T6	I	6.1	61 274 648 802	LQ0	E5		PP, EP, TOX, A	VE02				2	

UN No. or ID No.	Name and description	Class	Classi-fication Code	Packing group	Labels	Special provi-sions	Limited and excepted quantities		Carriage permitted	Equipment required	Venti-lation	Provisions concerning loading, unloading and carriage			Number of blue cones/ lights	Remarks
	3.1.2	2.2	2.2	2.1.1.3	5.2.2	3.3	3.4.6	3.5.1.2	3.2.1	8.1.5	7.1.6	7.1.6			7.1.5	3.2.1
(1)	(2)	(3a)	(3b)	(4)	(5)	(6)	(7a)	(7b)	(8)	(9)	(10)	(11)			(12)	(13)
3006	THIOCARBAMATE PESTICIDE, LIQUID, TOXIC	6.1	T6	II	6.1	61 274 648 802	LQ17	E4		PP, EP, TOX, A	VE02				2	
3006	THIOCARBAMATE PESTICIDE, LIQUID, TOXIC	6.1	T6	III	6.1	61 274 648 802	LQ7	E1		PP, EP, TOX, A	VE02				0	
3009	COPPER BASED PESTICIDE, LIQUID, TOXIC, FLAMMABLE, flash-point not less than 23 °C	6.1	TF2	I	6.1+3	61 274 802	LQ0	E5		PP, EP, EX, TOX, A	VE01, VE02				2	
3009	COPPER BASED PESTICIDE, LIQUID, TOXIC, FLAMMABLE, flash-point not less than 23 °C	6.1	TF2	II	6.1+3	61 274 802	LQ17	E4		PP, EP, EX, TOX, A	VE01, VE02				2	
3009	COPPER BASED PESTICIDE, LIQUID, TOXIC, FLAMMABLE, flash-point not less than 23 °C	6.1	TF2	III	6.1+3	61 274 802	LQ7	E1		PP, EP, EX, TOX, A	VE01, VE02				0	
3010	COPPER BASED PESTICIDE, LIQUID, TOXIC	6.1	T6	I	6.1	61 274 648 802	LQ0	E5		PP, EP, TOX, A	VE02				2	
3010	COPPER BASED PESTICIDE, LIQUID, TOXIC	6.1	T6	II	6.1	61 274 648 802	LQ17	E4		PP, EP, TOX, A	VE02				2	
3010	COPPER BASED PESTICIDE, LIQUID, TOXIC	6.1	T6	III	6.1	61 274 648 802	LQ7	E1		PP, EP, TOX, A	VE02				0	
3011	MERCURY BASED PESTICIDE, LIQUID, TOXIC, FLAMMABLE, flash-point not less than 23 °C	6.1	TF2	I	6.1+3	61 274 802	LQ0	E5		PP, EP, EX, TOX, A	VE01, VE02				2	
3011	MERCURY BASED PESTICIDE, LIQUID, TOXIC, FLAMMABLE, flash-point not less than 23 °C	6.1	TF2	II	6.1+3	61 274 802	LQ17	E4		PP, EP, EX, TOX, A	VE01, VE02				2	
3011	MERCURY BASED PESTICIDE, LIQUID, TOXIC, FLAMMABLE, flash-point not less than 23 °C	6.1	TF2	III	6.1+3	61 274 802	LQ7	E1		PP, EP, EX, TOX, A	VE01, VE02				0	
3012	MERCURY BASED PESTICIDE, LIQUID, TOXIC	6.1	T6	I	6.1	61 274 648 802	LQ0	E5		PP, EP, TOX, A	VE02				2	
3012	MERCURY BASED PESTICIDE, LIQUID, TOXIC	6.1	T6	II	6.1	61 274 648 802	LQ17	E4		PP, EP, TOX, A	VE02				2	

UN No. or ID No.	Name and description	Class	Classi-fication Code	Packing group	Labels	Special provi-sions	Limited and excepted quantities		Carriage permitted	Equipment required	Venti-lation	Provisions concerning loading, unloading and carriage			Number of blue cones/ lights	Remarks
	3.1.2	2.2	2.2	2.1.1.3	5.2.2	3.3	3.4.6	3.5.1.2	3.2.1	8.1.5	7.1.6	7.1.6			7.1.5	3.2.1
(1)	(2)	(3a)	(3b)	(4)	(5)	(6)	(7a)	(7b)	(8)	(9)	(10)	(11)			(12)	(13)
3012	MERCURY BASED PESTICIDE, LIQUID, TOXIC	6.1	T6	III	6.1	61 274 648 802	LQ7	E1		PP, EP, TOX, A	VE02				0	
3013	SUBSTITUTED NITROPHENOL PESTICIDE, LIQUID, TOXIC, FLAMMABLE, flash-point not less than 23 °C	6.1	TF2	I	6.1+3	61 274 802	LQ0	E5		PP, EP, EX, TOX, A	VE01, VE02				2	
3013	SUBSTITUTED NITROPHENOL PESTICIDE, LIQUID, TOXIC, FLAMMABLE, flash-point not less than 23 °C	6.1	TF2	II	6.1+3	61 274 802	LQ17	E4		PP, EP, EX, TOX, A	VE01, VE02				2	
3013	SUBSTITUTED NITROPHENOL PESTICIDE, LIQUID, TOXIC, FLAMMABLE, flash-point not less than 23 °C	6.1	TF2	III	6.1+3	61 274 802	LQ7	E1		PP, EP, EX, TOX, A	VE01, VE02				0	
3014	SUBSTITUTED NITROPHENOL PESTICIDE, LIQUID, TOXIC	6.1	T6	I	6.1	61 274 648 802	LQ0	E5		PP, EP, TOX, A	VE02				2	
3014	SUBSTITUTED NITROPHENOL PESTICIDE, LIQUID, TOXIC	6.1	T6	II	6.1	61 274 648 802	LQ17	E4		PP, EP, TOX, A	VE02				2	
3014	SUBSTITUTED NITROPHENOL PESTICIDE, LIQUID, TOXIC	6.1	T6	III	6.1	61 274 648 802	LQ7	E1		PP, EP, TOX, A	VE02				0	
3015	BIPYRIDILIUM PESTICIDE, LIQUID, TOXIC, FLAMMABLE, flash-point not less than 23 °C	6.1	TF2	I	6.1+3	61 274 802	LQ0	E5		PP, EP, EX, TOX, A	VE01, VE02				2	
3015	BIPYRIDILIUM PESTICIDE, LIQUID, TOXIC, FLAMMABLE, flash-point not less than 23 °C	6.1	TF2	II	6.1+3	61 274 802	LQ17	E4		PP, EP, EX, TOX, A	VE01, VE02				2	
3015	BIPYRIDILIUM PESTICIDE, LIQUID, TOXIC, FLAMMABLE, flash-point not less than 23 °C	6.1	TF2	III	6.1+3	61 274 802	LQ7	E1		PP, EP, EX, TOX, A	VE01, VE02				0	
3016	BIPYRIDILIUM PESTICIDE, LIQUID, TOXIC	6.1	T6	I	6.1	61 274 648 802	LQ0	E5		PP, EP, TOX, A	VE02				2	
3016	BIPYRIDILIUM PESTICIDE, LIQUID, TOXIC	6.1	T6	II	6.1	61 274 648 802	LQ17	E4		PP, EP, TOX, A	VE02				2	
3016	BIPYRIDILIUM PESTICIDE, LIQUID, TOXIC	6.1	T6	III	6.1	61 274 648 802	LQ7	E1		PP, EP, TOX, A	VE02				0	

UN No. or ID No.	Name and description	Class	Classi-fication Code	Packing group	Labels	Special provi-sions	Limited and excepted quantities		Carriage permitted	Equipment required	Venti-lation	Provisions concerning loading, unloading and carriage	Number of blue cones/ lights	Remarks
	3.1.2	2.2	2.2	2.1.1.3	5.2.2	3.3	3.4.6	3.5.1.2	3.2.1	8.1.5	7.1.6	7.1.6	7.1.5	3.2.1
(1)	(2)	(3a)	(3b)	(4)	(5)	(6)	(7a)	(7b)	(8)	(9)	(10)	(11)	(12)	(13)
3017	ORGANOPHOSPHORUS PESTICIDE, LIQUID, TOXIC, FLAMMABLE, flash-point not less than 23 °C	6.1	TF2	I	6.1+3	61 274 802	LQ0	E5		PP, EP, EX, TOX, A	VE01, VE02		2	
3017	ORGANOPHOSPHORUS PESTICIDE, LIQUID, TOXIC, FLAMMABLE, flash-point not less than 23 °C	6.1	TF2	II	6.1+3	61 274 802	LQ17	E4		PP, EP, EX, TOX, A	VE01, VE02		2	
3017	ORGANOPHOSPHORUS PESTICIDE, LIQUID, TOXIC, FLAMMABLE, flash-point not less than 23 °C	6.1	TF2	III	6.1+3	61 274 802	LQ7	E1		PP, EP, EX, TOX, A	VE01, VE02		0	
3018	ORGANOPHOSPHORUS PESTICIDE, LIQUID, TOXIC	6.1	T6	I	6.1	61 274 648 802	LQ0	E5		PP, EP, TOX, A	VE02		2	
3018	ORGANOPHOSPHORUS PESTICIDE, LIQUID, TOXIC	6.1	T6	II	6.1	61 274 648 802	LQ17	E4		PP, EP, TOX, A	VE02		2	
3018	ORGANOPHOSPHORUS PESTICIDE, LIQUID, TOXIC	6.1	T6	III	6.1	61 274 648 802	LQ7	E1		PP, EP, TOX, A	VE02		0	
3019	ORGANOTIN PESTICIDE, LIQUID, TOXIC, FLAMMABLE, flash-point not less than 23 °C	6.1	TF2	I	6.1+3	61 274 802	LQ0	E5		PP, EP, EX, TOX, A	VE01, VE02		2	
3019	ORGANOTIN PESTICIDE, LIQUID, TOXIC, FLAMMABLE, flash-point not less than 23 °C	6.1	TF2	II	6.1+3	61 274 802	LQ17	E4		PP, EP, EX, TOX, A	VE01, VE02		2	
3019	ORGANOTIN PESTICIDE, LIQUID, TOXIC, FLAMMABLE, flash-point not less than 23 °C	6.1	TF2	III	6.1+3	61 274 802	LQ7	E1		PP, EP, EX, TOX, A	VE01, VE02		0	
3020	ORGANOTIN PESTICIDE, LIQUID, TOXIC	6.1	T6	I	6.1	61 274 648 802	LQ0	E5		PP, EP, TOX, A	VE02		2	
3020	ORGANOTIN PESTICIDE, LIQUID, TOXIC	6.1	T6	II	6.1	61 274 648 802	LQ17	E4		PP, EP, TOX, A	VE02		2	
3020	ORGANOTIN PESTICIDE, LIQUID, TOXIC	6.1	T6	III	6.1	61 274 648 802	LQ7	E1		PP, EP, TOX, A	VE02		0	
3021	PESTICIDE, LIQUID, FLAMMABLE, TOXIC, N.O.S., flash-point less than 23 °C	3	FT2	I	3+6.1	61 274 802	LQ3	E0		PP, EP, EX, TOX, A	VE01, VE02		2	

UN No. or ID No.	Name and description	Class	Classi-fication Code	Packing group	Labels	Special provi-sions	Limited and excepted quantities		Carriage permitted	Equipment required	Venti-lation	Provisions concerning loading, unloading and carriage	Number of blue cones/ lights	Remarks
	3.1.2	2.2	2.2	2.1.1.3	5.2.2	3.3	3.4.6	3.5.1.2	3.2.1	8.1.5	7.1.6	7.1.6	7.1.5	3.2.1
(1)	(2)	(3a)	(3b)	(4)	(5)	(6)	(7a)	(7b)	(8)	(9)	(10)	(11)	(12)	(13)
3021	PESTICIDE, LIQUID, FLAMMABLE, TOXIC, N.O.S., flash-point less than 23 °C	3	FT2	II	3+6.1	61 274 802	LQ4	E2		PP, EP, EX, TOX, A	VE01, VE02		2	
3022	1,2-BUTYLENE OXIDE, STABILIZED	3	F1	II	3		LQ4	E2		PP, EX, A	VE01		1	
3023	2-METHYL-2-HEPTANETHIOL	6.1	TF1	I	6.1+3	802	LQ0	E5		PP, EP, EX, TOX, A	VE01, VE02		2	
3024	COUMARIN DERIVATIVE PESTICIDE, LIQUID, FLAMMABLE, TOXIC, flash-point less than 23 °C	3	FT2	I	3+6.1	61 274 802	LQ3	E0		PP, EP, EX, TOX, A	VE01, VE02		2	
3024	COUMARIN DERIVATIVE PESTICIDE, LIQUID, FLAMMABLE, TOXIC, flash-point less than 23 °C	3	FT2	II	3+6.1	61 274 802	LQ4	E2		PP, EP, EX, TOX, A	VE01, VE02		2	
3025	COUMARIN DERIVATIVE PESTICIDE, LIQUID, TOXIC, FLAMMABLE, flash-point not less than 23 °C	6.1	TF2	I	6.1+3	61 274 802	LQ0	E5		PP, EP, EX, TOX, A	VE01, VE02		2	
3025	COUMARIN DERIVATIVE PESTICIDE, LIQUID, TOXIC, FLAMMABLE, flash-point not less than 23 °C	6.1	TF2	II	6.1+3	61 274 802	LQ17	E4		PP, EP, EX, TOX, A	VE01, VE02		2	
3025	COUMARIN DERIVATIVE PESTICIDE, LIQUID, TOXIC, FLAMMABLE, flash-point not less than 23 °C	6.1	TF2	III	6.1+3	61 274 802	LQ7	E1		PP, EP, EX, TOX, A	VE01, VE02		0	
3026	COUMARIN DERIVATIVE PESTICIDE, LIQUID, TOXIC	6.1	T6	I	6.1	61 274 648 802	LQ0	E5		PP, EP, TOX, A	VE02		2	
3026	COUMARIN DERIVATIVE PESTICIDE, LIQUID, TOXIC	6.1	T6	II	6.1	61 274 648 802	LQ17	E4		PP, EP, TOX, A	VE02		2	
3026	COUMARIN DERIVATIVE PESTICIDE, LIQUID, TOXIC	6.1	T6	III	6.1	61 274 648 802	LQ7	E1		PP, EP, TOX, A	VE02		0	
3027	COUMARIN DERIVATIVE PESTICIDE, SOLID, TOXIC	6.1	T7	I	6.1	61 274 648 802	LQ0	E5		PP, EP			2	
3027	COUMARIN DERIVATIVE PESTICIDE, SOLID, TOXIC	6.1	T7	II	6.1	61 274 648 802	LQ18	E4		PP, EP			2	
3027	COUMARIN DERIVATIVE PESTICIDE, SOLID, TOXIC	6.1	T7	III	6.1	61 274 648 802	LQ9	E1		PP, EP			0	

UN No. or ID No.	Name and description	Class	Classi-fication Code	Packing group	Labels	Special provi-sions	Limited and excepted quantities		Carriage permitted	Equipment required	Venti-lation	Provisions concerning loading, unloading and carriage	Number of blue cones/ lights	Remarks
	3.1.2	2.2	2.2	2.1.1.3	5.2.2	3.3	3.4.6	3.5.1.2	3.2.1	8.1.5	7.1.6	7.1.6	7.1.5	3.2.1
(1)	(2)	(3a)	(3b)	(4)	(5)	(6)	(7a)	(7b)	(8)	(9)	(10)	(11)	(12)	(13)
3028	BATTERIES, DRY, CONTAINING POTASSIUM HYDROXIDE SOLID, electric storage	8	C11		8	295 304 598	LQ0	E0		PP, EP			0	
3048	ALUMINIUM PHOSPHIDE PESTICIDE	6.1	T7	I	6.1	153 648 802	LQ0	E5		PP, EP			2	
3054	CYCLOHEXYL MERCAPTAN	3	F1	III	3		LQ7	E1		PP, EX, A	VE01		0	
3055	2-(2-AMINOETHOXY)ETHANOL	8	C7	III	8		LQ7	E1		PP, EP			0	
3056	n-HEPTALDEHYDE	3	F1	III	3		LQ7	E1		PP, EX, A	VE01		0	
3057	TRIFLUOROACETYL CHLORIDE	2	2TC		2.3+8		LQ0	E0		PP, EP, TOX, A	VE02		2	
3064	NITROGLYCERIN, SOLUTION IN ALCOHOL with more than 1% but not more than 5% nitroglycerin	3	D	II	3		LQ0	E0		PP, EX, A	VE01		1	
3065	ALCOHOLIC BEVERAGES, with more than 70% alcohol by volume	3	F1	II	3		LQ5	E2		PP, EX, A	VE01		1	
3065	ALCOHOLIC BEVERAGES, with more than 24% but not more than 70% alcohol by volume	3	F1	III	3	144 145 247	LQ7	E1		PP, EX, A	VE01		0	
3066	PAINT (including paint, lacquer, enamel, stain, shellac, varnish, polish, liquid filler and liquid lacquer base) or PAINT RELATED MATERIAL (including paint thinning and reducing compound)	8	C9	II	8	163	LQ22	E2		PP, EP			0	
3066	PAINT (including paint, lacquer, enamel, stain, shellac, varnish, polish, liquid filler and liquid lacquer base) or PAINT RELATED MATERIAL (including paint thinning and reducing compound)	8	C9	III	8	163	LQ7	E1		PP, EP			0	
3070	ETHYLENE OXIDE AND DICHLORODIFLUOROMETHANE MIXTURE with not more than 12.5% ethylene oxide	2	2A		2.2		LQ1	E1		PP			0	
3071	MERCAPTANS, LIQUID, TOXIC, FLAMMABLE, N.O.S. or MERCAPTAN MIXTURE, LIQUID, TOXIC, FLAMMABLE, N.O.S.	6.1	TF1	II	6.1+3	274 802	LQ17	E4		PP, EP, EX, TOX, A	VE01, VE02		2	
3072	LIFE-SAVING APPLIANCES NOT SELF-INFLATING containing dangerous goods as equipment	9	M5		9	296 635	LQ0	E0		PP			0	
3073	VINYLPYRIDINES, STABILIZED	6.1	TFC	II	6.1+3+8	802	LQ17	E4		PP, EP, EX, TOX, A	VE01, VE02		2	

UN No. or ID No.	Name and description	Class	Classi-fication Code	Packing group	Labels	Special provi-sions	Limited and excepted quantities		Carriage permitted	Equipment required	Venti-lation	Provisions concerning loading, unloading and carriage	Number of blue cones/ lights	Remarks
	3.1.2	2.2	2.2	2.1.1.3	5.2.2	3.3	3.4.6	3.5.1.2	3.2.1	8.1.5	7.1.6	7.1.6	7.1.5	3.2.1
(1)	(2)	(3a)	(3b)	(4)	(5)	(6)	(7a)	(7b)	(8)	(9)	(10)	(11)	(12)	(13)
3077	ENVIRONMENTALLY HAZARDOUS SUBSTANCE, SOLID, N.O.S.	9	M7	III	9	274 335 601	LQ27	E1	T* B**	PP A***			0	* Only in the molten state. ** For carriage in bulk see also 7.1.4.1. ** * Only in the case of transport in bulk.
3078	CERIUM, turnings or gritty powder	4.3	W2	II	4.3	550	LQ11	E2		PP, EX, A	VE01	HA08	0	
3079	METHACRYLONITRILE, STABILIZED	3	FT1	I	3+6.1	802	LQ0	E0	**T**	PP, EP, EX, TOX, A	VE01, VE02		2	
3080	ISOCYANATES, TOXIC, FLAMMABLE, N.O.S. or ISOCYANATE SOLUTION, TOXIC, FLAMMABLE, N.O.S.	6.1	TF1	II	6.1+3	274 551 802	LQ17	E4		PP, EP, EX, TOX, A	VE01, VE02		2	
3082	ENVIRONMENTALLY HAZARDOUS SUBSTANCE, LIQUID, N.O.S.	9	M6	III	9	274 335 601	LQ7	E1	**T**	PP			0	
3083	PERCHLORYL FLUORIDE	2	2TO		2.3+5.1		LQ0	E0		PP, EP, TOX, A	VE02		2	
3084	CORROSIVE SOLID, OXIDIZING, N.O.S.	8	CO2	I	8+5.1	274	LQ0	E0		PP, EP			0	
3084	CORROSIVE SOLID, OXIDIZING, N.O.S.	8	CO2	II	8+5.1	274	LQ23	E2		PP, EP			0	
3085	OXIDIZING SOLID, CORROSIVE, N.O.S.	5.1	OC2	I	5.1+8	274	LQ0	E0		PP			0	
3085	OXIDIZING SOLID, CORROSIVE, N.O.S.	5.1	OC2	II	5.1+8	274	LQ11	E2		PP			0	
3085	OXIDIZING SOLID, CORROSIVE, N.O.S.	5.1	OC2	III	5.1+8	274	LQ12	E1		PP			0	
3086	TOXIC SOLID, OXIDIZING, N.O.S.	6.1	TO2	I	6.1+5.1	274 802	LQ0	E5		PP, EP			2	
3086	TOXIC SOLID, OXIDIZING, N.O.S.	6.1	TO2	II	6.1+5.1	274 802	LQ18	E4		PP, EP			2	
3087	OXIDIZING SOLID, TOXIC, N.O.S.	5.1	OT2	I	5.1+6.1	274 802	LQ0	E0		PP			2	
3087	OXIDIZING SOLID, TOXIC, N.O.S.	5.1	OT2	II	5.1+6.1	274 802	LQ11	E2		PP			2	
3087	OXIDIZING SOLID, TOXIC, N.O.S.	5.1	OT2	III	5.1+6.1	274 802	LQ12	E1		PP			0	
3088	SELF-HEATING SOLID, ORGANIC, N.O.S.	4.2	S2	II	4.2	274	LQ0	E2		PP			0	
3088	SELF-HEATING SOLID, ORGANIC, N.O.S.	4.2	S2	III	4.2	274	LQ0	E1		PP			0	
3089	METAL POWDER, FLAMMABLE, N.O.S.	4.1	F3	II	4.1	274 552	LQ8	E2		PP			1	
3089	METAL POWDER, FLAMMABLE, N.O.S.	4.1	F3	III	4.1	274 552	LQ9	E1		PP			0	
3090	LITHIUM METAL BATTERIES (including lithium alloy batteries)	9	M4	II	9	188 230 310 636	LQ0	E0		PP			0	

UN No. or ID No.	Name and description	Class	Classi-fication Code	Packing group	Labels	Special provi-sions	Limited and excepted quantities		Carriage permitted	Equipment required	Venti-lation	Provisions concerning loading, unloading and carriage	Number of blue cones/ lights	Remarks
	3.1.2	2.2	2.2	2.1.1.3	5.2.2	3.3	3.4.6	3.5.1.2	3.2.1	8.1.5	7.1.6	7.1.6	7.1.5	3.2.1
(1)	(2)	(3a)	(3b)	(4)	(5)	(6)	(7a)	(7b)	(8)	(9)	(10)	(11)	(12)	(13)
3091	LITHIUM METAL BATTERIES CONTAINED IN EQUIPMENT or LITHIUM METAL BATTERIES PACKED WITH EQUIPMENT (including lithium alloy batteries)	9	M4	II	9	188 230 636	LQ0	E0		PP			0	
3092	1-METHOXY-2-PROPANOL	3	F1	III	3		LQ7	E1	T	PP, EX, A	VE01		0	
3093	CORROSIVE LIQUID, OXIDIZING, N.O.S.	8	CO1	I	8+5.1	274	LQ0	E0		PP, EP			0	
3093	CORROSIVE LIQUID, OXIDIZING, N.O.S.	8	CO1	II	8+5.1	274	LQ22	E2		PP, EP			0	
3094	CORROSIVE LIQUID, WATER-REACTIVE, N.O.S.	8	CW1	I	8+4.3	274	LQ0	E0		PP, EP			0	
3094	CORROSIVE LIQUID, WATER-REACTIVE, N.O.S.	8	CW1	II	8+4.3	274	LQ22	E2		PP, EP			0	
3095	CORROSIVE SOLID, SELF-HEATING, N.O.S.	8	CS2	I	8+4.2	274	LQ0	E0		PP, EP			0	
3095	CORROSIVE SOLID, SELF-HEATING, N.O.S.	8	CS2	II	8+4.2	274	LQ23	E2		PP, EP			0	
3096	CORROSIVE SOLID, WATER-REACTIVE, N.O.S.	8	CW2	I	8+4.3	274	LQ0	E0		PP, EP			0	
3096	CORROSIVE SOLID, WATER-REACTIVE, N.O.S.	8	CW2	II	8+4.3	274	LQ23	E2		PP, EP			0	
3097	FLAMMABLE SOLID, OXIDIZING, N.O.S.	4.1	FO	CARRIAGE PROHIBITED										
3098	OXIDIZING LIQUID, CORROSIVE, N.O.S.	5.1	OC1	I	5.1+8	274	LQ0	E0		PP, EP			0	
3098	OXIDIZING LIQUID, CORROSIVE, N.O.S.	5.1	OC1	II	5.1+8	274	LQ10	E2		PP, EP			0	
3098	OXIDIZING LIQUID, CORROSIVE, N.O.S.	5.1	OC1	III	5.1+8	274	LQ13	E1		PP, EP			0	
3099	OXIDIZING LIQUID, TOXIC, N.O.S.	5.1	OT1	I	5.1+6.1	274 802	LQ0	E0		PP, EP, TOX, A	VE02		2	
3099	OXIDIZING LIQUID, TOXIC, N.O.S.	5.1	OT1	II	5.1+6.1	274 802	LQ10	E2		PP, EP, TOX, A	VE02		2	
3099	OXIDIZING LIQUID, TOXIC, N.O.S.	5.1	OT1	III	5.1+6.1	274 802	LQ13	E1		PP, EP, TOX, A	VE02		0	
3100	OXIDIZING SOLID, SELF-HEATING, N.O.S.	5.1	OS	CARRIAGE PROHIBITED										
3101	ORGANIC PEROXIDE TYPE B, LIQUID	5.2	P1		5.2+1	122 181 274	LQ14	E0		PP, EX, A	VE01	HA01, HA10	3	
3102	ORGANIC PEROXIDE TYPE B, SOLID	5.2	P1		5.2+1	122 181 274	LQ15	E0		PP, EX, A	VE01	HA01, HA10	3	
3103	ORGANIC PEROXIDE TYPE C, LIQUID	5.2	P1		5.2	122 274	LQ14	E0		PP, EX, A	VE01		0	
3104	ORGANIC PEROXIDE TYPE C, SOLID	5.2	P1		5.2	122 274	LQ15	E0		PP, EX, A	VE01		0	
3105	ORGANIC PEROXIDE TYPE D, LIQUID	5.2	P1		5.2	122 274	LQ16	E0		PP, EX, A	VE01		0	
3106	ORGANIC PEROXIDE TYPE D, SOLID	5.2	P1		5.2	122 274	LQ11	E0		PP, EX, A	VE01		0	
3107	ORGANIC PEROXIDE TYPE E, LIQUID	5.2	P1		5.2	122 274	LQ16	E0		PP, EX, A	VE01		0	
3108	ORGANIC PEROXIDE TYPE E, SOLID	5.2	P1		5.2	122 274	LQ11	E0		PP, EX, A	VE01		0	

UN No. or ID No.	Name and description	Class	Classi-fication Code	Packing group	Labels	Special provi-sions	Limited and excepted quantities		Carriage permitted	Equipment required	Venti-lation	Provisions concerning loading, unloading and carriage	Number of blue cones/ lights	Remarks
	3.1.2	2.2	2.2	2.1.1.3	5.2.2	3.3	3.4.6	3.5.1.2	3.2.1	8.1.5	7.1.6	7.1.6	7.1.5	3.2.1
(1)	(2)	(3a)	(3b)	(4)	(5)	(6)	(7a)	(7b)	(8)	(9)	(10)	(11)	(12)	(13)
3109	ORGANIC PEROXIDE TYPE F, LIQUID	5.2	P1		5.2	122 274	LQ16	E0		PP, EX, A	VE01		0	
3110	ORGANIC PEROXIDE TYPE F, SOLID	5.2	P1		5.2	122 274	LQ11	E0		PP, EX, A	VE01		0	
3111	ORGANIC PEROXIDE TYPE B, LIQUID, TEMPERATURE CONTROLLED	5.2	P2		5.2+1	122 181 274	LQ0	E0		PP, EX, A	VE01	HA01, HA10	3	
3112	ORGANIC PEROXIDE TYPE B, SOLID, TEMPERATURE CONTROLLED	5.2	P2		5.2+1	122 181 274	LQ0	E0		PP, EX, A	VE01	HA01, HA10	3	
3113	ORGANIC PEROXIDE TYPE C, LIQUID, TEMPERATURE CONTROLLED	5.2	P2		5.2	122 274	LQ0	E0		PP, EX, A	VE01		0	
3114	ORGANIC PEROXIDE TYPE C, SOLID, TEMPERATURE CONTROLLED	5.2	P2		5.2	122 274	LQ0	E0		PP, EX, A	VE01		0	
3115	ORGANIC PEROXIDE TYPE D, LIQUID, TEMPERATURE CONTROLLED	5.2	P2		5.2	122 274	LQ0	E0		PP, EX, A	VE01		0	
3116	ORGANIC PEROXIDE TYPE D, SOLID, TEMPERATURE CONTROLLED	5.2	P2		5.2	122 274	LQ0	E0		PP, EX, A	VE01		0	
3117	ORGANIC PEROXIDE TYPE E, LIQUID, TEMPERATURE CONTROLLED	5.2	P2		5.2	122 274	LQ0	E0		PP, EX, A	VE01		0	
3118	ORGANIC PEROXIDE TYPE E, SOLID, TEMPERATURE CONTROLLED	5.2	P2		5.2	122 274	LQ0	E0		PP, EX, A	VE01		0	
3119	ORGANIC PEROXIDE TYPE F, LIQUID, TEMPERATURE CONTROLLED	5.2	P2		5.2	122 274	LQ0	E0		PP, EX, A	VE01		0	
3120	ORGANIC PEROXIDE TYPE F, SOLID, TEMPERATURE CONTROLLED	5.2	P2		5.2	122 274	LQ0	E0		PP, EX, A	VE01		0	
3121	OXIDIZING SOLID, WATER-REACTIVE, N.O.S.	5.1	OW	CARRIAGE PROHIBITED										
3122	TOXIC LIQUID, OXIDIZING, N.O.S.	6.1	TO1	I	6.1+5.1	274 315 802	LQ0	E5		PP, EP, TOX, A	VE02		2	
3122	TOXIC LIQUID, OXIDIZING, N.O.S.	6.1	TO1	II	6.1+5.1	274 802	LQ17	E4		PP, EP, TOX, A	VE02		2	
3123	TOXIC LIQUID, WATER-REACTIVE, N.O.S.	6.1	TW1	I	6.1+4.3	274 315 802	LQ0	E5		PP, EP, TOX, A	VE02		2	
3123	TOXIC LIQUID, WATER-REACTIVE, N.O.S.	6.1	TW1	II	6.1+4.3	274 802	LQ17	E4		PP, EP, TOX, A	VE02		2	
3124	TOXIC SOLID, SELF-HEATING, N.O.S.	6.1	TS	I	6.1+4.2	274 802	LQ0	E5		PP, EP			2	
3124	TOXIC SOLID, SELF-HEATING, N.O.S.	6.1	TS	II	6.1+4.2	274 802	LQ18	E4		PP, EP			2	
3125	TOXIC SOLID, WATER-REACTIVE, N.O.S.	6.1	TW2	I	6.1+4.3	274 802	LQ0	E5		PP, EP			2	
3125	TOXIC SOLID, WATER-REACTIVE, N.O.S.	6.1	TW2	II	6.1+4.3	274 802	LQ18	E4		PP, EP			2	

UN No. or ID No.	Name and description	Class	Classi-fication Code	Packing group	Labels	Special provi-sions	Limited and excepted quantities		Carriage permitted	Equipment required	Venti-lation	Provisions concerning loading, unloading and carriage	Number of blue cones/ lights	Remarks
	3.1.2	2.2	2.2	2.1.1.3	5.2.2	3.3	3.4.6	3.5.1.2	3.2.1	8.1.5	7.1.6	7.1.6	7.1.5	3.2.1
(1)	(2)	(3a)	(3b)	(4)	(5)	(6)	(7a)	(7b)	(8)	(9)	(10)	(11)	(12)	(13)
3126	SELF-HEATING SOLID, CORROSIVE, ORGANIC, N.O.S.	4.2	SC2	II	4.2+8	274	LQ0	E2		PP			0	
3126	SELF-HEATING SOLID, CORROSIVE, ORGANIC, N.O.S.	4.2	SC2	III	4.2+8	274	LQ0	E1		PP			0	
3127	SELF-HEATING SOLID, OXIDIZING, N.O.S	4.2	SO	CARRIAGE PROHIBITED										
3128	SELF-HEATING SOLID, TOXIC, ORGANIC, N.O.S.	4.2	ST2	II	4.2+6.1	274 802	LQ0	E2		PP			2	
3128	SELF-HEATING SOLID, TOXIC, ORGANIC, N.O.S.	4.2	ST2	III	4.2+6.1	274 802	LQ0	E1		PP			0	
3129	WATER-REACTIVE LIQUID, CORROSIVE, N.O.S.	4.3	WC1	I	4.3+8	274	LQ0	E0		PP, EP, EX, A	VE01	HA08	0	
3129	WATER-REACTIVE LIQUID, CORROSIVE, N.O.S.	4.3	WC1	II	4.3+8	274	LQ10	E2		PP, EP, EX, A	VE01	HA08	0	
3129	WATER-REACTIVE LIQUID, CORROSIVE, N.O.S.	4.3	WC1	III	4.3+8	274	LQ13	E1		PP, EP, EX, A	VE01	HA08	0	
3130	WATER-REACTIVE LIQUID, TOXIC, N.O.S.	4.3	WT1	I	4.3+6.1	274 802	LQ0	E0		PP, EP, EX, TOX, A	VE01, VE02	HA08	2	
3130	WATER-REACTIVE LIQUID, TOXIC, N.O.S.	4.3	WT1	II	4.3+6.1	274 802	LQ10	E2		PP, EP, EX, TOX, A	VE01, VE02	HA08	2	
3130	WATER-REACTIVE LIQUID, TOXIC, N.O.S.	4.3	WT1	III	4.3+6.1	274 802	LQ13	E1		PP, EP, EX, TOX, A	VE01, VE02	HA08	0	
3131	WATER-REACTIVE SOLID, CORROSIVE, N.O.S.	4.3	WC2	I	4.3+8	274	LQ0	E0		PP, EP, EX, A	VE01	HA08	0	
3131	WATER-REACTIVE SOLID, CORROSIVE, N.O.S.	4.3	WC2	II	4.3+8	274	LQ11	E2		PP, EP, EX, A	VE01	HA08	0	
3131	WATER-REACTIVE SOLID, CORROSIVE, N.O.S.	4.3	WC2	III	4.3+8	274	LQ12	E1		PP, EP, EX, A	VE01	HA08	0	
3132	WATER-REACTIVE SOLID, FLAMMABLE, N.O.S.	4.3	WF2	I	4.3 + 4.1	274	LQ0	E0		PP,EX,A	VE01		1	
3132	WATER-REACTIVE SOLID, FLAMMABLE, N.O.S.	4.3	WF2	II	4.3 + 4.1	274	LQ11	E2		PP,EX,A	VE01		1	
3132	WATER-REACTIVE SOLID, FLAMMABLE, N.O.S.	4.3	WF2	III	4.3 + 4.1	274	LQ12	E1		PP,EX,A	VE01		0	
3133	WATER-REACTIVE SOLID, OXIDIZING, N.O.S.	4.3	WO	CARRIAGE PROHIBITED										
3134	WATER-REACTIVE SOLID, TOXIC, N.O.S.	4.3	WT2	I	4.3+6.1	274 802	LQ0	E0		PP, EP, EX, A	VE01	HA08	2	
3134	WATER-REACTIVE SOLID, TOXIC, N.O.S.	4.3	WT2	II	4.3+6.1	274 802	LQ11	E2		PP, EP, EX, A	VE01	HA08	2	
3134	WATER-REACTIVE SOLID, TOXIC, N.O.S.	4.3	WT2	III	4.3+6.1	274 802	LQ12	E1		PP, EP, EX, A	VE01	HA08	0	
3135	WATER-REACTIVE SOLID, SELF-HEATING, N.O.S.	4.3	WS	I	4.3 + 4.2	274	LQ0	E0		PP,EX,A	VE01		0	
3135	WATER-REACTIVE SOLID, SELF-HEATING, N.O.S.	4.3	WS	II	4.3 + 4.2	274	LQ11	E2		PP,EX,A	VE01		0	

UN No. or ID No.	Name and description 3.1.2	Class 2.2	Classi-fication Code 2.2	Packing group 2.1.1.3	Labels 5.2.2	Special provi-sions 3.3	Limited and excepted quantities 3.4.6	3.5.1.2	Carriage permitted 3.2.1	Equipment required 8.1.5	Venti-lation 7.1.6	Provisions concerning loading, unloading and carriage 7.1.6	Number of blue cones/ lights 7.1.5	Remarks 3.2.1
(1)	(2)	(3a)	(3b)	(4)	(5)	(6)	(7a)	(7b)	(8)	(9)	(10)	(11)	(12)	(13)
3135	WATER-REACTIVE SOLID, SELF-HEATING, N.O.S.	4.3	WS	III	4.3 + 4.2	274	LQ12	E1		PP,EX,A	VE01		0	
3136	TRIFLUOROMETHANE, REFRIGERATED LIQUID	2	3A		2.2	593	LQ1	E1		PP			0	
3137	OXIDIZING SOLID, FLAMMABLE, N.O.S.	5.1	OF	**CARRIAGE PROHIBITED**										
3138	ETHYLENE, ACETYLENE AND PROPYLENE MIXTURE, REFRIGERATED LIQUID containing at least 71.5% ethylene with not more than 22.5% acetylene and not more than 6% propylene	2	3F		2.1		LQ0	E0		PP, EX, A	VE01		1	
3139	OXIDIZING LIQUID, N.O.S.	5.1	O1	I	5.1	274	LQ0	E0		PP			0	
3139	OXIDIZING LIQUID, N.O.S.	5.1	O1	II	5.1	274	LQ10	E2		PP			0	
3139	OXIDIZING LIQUID, N.O.S.	5.1	O1	III	5.1	274	LQ13	E1		PP			0	
3140	ALKALOIDS, LIQUID, N.O.S. or ALKALOID SALTS, LIQUID, N.O.S.	6.1	T1	I	6.1	43 274 802	LQ0	E5		PP, EP, TOX, A	VE02		2	
3140	ALKALOIDS, LIQUID, N.O.S. or ALKALOID SALTS, LIQUID, N.O.S.	6.1	T1	II	6.1	43 274 802	LQ17	E4		PP, EP, TOX, A	VE02		2	
3140	ALKALOIDS, LIQUID, N.O.S. or ALKALOID SALTS, LIQUID, N.O.S.	6.1	T1	III	6.1	43 274 802	LQ7	E1		PP, EP, TOX, A	VE02		0	
3141	ANTIMONY COMPOUND, INORGANIC, LIQUID, N.O.S.	6.1	T4	III	6.1	45 274 512 802	LQ7	E1		PP, EP, TOX, A	VE02		0	
3142	DISINFECTANT, LIQUID, TOXIC, N.O.S.	6.1	T1	I	6.1	274 802	LQ0	E5		PP, EP, TOX, A	VE02		2	
3142	DISINFECTANT, LIQUID, TOXIC, N.O.S.	6.1	T1	II	6.1	274 802	LQ17	E4		PP, EP, TOX, A	VE02		2	
3142	DISINFECTANT, LIQUID, TOXIC, N.O.S.	6.1	T1	III	6.1	274 802	LQ7	E1		PP, EP, TOX, A	VE02		0	
3143	DYE, SOLID, TOXIC, N.O.S. or DYE INTERMEDIATE, SOLID, TOXIC, N.O.S.	6.1	T2	I	6.1	274 802	LQ0	E5		PP, EP			2	
3143	DYE, SOLID, TOXIC, N.O.S. or DYE INTERMEDIATE, SOLID, TOXIC, N.O.S.	6.1	T2	II	6.1	274 802	LQ18	E4		PP, EP			2	
3143	DYE, SOLID, TOXIC, N.O.S. or DYE INTERMEDIATE, SOLID, TOXIC, N.O.S.	6.1	T2	III	6.1	274 802	LQ9	E1		PP, EP			0	
3144	NICOTINE COMPOUND, LIQUID, N.O.S. or NICOTINE PREPARATION, LIQUID, N.O.S.	6.1	T1	I	6.1	43 274 802	LQ0	E5		PP, EP, TOX, A	VE02		2	
3144	NICOTINE COMPOUND, LIQUID, N.O.S. or NICOTINE PREPARATION, LIQUID, N.O.S.	6.1	T1	II	6.1	43 274 802	LQ17	E4		PP, EP, TOX, A	VE02		2	

UN No. or ID No.	Name and description	Class	Classi-fication Code	Packing group	Labels	Special provi-sions	Limited and excepted quantities		Carriage permitted	Equipment required	Venti-lation	Provisions concerning loading, unloading and carriage	Number of blue cones/ lights	Remarks
	3.1.2	2.2	2.2	2.1.1.3	5.2.2	3.3	3.4.6	3.5.1.2	3.2.1	8.1.5	7.1.6	7.1.6	7.1.5	3.2.1
(1)	(2)	(3a)	(3b)	(4)	(5)	(6)	(7a)	(7b)	(8)	(9)	(10)	(11)	(12)	(13)
3144	NICOTINE COMPOUND, LIQUID, N.O.S. or NICOTINE PREPARATION, LIQUID, N.O.S.	6.1	T1	III	6.1	43 274 802	LQ7	E1		PP, EP, TOX, A	VE02		0	
3145	ALKYLPHENOLS, LIQUID, N.O.S. (including C_2-C_{12} homologues)	8	C3	I	8	274	LQ0	E0		PP, EP			0	
3145	ALKYLPHENOLS, LIQUID, N.O.S. (including C_2-C_{12} homologues)	8	C3	II	8	274	LQ22	E2	**T**	PP, EP			0	
3145	ALKYLPHENOLS, LIQUID, N.O.S. (including C_2-C_{12} homologues)	8	C3	III	8	274	LQ7	E1	**T**	PP, EP			0	
3146	ORGANOTIN COMPOUND, SOLID, N.O.S.	6.1	T3	I	6.1	43 274 802	LQ0	E5		PP, EP			2	
3146	ORGANOTIN COMPOUND, SOLID, N.O.S.	6.1	T3	II	6.1	43 274 802	LQ18	E4		PP, EP			2	
3146	ORGANOTIN COMPOUND, SOLID, N.O.S.	6.1	T3	III	6.1	43 274 802	LQ9	E1		PP, EP			0	
3147	DYE, SOLID, CORROSIVE, N.O.S. or DYE INTERMEDIATE, SOLID, CORROSIVE, N.O.S.	8	C10	I	8	274	LQ0	E0		PP, EP			0	
3147	DYE, SOLID, CORROSIVE, N.O.S. or DYE INTERMEDIATE, SOLID, CORROSIVE, N.O.S.	8	C10	II	8	274	LQ23	E2		PP, EP			0	
3147	DYE, SOLID, CORROSIVE, N.O.S. or DYE INTERMEDIATE, SOLID, CORROSIVE, N.O.S.	8	C10	III	8	274	LQ24	E1		PP, EP			0	
3148	WATER-REACTIVE LIQUID, N.O.S.	4.3	W1	I	4.3	274	LQ0	E0		PP, EX, A	VE01	HA08	0	
3148	WATER-REACTIVE LIQUID, N.O.S.	4.3	W1	II	4.3	274	LQ10	E2		PP, EX, A	VE01	HA08	0	
3148	WATER-REACTIVE LIQUID, N.O.S.	4.3	W1	III	4.3	274	LQ13	E1		PP, EX, A	VE01	HA08	0	
3149	HYDROGEN PEROXIDE AND PEROXYACETIC ACID MIXTURE with acid(s), water and not more than 5% peroxyacetic acid, STABILIZED	5.1	OC1	II	5.1+8	196 553	LQ10	E2		PP, EP			0	
3150	DEVICES, SMALL, HYDROCARBON GAS POWERED or HYDROCARBON GAS REFILLS FOR SMALL DEVICES with release device	2	6F		2.1		LQ0	E0		PP, EX, A	VE01		1	
3151	POLYHALOGENATED BIPHENYLS, LIQUID or POLYHALOGENATED TERPHENYLS, LIQUID	9	M2	II	9	203 305 802	LQ26	E2		PP, EP			0	
3152	POLYHALOGENATED BIPHENYLS, SOLID or POLYHALOGENATED TERPHENYLS, SOLID	9	M2	II	9	203 305 802	LQ25	E2		PP, EP			0	
3153	PERFLUORO(METHYL VINYL ETHER)	2	2F		2.1		LQ0	E0		PP, EX, A	VE01		1	
3154	PERFLUORO(ETHYL VINYL ETHER)	2	2F		2.1		LQ0	E0		PP, EX, A	VE01		1	

UN No. or ID No.	Name and description	Class	Classi-fication Code	Packing group	Labels	Special provi-sions	Limited and excepted quantities		Carriage permitted	Equipment required	Venti-lation	Provisions concerning loading, unloading and carriage			Number of blue cones/ lights	Remarks
	3.1.2	2.2	2.2	2.1.1.3	5.2.2	3.3	3.4.6	3.5.1.2	3.2.1	8.1.5	7.1.6	7.1.6			7.1.5	3.2.1
(1)	(2)	(3a)	(3b)	(4)	(5)	(6)	(7a)	(7b)	(8)	(9)	(10)	(11)			(12)	(13)
3155	PENTACHLOROPHENOL	6.1	T2	II	6.1	43 802	LQ18	E4		PP, EP					2	
3156	COMPRESSED GAS, OXIDIZING, N.O.S.	2	1O		2.2+5.1	274	LQ0	E0		PP					0	
3157	LIQUEFIED GAS, OXIDIZING, N.O.S.	2	2O		2.2+5.1	274	LQ0	E0		PP					0	
3158	GAS, REFRIGERATED LIQUID, N.O.S.	2	3A		2.2	274 593	LQ1	E1		PP					0	
3159	1,1,1,2-TETRAFLUOROETHANE (REFRIGERANT GAS R 134a)	2	2A		2.2		LQ1	E1		PP					0	
3160	LIQUEFIED GAS, TOXIC, FLAMMABLE, N.O.S.	2	2TF		2.3+2.1	274	LQ0	E0		PP, EP, EX, TOX, A	VE01, VE02				2	
3161	LIQUEFIED GAS, FLAMMABLE, N.O.S.	2	2F		2.1	274	LQ0	E0		PP, EX, A	VE01				1	
3162	LIQUEFIED GAS, TOXIC, N.O.S.	2	2T		2.3	274	LQ0	E0		PP, EP, TOX, A	VE02				2	
3163	LIQUEFIED GAS, N.O.S.	2	2A		2.2	274	LQ1	E1		PP					0	
3164	ARTICLES, PRESSURIZED, PNEUMATIC or HYDRAULIC (containing non-flammable gas)	2	6A		2.2	283 594	LQ0	E0		PP					0	
3165	AIRCRAFT HYDRAULIC POWER UNIT FUEL TANK (containing a mixture of anhydrous hydrazine and methylhydrazine) (M86 fuel)	3	FTC	I	3+6.1+8	802	LQ0	E0		PP, EP, EX, TOX, A	VE01, VE02				2	
3166	Engine, internal combustion or vehicle, flammable gas powered or vehicle, flammable liquid powered	9	M11	NOT SUBJECT TO ADN												
3167	GAS SAMPLE, NON-PRESSURIZED, FLAMMABLE, N.O.S., not refrigerated liquid	2	7F		2.1	274	LQ0	E0		PP, EX, A	VE01				1	
3168	GAS SAMPLE, NON-PRESSURIZED, TOXIC, FLAMMABLE, N.O.S., not refrigerated liquid	2	7TF		2.3+2.1	274	LQ0	E0		PP, EP, EX, TOX, A	VE01, VE02				2	
3169	GAS SAMPLE, NON-PRESSURIZED, TOXIC, N.O.S., not refrigerated liquid	2	7T		2.3	274	LQ0	E0		PP, EP, TOX, A	VE02				2	
3170	ALUMINIUM SMELTING BY-PRODUCTS or ALUMINIUM REMELTING BY-PRODUCTS	4.3	W2	II	4.3	244	LQ11	E2		PP, EX, A	VE01		HA08		0	
3170	ALUMINIUM SMELTING BY-PRODUCTS or ALUMINIUM REMELTING BY-PRODUCTS	4.3	W2	III	4.3	244	LQ12	E1	**B**	PP, EX, A	VE01, VE03	LO03	HA07, HA08	IN01, IN02, IN03	0	VE03, LO03, HA07, IN01, IN02 and IN03 apply only when this substance is carried in bulk or without packaging
3171	Battery-powered vehicle or Battery-powered equipment	9	M11	NOT SUBJECT TO ADN												
3172	TOXINS, EXTRACTED FROM LIVING SOURCES, LIQUID, N.O.S.	6.1	T1	I	6.1	210 274 802	LQ0	E5		PP, EP, TOX, A	VE02				2	
3172	TOXINS, EXTRACTED FROM LIVING SOURCES, LIQUID, N.O.S.	6.1	T1	II	6.1	210 274 802	LQ17	E4		PP, EP, TOX, A	VE02				2	

UN No. or ID No.	Name and description	Class	Classi-fication Code	Packing group	Labels	Special provi-sions	Limited and excepted quantities		Carriage permitted	Equipment required	Venti-lation	Provisions concerning loading, unloading and carriage			Number of blue cones/ lights	Remarks
	3.1.2	2.2	2.2	2.1.1.3	5.2.2	3.3	3.4.6	3.5.1.2	3.2.1	8.1.5	7.1.6	7.1.6			7.1.5	3.2.1
(1)	(2)	(3a)	(3b)	(4)	(5)	(6)	(7a)	(7b)	(8)	(9)	(10)	(11)			(12)	(13)
3172	TOXINS, EXTRACTED FROM LIVING SOURCES, LIQUID, N.O.S.	6.1	T1	III	6.1	210 274 802	LQ7	E1		PP, EP, TOX, A	VE02				0	
3174	TITANIUM DISULPHIDE	4.2	S4	III	4.2		LQ0	E1		PP					0	
3175	SOLIDS or mixtures of solids (such as preparations and wastes) CONTAINING FLAMMABLE LIQUID, N.O.S. having a flash-point up to 60°C	4.1	F1	II	4.1	216 274 800	LQ8	E2	**B**	PP, EX, A	VE01, VE03			IN01, IN02	1	VE03, IN01and IN02 apply only when this substance is carried in bulk or without packaging
3175	SOLIDS CONTAINING FLAMMABLE LIQUID, MOLTEN, having a flash-point up to 60°C (DIALKYL-(C_{12}-C_{18})-DIMETHYL-AMMONIUM and 2-PROPANOL)	4.1	F1	II	4.1	216 274 800	LQ8	E2	**T**	PP, EX, A	VE01, VE03			IN01, IN02	1	VE03, IN01and IN02 apply only when this substance is carried in bulk or without packaging
3176	FLAMMABLE SOLID, ORGANIC, MOLTEN, N.O.S.	4.1	F2	II	4.1	274	LQ0	E0		PP					0	
3176	FLAMMABLE SOLID, ORGANIC, MOLTEN, N.O.S.	4.1	F2	III	4.1	274	LQ0	E0		PP					0	
3178	FLAMMABLE SOLID, INORGANIC, N.O.S.	4.1	F3	II	4.1	274	LQ8	E2		PP					1	
3178	FLAMMABLE SOLID, INORGANIC, N.O.S.	4.1	F3	III	4.1	274	LQ9	E1		PP					0	
3179	FLAMMABLE SOLID, TOXIC, INORGANIC, N.O.S.	4.1	FT2	II	4.1+6.1	274 802	LQ0	E2		PP					2	
3179	FLAMMABLE SOLID, TOXIC, INORGANIC, N.O.S.	4.1	FT2	III	4.1+6.1	274 802	LQ0	E1		PP					0	
3180	FLAMMABLE SOLID, CORROSIVE, INORGANIC, N.O.S.	4.1	FC2	II	4.1+8	274	LQ0	E2		PP					1	
3180	FLAMMABLE SOLID, CORROSIVE, INORGANIC, N.O.S.	4.1	FC2	III	4.1+8	274	LQ0	E1		PP					0	
3181	METAL SALTS OF ORGANIC COMPOUNDS, FLAMMABLE, N.O.S.	4.1	F3	II	4.1	274	LQ8	E2		PP					1	
3181	METAL SALTS OF ORGANIC COMPOUNDS, FLAMMABLE, N.O.S.	4.1	F3	III	4.1	274	LQ9	E1		PP					0	
3182	METAL HYDRIDES, FLAMMABLE, N.O.S.	4.1	F3	II	4.1	274 554	LQ8	E2		PP					1	
3182	METAL HYDRIDES, FLAMMABLE, N.O.S.	4.1	F3	III	4.1	274 554	LQ9	E1		PP					0	
3183	SELF-HEATING LIQUID, ORGANIC, N.O.S.	4.2	S1	II	4.2	274	LQ0	E2		PP					0	
3183	SELF-HEATING LIQUID, ORGANIC, N.O.S.	4.2	S1	III	4.2	274	LQ0	E1		PP					0	
3184	SELF-HEATING LIQUID, TOXIC, ORGANIC, N.O.S.	4.2	ST1	II	4.2+6.1	274 802	LQ0	E2		PP, EP, TOX, A	VE02				2	
3184	SELF-HEATING LIQUID, TOXIC, ORGANIC, N.O.S.	4.2	ST1	III	4.2+6.1	274 802	LQ0	E1		PP, EP, TOX, A	VE02				0	
3185	SELF-HEATING LIQUID, CORROSIVE, ORGANIC, N.O.S.	4.2	SC1	II	4.2+8	274	LQ0	E2		PP, EP					0	
3185	SELF-HEATING LIQUID, CORROSIVE, ORGANIC, N.O.S.	4.2	SC1	III	4.2+8	274	LQ0	E1		PP, EP					0	

UN No. or ID No.	Name and description	Class	Classi-fication Code	Packing group	Labels	Special provi-sions	Limited and excepted quantities		Carriage permitted	Equipment required	Venti-lation	Provisions concerning loading, unloading and carriage	Number of blue cones/ lights	Remarks
	3.1.2	2.2	2.2	2.1.1.3	5.2.2	3.3	3.4.6	3.5.1.2	3.2.1	8.1.5	7.1.6	7.1.6	7.1.5	3.2.1
(1)	(2)	(3a)	(3b)	(4)	(5)	(6)	(7a)	(7b)	(8)	(9)	(10)	(11)	(12)	(13)
3186	SELF-HEATING LIQUID, INORGANIC, N.O.S.	4.2	S3	II	4.2	274	LQ0	E2		PP			0	
3186	SELF-HEATING LIQUID, INORGANIC, N.O.S.	4.2	S3	III	4.2	274	LQ0	E1		PP			0	
3187	SELF-HEATING LIQUID, TOXIC, INORGANIC, N.O.S.	4.2	ST3	II	4.2+6.1	274 802	LQ0	E2		PP, EP, TOX, A	VE02		2	
3187	SELF-HEATING LIQUID, TOXIC, INORGANIC, N.O.S.	4.2	ST3	III	4.2+6.1	274 802	LQ0	E1		PP, EP, TOX, A	VE02		0	
3188	SELF-HEATING LIQUID, CORROSIVE, INORGANIC, N.O.S.	4.2	SC3	II	4.2+8	274	LQ0	E2		PP, EP			0	
3188	SELF-HEATING LIQUID, CORROSIVE, INORGANIC, N.O.S.	4.2	SC3	III	4.2+8	274	LQ0	E1		PP, EP			0	
3189	METAL POWDER, SELF-HEATING, N.O.S.	4.2	S4	II	4.2	274 555	LQ0	E2		PP			0	
3189	METAL POWDER, SELF-HEATING, N.O.S.	4.2	S4	III	4.2	274 555	LQ0	E1		PP			0	
3190	SELF-HEATING SOLID, INORGANIC, N.O.S.	4.2	S4	II	4.2	274	LQ0	E2		PP			0	
3190	SELF-HEATING SOLID, INORGANIC, N.O.S.	4.2	S4	III	4.2	274	LQ0	E1	B	PP			0	
3191	SELF-HEATING SOLID, TOXIC, INORGANIC, N.O.S.	4.2	ST4	II	4.2+6.1	274 802	LQ0	E2		PP			2	
3191	SELF-HEATING SOLID, TOXIC, INORGANIC, N.O.S.	4.2	ST4	III	4.2+6.1	274 802	LQ0	E1		PP			0	
3192	SELF-HEATING SOLID, CORROSIVE, INORGANIC, N.O.S.	4.2	SC4	II	4.2+8	274	LQ0	E2		PP			0	
3192	SELF-HEATING SOLID, CORROSIVE, INORGANIC, N.O.S.	4.2	SC4	III	4.2+8	274	LQ0	E1		PP			0	
3194	PYROPHORIC LIQUID, INORGANIC, N.O.S.	4.2	S3	I	4.2	274	LQ0	E0		PP			0	
3200	PYROPHORIC SOLID, INORGANIC, N.O.S.	4.2	S4	I	4.2	274	LQ0	E0		PP			0	
3205	ALKALINE EARTH METAL ALCOHOLATES, N.O.S.	4.2	S4	II	4.2	183 274	LQ0	E2		PP			0	
3205	ALKALINE EARTH METAL ALCOHOLATES, N.O.S.	4.2	S4	III	4.2	183 274	LQ0	E1		PP			0	
3206	ALKALI METAL ALCOHOLATES, SELF-HEATING, CORROSIVE, N.O.S.	4.2	SC4	II	4.2+8	182 274	LQ0	E2		PP			0	
3206	ALKALI METAL ALCOHOLATES, SELF-HEATING, CORROSIVE, N.O.S.	4.2	SC4	III	4.2+8	183 274	LQ0	E1		PP			0	
3208	METALLIC SUBSTANCE, WATER-REACTIVE, N.O.S.	4.3	W2	I	4.3	274 557	LQ0	E0		PP, EX, A	VE01	HA08	0	
3208	METALLIC SUBSTANCE, WATER-REACTIVE, N.O.S.	4.3	W2	II	4.3	274 557	LQ11	E2		PP, EX, A	VE01	HA08	0	
3208	METALLIC SUBSTANCE, WATER-REACTIVE, N.O.S.	4.3	W2	III	4.3	274 557	LQ12	E1		PP, EX, A	VE01	HA08	0	
3209	METALLIC SUBSTANCE, WATER-REACTIVE, SELF-HEATING, N.O.S.	4.3	WS	I	4.3+4.2	274 558	LQ0	E0		PP, EX, A	VE01	HA08	0	
3209	METALLIC SUBSTANCE, WATER-REACTIVE, SELF-HEATING, N.O.S.	4.3	WS	II	4.3+4.2	274 558	LQ11	E2		PP, EX, A	VE01	HA08	0	
3209	METALLIC SUBSTANCE, WATER-REACTIVE, SELF-HEATING, N.O.S.	4.3	WS	III	4.3+4.2	274 558	LQ12	E1		PP, EX, A	VE01	HA08	0	

UN No. or ID No.	Name and description	Class	Classi-fication Code	Packing group	Labels	Special provi-sions	Limited and excepted quantities		Carriage permitted	Equipment required	Venti-lation	Provisions concerning loading, unloading and carriage	Number of blue cones/ lights	Remarks
	3.1.2	2.2	2.2	2.1.1.3	5.2.2	3.3	3.4.6	3.5.1.2	3.2.1	8.1.5	7.1.6	7.1.6	7.1.5	3.2.1
(1)	(2)	(3a)	(3b)	(4)	(5)	(6)	(7a)	(7b)	(8)	(9)	(10)	(11)	(12)	(13)
3210	CHLORATES, INORGANIC, AQUEOUS SOLUTION, N.O.S.	5.1	O1	II	5.1	274 605	LQ10	E2		PP			0	
3210	CHLORATES, INORGANIC, AQUEOUS SOLUTION, N.O.S.	5.1	O1	III	5.1	274 605	LQ13	E1		PP			0	
3211	PERCHLORATES, INORGANIC, AQUEOUS SOLUTION, N.O.S.	5.1	O1	II	5.1	274	LQ10	E2		PP			0	
3211	PERCHLORATES, INORGANIC, AQUEOUS SOLUTION, N.O.S.	5.1	O1	III	5.1	274	LQ13	E1		PP			0	
3212	HYPOCHLORITES, INORGANIC, N.O.S.	5.1	O2	II	5.1	274 559	LQ11	E2		PP			0	
3213	BROMATES, INORGANIC, AQUEOUS SOLUTION, N.O.S.	5.1	O1	II	5.1	274 604	LQ10	E2		PP			0	
3213	BROMATES, INORGANIC, AQUEOUS SOLUTION, N.O.S.	5.1	O1	III	5.1	274 604	LQ13	E1		PP			0	
3214	PERMANGANATES, INORGANIC, AQUEOUS SOLUTION, N.O.S.	5.1	O1	II	5.1	274 608	LQ10	E2		PP			0	
3215	PERSULPHATES, INORGANIC, N.O.S.	5.1	O2	III	5.1	274	LQ12	E1		PP			0	
3216	PERSULPHATES, INORGANIC, AQUEOUS SOLUTION, N.O.S.	5.1	O1	III	5.1	274	LQ13	E1		PP			0	
3218	NITRATES, INORGANIC, AQUEOUS SOLUTION, N.O.S.	5.1	O1	II	5.1	270 274 511	LQ10	E2		PP			0	
3218	NITRATES, INORGANIC, AQUEOUS SOLUTION, N.O.S.	5.1	O1	III	5.1	270 274 511	LQ13	E1		PP			0	
3219	NITRITES, INORGANIC, AQUEOUS SOLUTION, N.O.S.	5.1	O1	II	5.1	103 274	LQ10	E2		PP			0	
3219	NITRITES, INORGANIC, AQUEOUS SOLUTION, N.O.S.	5.1	O1	III	5.1	103 274	LQ13	E1		PP			0	
3220	PENTAFLUOROETHANE (REFRIGERANT GAS R 125)	2	2A		2.2		LQ1	E1		PP			0	
3221	SELF-REACTIVE LIQUID TYPE B	4.1	SR1		4.1+1	181 194 274	LQ14	E0		PP		HA01, HA10	3	
3222	SELF-REACTIVE SOLID TYPE B	4.1	SR1		4.1+1	181 194 274	LQ15	E0		PP		HA01, HA10	3	
3223	SELF-REACTIVE LIQUID TYPE C	4.1	SR1		4.1	194 274	LQ14	E0		PP			0	
3224	SELF-REACTIVE SOLID TYPE C	4.1	SR1		4.1	194 274	LQ15	E0		PP			0	
3225	SELF-REACTIVE LIQUID TYPE D	4.1	SR1		4.1	194 274	LQ16	E0		PP			0	
3226	SELF-REACTIVE SOLID TYPE D	4.1	SR1		4.1	194 274	LQ11	E0		PP			0	

UN No. or ID No.	Name and description	Class	Classi-fication Code	Packing group	Labels	Special provi-sions	Limited and excepted quantities		Carriage permitted	Equipment required	Venti-lation	Provisions concerning loading, unloading and carriage	Number of blue cones/ lights	Remarks
	3.1.2	2.2	2.2	2.1.1.3	5.2.2	3.3	3.4.6	3.5.1.2	3.2.1	8.1.5	7.1.6	7.1.6	7.1.5	3.2.1
(1)	(2)	(3a)	(3b)	(4)	(5)	(6)	(7a)	(7b)	(8)	(9)	(10)	(11)	(12)	(13)
3227	SELF-REACTIVE LIQUID TYPE E	4.1	SR1		4.1	194 274	LQ16	E0		PP			0	
3228	SELF-REACTIVE SOLID TYPE E	4.1	SR1		4.1	194 274	LQ11	E0		PP			0	
3229	SELF-REACTIVE LIQUID TYPE F	4.1	SR1		4.1	194 274	LQ16	E0		PP			0	
3230	SELF-REACTIVE SOLID TYPE F	4.1	SR1		4.1	194 274	LQ11	E0		PP			0	
3231	SELF-REACTIVE LIQUID TYPE B, TEMPERATURE CONTROLLED	4.1	SR2		4.1+1	181 194 274	LQ0	E0		PP		HA01, HA10	3	
3232	SELF-REACTIVE SOLID TYPE B, TEMPERATURE CONTROLLED	4.1	SR2		4.1+1	181 194 274	LQ0	E0		PP		HA01, HA10	3	
3233	SELF-REACTIVE LIQUID TYPE C, TEMPERATURE CONTROLLED	4.1	SR2		4.1	194 274	LQ0	E0		PP			0	
3234	SELF-REACTIVE SOLID TYPE C, TEMPERATURE CONTROLLED	4.1	SR2		4.1	194 274	LQ0	E0		PP			0	
3235	SELF-REACTIVE LIQUID TYPE D, TEMPERATURE CONTROLLED	4.1	SR2		4.1	194 274	LQ0	E0		PP			0	
3236	SELF-REACTIVE SOLID TYPE D, TEMPERATURE CONTROLLED	4.1	SR2		4.1	194 274	LQ0	E0		PP			0	
3237	SELF-REACTIVE LIQUID TYPE E, TEMPERATURE CONTROLLED	4.1	SR2		4.1	194 274	LQ0	E0		PP			0	
3238	SELF-REACTIVE SOLID TYPE E, TEMPERATURE CONTROLLED	4.1	SR2		4.1	194 274	LQ0	E0		PP			0	
3239	SELF-REACTIVE LIQUID TYPE F, TEMPERATURE CONTROLLED	4.1	SR2		4.1	194 274	LQ0	E0		PP			0	
3240	SELF-REACTIVE SOLID TYPE F, TEMPERATURE CONTROLLED	4.1	SR2		4.1	194 274	LQ0	E0		PP			0	
3241	2-BROMO-2-NITROPROPANE-1,3-DIOL	4.1	SR1	III	4.1	638	LQ0	E1		PP			0	
3242	AZODICARBONAMIDE	4.1	SR1	II	4.1	215 638	LQ0	E2		PP			0	
3243	SOLIDS CONTAINING TOXIC LIQUID, N.O.S.	6.1	T9	II	6.1	217 274 802	LQ18	E4		PP, EP, TOX, A	VE02		2	
3244	SOLIDS CONTAINING CORROSIVE LIQUID, N.O.S.	8	C10	II	8	218 274	LQ23	E2		PP, EP			0	
3245	GENETICALLY MODIFIED MICROORGANISMS or GENETICALLY MODIFIED ORGANISMS	9	M8		9	219 637 802	LQ0	E0		PP			0	
3245	GENETICALLY MODIFIED MICRO-ORGANISMS or GENETICALLY MODIFIED ORGANISMS, in refrigerated liquid nitrogen	9	M8		9 +2.2	219 637 802	LQ0	E0		PP			0	
3246	METHANESULPHONYL CHLORIDE	6.1	TC1	I	6.1+8	802	LQ0	E5		PP, EP, TOX, A	VE02		2	

UN No. or ID No.	Name and description	Class	Classi-fication Code	Packing group	Labels	Special provi-sions	Limited and excepted quantities		Carriage permitted	Equipment required	Venti-lation	Provisions concerning loading, unloading and carriage	Number of blue cones/ lights	Remarks
	3.1.2	2.2	2.2	2.1.1.3	5.2.2	3.3	3.4.6	3.5.1.2	3.2.1	8.1.5	7.1.6	7.1.6	7.1.5	3.2.1
(1)	(2)	(3a)	(3b)	(4)	(5)	(6)	(7a)	(7b)	(8)	(9)	(10)	(11)	(12)	(13)
3247	SODIUM PEROXOBORATE, ANHYDROUS	5.1	O2	II	5.1		LQ11	E2		PP			0	
3248	MEDICINE, LIQUID, FLAMMABLE, TOXIC, N.O.S.	3	FT1	II	3+6.1	220 221 274 601 802	LQ0	E2		PP, EP, EX, TOX, A	VE01, VE02		2	
3248	MEDICINE, LIQUID, FLAMMABLE, TOXIC, N.O.S.	3	FT1	III	3+6.1	220 221 274 601 802	LQ7	E1		PP, EP, EX, TOX, A	VE01, VE02		0	
3249	MEDICINE, SOLID, TOXIC, N.O.S.	6.1	T2	II	6.1	221 274 601 802	LQ18	E4		PP, EP			2	
3249	MEDICINE, SOLID, TOXIC, N.O.S.	6.1	T2	III	6.1	221 274 601 802	LQ9	E1		PP, EP			0	
3250	CHLOROACETIC ACID, MOLTEN	6.1	TC1	II	6.1+8	802	LQ0	E0		PP, EP, TOX, A	VE02		2	
3251	ISOSORBIDE-5-MONONITRATE	4.1	SR1	III	4.1	226 638	LQ0	E1		PP			0	
3252	DIFLUOROMETHANE (REFRIGERANT GAS R 32)	2	2F		2.1		LQ0	E0		PP, EX, A	VE01		1	
3253	DISODIUM TRIOXOSILICATE	8	C6	III	8		LQ24	E1		PP, EP			0	
3254	TRIBUTYLPHOSPHANE	4.2	S1	I	4.2		LQ0	E0		PP			0	
3255	tert-BUTYL HYPOCHLORITE	4.2	SC1	**CARRIAGE PROHIBITED**										
3256	ELEVATED TEMPERATURE LIQUID, FLAMMABLE, N.O.S. with flash-point above 60 °C, at or above its flash-point	3	F2	III	3	274 560	LQ0	E0	**T**	PP, EX, A	VE01		0	
3257	ELEVATED TEMPERATURE LIQUID, N.O.S., at or above 100 °C and below its flash-point (including molten metals, molten salts, etc.)	9	M9	III	9	274 580 643	LQ0	E0	**T**	PP			0	
3258	ELEVATED TEMPERATURE SOLID, N.O.S., at or above 240 °C	9	M10	III	9	274 580 643	LQ0	E0		PP			0	
3259	AMINES, SOLID, CORROSIVE, N.O.S. or POLYAMINES, SOLID, CORROSIVE, N.O.S.	8	C8	I	8	274	LQ0	E0		PP, EP			0	
3259	AMINES, SOLID, CORROSIVE, N.O.S. or POLYAMINES, SOLID, CORROSIVE, N.O.S.	8	C8	II	8	274	LQ23	E2		PP, EP			0	
3259	AMINES, SOLID, CORROSIVE, N.O.S. or POLYAMINES, SOLID, CORROSIVE, N.O.S.	8	C8	III	8	274	LQ24	E1	**T**	PP, EP			0	
3260	CORROSIVE SOLID, ACIDIC, INORGANIC, N.O.S.	8	C2	I	8	274	LQ0	E0		PP, EP			0	

UN No. or ID No.	Name and description	Class	Classi-fication Code	Packing group	Labels	Special provi-sions	Limited and excepted quantities		Carriage permitted	Equipment required	Venti-lation	Provisions concerning loading, unloading and carriage	Number of blue cones/ lights	Remarks
	3.1.2	2.2	2.2	2.1.1.3	5.2.2	3.3	3.4.6	3.5.1.2	3.2.1	8.1.5	7.1.6	7.1.6	7.1.5	3.2.1
(1)	(2)	(3a)	(3b)	(4)	(5)	(6)	(7a)	(7b)	(8)	(9)	(10)	(11)	(12)	(13)
3260	CORROSIVE SOLID, ACIDIC, INORGANIC, N.O.S.	8	C2	II	8	274	LQ23	E2		PP, EP			0	
3260	CORROSIVE SOLID, ACIDIC, INORGANIC, N.O.S.	8	C2	III	8	274	LQ24	E1		PP, EP			0	
3261	CORROSIVE SOLID, ACIDIC, ORGANIC, N.O.S.	8	C4	I	8	274	LQ0	E0		PP, EP			0	
3261	CORROSIVE SOLID, ACIDIC, ORGANIC, N.O.S.	8	C4	II	8	274	LQ23	E2		PP, EP			0	
3261	CORROSIVE SOLID, ACIDIC, ORGANIC, N.O.S.	8	C4	III	8	274	LQ24	E1		PP, EP			0	
3262	CORROSIVE SOLID, BASIC, INORGANIC, N.O.S.	8	C6	I	8	274	LQ0	E0		PP, EP			0	
3262	CORROSIVE SOLID, BASIC, INORGANIC, N.O.S.	8	C6	II	8	274	LQ23	E2		PP, EP			0	
3262	CORROSIVE SOLID, BASIC, INORGANIC, N.O.S.	8	C6	III	8	274	LQ24	E1		PP, EP			0	
3263	CORROSIVE SOLID, BASIC, ORGANIC, N.O.S.	8	C8	I	8	274	LQ0	E0		PP, EP			0	
3263	CORROSIVE SOLID, BASIC, ORGANIC, N.O.S.	8	C8	II	8	274	LQ23	E2		PP, EP			0	
3263	CORROSIVE SOLID, BASIC, ORGANIC, N.O.S.	8	C8	III	8	274	LQ24	E1		PP, EP			0	
3264	CORROSIVE LIQUID, ACIDIC, INORGANIC, N.O.S.	8	C1	I	8	274	LQ0	E0	**T**	PP, EP			0	
3264	CORROSIVE LIQUID, ACIDIC, INORGANIC, N.O.S.	8	C1	II	8	274	LQ22	E2	**T**	PP, EP			0	
3264	CORROSIVE LIQUID, ACIDIC, INORGANIC, N.O.S.	8	C1	III	8	274	LQ7	E1	**T**	PP, EP			0	
3265	CORROSIVE LIQUID, ACIDIC, ORGANIC, N.O.S.	8	C3	I	8	274	LQ0	E0	**T**	PP, EP			0	
3265	CORROSIVE LIQUID, ACIDIC, ORGANIC, N.O.S.	8	C3	II	8	274	LQ22	E2	**T**	PP, EP			0	
3265	CORROSIVE LIQUID, ACIDIC, ORGANIC, N.O.S.	8	C3	III	8	274	LQ7	E1	**T**	PP, EP			0	
3266	CORROSIVE LIQUID, BASIC, INORGANIC, N.O.S.	8	C5	I	8	274	LQ0	E0	**T**	PP, EP			0	
3266	CORROSIVE LIQUID, BASIC, INORGANIC, N.O.S.	8	C5	II	8	274	LQ22	E2	**T**	PP, EP			0	
3266	CORROSIVE LIQUID, BASIC, INORGANIC, N.O.S.	8	C5	III	8	274	LQ7	E1	**T**	PP, EP			0	
3267	CORROSIVE LIQUID, BASIC, ORGANIC, N.O.S.	8	C7	I	8	274	LQ0	E0	**T**	PP, EP			0	
3267	CORROSIVE LIQUID, BASIC, ORGANIC, N.O.S.	8	C7	II	8	274	LQ22	E2	**T**	PP, EP			0	
3267	CORROSIVE LIQUID, BASIC, ORGANIC, N.O.S.	8	C7	III	8	274	LQ7	E1	**T**	PP, EP			0	

UN No. or ID No.	Name and description	Class	Classi-fication Code	Packing group	Labels	Special provi-sions	Limited and excepted quantities		Carriage permitted	Equipment required	Venti-lation	Provisions concerning loading, unloading and carriage			Number of blue cones/ lights	Remarks
	3.1.2	2.2	2.2	2.1.1.3	5.2.2	3.3	3.4.6	3.5.1.2	3.2.1	8.1.5	7.1.6	7.1.6			7.1.5	3.2.1
(1)	(2)	(3a)	(3b)	(4)	(5)	(6)	(7a)	(7b)	(8)	(9)	(10)	(11)			(12)	(13)
3268	AIR BAG INFLATORS or AIR BAG MODULES or SEAT-BELT PRETENSIONERS	9	M5	III	9	280 289	LQ0	E0		PP					0	
3269	POLYESTER RESIN KIT	3	F1	II	3	236 340	LQ6	E0		PP, EX, A	VE01				1	
3269	POLYESTER RESIN KIT	3	F1	III	3	236 340	LQ7	E0		PP, EX, A	VE01				0	
3270	NITROCELLULOSE MEMBRANE FILTERS, with not more than 12.6% nitrogen, by dry mass	4.1	F1	II	4.1	237 286	LQ8	E2		PP					1	
3271	ETHERS, N.O.S.	3	F1	II	3	274	LQ4	E2	**T**	PP, EX, A	VE01				1	
3271	ETHERS, N.O.S.	3	F1	III	3	274	LQ7	E1	**T**	PP, EX, A	VE01				0	
3272	ESTERS, N.O.S.	3	F1	II	3	274 601	LQ4	E2	**T**	PP, EX, A	VE01				1	
3272	ESTERS, N.O.S.	3	F1	III	3	274 601	LQ7	E1	**T**	PP, EX, A	VE01				0	
3273	NITRILES, FLAMMABLE, TOXIC, N.O.S.	3	FT1	I	3+6.1	274 802	LQ0	E0		PP, EP, EX, TOX, A	VE01, VE02				2	
3273	NITRILES, FLAMMABLE, TOXIC, N.O.S.	3	FT1	II	3+6.1	274 802	LQ0	E2		PP, EP, EX, TOX, A	VE01, VE02				2	
3274	ALCOHOLATES SOLUTION, N.O.S., in alcohol	3	FC	II	3+8	274	LQ4	E2		PP, EP, EX, A	VE01				1	
3275	NITRILES, TOXIC, FLAMMABLE, N.O.S.	6.1	TF1	I	6.1+3	274 315 802	LQ0	E5		PP, EP, EX, TOX, A	VE01, VE02				2	
3275	NITRILES, TOXIC, FLAMMABLE, N.O.S.	6.1	TF1	II	6.1+3	274 802	LQ17	E4		PP, EP, EX, TOX, A	VE01, VE02				2	
3276	NITRILES, TOXIC, LIQUID, N.O.S.	6.1	T1	I	6.1	274 315 802	LQ0	E5		PP, EP, TOX, A	VE02				2	
3276	NITRILES, TOXIC, LIQUID, N.O.S.	6.1	T1	II	6.1	274 802	LQ17	E4	**T**	PP, EP, TOX, A	VE02				2	
3276	NITRILES, TOXIC, LIQUID, N.O.S.	6.1	T1	III	6.1	274 802	LQ7	E1		PP, EP, TOX, A	VE02				0	
3277	CHLOROFORMATES, TOXIC, CORROSIVE, N.O.S.	6.1	TC1	II	6.1+8	274 561 802	LQ17	E4		PP, EP, TOX, A	VE02				2	
3278	ORGANOPHOSPHORUS COMPOUND, TOXIC, LIQUID, N.O.S.	6.1	T1	I	6.1	43 274 315 802	LQ0	E5		PP, EP, TOX, A	VE02				2	
3278	ORGANOPHOSPHORUS COMPOUND, TOXIC, LIQUID, N.O.S.	6.1	T1	II	6.1	43 274 802	LQ17	E4		PP, EP, TOX, A	VE02				2	
3278	ORGANOPHOSPHORUS COMPOUND, TOXIC, LIQUID, N.O.S.	6.1	T1	III	6.1	43 274 802	LQ7	E1		PP, EP, TOX, A	VE02				0	

UN No. or ID No.	Name and description	Class	Classi-fication Code	Packing group	Labels	Special provi-sions	Limited and excepted quantities		Carriage permitted	Equipment required	Venti-lation	Provisions concerning loading, unloading and carriage	Number of blue cones/ lights	Remarks
	3.1.2	2.2	2.2	2.1.1.3	5.2.2	3.3	3.4.6	3.5.1.2	3.2.1	8.1.5	7.1.6	7.1.6	7.1.5	3.2.1
(1)	(2)	(3a)	(3b)	(4)	(5)	(6)	(7a)	(7b)	(8)	(9)	(10)	(11)	(12)	(13)
3279	ORGANOPHOSPHORUS COMPOUND, TOXIC, FLAMMABLE, N.O.S.	6.1	TF1	I	6.1+3	43 274 315 802	LQ0	E5		PP, EP, EX, TOX, A	VE01, VE02		2	
3279	ORGANOPHOSPHORUS COMPOUND, TOXIC, FLAMMABLE, N.O.S.	6.1	TF1	II	6.1+3	43 274 802	LQ17	E4		PP, EP, EX, TOX, A	VE01, VE02		2	
3280	ORGANOARSENIC COMPOUND, LIQUID, N.O.S.	6.1	T3	I	6.1	274 315 802	LQ0	E5		PP, EP, TOX, A	VE02		2	
3280	ORGANOARSENIC COMPOUND, LIQUID, N.O.S.	6.1	T3	II	6.1	274 802	LQ17	E4		PP, EP, TOX, A	VE02		2	
3280	ORGANOARSENIC COMPOUND, LIQUID, N.O.S.	6.1	T3	III	6.1	274 802	LQ7	E1		PP, EP, TOX, A	VE02		0	
3281	METAL CARBONYLS, LIQUID, N.O.S.	6.1	T3	I	6.1	274 315 562 802	LQ0	E5		PP, EP, TOX, A	VE02		2	
3281	METAL CARBONYLS, LIQUID, N.O.S.	6.1	T3	II	6.1	274 562 802	LQ17	E4		PP, EP, TOX, A	VE02		2	
3281	METAL CARBONYLS, LIQUID, N.O.S.	6.1	T3	III	6.1	274 562 802	LQ7	E1		PP, EP, TOX, A	VE02		0	
3282	ORGANOMETALLIC COMPOUND, TOXIC, LIQUID, N.O.S.	6.1	T3	I	6.1	274 562 802	LQ0	E5		PP, EP, TOX, A	VE02		2	
3282	ORGANOMETALLIC COMPOUND, TOXIC, LIQUID, N.O.S.	6.1	T3	II	6.1	274 562 802	LQ17	E4		PP, EP, TOX, A	VE02		2	
3282	ORGANOMETALLIC COMPOUND, TOXIC, LIQUID, N.O.S.	6.1	T3	III	6.1	274 562 802	LQ7	E1		PP, EP, TOX, A	VE02		0	
3283	SELENIUM COMPOUND, SOLID, N.O.S.	6.1	T5	I	6.1	274 563 802	LQ0	E5		PP, EP			2	
3283	SELENIUM COMPOUND, SOLID, N.O.S.	6.1	T5	II	6.1	274 563 802	LQ18	E4		PP, EP			2	
3283	SELENIUM COMPOUND, SOLID, N.O.S.	6.1	T5	III	6.1	274 563 802	LQ9	E1		PP, EP			0	
3284	TELLURIUM COMPOUND, N.O.S.	6.1	T5	I	6.1	274 802	LQ0	E5		PP, EP			2	
3284	TELLURIUM COMPOUND, N.O.S.	6.1	T5	II	6.1	274 802	LQ18	E4		PP, EP			2	

UN No. or ID No.	Name and description	Class	Classi-fication Code	Packing group	Labels	Special provi-sions	Limited and excepted quantities		Carriage permitted	Equipment required	Venti-lation	Provisions concerning loading, unloading and carriage	Number of blue cones/ lights	Remarks
	3.1.2	2.2	2.2	2.1.1.3	5.2.2	3.3	3.4.6	3.5.1.2	3.2.1	8.1.5	7.1.6	7.1.6	7.1.5	3.2.1
(1)	(2)	(3a)	(3b)	(4)	(5)	(6)	(7a)	(7b)	(8)	(9)	(10)	(11)	(12)	(13)
3284	TELLURIUM COMPOUND, N.O.S.	6.1	T5	III	6.1	274 802	LQ9	E1		PP, EP			0	
3285	VANADIUM COMPOUND, N.O.S.	6.1	T5	I	6.1	274 564 802	LQ0	E5		PP, EP			2	
3285	VANADIUM COMPOUND, N.O.S.	6.1	T5	II	6.1	274 564 802	LQ18	E4		PP, EP			2	
3285	VANADIUM COMPOUND, N.O.S.	6.1	T5	III	6.1	274 564 802	LQ9	E1		PP, EP			0	
3286	FLAMMABLE LIQUID, TOXIC, CORROSIVE, N.O.S.	3	FTC	I	3+6.1+8	274 802	LQ0	E0	**T**	PP, EP, EX, TOX, A	VE01, VE02		2	
3286	FLAMMABLE LIQUID, TOXIC, CORROSIVE, N.O.S.	3	FTC	II	3+6.1+8	274 802	LQ0	E2	**T**	PP, EP, EX, TOX, A	VE01, VE02		2	
3287	TOXIC LIQUID, INORGANIC, N.O.S.	6.1	T4	I	6.1	274 315 802	LQ0	E5	**T**	PP, EP, TOX, A	VE02		2	
3287	TOXIC LIQUID, INORGANIC, N.O.S.	6.1	T4	II	6.1	274 802	LQ17	E4	**T**	PP, EP, TOX, A	VE02		2	
3287	TOXIC LIQUID, INORGANIC, N.O.S.	6.1	T4	III	6.1	274 802	LQ7	E1	**T**	PP, EP, TOX, A	VE02		0	
3288	TOXIC SOLID, INORGANIC, N.O.S.	6.1	T5	I	6.1	274 802	LQ0	E5		PP, EP			2	
3288	TOXIC SOLID, INORGANIC, N.O.S.	6.1	T5	II	6.1	274 802	LQ18	E4		PP, EP			2	
3288	TOXIC SOLID, INORGANIC, N.O.S.	6.1	T5	III	6.1	274 802	LQ9	E1		PP, EP			0	
3289	TOXIC LIQUID, CORROSIVE, INORGANIC, N.O.S.	6.1	TC3	I	6.1+8	274 315 802	LQ0	E5	**T**	PP, EP, TOX, A	VE02		2	
3289	TOXIC LIQUID, CORROSIVE, INORGANIC, N.O.S.	6.1	TC3	II	6.1+8	274 802	LQ17	E4	**T**	PP, EP, TOX, A	VE02		2	
3290	TOXIC SOLID, CORROSIVE, INORGANIC, N.O.S.	6.1	TC4	I	6.1+8	274 802	LQ0	E5		PP, EP			2	
3290	TOXIC SOLID, CORROSIVE, INORGANIC, N.O.S.	6.1	TC4	II	6.1+8	274 802	LQ18	E4		PP, EP			2	
3291	CLINICAL WASTE, UNSPECIFIED, N.O.S. or (BIO) MEDICAL WASTE, N.O.S. or REGULATED MEDICAL WASTE, N.O.S.	6.2	I3	II	6.2	565 802	LQ0	E0		PP			0	
3291	CLINICAL WASTE, UNSPECIFIED, N.O.S. or (BIO) MEDICAL WASTE, N.O.S. or REGULATED MEDICAL WASTE, N.O.S., in refrigerated liquid nitrogen	6.2	I3	II	6.2 +2.2	565 802	LQ0	E0		PP			0	
3292	BATTERIES, CONTAINING SODIUM, or CELLS, CONTAINING SODIUM	4.3	W3	II	4.3	239 295	LQ0	E0		PP, EX, A	VE01	HA08	0	

UN No. or ID No.	Name and description	Class	Classi-fication Code	Packing group	Labels	Special provi-sions	Limited and excepted quantities		Carriage permitted	Equipment required	Venti-lation	Provisions concerning loading, unloading and carriage	Number of blue cones/ lights	Remarks
	3.1.2	2.2	2.2	2.1.1.3	5.2.2	3.3	3.4.6	3.5.1.2	3.2.1	8.1.5	7.1.6	7.1.6	7.1.5	3.2.1
(1)	(2)	(3a)	(3b)	(4)	(5)	(6)	(7a)	(7b)	(8)	(9)	(10)	(11)	(12)	(13)
3293	HYDRAZINE, AQUEOUS SOLUTION with not more than 37% hydrazine, by mass	6.1	T4	III	6.1	566 802	LQ7	E1		PP, EP, TOX, A	VE02		0	
3294	HYDROGEN CYANIDE, SOLUTION IN ALCOHOL with not more than 45% hydrogen cyanide	6.1	TF1	I	6.1+3	610 802	LQ0	E5		PP, EP, EX, TOX, A	VE01, VE02		2	
3295	HYDROCARBONS, LIQUID, N.O.S.	3	F1	I	3	649	LQ3	E3	T	PP, EX, A	VE01		1	
3295	HYDROCARBONS, LIQUID, N.O.S. (vapour pressure at 50 °C more than 110 kPa)	3	F1	II	3	640C 649	LQ4	E2	T	PP, EX, A	VE01		1	
3295	HYDROCARBONS, LIQUID, N.O.S. (vapour pressure at 50 °C not more than 110 kPa)	3	F1	II	3	640D 649	LQ4	E2	T	PP, EX, A	VE01		1	
3295	HYDROCARBONS, LIQUID, N.O.S.	3	F1	III	3		LQ7	E1	T	PP, EX, A	VE01		0	
3296	HEPTAFLUOROPROPANE (REFRIGERANT GAS R 227)	2	2A		2.2		LQ1	E1		PP			0	
3297	ETHYLENE OXIDE AND CHLOROTETRAFLUOROETHANE MIXTURE with not more than 8.8% ethylene oxide	2	2A		2.2		LQ1	E1		PP			0	
3298	ETHYLENE OXIDE AND PENTAFLUOROETHANE MIXTURE with not more than 7.9% ethylene oxide	2	2A		2.2		LQ1	E1		PP			0	
3299	ETHYLENE OXIDE AND TETRAFLUOROETHANE MIXTURE with not more than 5.6% ethylene oxide	2	2A		2.2		LQ1	E1		PP			0	
3300	ETHYLENE OXIDE AND CARBON DIOXIDE MIXTURE with more than 87% ethylene oxide	2	2TF		2.3+2.1		LQ0	E0		PP, EP, EX, TOX, A	VE01, VE02		2	
3301	CORROSIVE LIQUID, SELF-HEATING, N.O.S.	8	CS1	I	8+4.2	274	LQ0	E0		PP, EP			0	
3301	CORROSIVE LIQUID, SELF-HEATING, N.O.S.	8	CS1	II	8+4.2	274	LQ22	E2		PP, EP			0	
3302	2-DIMETHYLAMINOETHYL ACRYLATE	6.1	T1	II	6.1	802	LQ17	E4		PP, EP, TOX, A	VE02		2	
3303	COMPRESSED GAS, TOXIC, OXIDIZING, N.O.S.	2	1TO		2.3+5.1	274	LQ0	E0		PP, EP, TOX, A	VE02		2	
3304	COMPRESSED GAS, TOXIC, CORROSIVE, N.O.S.	2	1TC		2.3+8	274	LQ0	E0		PP, EP, TOX, A	VE02		2	
3305	COMPRESSED GAS, TOXIC, FLAMMABLE, CORROSIVE, N.O.S.	2	1TFC		2.3+2.1+8	274	LQ0	E0		PP, EP, EX, TOX, A	VE01, VE02		2	
3306	COMPRESSED GAS, TOXIC, OXIDIZING, CORROSIVE, N.O.S.	2	1TOC		2.3+5.1+8	274	LQ0	E0		PP, EP, TOX, A	VE02		2	
3307	LIQUEFIED GAS, TOXIC, OXIDIZING, N.O.S.	2	2TO		2.3+5.1	274	LQ0	E0		PP, EP, TOX, A	VE02		2	
3308	LIQUEFIED GAS, TOXIC, CORROSIVE, N.O.S.	2	2TC		2.3+8	274	LQ0	E0		PP, EP, TOX, A	VE02		2	
3309	LIQUEFIED GAS, TOXIC, FLAMMABLE, CORROSIVE, N.O.S.	2	2TFC		2.3+2.1+8	274	LQ0	E0		PP, EP, EX, TOX, A	VE01, VE02		2	
3310	LIQUEFIED GAS, TOXIC, OXIDIZING, CORROSIVE, N.O.S.	2	2TOC		2.3+5.1+8	274	LQ0	E0		PP, EP, TOX, A	VE02		2	

UN No. or ID No.	Name and description	Class	Classi-fication Code	Packing group	Labels	Special provi-sions	Limited and excepted quantities		Carriage permitted	Equipment required	Venti-lation	Provisions concerning loading, unloading and carriage			Number of blue cones/ lights	Remarks
	3.1.2	2.2	2.2	2.1.1.3	5.2.2	3.3	3.4.6	3.5.1.2	3.2.1	8.1.5	7.1.6	7.1.6			7.1.5	3.2.1
(1)	(2)	(3a)	(3b)	(4)	(5)	(6)	(7a)	(7b)	(8)	(9)	(10)	(11)			(12)	(13)
3311	GAS, REFRIGERATED LIQUID, OXIDIZING, N.O.S.	2	3O		2.2+5.1	274	LQ0	E0		PP					0	
3312	GAS, REFRIGERATED LIQUID, FLAMMABLE, N.O.S.	2	3F		2.1	274	LQ0	E0		PP, EX, A	VE01				1	
3313	ORGANIC PIGMENTS, SELF-HEATING	4.2	S2	II	4.2		LQ0	E2		PP					0	
3313	ORGANIC PIGMENTS, SELF-HEATING	4.2	S2	III	4.2		LQ0	E1		PP					0	
3314	PLASTICS MOULDING COMPOUND in dough, sheet or extruded rope form evolving flammable vapour	9	M3	III	**none**	207 633	LQ27	E1		PP, EP, EX, A	VE01				0	
3315	CHEMICAL SAMPLE, TOXIC	6.1	T8	I	6.1	250 802	LQ0	E5		PP, EP, TOX, A	VE02				2	
3316	CHEMICAL KIT or FIRST AID KIT	9	M11	II	9	251 340	LQ0	E0		PP					0	
3316	CHEMICAL KIT or FIRST AID KIT	9	M11	III	9	251 340	LQ0	E0		PP					0	
3317	2-AMINO-4,6-DINITROPHENOL, WETTED with not less than 20% water, by mass	4.1	D	I	4.1		LQ0	E0		PP					1	
3318	AMMONIA SOLUTION, relative density less than 0.880 at 15 °C in water, with more than 50% ammonia	2	4TC		2.3+8	23	LQ0	E0		PP, EP, TOX, A	VE02				2	
3319	NITROGLYCERIN MIXTURE, DESENSITIZED, SOLID, N.O.S. with more than 2% but not more than 10% nitroglycerin, by mass	4.1	D	II	4.1	272 274	LQ0	E0		PP					0	
3320	SODIUM BOROHYDRIDE AND SODIUM HYDROXIDE SOLUTION, with not more than 12% sodium borohydride and not more than 40% sodium hydroxide by mass	8	C5	II	8		LQ22	E2		PP, EP					0	
3320	SODIUM BOROHYDRIDE AND SODIUM HYDROXIDE SOLUTION, with not more than 12% sodium borohydride and not more than 40% sodium hydroxide by mass	8	C5	III	8		LQ7	E1		PP, EP					0	
3321	RADIOACTIVE MATERIAL, LOW SPECIFIC ACTIVITY (LSA-II), non fissile or fissile-excepted	7			7X	172 317 325 336	LQ0	E0		PP					2	
3322	RADIOACTIVE MATERIAL, LOW SPECIFIC ACTIVITY (LSA-III), non fissile or fissile-excepted	7			7X	172 317 325 336	LQ0	E0		PP					2	
3323	RADIOACTIVE MATERIAL, TYPE C PACKAGE, non fissile or fissile-excepted	7			7X	172 317	LQ0	E0		PP					2	
3324	RADIOACTIVE MATERIAL, LOW SPECIFIC ACTIVITY (LSA-II), FISSILE	7			7X+7E	172 326 336	LQ0	E0		PP					2	

UN No. or ID No.	Name and description	Class	Classi-fication Code	Packing group	Labels	Special provi-sions	Limited and excepted quantities		Carriage permitted	Equipment required	Venti-lation	Provisions concerning loading, unloading and carriage	Number of blue cones/ lights	Remarks
	3.1.2	2.2	2.2	2.1.1.3	5.2.2	3.3	3.4.6	3.5.1.2	3.2.1	8.1.5	7.1.6	7.1.6	7.1.5	3.2.1
(1)	(2)	(3a)	(3b)	(4)	(5)	(6)	(7a)	(7b)	(8)	(9)	(10)	(11)	(12)	(13)
3325	RADIOACTIVE MATERIAL, LOW SPECIFIC ACTIVITY, (LSA-III), FISSILE	7			7X+7E	172 326 336	LQ0	E0		PP			2	
3326	RADIOACTIVE MATERIAL, SURFACE CONTAMINATED OBJECTS (SCO-I or SCO-II), FISSILE	7			7X+7E	172 336	LQ0	E0		PP			2	
3327	RADIOACTIVE MATERIAL, TYPE A PACKAGE, FISSILE, non-special form	7			7X+7E	172 326	LQ0	E0		PP			2	
3328	RADIOACTIVE MATERIAL, TYPE B(U) PACKAGE, FISSILE	7			7X+7E	172 337	LQ0	E0		PP			2	
3329	RADIOACTIVE MATERIAL, TYPE B(M) PACKAGE, FISSILE	7			7X+7E	172 337	LQ0	E0		PP			2	
3330	RADIOACTIVE MATERIAL, TYPE C PACKAGE, FISSILE	7			7X+7E	172	LQ0	E0		PP			2	
3331	RADIOACTIVE MATERIAL, TRANSPORTED UNDER SPECIAL ARRANGEMENT, FISSILE	7			7X+7E	172	LQ0	E0		PP			2	
3332	RADIOACTIVE MATERIAL, TYPE A PACKAGE, SPECIAL FORM, non fissile or fissile-excepted	7			7X	172 317	LQ0	E0		PP			2	
3333	RADIOACTIVE MATERIAL, TYPE A PACKAGE, SPECIAL FORM, FISSILE	7			7X+7E	172	LQ0	E0		PP			2	
3334	Aviation regulated liquid, n.o.s.	9	M11	NOT SUBJECT TO ADN										
3335	Aviation regulated solid, n.o.s.	9	M11	NOT SUBJECT TO ADN										
3336	MERCAPTANS, LIQUID, FLAMMABLE, N.O.S. or MERCAPTAN MIXTURE, LIQUID, FLAMMABLE, N.O.S.	3	F1	I	3	274	LQ3	E3		PP, EX, A	VE01		1	
3336	MERCAPTANS, LIQUID, FLAMMABLE, N.O.S. or MERCAPTAN MIXTURE, LIQUID, FLAMMABLE, N.O.S. (vapour pressure at 50 °C more than 110 kPa)	3	F1	II	3	274 640C	LQ4	E2		PP, EX, A	VE01		1	
3336	MERCAPTANS, LIQUID, FLAMMABLE, N.O.S. or MERCAPTAN MIXTURE, LIQUID, FLAMMABLE, N.O.S. (vapour pressure at 50 °C not more than 110 kPa)	3	F1	II	3	274 640D	LQ4	E2		PP, EX, A	VE01		1	
3336	MERCAPTANS, LIQUID, FLAMMABLE, N.O.S. or MERCAPTAN MIXTURE, LIQUID, FLAMMABLE, N.O.S.	3	F1	III	3	274	LQ7	E1		PP, EX, A	VE01		0	
3337	REFRIGERANT GAS R 404A (Pentafluoroethane, 1,1,1-trifluoroethane, and 1,1,1,2-tetrafluoroethane zeotropic mixture with approximately 44% pentafluoroethane and 52% 1,1,1-trifluoroethane)	2	2A		2.2		LQ1	E1		PP			0	

UN No. or ID No.	Name and description	Class	Classi-fication Code	Packing group	Labels	Special provi-sions	Limited and excepted quantities		Carriage permitted	Equipment required	Venti-lation	Provisions concerning loading, unloading and carriage	Number of blue cones/ lights	Remarks
	3.1.2	2.2	2.2	2.1.1.3	5.2.2	3.3	3.4.6	3.5.1.2	3.2.1	8.1.5	7.1.6	7.1.6	7.1.5	3.2.1
(1)	(2)	(3a)	(3b)	(4)	(5)	(6)	(7a)	(7b)	(8)	(9)	(10)	(11)	(12)	(13)
3338	REFRIGERANT GAS R 407A (Difluoromethane, pentafluoroethane, and 1,1,1,2-tetrafluoroethane zeotropic mixture with approximately 20% difluoromethane and 40% pentafluoroethane)	2	2A		2.2		LQ1	E1		PP			0	
3339	REFRIGERANT GAS R 407B (Difluoromethane, pentafluoroethane, and 1,1,1,2-tetrafluoroethane zeotropic mixture with approximately 10% difluoromethane and 70% pentafluoroethane	2	2A		2.2		LQ1	E1		PP			0	
3340	REFRIGERANT GAS R 407C (Difluoromethane, pentafluoroethane, and 1,1,1,2-tetrafluoroethane zeotropic mixture with approximately 23%difluoromethane and 25% pentafluoroethane)	2	2A		2.2		LQ1	E1		PP			0	
3341	THIOUREA DIOXIDE	4.2	S2	II	4.2		LQ0	E2		PP			0	
3341	THIOUREA DIOXIDE	4.2	S2	III	4.2		LQ0	E1		PP			0	
3342	XANTHATES	4.2	S2	II	4.2		LQ0	E2		PP			0	
3342	XANTHATES	4.2	S2	III	4.2		LQ0	E1		PP			0	
3343	NITROGLYCERIN MIXTURE, DESENSITIZED, LIQUID, FLAMMABLE, N.O.S. with not more than 30% nitroglycerin, by mass	3	D		3	274 278	LQ0	E0		PP, EX, A	VE01		0	
3344	PENTAERYTHRITE TETRANITRATE (PENTAERYTHRITOL TETRANITRATE; PETN) MIXTURE, DESENSITIZED, SOLID, N.O.S. with more than 10% but not more than 20% PETN, by mass	4.1	D	II	4.1	272 274	LQ0	E0		PP			1	
3345	PHENOXYACETIC ACID DERIVATIVE PESTICIDE, SOLID, TOXIC	6.1	T7	I	6.1	61 274 648 802	LQ0	E5		PP, EP			2	
3345	PHENOXYACETIC ACID DERIVATIVE PESTICIDE, SOLID, TOXIC	6.1	T7	II	6.1	61 274 648 802	LQ18	E4		PP, EP			2	
3345	PHENOXYACETIC ACID DERIVATIVE PESTICIDE, SOLID, TOXIC	6.1	T7	III	6.1	61 274 648 802	LQ9	E1		PP, EP			0	
3346	PHENOXYACETIC ACID DERIVATIVE PESTICIDE, LIQUID, FLAMMABLE, TOXIC, flash-point less than 23 °C	3	FT2	I	3+6.1	61 274 802	LQ3	E0		PP, EP, EX, TOX, A	VE01, VE02		2	
3346	PHENOXYACETIC ACID DERIVATIVE PESTICIDE, LIQUID, FLAMMABLE, TOXIC, flash-point less than 23 °C	3	FT2	II	3+6.1	61 274 802	LQ4	E2		PP, EP, EX, TOX, A	VE01, VE02		2	

UN No. or ID No.	Name and description	Class	Classi-fication Code	Packing group	Labels	Special provi-sions	Limited and excepted quantities		Carriage permitted	Equipment required	Venti-lation	Provisions concerning loading, unloading and carriage	Number of blue cones/ lights	Remarks
	3.1.2	2.2	2.2	2.1.1.3	5.2.2	3.3	3.4.6	3.5.1.2	3.2.1	8.1.5	7.1.6	7.1.6	7.1.5	3.2.1
(1)	(2)	(3a)	(3b)	(4)	(5)	(6)	(7a)	(7b)	(8)	(9)	(10)	(11)	(12)	(13)
3347	PHENOXYACETIC ACID DERIVATIVE PESTICIDE, LIQUID, TOXIC, FLAMMABLE, flash-point not less than 23 °C	6.1	TF2	I	6.1+3	61 274 802	LQ0	E5		PP, EP, EX, TOX, A	VE01, VE02		2	
3347	PHENOXYACETIC ACID DERIVATIVE PESTICIDE, LIQUID, TOXIC, FLAMMABLE, flash-point not less than 23 °C	6.1	TF2	II	6.1+3	61 274 802	LQ17	E4		PP, EP, EX, TOX, A	VE01, VE02		2	
3347	PHENOXYACETIC ACID DERIVATIVE PESTICIDE, LIQUID, TOXIC, FLAMMABLE, flash-point not less than 23 °C	6.1	TF2	III	6.1+3	61 274 802	LQ7	E1		PP, EP, EX, TOX, A	VE01, VE02		0	
3348	PHENOXYACETIC ACID DERIVATIVE PESTICIDE, LIQUID, TOXIC	6.1	T6	I	6.1	61 274 648 802	LQ0	E5		PP, EP, TOX, A	VE02		2	
3348	PHENOXYACETIC ACID DERIVATIVE PESTICIDE, LIQUID, TOXIC	6.1	T6	II	6.1	61 274 648 802	LQ17	E4		PP, EP, TOX, A	VE02		2	
3348	PHENOXYACETIC ACID DERIVATIVE PESTICIDE, LIQUID, TOXIC	6.1	T6	III	6.1	61 274 648 802	LQ7	E1		PP, EP, TOX, A	VE02		0	
3349	PYRETHROID PESTICIDE, SOLID, TOXIC	6.1	T7	I	6.1	61 274 648 802	LQ0	E5		PP, EP			2	
3349	PYRETHROID PESTICIDE, SOLID, TOXIC	6.1	T7	II	6.1	61 274 648 802	LQ18	E4		PP, EP			2	
3349	PYRETHROID PESTICIDE, SOLID, TOXIC	6.1	T7	III	6.1	61 274 648 802	LQ9	E1		PP, EP			0	
3350	PYRETHROID PESTICIDE, LIQUID, FLAMMABLE, TOXIC, flash-point less than 23 °C	3	FT2	I	3+6.1	61 274 802	LQ3	E0		PP, EP, EX, TOX, A	VE01, VE02		2	
3350	PYRETHROID PESTICIDE, LIQUID, FLAMMABLE, TOXIC, flash-point less than 23 °C	3	FT2	II	3+6.1	61 274 802	LQ4	E2		PP, EP, EX, TOX, A	VE01, VE02		2	
3351	PYRETHROID PESTICIDE, LIQUID, TOXIC, FLAMMABLE, flash-point not less than 23 °C	6.1	TF2	I	6.1+3	61 274 802	LQ0	E5		PP, EP, EX, TOX, A	VE01, VE02		2	
3351	PYRETHROID PESTICIDE, LIQUID, TOXIC, FLAMMABLE, flash-point not less than 23 °C	6.1	TF2	II	6.1+3	61 274 802	LQ17	E4		PP, EP, EX, TOX, A	VE01, VE02		2	

UN No. or ID No.	Name and description	Class	Classi-fication Code	Packing group	Labels	Special provi-sions	Limited and excepted quantities		Carriage permitted	Equipment required	Venti-lation	Provisions concerning loading, unloading and carriage	Number of blue cones/ lights	Remarks
	3.1.2	2.2	2.2	2.1.1.3	5.2.2	3.3	3.4.6	3.5.1.2	3.2.1	8.1.5	7.1.6	7.1.6	7.1.5	3.2.1
(1)	(2)	(3a)	(3b)	(4)	(5)	(6)	(7a)	(7b)	(8)	(9)	(10)	(11)	(12)	(13)
3351	PYRETHROID PESTICIDE, LIQUID, TOXIC, FLAMMABLE, flash-point not less than 23 °C	6.1	TF2	III	6.1+3	61 274 802	LQ7	E1		PP, EP, EX, TOX, A	VE01, VE02		0	
3352	PYRETHROID PESTICIDE, LIQUID, TOXIC	6.1	T6	I	6.1	61 274 648 802	LQ0	E5		PP, EP, TOX, A	VE02		2	
3352	PYRETHROID PESTICIDE, LIQUID, TOXIC	6.1	T6	II	6.1	61 274 648 802	LQ17	E4		PP, EP, TOX, A	VE02		2	
3352	PYRETHROID PESTICIDE, LIQUID, TOXIC	6.1	T6	III	6.1	61 274 648 802	LQ7	E1		PP, EP, TOX, A	VE02		0	
3354	INSECTICIDE GAS, FLAMMABLE, N.O.S.	2	2F		2.1	274	LQ0	E0		PP, EX, A	VE01		1	
3355	INSECTICIDE GAS, TOXIC, FLAMMABLE, N.O.S.	2	2TF		2.3+2.1	274	LQ0	E0		PP, EP, EX, TOX, A	VE01, VE02		2	
3356	OXYGEN GENERATOR, CHEMICAL	5.1	O3	II	5.1	284	LQ0	E0		PP			0	
3357	NITROGLYCERIN MIXTURE, DESENSITIZED, LIQUID, N.O.S. with not more than 30% nitroglycerin, by mass	3	D	II	3	274 288	LQ0	E0		PP, EX, A	VE01		1	
3358	REFRIGERATING MACHINES containing flammable, non-toxic, liquefied gas	2	6F		2.1	291	LQ0	E0		PP, EX, A	VE01		1	
3359	FUMIGATED UNIT	9	M11			302				PP				
3360	Fibres, vegetable, dry	4.1	F1	**NOT SUBJECT TO ADN**										
3361	CHLOROSILANES, TOXIC, CORROSIVE, N.O.S.	6.1	TC1	II	6.1+8	274 802	LQ0	E4		PP, EP, TOX, A	VE02		2	
3362	CHLOROSILANES, TOXIC, CORROSIVE, FLAMMABLE, N.O.S.	6.1	TFC	II	6.1+3+8	274	LQ0	E4		PP, EP, EX, TOX, A	VE01, VE 02		2	
3363	Dangerous goods in machinery or dangerous goods in apparatus	9	M11	**NOT SUBJECT TO ADN** [see also 1.1.3.1 (b)]										
3364	TRINITROPHENOL (PICRIC ACID) WETTED with not less than 10% water, by mass	4.1	D	I	4.1		LQ0	E0		PP			1	
3365	TRINITROCHLOROBENZENE (PICRYL CHLORIDE) WETTED with not less than 10% water, by mass	4.1	D	I	4.1		LQ0	E0		PP			1	
3366	TRINITROTOLUENE (TNT), WETTED with not less than 10% water, by mass	4.1	D	I	4.1		LQ0	E0		PP			1	
3367	TRINITROBENZENE, WETTED with not less than 10% water, by mass	4.1	D	I	4.1		LQ0	E0		PP			1	
3368	TRINITROBENZOIC ACID, WETTED with not less than 10% water, by mass	4.1	D	I	4.1		LQ0	E0		PP			1	
3369	SODIUM DINITRO-o-CRESOLATE, WETTED with not less than 10% water, by mass	4.1	DT	I	4.1+6.1	802	LQ0	E0		PP			2	

UN No. or ID No.	Name and description	Class	Classi-fication Code	Packing group	Labels	Special provi-sions	Limited and excepted quantities		Carriage permitted	Equipment required	Venti-lation	Provisions concerning loading, unloading and carriage	Number of blue cones/ lights	Remarks
	3.1.2	2.2	2.2	2.1.1.3	5.2.2	3.3	3.4.6	3.5.1.2	3.2.1	8.1.5	7.1.6	7.1.6	7.1.5	3.2.1
(1)	(2)	(3a)	(3b)	(4)	(5)	(6)	(7a)	(7b)	(8)	(9)	(10)	(11)	(12)	(13)
3370	UREA NITRATE, WETTED with not less than 10% water, by mass	4.1	D	I	4.1		LQ0	E0		PP			1	
3371	2-METHYLBUTANAL	3	F1	II	3		LQ4	E2		PP, EX, A	VE01		1	
3373	BIOLOGICAL SUBSTANCE, CATEGORY B	6.2	I4		6.2	319	LQ0	E0		PP			0	
3373	BIOLOGICAL SUBSTANCE, CATEGORY B (animal material only)	6.2	I4		6.2	319	LQ0	E0		PP			0	
3374	ACETYLENE, SOLVENT FREE	2	2F		2.1		LQ0	E0		PP, EX, A	VE01		1	
3375	AMMONIUM NITRATE EMULSION, or SUSPENSION or GEL, intermediate for blasting explosives, liquid	5.1	O1	II	5.1	309	LQ0	E2		PP			0	
3375	AMMONIUM NITRATE EMULSION, or SUSPENSION or GEL, intermediate for blasting explosives, solid	5.1	O2	II	5.1	309	LQ0	E2		PP			0	
3376	4-NITROPHENYLHYDRAZINE, with not less than 30% water, by mass	4.1	D	I	4.1		LQ0	E0		PP			1	
3377	SODIUM PERBORATE MONOHYDRATE	5.1	O2	III	5.1		LQ12	E1		PP			0	
3378	SODIUM CARBONATE PEROXYHYDRATE	5.1	O2	II	5.1		LQ11	E2		PP			0	
3378	SODIUM CARBONATE PEROXYHYDRATE	5.1	O2	III	5.1		LQ12	E1		PP			0	
3379	DESENSITIZED EXPLOSIVE, LIQUID, N.O.S.	3	D	I	3	274 311	LQ0	E0		PP, EX, A	VE01		1	
3380	DESENSITIZED EXPLOSIVE, SOLID, N.O.S.	4.1	D	I	4.1	274 311	LQ0	E0		PP			1	
3381	TOXIC BY INHALATION LIQUID, N.O.S. with an inhalation toxicity lower than or equal to 200 ml/m^3 and saturated vapour concentration greater than or equal to 500 LC_{50}	6.1	T1 or T4	I	6.1	274 802	LQ0	E5		PP, EP, TOX, A	VE02		2	
3382	TOXIC BY INHALATION LIQUID, N.O.S. with an inhalation toxicity lower than or equal to 1000 ml/m^3 and saturated vapour concentration greater than or equal to 10 LC_{50}	6.1	T1 or T4	I	6.1	274 802	LQ0	E5		PP, EP, TOX, A	VE02		2	
3383	TOXIC BY INHALATION LIQUID, FLAMMABLE, N.O.S. with an inhalation toxicity lower than or equal to 200 ml/m^3 and saturated vapour concentration greater than or equal to 500 LC_{50}	6.1	TF1	I	6.1 +3	274 802	LQ0	E5		PP, EP, EX, TOX, A	VE01, VE02		2	
3384	TOXIC BY INHALATION LIQUID, FLAMMABLE, N.O.S. with an inhalation toxicity lower than or equal to 1000 ml/m^3 and saturated vapour concentration greater than or equal to 10 LC_{50}	6.1	TF1	I	6.1 +3	274 802	LQ0	E5		PP, EP, EX, TOX, A	VE01, VE02		2	

UN No. or ID No.	Name and description	Class	Classi-fication Code	Packing group	Labels	Special provi-sions	Limited and excepted quantities		Carriage permitted	Equipment required	Venti-lation	Provisions concerning loading, unloading and carriage	Number of blue cones/ lights	Remarks
	3.1.2	2.2	2.2	2.1.1.3	5.2.2	3.3	3.4.6	3.5.1.2	3.2.1	8.1.5	7.1.6	7.1.6	7.1.5	3.2.1
(1)	(2)	(3a)	(3b)	(4)	(5)	(6)	(7a)	(7b)	(8)	(9)	(10)	(11)	(12)	(13)
3385	TOXIC BY INHALATION LIQUID, WATER-REACTIVE, N.O.S. with an inhalation toxicity lower than or equal to 200 ml/m^3 and saturated vapour concentration greater than or equal to 500 LC_{50}	6.1	TW1	I	6.1 +4.3	274 802	LQ0	E5		PP, EP, TOX, A	VE02		2	
3386	TOXIC BY INHALATION LIQUID, WATER-REACTIVE, N.O.S. with an inhalation toxicity lower than or equal to 1000 ml/m^3 and saturated vapour concentration greater than or equal to 10 LC_{50}	6.1	TW1	I	6.1 +4.3	274 802	LQ0	E5		PP, EP, TOX, A	VE02		2	
3387	TOXIC BY INHALATION LIQUID, OXIDIZING, N.O.S. with an inhalation toxicity lower than or equal to 200 ml/m^3 and saturated vapour concentration greater than or equal to 500 LC_{50}	6.1	TO1	I	6.1 +5.1	274 802	LQ0	E5		PP, EP, TOX, A	VE02		2	
3388	TOXIC BY INHALATION LIQUID, OXIDIZING, N.O.S. with an inhalation toxicity lower than or equal to 1000 ml/m^3 and saturated vapour concentration greater than or equal to 10 LC_{50}	6.1	TO1	I	6.1 +5.1	274 802	LQ0	E5		PP, EP, TOX, A	VE02		2	
3389	TOXIC BY INHALATION LIQUID, CORROSIVE, N.O.S. with an inhalation toxicity lower than or equal to 200 ml/m^3 and saturated vapour concentration greater than or equal to 500 LC_{50}	6.1	TC1 or TC3	I	6.1 +8	274 802	LQ0	E5		PP, EP, TOX, A	VE02		2	
3390	TOXIC BY INHALATION LIQUID, CORROSIVE, N.O.S. with an inhalation toxicity lower than or equal to 1000 ml/m^3 and saturated vapour concentration greater than or equal to 10 LC_{50}	6.1	TC1 or TC3	I	6.1 +8	274 802	LQ0	E5		PP, EP, TOX, A	VE02		2	
3391	ORGANOMETALLIC SUBSTANCE, SOLID, PYROPHORIC	4.2	S5	I	4.2	274	LQ0	E0		PP			0	
3392	ORGANOMETALLIC SUBSTANCE, LIQUID, PYROPHORIC	4.2	S5	I	4.2	274	LQ0	E0		PP			0	
3393	ORGANOMETALLIC SUBSTANCE, SOLID, PYROPHORIC, WATER REACTIVE	4.2	SW	I	4.2 +4.3	274	LQ0	E0		PP, EX, A	VE01		0	
3394	ORGANOMETALLIC SUBSTANCE, LIQUID, PYROPHORIC, WATER REACTIVE	4.2	SW	I	4.2 +4.3	274	LQ0	E0		PP, EX, A	VE01		0	
3395	ORGANOMETALLIC SUBSTANCE, SOLID, WATER REACTIVE	4.3	W2	I	4.3	274	LQ0	E0		PP, EX, A	VE01	HA08	0	
3395	ORGANOMETALLIC SUBSTANCE, SOLID, WATER REACTIVE	4.3	W2	II	4.3	274	LQ11	E2		PP, EX, A	VE01	HA08	0	
3395	ORGANOMETALLIC SUBSTANCE, SOLID, WATER REACTIVE	4.3	W2	III	4.3	274	LQ12	E1		PP, EX, A	VE01	HA08	0	

UN No. or ID No.	Name and description	Class	Classi-fication Code	Packing group	Labels	Special provi-sions	Limited and excepted quantities		Carriage permitted	Equipment required	Venti-lation	Provisions concerning loading, unloading and carriage	Number of blue cones/ lights	Remarks
	3.1.2	2.2	2.2	2.1.1.3	5.2.2	3.3	3.4.6	3.5.1.2	3.2.1	8.1.5	7.1.6	7.1.6	7.1.5	3.2.1
(1)	(2)	(3a)	(3b)	(4)	(5)	(6)	(7a)	(7b)	(8)	(9)	(10)	(11)	(12)	(13)
3396	ORGANOMETALLIC SUBSTANCE, SOLID, WATER REACTIVE, FLAMMABLE	4.3	WF2	I	4.3 +4.1	274	LQ0	E0		PP, EX, A	VE01		1	
3396	ORGANOMETALLIC SUBSTANCE, SOLID, WATER REACTIVE, FLAMMABLE	4.3	WF2	II	4.3 +4.1	274	LQ11	E2		PP, EX, A	VE01		1	
3396	ORGANOMETALLIC SUBSTANCE, SOLID, WATER REACTIVE, FLAMMABLE	4.3	WF2	III	4.3 +4.1	274	LQ12	E1		PP, EX, A	VE01		0	
3397	ORGANOMETALLIC SUBSTANCE, SOLID, WATER REACTIVE, SELF-HEATING	4.3	WS	I	4.3 +4.2	274	LQ0	E0		PP, EX, A	VE01	HA08	0	
3397	ORGANOMETALLIC SUBSTANCE, SOLID, WATER REACTIVE, SELF-HEATING	4.3	WS	II	4.3 +4.2	274	LQ11	E2		PP, EX, A	VE01	HA08	0	
3397	ORGANOMETALLIC SUBSTANCE, SOLID, WATER REACTIVE, SELF-HEATING	4.3	WS	III	4.3 +4.2	274	LQ12	E1		PP, EX, A	VE01	HA08	0	
3398	ORGANOMETALLIC SUBSTANCE, LIQUID, WATER REACTIVE	4.3	W1	I	4.3	274	LQ0	E0		PP, EX, A	VE01	HA08	0	
3398	ORGANOMETALLIC SUBSTANCE, LIQUID, WATER REACTIVE	4.3	W1	II	4.3	274	LQ10	E2		PP, EX, A	VE01	HA08	0	
3398	ORGANOMETALLIC SUBSTANCE, LIQUID, WATER REACTIVE	4.3	W1	III	4.3	274	LQ13	E1		PP, EX, A	VE01	HA08	0	
3399	ORGANOMETALLIC SUBSTANCE, LIQUID, WATER REACTIVE, FLAMMABLE	4.3	WF1	I	4.3 +3	274	LQ0	E0		PP, EX, A	VE01	HA08	1	
3399	ORGANOMETALLIC SUBSTANCE, LIQUID, WATER REACTIVE, FLAMMABLE	4.3	WF1	II	4.3 +3	274	LQ10	E2		PP, EX, A	VE01	HA08	1	
3399	ORGANOMETALLIC SUBSTANCE, LIQUID, WATER REACTIVE, FLAMMABLE	4.3	WF1	III	4.3 +3	274	LQ13	E1		PP, EX, A	VE01	HA08	0	
3400	ORGANOMETALLIC SUBSTANCE, SOLID, SELF-HEATING	4.2	S5	II	4.2	274	LQ18	E2		PP			0	
3400	ORGANOMETALLIC SUBSTANCE, SOLID, SELF-HEATING	4.2	S5	III	4.2	274	LQ11	E1		PP			0	
3401	ALKALI METAL AMALGAM, SOLID	4.3	W2	I	4.3	182 274	LQ0	E0		PP, EX, A	VE01	HA08	0	
3402	ALKALINE EARTH METAL AMALGAM, SOLID	4.3	W2	I	4.3	183 274 506	LQ0	E0		PP, EX, A	VE01	HA08	0	
3403	POTASSIUM METAL ALLOYS, SOLID	4.3	W2	I	4.3		LQ0	E0		PP, EX, A	VE01	HA08	0	
3404	Potassium sodium alloys, SOLID	4.3	W2	I	4.3		LQ0	E0		PP, EX, A	VE01	HA08	0	
3405	BARIUM CHLORATE SOLUTION	5.1	OT1	II	5.1 +6.1	802	LQ10	E2		PP, EP, TOX, A	VE02		2	
3405	BARIUM CHLORATE SOLUTION	5.1	OT1	III	5.1 +6.1	802	LQ13	E1		PP, EP, TOX, A	VE02		0	
3406	BARIUM PERCHLORATE SOLUTION	5.1	OT1	II	5.1 +6.1	802	LQ10	E2		PP, EP, TOX, A	VE02		2	
3406	BARIUM PERCHLORATE SOLUTION	5.1	OT1	III	5.1 +6.1	802	LQ13	E1		PP, EP, TOX, A	VE02		0	
3407	CHLORATE AND MAGNESIUM CHLORIDE MIXTURE SOLUTION	5.1	O1	II	5.1		LQ10	E2		PP			0	

UN No. or ID No.	Name and description	Class	Classi-fication Code	Packing group	Labels	Special provi-sions	Limited and excepted quantities		Carriage permitted	Equipment required	Venti-lation	Provisions concerning loading, unloading and carriage	Number of blue cones/ lights	Remarks
	3.1.2	2.2	2.2	2.1.1.3	5.2.2	3.3	3.4.6	3.5.1.2	3.2.1	8.1.5	7.1.6	7.1.6	7.1.5	3.2.1
(1)	(2)	(3a)	(3b)	(4)	(5)	(6)	(7a)	(7b)	(8)	(9)	(10)	(11)	(12)	(13)
3407	CHLORATE AND MAGNESIUM CHLORIDE MIXTURE SOLUTION	5.1	O1	III	5.1		LQ13	E1		PP			0	
3408	LEAD PERCHLORATE SOLUTION	5.1	OT1	II	5.1 +6.1		LQ10	E2		PP			0	
3408	LEAD PERCHLORATE SOLUTION	5.1	OT1	III	5.1 +6.1		LQ13	E1		PP			0	
3409	CHLORONITROBENZENES, LIQUID	6.1	T1	II	6.1	279 802	LQ17	E4		PP, EP			2	
3410	4-CHLORO-o-TOLUIDINE HYDROCHLORIDE SOLUTION	6.1	T1	III	6.1	802	LQ7	E1		PP, EP, TOX, A	VE02		0	
3411	beta-NAPHTHYLAMINE SOLUTION	6.1	T1	II	6.1	802	LQ17	E4		PP, EP, TOX, A	VE02		2	
3411	beta-NAPHTHYLAMINE SOLUTION	6.1	T1	III	6.1	802	LQ7	E1		PP, EP, TOX, A	VE02		0	
3412	FORMIC ACID with not less than 10% but not more than 85% acid by mass	8	C3	II	8		LQ22	E2	T	PP, EP			0	
3412	FORMIC ACID with not less than 5% but less than 10% acid by mass	8	C3	III	8		LQ7	E1	T	PP, EP			0	
3413	POTASSIUM CYANIDE SOLUTION	6.1	T4	I	6.1	802	LQ0	E5		PP, EP, TOX, A	VE02		2	
3413	POTASSIUM CYANIDE SOLUTION	6.1	T4	II	6.1	802	LQ17	E4		PP, EP, TOX, A	VE02		2	
3413	POTASSIUM CYANIDE SOLUTION	6.1	T4	III	6.1	802	LQ7	E1		PP, EP, TOX, A	VE02		0	
3414	SODIUM CYANIDE SOLUTION	6.1	T4	I	6.1	802	LQ0	E5		PP, EP, TOX, A	VE02		2	
3414	SODIUM CYANIDE SOLUTION	6.1	T4	II	6.1	802	LQ17	E4		PP, EP, TOX, A	VE02		2	
3414	SODIUM CYANIDE SOLUTION	6.1	T4	III	6.1	802	LQ7	E1		PP, EP, TOX, A	VE02		0	
3415	SODIUM FLUORIDE SOLUTION	6.1	T4	III	6.1	802	LQ7	E1		PP, EP, TOX, A	VE02		0	
3416	CHLOROACETOPHENONE, LIQUID	6.1	T1	II	6.1	802	LQ17	E4		PP, EP, TOX, A	VE02		2	
3417	XYLYL BROMIDE, SOLID	6.1	T2	II	6.1	802	LQ18	E4		PP, EP			2	
3418	2,4-TOLUYLENEDIAMINE SOLUTION	6.1	T1	III	6.1	802	LQ7	E1		PP, EP, TOX, A	VE02		0	
3419	BORON TRIFLUORIDE ACETIC ACID COMPLEX, SOLID	8	C4	II	8		LQ23	E2		PP, EP			0	
3420	BORON TRIFLUORIDE PROPIONIC ACID COMPLEX, SOLID	8	C4	II	8		LQ23	E2		PP, EP			0	
3421	POTASSIUM HYDROGENDI-FLUORIDE SOLUTION	8	CT1	II	8 +6.1	802	LQ22	E2		PP, EP, TOX, A	VE02		2	
3421	POTASSIUM HYDROGENDI-FLUORIDE SOLUTION	8	CT1	III	8 +6.1	802	LQ7	E1		PP, EP, TOX, A	VE02		0	
3422	POTASSIUM FLUORIDE SOLUTION	6.1	T4	III	6.1	802	LQ7	E1		PP, EP, TOX, A	VE02		0	

UN No. or ID No.	Name and description	Class	Classi-fication Code	Packing group	Labels	Special provi-sions	Limited and excepted quantities		Carriage permitted	Equipment required	Venti-lation	Provisions concerning loading, unloading and carriage	Number of blue cones/ lights	Remarks
	3.1.2	2.2	2.2	2.1.1.3	5.2.2	3.3	3.4.6	3.5.1.2	3.2.1	8.1.5	7.1.6	7.1.6	7.1.5	3.2.1
(1)	(2)	(3a)	(3b)	(4)	(5)	(6)	(7a)	(7b)	(8)	(9)	(10)	(11)	(12)	(13)
3423	TETRAMETHYLAMMONIUM HYDROXIDE, SOLID	8	C8	II	8		LQ24	E2		PP, EP			0	
3424	AMMONIUM DINITRO -o-CRESOLATE SOLUTION	6.1	T1	II	6.1	802	LQ17	E4		PP, EP, TOX, A	VE02		2	
3424	AMMONIUM DINITRO -o-CRESOLATE SOLUTION	6.1	T1	III	6.1	802	LQ7	E1		PP, EP, TOX, A	VE02		0	
3425	BROMOACETIC ACID, SOLID	8	C4	II	8		LQ23	E2		PP, EP			0	
3426	ACRYLAMIDE SOLUTION	6.1	T1	III	6.1		LQ7	E1	T	PP, EP, TOX, A	VE02		0	
3427	CHLOROBENZYL CHLORIDES, SOLID	6.1	T2	III	6.1	802	LQ9	E1		PP, EP			0	
3428	3-CHLORO-4-METHYLPHENYL ISOCYANATE, SOLID	6.1	T2	II	6.1	802	LQ18	E4		PP, EP			2	
3429	CHLOROTOLUIDINES, LIQUID	6.1	T1	III	6.1	802	LQ7	E1		PP, EP, TOX, A	VE02		0	
3430	XYLENOLS, liquid	6.1	T1	II	6.1	802	LQ17	E4		PP, EP, TOX, A	VE02		2	
3431	NITROBENZOTRIFLUORIDES, solid	6.1	T2	II	6.1	802	LQ18	E4		PP, EP			2	
3432	POLYCHLORINATED BIPHENYLS, SOLID	9	M2	II	9	305 802	LQ25	E2		PP, EP			0	
3434	NITROCRESOLS, liquid	6.1	T1	III	6.1	802	LQ7	E1		PP, EP, TOX, A	VE02		0	
3436	HEXAFLUOROACETONE HYDRATE, SOLID	6.1	T2	II	6.1	802	LQ18	E4		PP, EP			2	
3437	CHLOROCRESOLS, solid	6.1	T2	II	6.1	802	LQ18	E4		PP, EP			2	
3438	alpha-METHYLBENZYL ALCOHOL, SOLID	6.1	T2	III	6.1	802	LQ9	E1		PP, EP			0	
3439	NITRILES, TOXIC, SOLID, N.O.S.	6.1	T2	I	6.1	274 802	LQ0	E5		PP, EP			2	
3439	NITRILES, TOXIC, SOLID, N.O.S.	6.1	T2	II	6.1	274 802	LQ18	E4		PP, EP			2	
3439	NITRILES, TOXIC, SOLID, N.O.S.	6.1	T2	III	6.1	274 802	LQ9	E1		PP, EP			0	
3440	SELENIUM COMPOUND, LIQUID, N.O.S.	6.1	T4	I	6.1	274 802	LQ0	E5		PP, EP, TOX, A	VE02		2	
3440	SELENIUM COMPOUND, LIQUID, N.O.S.	6.1	T4	II	6.1	274 802	LQ17	E4		PP, EP, TOX, A	VE02		2	
3440	SELENIUM COMPOUND, LIQUID, N.O.S.	6.1	T4	III	6.1	274 802	LQ7	E1		PP, EP, TOX, A	VE02		0	
3441	CHLORODINITROBENZENES, SOLID	6.1	T2	II	6.1	279 802	LQ18	E4		PP, EP			2	
3442	DICHLOROANILINES, SOLID	6.1	T2	II	6.1	279 802	LQ18	E4		PP, EP			2	
3443	DINITROBENZENES, SOLID	6.1	T2	II	6.1	802	LQ18	E4		PP, EP			2	
3444	NICOTINE HYDROCHLORIDE, SOLID	6.1	T2	II	6.1	43 802	LQ18	E4		PP, EP			2	
3445	NICOTINE SULPHATE, SOLID	6.1	T2	II	6.1	802	LQ18	E4		PP, EP			2	
3446	NITROTOLUENES, SOLID	6.1	T2	II	6.1	802	LQ18	E4	T	PP, EP			2	
3447	NITROXYLENES, SOLID	6.1	T2	II	6.1	802	LQ18	E4		PP, EP			2	

UN No. or ID No.	Name and description	Class	Classi-fication Code	Packing group	Labels	Special provi-sions	Limited and excepted quantities		Carriage permitted	Equipment required	Venti-lation	Provisions concerning loading, unloading and carriage	Number of blue cones/ lights	Remarks
	3.1.2	2.2	2.2	2.1.1.3	5.2.2	3.3	3.4.6	3.5.1.2	3.2.1	8.1.5	7.1.6	7.1.6	7.1.5	3.2.1
(1)	(2)	(3a)	(3b)	(4)	(5)	(6)	(7a)	(7b)	(8)	(9)	(10)	(11)	(12)	(13)
3448	TEAR GAS SUBSTANCE, SOLID, N.O.S.	6.1	T2	I	6.1	274 802	LQ0	E5		PP, EP			2	
3448	TEAR GAS SUBSTANCE, SOLID, N.O.S.	6.1	T2	II	6.1	274 802	LQ18	E4		PP, EP			2	
3449	BROMOBENZYL CYANIDES, SOLID	6.1	T2	I	6.1	138 802	LQ0	E5		PP, EP			2	
3450	DIPHENYLCHLOROARSINE, SOLID	6.1	T3	I	6.1	802	LQ0	E5		PP, EP			2	
3451	TOLUIDINES, SOLID	6.1	T2	II	6.1	279 802	LQ18	E4	T	PP, EP			2	
3452	XYLIDINES, SOLID	6.1	T2	II	6.1	802	LQ18	E4		PP, EP			2	
3453	PHOSPHORIC ACID, SOLID	8	C2	III	8		LQ24	E1		PP, EP			0	
3454	DINITROTOLUENES, SOLID	6.1	T2	II	6.1	802	LQ18	E4		PP, EP			2	
3455	CRESOLS, SOLID	6.1	TC2	II	6.1+8	802	LQ18	E4	T	PP, EP			2	
3456	NITROSYLSULPHURIC ACID, SOLID	8	C2	II	8		LQ23	E2	T3	PP, EP			0	
3457	CHLORONITROTOLUENES, SOLID	6.1	T2	III	6.1	802	LQ9	E1		PP, EP			0	
3458	NITROANISOLES, SOLID	6.1	T2	III	6.1	279 802	LQ9	E1		PP, EP			0	
3459	NITROBROMOBENZENES, SOLID	6.1	T2	III	6.1	802	LQ9	E1		PP, EP			0	
3460	N-ETHYLBENZYLTOLUIDINES, SOLID	6.1	T2	III	6.1	802	LQ9	E1		PP, EP			0	
3462	TOXINS, EXTRACTED FROM LIVING SOURCES, SOLID, N.O.S.	6.1	T2	I	6.1	210 274 802	LQ0	E5		PP, EP			2	
3462	TOXINS, EXTRACTED FROM LIVING SOURCES, SOLID, N.O.S.	6.1	T2	II	6.1	210 274 802	LQ18	E4		PP, EP			2	
3462	TOXINS, EXTRACTED FROM LIVING SOURCES, SOLID, N.O.S.	6.1	T2	III	6.1	210 274 802	LQ9	E1		PP, EP			0	
3463	PROPIONIC ACID with not less than 90% acid by mass	8	CF1	II	8 +3		LQ22	E2	T	PP, EP, EX, A			0	
3464	ORGANOPHOSPHORUS COMPOUND, TOXIC, SOLID, N.O.S.	6.1	T2	I	6.1	43 274 802	LQ0	E5		PP, EP			2	
3464	ORGANOPHOSPHORUS COMPOUND, TOXIC, SOLID, N.O.S.	6.1	T2	II	6.1	43 274 802	LQ18	E4		PP, EP			2	
3464	ORGANOPHOSPHORUS COMPOUND, TOXIC, SOLID, N.O.S.	6.1	T2	III	6.1	43 274 802	LQ9	E1		PP, EP			0	
3465	ORGANOARSENIC COMPOUND, SOLID, N.O.S.	6.1	T3	I	6.1	274 802	LQ0	E5		PP, EP			2	
3465	ORGANOARSENIC COMPOUND, SOLID, N.O.S.	6.1	T3	II	6.1	274 802	LQ18	E4		PP, EP			2	
3465	ORGANOARSENIC COMPOUND, SOLID, N.O.S.	6.1	T3	III	6.1	274 802	LQ9	E1		PP, EP			0	

UN No. or ID No.	Name and description	Class	Classi-fication Code	Packing group	Labels	Special provi-sions	Limited and excepted quantities		Carriage permitted	Equipment required	Venti-lation	Provisions concerning loading, unloading and carriage	Number of blue cones/ lights	Remarks
	3.1.2	2.2	2.2	2.1.1.3	5.2.2	3.3	3.4.6	3.5.1.2	3.2.1	8.1.5	7.1.6	7.1.6	7.1.5	3.2.1
(1)	(2)	(3a)	(3b)	(4)	(5)	(6)	(7a)	(7b)	(8)	(9)	(10)	(11)	(12)	(13)
3466	METAL CARBONYLS, SOLID, N.O.S	6.1	T3	I	6.1	274 562 802	LQ0	E5		PP, EP			2	
3466	METAL CARBONYLS, SOLID, N.O.S	6.1	T3	II	6.1	274 562 802	LQ18	E4		PP, EP			2	
3466	METAL CARBONYLS, SOLID, N.O.S	6.1	T3	III	6.1	274 562 802	LQ9	E1		PP, EP			0	
3467	ORGANOMETALLIC COMPOUND, TOXIC, SOLID, N.O.S.	6.1	T3	I	6.1	274 562 802	LQ0	E5		PP, EP			2	
3467	ORGANOMETALLIC COMPOUND, TOXIC, SOLID, N.O.S.	6.1	T3	II	6.1	274 562 802	LQ18	E4		PP, EP			2	
3467	ORGANOMETALLIC COMPOUND, TOXIC, SOLID, N.O.S	6.1	T3	III	6.1	274 562 802	LQ9	E1		PP, EP			0	
3468	HYDROGEN IN A METAL HYDRIDE STORAGE SYSTEM or HYDROGEN IN A METAL HYDRIDE STORAGE SYSTEM CONTAINED IN EQUIPMENT or HYDROGEN IN A METAL HYDRIDE STORAGE SYSTEM PACKED WITH EQUIPMENT	2	2F		2.1	321	LQ0	E0	T	PP, EX, A	VE01		1	
3469	PAINT, FLAMMABLE, CORROSIVE (including paint, lacquer, enamel, stain, shellac, varnish, polish, liquid filler and liquid lacquer base) or PAINT RELATED MATERIAL, FLAMMABLE, CORROSIVE (including paint thinning or reducing compound)	3	FC	I	3 +8	163	LQ3	E0		PP, EX, A	VE01		1	
3469	PAINT, FLAMMABLE, CORROSIVE (including paint, lacquer, enamel, stain, shellac, varnish, polish, liquid filler and liquid lacquer base) or PAINT RELATED MATERIAL, FLAMMABLE, CORROSIVE (including paint thinning or reducing compound)	3	FC	II	3 +8	163	LQ4	E2		PP, EX, A	VE01		1	
3469	PAINT, FLAMMABLE, CORROSIVE (including paint, lacquer, enamel, stain, shellac, varnish, polish, liquid filler and liquid lacquer base) or PAINT RELATED MATERIAL, FLAMMABLE, CORROSIVE (including paint thinning or reducing compound)	3	FC	III	3 +8	163	LQ7	E1		PP, EX, A	VE01		0	

UN No. or ID No.	Name and description	Class	Classi-fication Code	Packing group	Labels	Special provi-sions	Limited and excepted quantities		Carriage permitted	Equipment required	Venti-lation	Provisions concerning loading, unloading and carriage	Number of blue cones/ lights	Remarks
	3.1.2	2.2	2.2	2.1.1.3	5.2.2	3.3	3.4.6	3.5.1.2	3.2.1	8.1.5	7.1.6	7.1.6	7.1.5	3.2.1
(1)	(2)	(3a)	(3b)	(4)	(5)	(6)	(7a)	(7b)	(8)	(9)	(10)	(11)	(12)	(13)
3470	PAINT, CORROSIVE, FLAMMABLE (including paint, lacquer, enamel, stain, shellac, varnish, polish, liquid filler and liquid lacquer base) or PAINT RELATED MATERIAL CORROSIVE, FLAMMABLE (including paint thinning or reducing compound)	8	CF1	II	8 +3	163	LQ22	E2		PP, EP, EX, A	VE01		0	
3471	HYDROGENDIFLUORIDES SOLUTION, N.O.S.	8	CT1	II	8 +6.1		LQ22	E2		PP, EP			0	
3471	HYDROGENDIFLUORIDES SOLUTION, N.O.S.	8	CT1	III	8 +6.1		LQ7	E1		PP, EP			0	
3472	CROTONIC ACID, LIQUID	8	C3	III	8		LQ7	E1		PP, EP			0	
3473	FUEL CELL CARTRIDGES or FUEL CELL CARTRIDGES CONTAINED IN EQUIPMENT or FUEL CELL CARTRIDGES PACKED WITH EQUIPMENT containing flammable liquids	3	F1		3	328	LQ13	E0						
3474	1-HYDROXYBENZOTRIAZOLE, ANHYDROUS, WETTED with not less than 20% water, by mass	4.1	D	I	4.1		LQ0	E0		PP			1	
3475	ETHANOL AND GASOLINE MIXTURE or ETHANOL AND MOTOR SPIRIT MIXTURE or ETHANOL AND PETROL MIXTURE, with more than 10% ethanol	3	F1	II	3	333	LQ4	E2	T	PP,EX, A	VE01		1	
3476	FUEL CELL CARTRIDGES or FUEL CELL CARTRIDGES CONTAINED IN EQUIPMENT or FUEL CELL CARTRIDGES PACKED WITH EQUIPMENT, containing water-reactive substances	4.3	W3		4.3	328 334	LQ10 LQ11	E0		PP,EX, A	VE01	HA08	0	
3477	FUEL CELL CARTRIDGES or FUEL CELL CARTRIDGES CONTAINED IN EQUIPMENT or FUEL CELL CARTRIDGES PACKED WITH EQUIPMENT, containing corrosive substances	8	C11		8	328 334	LQ12 LQ13	E0		PP,EX, A			0	
3478	FUEL CELL CARTRIDGES or FUEL CELL CARTRIDGES CONTAINED IN EQUIPMENT or FUEL CELL CARTRIDGES PACKED WITH EQUIPMENT, containing liquefied flammable gas	2	6F		2.1	328 338	LQ1	E0		PP,EX, A	VE01			
3479	FUEL CELL CARTRIDGES or FUEL CELL CARTRIDGES CONTAINED IN EQUIPMENT or FUEL CELL CARTRIDGES PACKED WITH EQUIPMENT, containing hydrogen in metal hydride	2	6F		2.1	328 339	LQ1	E0		PP,EX, A	VE01		1	

UN No. or ID No.	Name and description	Class	Classi-fication Code	Packing group	Labels	Special provi-sions	Limited and excepted quantities		Carriage permitted	Equipment required	Venti-lation	Provisions concerning loading, unloading and carriage	Number of blue cones/ lights	Remarks
	3.1.2	2.2	2.2	2.1.1.3	5.2.2	3.3	3.4.6	3.5.1.2	3.2.1	8.1.5	7.1.6	7.1.6	7.1.5	3.2.1
(1)	(2)	(3a)	(3b)	(4)	(5)	(6)	(7a)	(7b)	(8)	(9)	(10)	(11)	(12)	(13)
3480	LITHIUM ION BATTERIES (including lithium ion polymer batteries)	9	M4	II	9	188 230 310 636	LQ0	E0		PP			0	
3481	LITHIUM ION BATTERIES CONTAINED IN EQUIPMENT or LITHIUM ION BATTERIES PACKED WITH EQUIPMENT (including lithium ion polymer batteries)	9	M4	II	9	188 230 636	LQ0	E0		PP			0	
9000	AMMONIA, DEEPLY REFRIGERATED	2	3TC		2.3+8				**T**	PP			2	Admitted only for carriage in tank vessels
9001	SUBSTANCES WITH A FLASH-POINT ABOVE 60 °C which are carried heated within a limiting range of 15K below their flash-point	3	F3		none				**T**	PP			0	Dangerous only when carried in tank vessels
9002	SUBSTANCES WITH A SELF-IGNITION TEMPERATURE OF 200 °C AND BELOW, n.o.s.	3	F4		none				**T**	PP			0	Dangerous only when carried in tank vessels
9003	SUBSTANCES WITH A FLASH-POINT ABOVE 60 °C AND NOT MORE THAN 100 °C, which do not belong to another Class	9			none				**T**	PP			0	Dangerous only when carried in tank vessels
9004	DIPHENYLMETHANE-4, 4'-DIISOCYANATE	9			none				**T**	PP			0	Dangerous only when carried in tank vessels
9005	ENVIRONMENTALLY HAZARDOUS SUBSTANCE, SOLID, N.O.S., MOLTEN	9			none				**T**	PP			0	Dangerous only when carried in tank vessels
9006	ENVIRONMENTALLY HAZARDOUS SUBSTANCE, LIQUID, N.O.S.	9			none				**T**	PP			0	Dangerous only when carried in tank vessels

3.2.2 Table B: List of dangerous goods in alphabetical order

The following Table B is an alphabetical list of the substances and articles which are listed in the UN numerical order in Table A of 3.2.1. It does not form an integral part of ADN. It has been prepared, with all necessary care by the Secretariat of the United Nations Economic Commission for Europe, in order to facilitate the consultation of Annexes A and B, but it cannot be relied upon as a substitute for the careful study and observance of the actual provisions of those annexed Regulations which, in case of conflict, are deemed to be authoritative.

NOTE 1: *For the purpose of determining the alphabetical order the following information has been ignored, even when it forms part of the proper shipping name: numbers; Greek letters; the abbreviations "sec" and "tert"; and the letters "N" (nitrogen), "n" (normal), "o" (ortho) "m" (meta), "p" (para) and "N.O.S." (not otherwise specified).*

NOTE 2: *The name of a substance or article in block capital letters indicates a proper shipping name (see 3.1.2).*

NOTE 3: *The name of a substance or article in block capital letters followed by the word "see" indicates an alternative proper shipping name or part of a proper shipping name (except for PCBs) (see 3.1.2.1).*

NOTE 4: *An entry in lower case letters followed by the word "see" indicates that the entry is not a proper shipping name; it is a synonym.*

NOTE 5: *Where an entry is partly in block capital letters and partly in lower case letters, the latter part is considered not to be part of the proper shipping name (see 3.1.2.1).*

NOTE 6: *A proper shipping name may be used in the singular or plural, as appropriate, for the purposes of documentation and package marking (see 3.1.2.3).*

NOTE 7: *For the exact determination of a proper shipping name, see 3.1.2.*

Name and description	UN No.	Class	Remarks
Accumulators, electric, see	2794	8	
	2795	8	
	2800	8	
	3028	8	
	3292	4.3	
ACETAL	1088	3	
ACETALDEHYDE	1089	3	
ACETALDEHYDE AMMONIA	1841	9	
ACETALDEHYDE OXIME	2332	3	
ACETIC ACID, GLACIAL	2789	8	
ACETIC ACID SOLUTION, more than 10% but not more than 80% acid, by mass	2790	8	
ACETIC ACID SOLUTION, more than 80% acid, by mass	2789	8	
ACETIC ANHYDRIDE	1715	8	
Acetoin, see	2621	3	
ACETONE	1090	3	
ACETONE CYANOHYDRIN, STABILIZED	1541	6.1	
ACETONE OILS	1091	3	
ACETONITRILE	1648	3	
ACETYL BROMIDE	1716	8	
ACETYL CHLORIDE	1717	3	
ACETYLENE, DISSOLVED	1001	2	
ACETYLENE, SOLVENT FREE	3374	2	
Acetylene tetrabromide, see	2504	6.1	
Acetylene tetrachloride, see	1702	6.1	
ACETYL IODIDE	1898	8	
ACETYL METHYL CARBINOL	2621	3	
Acid butyl phosphate, see	1718	8	
Acid mixture, hydrofluoric and sulphuric, see	1786	8	
Acid mixture, nitrating acid, see	1796	8	
Acid mixture, spent, nitrating acid, see	1826	8	
Acraldehyde, inhibited, see	1092	6.1	
ACRIDINE	2713	6.1	
ACROLEIN DIMER, STABILIZED	2607	3	
ACROLEIN, STABILIZED	1092	6.1	
ACRYLAMIDE, SOLID	2074	6.1	
ACRYLAMIDE, SOLUTION	3426	6.1	
ACRYLIC ACID, STABILIZED	2218	8	
ACRYLONITRILE, STABILIZED	1093	3	
Actinolite, see	2590	9	
Activated carbon, see	1362	4.2	
Activated charcoal, see	1362	4.2	
ADHESIVES containing flammable liquid	1133	3	
ADIPONITRILE	2205	6.1	
Aeroplane flares, see	0093	1	
	0403	1	
	0404	1	
	0420	1	
	0421	1	
AEROSOLS	1950	2	
AGENT, BLASTING, TYPE B	0331	1	
AGENT, BLASTING, TYPE E	0332	1	
AIR BAG INFLATORS	0503	1	
	3268	9	
AIR BAG MODULES	0503	1	
	3268	9	
AIR, COMPRESSED	1002	2	
Aircraft evacuation slides, see	2990	9	
AIRCRAFT HYDRAULIC POWER UNIT FUEL TANK (containing a mixture of anhydrous hydrazine and methylhydrazine) (M86 fuel)	3165	3	
Aircraft survival kits, see	2990	9	
AIR, REFRIGERATED LIQUID	1003	2	
ALCOHOLATES SOLUTION, N.O.S., in alcohol	3274	3	
Alcohol, denaturated, see	1986	3	
	1987	3	
Alcohol, industrial, see	1986	3	
	1987	3	
ALCOHOLS, N.O.S.	1987	3	

Name and description	UN No.	Class	Remarks
ALCOHOLS, FLAMMABLE, TOXIC, N.O.S.	1986	3	
ALCOHOLIC BEVERAGES, with more than 24% but not more than 70% alcohol by volume	3065	3	
ALCOHOLIC BEVERAGES, with more than 70% alcohol by volume	3065	3	
Aldehyde, see	1989	3	
ALDEHYDES, N.O.S.	1989	3	
ALDEHYDES, FLAMMABLE, TOXIC, N.O.S.	1988	3	
ALDOL	2839	6.1	
ALKALI METAL ALCOHOLATES, SELF-HEATING, CORROSIVE, N.O.S.	3206	4.2	
ALKALI METAL ALLOY, LIQUID, N.O.S.	1421	4.3	
ALKALI METAL AMALGAM, LIQUID	1389	4.3	
ALKALI METAL AMALGAM, SOLID	3401	4.3	
ALKALI METAL AMIDES	1390	4.3	
ALKALI METAL DISPERSION	1391	4.3	
Alkaline corrosive battery fluid, see	2797	8	
ALKALINE EARTH METAL ALCOHOLATES, N.O.S.	3205	4.2	
ALKALINE EARTH METAL ALLOY, N.O.S.	1393	4.3	
ALKALINE EARTH METAL AMALGAM, LIQUID	1392	4.3	
ALKALINE EARTH METAL AMALGAM, SOLID	3402	4.3	
ALKALINE EARTH METAL DISPERSION	1391	4.3	
ALKALOIDS, LIQUID, N.O.S.	3140	6.1	
ALKALOIDS, SOLID, N.O.S.	1544	6.1	
ALKALOID SALTS, LIQUID, N.O.S.	3140	6.1	
ALKALOID SALTS, SOLID, N.O.S.	1544	6.1	
Alkyl aluminium halides, see	3394	4.2	
ALKYLPHENOLS, LIQUID, N.O.S. (including C_2-C_{12} homologues)	3145	8	
ALKYLPHENOLS, SOLID, N.O.S. (including C_2-C_{12} homologues)	2430	8	
ALKYLSULPHONIC ACIDS, LIQUID with more than 5% free sulphuric acid	2584	8	
ALKYLSULPHONIC ACIDS, LIQUID with not more than 5% free sulphuric acid	2586	8	
ALKYLSULPHONIC ACIDS, SOLID with more than 5% free sulphuric acid	2583	8	
ALKYLSULPHONIC ACIDS, SOLID with not more than 5% free sulphuric acid	2585	8	
ALKYLSULPHURIC ACIDS	2571	8	
Allene, see	2200	2	
ALLYL ACETATE	2333	3	
ALLYL ALCOHOL	1098	6.1	
ALLYLAMINE	2334	6.1	
ALLYL BROMIDE	1099	3	
ALLYL CHLORIDE	1100	3	
Allyl chlorocarbonate, see	1722	6.1	
ALLYL CHLOROFORMATE	1722	6.1	
ALLYL ETHYL ETHER	2335	3	
ALLYL FORMATE	2336	3	
ALLYL GLYCIDYL ETHER	2219	3	
ALLYL IODIDE	1723	3	
ALLYL ISOTHIOCYANATE, STABILIZED	1545	6.1	
ALLYLTRICHLOROSILANE, STABILIZED	1724	8	
Aluminium alkyls, see	3394	4.2	
Aluminium alkyl halides, liquid, see	3394	4.2	
Aluminium alkyl halides, solid, see	3393	4.2	
Aluminium alkyl hydrides, see	3394	4.2	
ALUMINIUM BOROHYDRIDE	2870	4.2	
ALUMINIUM BOROHYDRIDE IN DEVICES	2870	4.2	
ALUMINIUM BROMIDE, ANHYDROUS	1725	8	

Name and description	UN No.	Class	Remarks
ALUMINIUM BROMIDE SOLUTION	2580	8	
ALUMINIUM CARBIDE	1394	4.3	
ALUMINIUM CHLORIDE, ANHYDROUS	1726	8	
ALUMINIUM CHLORIDE SOLUTION	2581	8	
Aluminium dross, see	3170	4.3	
ALUMINIUM FERROSILICON POWDER	1395	4.3	
ALUMINIUM HYDRIDE	2463	4.3	
ALUMINIUM NITRATE	1438	5.1	
ALUMINIUM PHOSPHIDE	1397	4.3	
ALUMINIUM PHOSPHIDE PESTICIDE	3048	6.1	
ALUMINIUM POWDER, COATED	1309	4.1	
ALUMINIUM POWDER, UNCOATED	1396	4.3	
ALUMINIUM REMELTING BY-PRODUCTS	3170	4.3	
ALUMINIUM RESINATE	2715	4.1	
ALUMINIUM SILICON POWDER, UNCOATED	1398	4.3	
ALUMINIUM SMELTING BY-PRODUCTS	3170	4.3	
Amatols, see	0082	1	
AMINES, FLAMMABLE, CORROSIVE, N.O.S.	2733	3	
AMINES, LIQUID, CORROSIVE, N.O.S.	2735	8	
AMINES, LIQUID, CORROSIVE, FLAMMABLE, N.O.S.	2734	8	
AMINES, SOLID, CORROSIVE, N.O.S.	3259	8	
Aminobenzene, see	1547	6.1	
2-Aminobenzotrifluoruride, see	2942	6.1	
3-Aminobenzotrifluoruride, see	2948	6.1	
Aminobutane, see	1125	3	
2-AMINO-4-CHLOROPHENOL	2673	6.1	
2-AMINO-5-DIETHYLAMINOPENTANE	2946	6.1	
2-AMINO-4,6-DINITROPHENOL, WETTED with not less than 20% water, by mass	3317	4.1	
2-(2-AMINOETHOXY) ETHANOL	3055	8	
N-AMINOETHYLPIPERAZINE	2815	8	
1-Amino-2-nitrobenzene, see	1661	6.1	
1-Amino-3-nitrobenzene, see	1661	6.1	
1-Amino-4-nitrobenzene, see	1661	6.1	
AMINOPHENOLS (o-, m-, p-)	2512	6.1	
AMINOPYRIDINES (o-, m-, p-)	2671	6.1	
AMMONIA, ANHYDROUS	1005	2	
AMMONIA, DEEPLY REFRIGERATED	9000	2	Admitted only for carriage in tank vessel
AMMONIA SOLUTION relative density between 0.880 and 0.957 at 15 °C in water, with more than 10% but not more than 35% ammonia	2672	8	
AMMONIA SOLUTION, relative density less than 0.880 at 15 °C in water, with more than 35% but not more than 50% ammonia	2073	2	
AMMONIA SOLUTION, relative density less than 0.880 at 15 °C in water, with more than 50% ammonia	3318	2	
AMMONIUM ARSENATE	1546	6.1	
Ammonium bichromate, see	1439	5.1	
Ammonium bifluoride solid, see	1727	8	
Ammonium bifluoride solution, see	2817	8	
Ammonium bisulphate, see	2506	8	
Ammonium bisulphite solution, see	2693	8	
AMMONIUM DICHROMATE	1439	5.1	
AMMONIUM DINITRO-o-CRESOLATE, SOLID	1843	6.1	
AMMONIUM DINITRO-o-CRESOLATE, SOLUTION	3424	6.1	
AMMONIUM FLUORIDE	2505	6.1	

Name and description	UN No.	Class	Remarks
AMMONIUM FLUOROSILICATE	2854	6.1	
Ammonium hexafluorosilicate, see	2854	6.1	
AMMONIUM HYDROGENDIFLUORIDE, SOLID	1727	8	
AMMONIUM HYDROGENDIFLUORIDE SOLUTION	2817	8	
AMMONIUM HYDROGEN SULPHATE	2506	8	
Ammonium hydrosulphide solution (treat as ammonium sulphide solution), see	2683	8	
AMMONIUM METAVANADATE	2859	6.1	
AMMONIUM NITRATE with more than 0.2% combustible substances, including any organic substance calculated as carbon, to the exclusion of any other added substance	0222	1	
AMMONIUM NITRATE with not more than 0.2% total combustible material, including any organic substance calculated as carbon, to the exclusion of any other added substance	1942	5.1	
AMMONIUM NITRATE EMULSION, intermediate for blasting explosives, liquid	3375	5.1	
AMMONIUM NITRATE EMULSION, intermediate for blasting explosives, solid	3375	5.1	
Ammonium nitrate explosive, see	0082 0331	1 1	
AMMONIUM NITRATE BASED FERTILIZER	2067	5.1	
AMMONIUM NITRATE BASED FERTILIZER, uniform mixtures of the nitrogen/phosphate, nitrogen/potash or nitrogen/phosphate/potash type, containing not more than 70% ammonium nitrate and not more than 0.4% total combustible/organic material calculated as carbon or with not more than 45% ammonium nitrate and unrestricted combustible material	2071	9	
AMMONIUM NITRATE GEL, intermediate for blasting explosives, liquid	3375	5.1	
AMMONIUM NITRATE GEL, intermediate for blasting explosives, solid	3375	5.1	
AMMONIUM NITRATE, LIQUID hot concentrated solution, in a concentration of more than 80% but not more than 93%	2426	5.1	
AMMONIUM NITRATE SUSPENSION, intermediate for blasting explosives, liquid	3375	5.1	
AMMONIUM NITRATE SUSPENSION, intermediate for blasting explosives, solid	3375	5.1	
AMMONIUM PERCHLORATE	0402 1442	1 5.1	
Ammonium permanganate, see	1482	5.1	
AMMONIUM PERSULPHATE	1444	5.1	
AMMONIUM PICRATE dry or wetted with less than 10% water, by mass	0004	1	
AMMONIUM PICRATE, WETTED with not less than 10% water, by mass	1310	4.1	
AMMONIUM POLYSULPHIDE SOLUTION	2818	8	
AMMONIUM POLYVANADATE	2861	6.1	
Ammonium silicofluoride, see	2854	6.1	
AMMONIUM SULPHIDE SOLUTION	2683	8	
Ammunition, blank, see	0014 0326 0327 0338 0413	1 1 1 1 1	
Ammunition, fixed Ammunition, semi-fixed Ammunition, separate loading, see	0005 0006 0007 0321 0348 0412	1 1 1 1 1 1	
AMMUNITION, ILLUMINATING with or without burster, expelling charge or propelling charge	0171 0254 0297	1 1 1	
AMMUNITION, INCENDIARY, liquid or gel, with burster, expelling charge or propelling charge	0247	1	

Name and description	UN No.	Class	Remarks
AMMUNITION, INCENDIARY with or without burster, expelling charge or propelling charge	0009 0010 0300	1 1 1	
Ammunition, incendiary (water-activated contrivances) with burster, expelling charge or propelling charge, see	0248 0249	1 1	
AMMUNITION, INCENDIARY, WHITE PHOSPHORUS with burster, expelling charge or propelling charge	0243 0244	1 1	
Ammunition, industrial, see	0275 0276 0277 0278 0323 0381	1 1 1 1 1 1	
Ammunition, lachrymatory, see	0018 0019 0301 2017	1 1 1 1	
AMMUNITION, PRACTICE	0362 0488	1 1	
AMMUNITION, PROOF	0363	1	
AMMUNITION, SMOKE with or without burster, expelling charge or propelling charge	0015 0016 0303	1 1 1	
Ammunition, smoke (water-activated contrivances), white phosphorus with burster, expelling charge or propelling charge, see	0248	1	
Ammunition, smoke (water-activated contrivances), without white phosphorus or phosphides with burster, expelling charge or propelling charge, see	0249	1	
AMMUNITION, SMOKE, WHITE PHOSPHORUS with burster, expelling charge or propelling charge	0245 0246	1 1	
Ammunition, sporting, see	0012 0328 0339 0417	1 1 1 1	
AMMUNITION, TEAR-PRODUCING, NON-EXPLOSIVE without burster or expelling charge, non-fuzed	2017	6.1	
AMMUNITION, TEAR-PRODUCING with burster, expelling charge or propelling charge	0018 0019 0301	1 1 1	
AMMUNITION, TOXIC with burster, expelling charge or propelling charge	0020	1	Carriage prohibited
AMMUNITION, TOXIC with burster, expelling charge or propelling charge	0021	1	Carriage prohibited
Ammunition, toxic (water-activated contrivances) with burster, expelling charge or propelling charge, see	0248 0249	1 1	
AMMUNITION, TOXIC, NON-EXPLOSIVE without burster or expelling charge, non-fuzed	2016	6.1	
Amosite, see	2212	9	
AMYL ACETATES	1104	3	
AMYL ACID PHOSPHATE	2819	8	
Amyl aldehyde, see	2058	3	
AMYLAMINE	1106	3	
n-Amylamine, see	1106	3	
AMYL BUTYRATES	2620	3	
AMYL CHLORIDE	1107	3	
n-AMYLENE, see	1108	3	
AMYL FORMATES	1109	3	
AMYL MERCAPTAN	1111	3	
n-AMYL METHYL KETONE	1110	3	
AMYL NITRATE	1112	3	
AMYL NITRITE	1113	3	
AMYLTRICHLOROSILANE	1728	8	
Anaesthetic ether, see	1155	3	
ANILINE	1547	6.1	
Aniline chloride, see	1548	6.1	
ANILINE HYDROCHLORIDE	1548	6.1	
Aniline oil, see	1547	6.1	
Aniline salt, see	1548	6.1	
ANISIDINES	2431	6.1	
ANISOLE	2222	3	
ANISOYL CHLORIDE	1729	8	
Anthophyllite, see	2590	9	

Name and description	UN No.	Class	Remarks
Antimonous chloride, see	1733	8	
ANTIMONY COMPOUND, INORGANIC, LIQUID, N.O.S.	3141	6.1	
ANTIMONY COMPOUND, INORGANIC, SOLID, N.O.S.	1549	6.1	
Antimony hydride, see	2676	2	
ANTIMONY LACTATE	1550	6.1	
Antimony (III) lactate, see	1550	6.1	
ANTIMONY PENTACHLORIDE, LIQUID	1730	8	
ANTIMONY PENTACHLORIDE SOLUTION	1731	8	
ANTIMONY PENTAFLUORIDE	1732	8	
Antimony perchloride, liquid, see	1730	8	
ANTIMONY POTASSIUM TARTRATE	1551	6.1	
ANTIMONY POWDER	2871	6.1	
ANTIMONY TRICHLORIDE	1733	8	
A.n.t.u., see	1651	6.1	
ARGON, COMPRESSED	1006	2	
ARGON, REFRIGERATED LIQUID	1951	2	
Arsenates, n.o.s., see	1556	6.1	
	1557	6.1	
ARSENIC	1558	6.1	
ARSENIC ACID, LIQUID	1553	6.1	
ARSENIC ACID, SOLID	1554	6.1	
ARSENICAL DUST	1562	6.1	
Arsenical flue dust, see	1562	6.1	
ARSENICAL PESTICIDE, LIQUID, FLAMMABLE, TOXIC, flash-point less than 23 °C	2760	3	
ARSENICAL PESTICIDE, LIQUID, TOXIC	2994	6.1	
ARSENICAL PESTICIDE, LIQUID, TOXIC, FLAMMABLE, flash-point not less than 23 °C	2993	6.1	
ARSENICAL PESTICIDE, SOLID, TOXIC	2759	6.1	
ARSENIC BROMIDE	1555	6.1	
Arsenic (III) bromide, see	1555	6.1	
Arsenic chloride, see	1560	6.1	
ARSENIC COMPOUND, LIQUID, N.O.S., inorganic, including: Arsenates, n.o.s., Arsenites, n.o.s.; and Arsenic sulphides, n.o.s.	1556	6.1	
ARSENIC COMPOUND, SOLID, N.O.S., inorganic, including: Arsenates, n.o.s.; Arsenites, n.o.s.; and Arsenic sulphides, n.o.s.	1557	6.1	
Arsenic (III) oxide, see	1561	6.1	
Arsenic (V) oxide, see	1559	6.1	
ARSENIC PENTOXIDE	1559	6.1	
Arsenic sulphides, see	1556	6.1	
	1557	6.1	
ARSENIC TRICHLORIDE	1560	6.1	
ARSENIC TRIOXIDE	1561	6.1	
Arsenious chloride, see	1560	6.1	
Arsenites, n.o.s., see	1556	6.1	
	1557	6.1	
Arsenous chloride, see	1560	6.1	
ARSINE	2188	2	
ARTICLES, EEI, see	0486	1	
ARTICLES, EXPLOSIVE, EXTREMELY INSENSITIVE	0486	1	
ARTICLES, EXPLOSIVE, N.O.S.	0349	1	
	0350	1	
	0351	1	
	0352	1	
	0353	1	
	0354	1	
	0355	1	
	0356	1	
	0462	1	
	0463	1	
	0464	1	
	0465	1	
	0466	1	
	0467	1	
	0468	1	
	0469	1	
	0470	1	
	0471	1	
	0472	1	

Name and description	UN No.	Class	Remarks
ARTICLES, PRESSURIZED, HYDRAULIC (containing non-flammable gas)	3164	2	
ARTICLES, PRESSURIZED, PNEUMATIC (containing non-flammable gas)	3164	2	
ARTICLES, PYROPHORIC	0380	1	
ARTICLES, PYROTECHNIC for technical purposes	0428	1	
	0429	1	
	0430	1	
	0431	1	
	0432	1	
ARYLSULPHONIC ACIDS, LIQUID with more than 5% free sulphuric acid	2584	8	
ARYLSULPHONIC ACIDS, LIQUID with not more than 5% free sulphuric acid	2586	8	
ARYLSULPHONIC ACIDS, SOLID with more than 5% free sulphuric acid	2583	8	
ARYLSULPHONIC ACIDS, SOLID with not more than 5% free sulphuric acid	2585	8	
Asbestos, blue or brown, see	2212	9	
Asbestos, white, see	2590	9	
Asphalt, with a flash-point not greater than 60 °C, see	1999	3	
Asphalt, with a flash-point above 60 °C, at or above its flash-point, see	3256	3	
Asphalt, at or above 100 °C and below its flash-point, see	3257	9	
Aviation regulated liquid, n.o.s.	3334	9	Not subject to ADN
Aviation regulated solid, n.o.s.	3335	9	Not subject to ADN
AZODICARBONAMIDE	3242	4.1	
Bag charges, see	0242	1	
	0279	1	
	0414	1	
Ballistite, see	0160	1	
	0161	1	
Bangalore torpedoes, see	0136	1	
	0137	1	
	0138	1	
	0294	1	
BARIUM	1400	4.3	
BARIUM ALLOYS, PYROPHORIC	1854	4.2	
BARIUM AZIDE, dry or wetted with less than 50% water, by mass	0224	1	
BARIUM AZIDE, WETTED with not less than 50% water, by mass	1571	4.1	
Barium binoxide, see	1449	5.1	
BARIUM BROMATE	2719	5.1	
BARIUM CHLORATE, SOLID	1445	5.1	
BARIUM CHLORATE, SOLUTION	3405	5.1	
BARIUM COMPOUND, N.O.S.	1564	6.1	
BARIUM CYANIDE	1565	6.1	
Barium dioxide, see	1449	5.1	
BARIUM HYPOCHLORITE with more than 22% available chlorine	2741	5.1	
BARIUM NITRATE	1446	5.1	
BARIUM OXIDE	1884	6.1	
BARIUM PERCHLORATE, SOLID	1447	5.1	
BARIUM PERCHLORATE, SOLUTION	3406	5.1	
BARIUM PERMANGANATE	1448	5.1	
BARIUM PEROXIDE	1449	5.1	
Barium selenate, see	2630	6.1	
Barium selenite, see	2630	6.1	
Barium superoxide, see	1449	5.1	
BATTERIES, CONTAINING SODIUM	3292	4.3	
BATTERIES, DRY, CONTAINING POTASSIUM HYDROXIDE SOLID, electric storage	3028	8	
BATTERIES, WET, FILLED WITH ACID, electric storage	2794	8	
BATTERIES, WET, FILLED WITH ALKALI, electric storage	2795	8	
BATTERIES, WET, NON-SPILLABLE, electric storage	2800	8	
BATTERY FLUID, ACID	2796	8	
BATTERY FLUID, ALKALI	2797	8	

Name and description	UN No.	Class	Remarks
Battery-powered vehicle or Battery-powered equipment	3171	9	Not subject to ADN
BENZALDEHYDE	1990	9	
BENZENE	1114	3	
BENZENESULPHONYL CHLORIDE	2225	8	
Benzenethiol, see	2337	6.1	
BENZIDINE	1885	6.1	
Benzol, see	1114	3	
Benzolene, see	1268	3	
BENZONITRILE	2224	6.1	
BENZOQUINONE	2587	6.1	
Benzosulphochloride, see	2225	8	
BENZOTRICHLORIDE	2226	8	
BENZOTRIFLUORIDE	2338	3	
BENZOYL CHLORIDE	1736	8	
BENZYL BROMIDE	1737	6.1	
BENZYL CHLORIDE	1738	6.1	
Benzyl chlorocarbonate, see	1739	8	
BENZYL CHLOROFORMATE	1739	8	
Benzyl cyanide, see	2470	6.1	
BENZYLDIMETHYLAMINE	2619	8	
BENZYLIDENE CHLORIDE	1886	6.1	
BENZYL IODIDE	2653	6.1	
BERYLLIUM COMPOUND, N.O.S.	1566	6.1	
BERYLLIUM NITRATE	2464	5.1	
BERYLLIUM POWDER	1567	6.1	
Bhusa	1327	4.1	Not subject to ADN
BICYCLO[2.2.1]HEPTA-2,5-DIENE, STABILIZED	2251	3	
Bifluorides, n.o.s., see	1740	8	
BIOLOGICAL SUBSTANCE, CATEGORY B	3373	6.2	
BIOLOGICAL SUBSTANCE, CATEGORY B (animal material only)	3373	6.2	
(BIO) MEDICAL WASTE, N.O.S.	3291	6.2	
BIPYRIDILIUM PESTICIDE, LIQUID, FLAMMABLE, TOXIC, flash-point less than 23 °C	2782	3	
BIPYRIDILIUM PESTICIDE, LIQUID, TOXIC	3016	6.1	
BIPYRIDILIUM PESTICIDE, LIQUID, TOXIC, FLAMMABLE, flash-point not less than 23 °C	3015	6.1	
BIPYRIDILIUM PESTICIDE, SOLID, TOXIC	2781	6.1	
BISULPHATES, AQUEOUS SOLUTION	2837	8	
BISULPHITES, AQUEOUS SOLUTION, N.O.S.	2693	8	
Bitumen, with a flash-point not greater than 60 °C,see	1999	3	
Bitumen, with a flash-point not greater than 60 °C, see	1999	3	
Bitumen, with a flash-point above 60 °C, at or above its flash-point, see	3256	3	
Bitumen, at or above 100 °C and below its flash-point, see	3257	9	
BLACK POWDER, COMPRESSED	0028	1	
BLACK POWDER, granular or as a meal	0027	1	
BLACK POWDER, IN PELLETS	0028	1	
Blasting cap assemblies, see	0360 0361	1 1	
Blasting caps, electric, see	0030 0255 0456	1 1 1	
Bleaching powder, see	2208	5.1	
BLUE ASBESTOS (crocidolite)	2212	9	
BOMBS with bursting charge	0033 0034 0035 0291	1 1 1 1	
Bombs, illuminating, see	0254	1	

Name and description	UN No.	Class	Remarks
BOMBS, PHOTO-FLASH	0037 0038 0039 0299	1 1 1 1	
BOMBS, SMOKE, NON-EXPLOSIVE with corrosive liquid, without initiating device	2028	8	
Bombs, target identification, see	0171 0254 0297	1 1 1	
BOMBS WITH FLAMMABLE LIQUID with bursting charge	0399 0400	1 1	
BOOSTERS WITH DETONATOR	0225 0268	1 1	
BOOSTERS without detonator	0042 0283	1 1	
Borate and chlorate mixture, see	1458	5.1	
BORNEOL	1312	4.1	
BORON TRIBROMIDE	2692	8	
BORON TRICHLORIDE	1741	2	
BORON TRIFLUORIDE ACETIC ACID COMPLEX, LIQUID	1742	8	
BORON TRIFLUORIDE ACETIC ACID COMPLEX, SOLID	3419	8	
BORON TRIFLUORIDE	1008	2	
BORON TRIFLUORIDE DIETHYL ETHERATE	2604	8	
BORON TRIFLUORIDE DIHYDRATE	2851	8	
BORON TRIFLUORIDE DIMETHYL ETHERATE	2965	4.3	
BORON TRIFLUORIDE PROPIONIC ACID COMPLEX, LIQUID	1743	8	
BORON TRIFLUORIDE PROPIONIC ACID COMPLEX, SOLID	3420	8	
BROMATES, INORGANIC, N.O.S.	1450	5.1	
BROMATES, INORGANIC, AQUEOUS SOLUTION, N.O.S	3213	5.1	
BROMINE	1744	8	
BROMINE CHLORIDE	2901	2	
BROMINE PENTAFLUORIDE	1745	5.1	
BROMINE SOLUTION	1744	8	
BROMINE TRIFLUORIDE	1746	5.1	
BROMOACETIC ACID, SOLID	3425	8	
BROMOACETIC ACID, SOLUTION	1938	8	
BROMOACETONE	1569	6.1	
omega-Bromoacetone, see	2645	6.4	
BROMOACETYL BROMIDE	2513	8	
BROMOBENZENE	2514	3	
BROMOBENZYL CYANIDES, LIQUID	1694	6.1	
BROMOBENZYL CYANIDES, SOLID	3449	6.1	
1-BROMOBUTANE	1126	3	
2-BROMOBUTANE	2339	3	
BROMOCHLORO-METHANE	1887	6.1	
1-BROMO-3-CHLOROPROPANE	2688	6.1	
1-Bromo-2,3-epoxypropane, see	2558	6.1	
Bromoethane, see	1891	6.1	
2-BROMOETHYL ETHYL ETHER	2340	3	
BROMOFORM	2515	6.1	
Bromomethane, see	1062	2	
1-BROMO-3-METHYLBUTANE	2341	3	
BROMOMETHYLPROPANES	2342	3	
2-BROMO-2-NITROPROPANE-1,3-DIOL	3241	4.1	
2-BROMOPENTANE	2343	3	
BROMOPROPANES	2344	3	
3-BROMOPROPYNE	2345	3	
BROMOTRIFLUOROETHYLENE	2419	2	
BROMOTRIFLUOROMETHANE	1009	2	
BROWN ASBESTOS (amosite, mysorite)	2212	9	
BRUCINE	1570	6.1	
BURSTERS, explosive	0043	1	

Name and description	UN No.	Class	Remarks
BUTADIENES AND HYDROCARBON MIXTURE, STABILIZED, having a vapour pressure at 70 °C not exceeding 1.1 MPa (11 bar) and a density at 50 °C not lower than 0.525 kg/l	1010	2	
BUTADIENES, STABILIZED, (1,2-butadiene)	1010	2	
BUTADIENES, STABILIZED, (1,3-butadiene)	1010	2	
BUTANE	1011	2	
BUTANEDIONE	2346	3	
Butane-1-thiol, see	2347	3	
BUTANOLS	1120	3	
1-Butanol, see	1120	3	
Butan-2-ol, see	1120	3	
Butanol, secondary, see	1120	3	
Butanol, tertiary, see	1120	3	
Butanone, see	1193	3	
2-Butenal, see	1143	6.1	
Butene, see	1012	2	
Bute-1-ene-3-one, see	1251	3	
1,2-Buteneoxide, see	3022	3	
2-Buten-1-ol, see	2614	3	
BUTYL ACETATES	1123	3	
Butyl acetate, secondary, see	1123	3	
BUTYL ACID PHOSPHATE	1718	8	
BUTYL ACRYLATES, STABILIZED	2348	3	
n-Butyl alcohol, see	1120	3	
Butyl alcohols, see	1120	3	
n-BUTYLAMINE	1125	3	
N-BUTYLANILINE	2738	6.1	
sec-Butyl benzene, see	2709	3	
BUTYLBENZENES	2709	3	
n-Butyl bromide, see	1126	3	
n-Butyl chloride, see	1127	3	
n-BUTYL CHLOROFORMATE	2743	6.1	
tert-BUTYLCYCLOHEXYL CHLOROFORMATE	2747	6.1	
BUTYLENES MIXTURE or 1-BUTYLENE or CIS-2-BUTYLENE or TRANS-2-BUTYLENE	1012	2	
1,2-BUTYLENE OXIDE, STABILIZED	3022	3	
Butyl ethers, see	1149	3	
Butyl ethyl ether, see	1179	3	
n-BUTYL FORMATE	1128	3	
tert-BUTYL HYPOCHLORITE	3255	4.2	Carriage prohibited
N,n-BUTYLIMIDAZOLE	2690	6.1	
N,n-Butyliminazole, see	2690	6.1	
n-BUTYL ISOCYANATE	2485	6.1	
tert-BUTYL ISOCYANATE	2484	6.1	
Butyl lithium, see	3394	4.2	
BUTYL MERCAPTAN	2347	3	
n-BUTYL METHACRYLATE, STABILIZED	2227	3	
BUTYL METHYL ETHER	2350	3	
BUTYL NITRITES	2351	3	
Butylphenols, liquid, see	3145	8	
Butylphenols, solid, see	2430	8	
BUTYL PROPIONATES	1914	3	
p-tert-Butyltoluene, see	2667	6.1	
BUTYLTOLUENES	2667	6.1	
BUTYLTRICHLOROSILANE	1747	8	
5-tert-BUTYL-2,4,6-TRINITRO-m-XYLENE	2956	4.1	
BUTYL VINYL ETHER, STABILIZED	2352	3	
But-1-yne, see	2452	2	
1,4-BUTYNEDIOL	2716	6.1	
2-Butyne-1,4-diol, see	2716	6.1	

Name and description	UN No.	Class	Remarks
BUTYRALDEHYDE	1129	3	
n-Butyraldehyde, see	1129	3	
BUTYRALDOXIME	2840	3	
BUTYRIC ACID	2820	8	
BUTYRIC ANHYDRIDE	2739	8	
Butyrone, see	2710	3	
BUTYRONITRILE	2411	3	
Butyroyl chloride, see	2353	3	
BUTYRYL CHLORIDE	2353	3	
Cable cutters, explosive, see	0070	1	
CACODYLIC ACID	1572	6.1	
CADMIUM COMPOUND	2570	6.1	
CAESIUM	1407	4.3	
CAESIUM HYDROXIDE	2682	8	
CAESIUM HYDROXIDE SOLUTION	2681	8	
CAESIUM NITRATE	1451	5.1	
Caffeine, see	1544	6.1	
Cajeputene, see	2052	3	
CALCIUM	1401	4.3	
CALCIUM ALLOYS, PYROPHORIC	1855	4.2	
CALCIUM ARSENATE	1573	6.1	
CALCIUM ARSENATE AND CALCIUM ARSENITE MIXTURE, SOLID	1574	6.1	
Calcium bisulphite solution, see	2693	8	
CALCIUM CARBIDE	1402	4.3	
CALCIUM CHLORATE	1452	5.1	
CALCIUM CHLORATE, AQUEOUS SOLUTION	2429	5.1	
CALCIUM CHLORITE	1453	5.1	
CALCIUM CYANAMIDE with more than 0.1% calcium carbide	1403	4.3	
CALCIUM CYANIDE	1575	6.1	
CALCIUM DITHIONITE	1923	4.2	
CALCIUM HYDRIDE	1404	4.3	
CALCIUM HYDROSULPHITE, see	1923	4.2	
CALCIUM HYPOCHLORITE, DRY	1748	5.1	
CALCIUM HYPOCHLORITE, HYDRATED with not less than 5.5% but not more than 16% water	2880	5.1	
CALCIUM HYPOCHLORITE, HYDRATED MIXTURE with not less than 5.5% but not more than 16% water	2880	5.1	
CALCIUM HYPOCHLORITE MIXTURE, DRY with more than 10% but not more than 39% available chlorine	2208	5.1	
CALCIUM HYPOCHLORITE MIXTURE, DRY with more than 39% available chlorine (8.8% available oxygen)	1748	5.1	
CALCIUM MANGANESE SILICON	2844	4.3	
CALCIUM NITRATE	1454	5.1	
Calcium oxide	1910	8	Not subject to ADN
CALCIUM PERCHLORATE	1455	5.1	
CALCIUM PERMANGANATE	1456	5.1	
CALCIUM PEROXIDE	1457	5.1	
CALCIUM PHOSPHIDE	1360	4.3	
CALCIUM, PYROPHORIC	1855	4.2	
CALCIUM RESINATE	1313	4.1	
CALCIUM RESINATE, FUSED	1314	4.1	
Calcium selenate, see	2630	6.1	
CALCIUM SILICIDE	1405	4.3	
Calcium silicon, see	1405	4.3	
Calcium superoxide, see	1457	5.1	
Camphanone, see	2717	4.1	
CAMPHOR OIL	1130	3	
CAMPHOR, synthetic	2717	4.1	
CAPROIC ACID	2829	8	

Name and description	UN No.	Class	Remarks
CARBAMATE PESTICIDE, LIQUID, FLAMMABLE, TOXIC, flash-point less than 23 °C	2758	3	
CARBAMATE PESTICIDE, LIQUID, TOXIC	2992	6.1	
CARBAMATE PESTICIDE, LIQUID, TOXIC, FLAMMABLE, flash-point not less than 23 °C	2991	6.1	
CARBAMATE PESTICIDE, SOLID, TOXIC	2757	6.1	
Carbolic acid, see	1671 2312 2821	6.1 6.1 6.1	
CARBON, animal or vegetable origin	1361	4.2	
CARBON, ACTIVATED	1362	4.2	
Carbon bisulphide, see	1131	3	
Carbon black (animal or vegetable origin), see	1361	4.2	
CARBON DIOXIDE	1013	2	
Carbon dioxide and ethylene oxide mixture, see	1041 1952 3300	2 2 2	
CARBON DIOXIDE, REFRIGERATED LIQUID	2187	2	
Carbon dioxide, solid	1845	9	Not subject to ADN
CARBON DISULPHIDE	1131	3	
Carbonic anhydride, see	1013 1845 2187	2 9 2	
CARBON MONOXIDE, COMPRESSED	1016	2	
Carbon oxysulphide, see	2204	2.3	
Carbon sulphide, see	1131	3	
CARBON TETRABROMIDE	2516	6.1	
CARBON TETRACHLORIDE	1846	6.1	
Carbonyl chloride, see	1076	2	
CARBONYL FLUORIDE	2417	2	
CARBONYL SULPHIDE	2204	2	
Cartridge cases, empty, primed, see	0055 0379	1 1	

Name and description	UN No.	Class	Remarks
Cartridges, actuating, for fire extinguisher or apparatus valve, see	0275 0276 0323 0381	1 1 1 1	
Cartridges, explosive, see	0048	1	
CARTRIDGES, FLASH	0049 0050	1 1	
CARTRIDGES FOR WEAPONS with bursting charge	0005 0006 0007 0321 0348 0412	1 1 1 1 1 1	
CARTRIDGES FOR WEAPONS, BLANK	0014 0326 0327 0338 0413	1 1 1 1 1	
CARTRIDGES FOR WEAPONS, INERT PROJECTILE	0012 0328 0339 0417	1 1 1 1	
Cartridges, illuminating, see	0171 0254 0297	1 1 1	
CARTRIDGES, OIL WELL	0277 0278	1 1	
CARTRIDGES, POWER DEVICE	0275 0276 0323 0381	1 1 1 1	
CARTRIDGES, SIGNAL	0054 0312 0405	1 1 1	
CARTRIDGES, SMALL ARMS	0012 0339 0417	1 1 1	
CARTRIDGES, SMALL ARMS, BLANK	0014 0327 0338	1 1 1	
Cartridges, starter, jet engine, see	0275 0276 0323 0381	1 1 1 1	
CASES, CARTRIDGE, EMPTY, WITH PRIMER	0055 0379	1 1	
CASES, COMBUSTIBLE, EMPTY, WITHOUT PRIMER	0446 0447	1 1	
Casinghead gasoline, see	1203	3	

Name and description	UN No.	Class	Remarks
CASTOR BEANS	2969	9	
CASTOR FLAKE	2969	9	
CASTOR MEAL	2969	9	
CASTOR POMACE	2969	9	
CAUSTIC ALKALI LIQUID, N.O.S.	1719	8	
Caustic potash, see	1814	8	
Caustic soda, see	1824	8	
Caustic soda liquor, see	1824	8	
CELLS, CONTAINING SODIUM	3292	4.3	
CELLULOID in block, rods, rolls, sheets, tubes, etc., except scrap	2000	4.1	
CELLULOID, SCRAP	2002	4.2	
Cement, see	1133	3	
CERIUM, slabs, ingots or rods	1333	4.1	
CERIUM, turnings or gritty powder	3078	4.3	
Cer mishmetall, see	1323	4.1	
Charcoal, activated, see	1362	4.1	
Charcoal, non-activated, see	1361	4.2	
CHARGES, BURSTING, PLASTICS BONDED	0457 0458 0459 0460	1 1 1 1	
CHARGES, DEMOLITION	0048	1	
CHARGES, DEPTH	0056	1	
Charges, expelling, explosive, for fire extinguishers, see	0275 0276 0323 0381	1 1 1 1	
CHARGES, EXPLOSIVE, COMMERCIAL without detonator	0442 0443 0444 0445	1 1 1 1	
CHARGES, PROPELLING	0271 0272 0415 0491	1 1 1 1	
CHARGES, PROPELLING, FOR CANNON	0242 0279 0414	1 1 1	
CHARGES, SHAPED, FLEXIBLE, LINEAR	0237 0288	1 1	
CHARGES, SHAPED, without detonator	0059 0439 0440 0441	1 1 1 1	
CHARGES, SUPPLEMENTARY, EXPLOSIVE	0060	1	
CHEMICAL KIT	3316	9	
CHEMICAL SAMPLE, TOXIC	3315	6.1	
Chile saltpetre, see	1498	5.1	
CHLORAL, ANHYDROUS, STABILIZED	2075	6.1	
CHLORATE AND BORATE MIXTURE	1458	5.1	
CHLORATE AND MAGNESIUM CHLORIDE MIXTURE, SOLID	1459	5.1	
CHLORATE AND MAGNESIUM CHLORIDE MIXTURE, SOLUTION	3407	5.1	
CHLORATES, INORGANIC, N.O.S.	1461	5.1	
CHLORATES, INORGANIC, AQUEOUS SOLUTION, N.O.S.	3210	5.1	
CHLORIC ACID, AQUEOUS SOLUTION with not more than 10% chloric acid	2626	5.1	
CHLORINE	1017	2	
CHLORINE PENTAFLUORIDE	2548	2	
CHLORINE TRIFLUORIDE	1749	2	
CHLORITES, INORGANIC, N.O.S.	1462	5.1	
CHLORITE SOLUTION	1908	8	
Chloroacetaldehyde, see	2232	6.1	
CHLOROACETIC ACID, MOLTEN	3250	6.1	
CHLOROACETIC ACID, SOLID	1751	6.1	
CHLOROACETIC ACID SOLUTION	1750	6.1	
CHLOROACETONE, STABILIZED	1695	6.1	
CHLOROACETONITRILE	2668	6.1	
CHLOROACETOPHENONE, LIQUID	3416	6.1	

Name and description	UN No.	Class	Remarks
CHLOROACETOPHENONE, SOLID	1697	6.1	
CHLOROACETYL CHLORIDE	1752	6.1	
CHLOROANILINES, LIQUID	2019	6.1	
CHLOROANILINES, SOLID	2018	6.1	
CHLOROANISIDINES	2233	6.1	
CHLOROBENZENE	1134	3	
CHLOROBENZOTRIFLUORIDES	2234	3	
CHLOROBENZYL CHLORIDES, LIQUID	2235	6.1	
CHLOROBENZYL CHLORIDES, SOLID	3427	6.1	
1-Chloro-3-bromopropane, see	2688	6.1	
1-Chlorobutane, see	1127	3	
2-Chlorobutane, see	1127	3	
CHLOROBUTANES	1127	3	
CHLOROCRESOLS, SOLUTION	2669	6.1	
CHLOROCRESOLS, SOLID	3437	6.1	
CHLORODIFLUORO-BROMOMETHANE	1974	2	
1-CHLORO-1,1-DIFLUORO-ETHANE	2517	2	
CHLORODIFLUOROMETHANE	1018	2	
CHLORODIFLUORO-METHANE AND CHLORO-PENTAFLUOROETHANE MIXTURE with fixed boiling point, with approximately 49% chlorodifluoromethane	1973	2	
3-Chloro-1,2-dihydroxypropane, see	2689	6.1	
Chlorodimethyl ether, see	1239	6.1	
1Chloro-2,2-dimethylpropane, see	1107	3	
CHLORODINITROBENZENES, LIQUID	1577	6.1	
CHLORODINITROBENZENES, SOLID	3441	6.1	
2-CHLOROETHANAL	2232	6.1	
Chloroethane, see	1037	2	

Name and description	UN No.	Class	Remarks
Chloroethane nitrile, see	2668	6.1	
2-Chloroethanol, see	1135	6.1	
CHLOROFORM	1888	6.1	
CHLOROFORMATES, TOXIC, CORROSIVE, N.O.S.	3277	6.1	
CHLOROFORMATES, TOXIC, CORROSIVE, FLAMMABLE, N.O.S.	2742	6.1	
Chloromethane, see	1063	2	
1-Chloro-3-methylbutane, see	1107	3	
2-Chloro-2-methylbutane, see	1107	3	
CHLOROMETHYL CHLOROFORMATE	2745	6.1	
Chloromethyl cyanide, see	2668	6.1	
CHLOROMETHYL ETHYL ETHER	2354	3	
1-Chloro-3-methylbutane, see	1107	3	
1-Chloro-3-methylbutane, see	1107	3	
Chloromethyl methyl ether, see	1239	6.1	
3-CHLORO-4-METHYLPHENYL ISOCYANATE, LIQUID	2236	6.1	
3-CHLORO-4-METHYLPHENYL ISOCYANATE, SOLID	3428	6.1	
1-Chloro-2-methylpropane, see	1127	3	
2-Chloro-2-methylpropane, see	1127	3	
3-Chloro-2-methylprop-1-ene, see	2554	3	
CHLORONITROANILINES	2237	6.1	
CHLORONITROBENZENES LIQUID	3409	6.1	
CHLORONITROBENZENES SOLID	1578	6.1	
CHLORONITROTOLUENES, LIQUID	2433	6.1	
CHLORONITROTOLUENES, SOLID	3457	6.1	
CHLOROPENTAFLUORO-ETHANE	1020	2	
1-Chloropentane	1107	3	
CHLOROPHENOLATES, LIQUID	2904	8	

Name and description	UN No.	Class	Remarks
CHLOROPHENOLATES, SOLID	2905	8	
CHLOROPHENOLS, LIQUID	2021	6.1	
CHLOROPHENOLS, SOLID	2020	6.1	
CHLOROPHENYL-TRICHLOROSILANE	1753	8	
CHLOROPICRIN	1580	6.1	
CHLOROPICRIN AND METHYL BROMIDE MIXTURE, with more than 2% chloropicrin	1581	2	
CHLOROPICRIN AND METHYL CHLORIDE MIXTURE	1582	2	
CHLOROPICRIN MIXTURE, N.O.S.	1583	6.1	
CHLOROPLATINIC ACID, SOLID	2507	8	
CHLOROPRENE, STABILIZED	1991	3	
1-CHLOROPROPANE	1278	3	
2-CHLOROPROPANE	2356	3	
3-Chloro-propanediol-1,2, see	2689	6.1	
3-CHLOROPROPANOL-1	2849	6.1	
2-CHLOROPROPENE	2456	3	
3-Chloropropene, see	1100	3	
3-Chloroprop-1-ene, see	1100	3	
2-CHLOROPROPIONIC ACID	2511	8	
2-CHLOROPYRIDINE	2822	6.1	
CHLOROSILANES, CORROSIVE, N.O.S.	2987	8	
CHLOROSILANES, CORROSIVE, FLAMMABLE, N.O.S.	2986	8	
CHLOROSILANES, FLAMMABLE, CORROSIVE, N.O.S.	2985	3	
CHLOROSILANES, TOXIC, CORROSIVE, N.O.S.	3361	6.1	
CHLOROSILANES, TOXIC, CORROSIVE, FLAMMABLE, N.O.S.	3362	6.1	
CHLOROSILANES, WATER-REACTIVE, FLAMMABLE, CORROSIVE, N.O.S.	2988	4.3	
CHLOROSULPHONIC ACID (with or without sulphur trioxide)	1754	8	
1-CHLORO-1,2,2,2-TETRAFLUOROETHANE	1021	2	
CHLOROTOLUENES	2238	3	
4-CHLORO-o-TOLUIDINE HYDROCHLORIDE, SOLID	1579	6.1	
4-CHLORO-o-TOLUIDINE HYDROCHLORIDE, SOLUTION	3410	6.1	
CHLOROTOLUIDINES LIQUID	3429	6.1	
CHLOROTOLUIDINES SOLID	2239	6.1	
1-CHLORO-2,2,2-TRIFLUOROETHANE	1983	2	
Chlorotrifluoroethylene, see	1082	2	
CHLOROTRIFLUOROMETHANE	1022	2	
CHLOROTRIFLUOROMETHANE AND TRIFLUOROMETHANE AZEOTROPIC MIXTURE with approximately 60% chlorotrifluoromethane	2599	2	
Chromic acid, solid, see	1463	5.1	
CHROMIC ACID SOLUTION	1755	8	
Chromic anhydride, solid, see	1463	5.1	
CHROMIC FLUORIDE, SOLID	1756	8	
CHROMIC FLUORIDE SOLUTION	1757	8	
Chromic nitrate, see	2720	5.1	
Chromium (VI) dichloride dioxide, see	1758	8	
Chromium (III) fluoride, solid, see	1756	8	
CHROMIUM NITRATE	2720	5.1	
Chromium (III) nitrate, see	2720	5.1	
CHROMIUM OXYCHLORIDE	1758	8	
CHROMIUM TRIOXIDE, ANHYDROUS	1463	5.1	
CHROMOSULPHURIC ACID	2240	8	
Chryosotile, see	2590	9	
Cinene, see	2052	3	
Cinnamene, see	2055	3	

Name and description	UN No.	Class	Remarks
Cinnamol, see	2055	3	
CLINICAL WASTE, UNSPECIFIED, N.O.S.	3291	6.2	
COAL GAS, COMPRESSED	1023	2	
COAL TAR DISTILLATES, FLAMMABLE	1136	3	
Coal tar naphtha, see	1268	3	
Coal tar oil, see	1136	3	
COATING SOLUTION (includes surface treatments or coatings used for industrial or other purposes such as vehicle under coating, drum or barrel lining)	1139	3	
COBALT NAPHTHENATES, POWDER	2001	4.1	
COBALT RESINATE, PRECIPITATED	1318	4.1	
Cocculus, see	3172 3462	6.1 6.1	
Collodion cottons, see	0340 0341 0342 2059 2555 2556 2557	1 1 1 3 4.1 4.1 4.1	
COMPONENTS, EXPLOSIVE TRAIN, N.O.S.	0382 0383 0384 0461	1 1 1 1	
Composition B, see	0118	1	
COMPRESSED GAS, N.O.S.	1956	2	
COMPRESSED GAS, FLAMMABLE, N.O.S.	1954	2	
COMPRESSED GAS, OXIDIZING, N.O.S.	3156	2	
COMPRESSED GAS, TOXIC, N.O.S.	1955	2	
COMPRESSED GAS, TOXIC, CORROSIVE, N.O.S.	3304	2	
COMPRESSED GAS, TOXIC, FLAMMABLE, N.O.S.	1953	2	
COMPRESSED GAS, TOXIC, FLAMMABLE, CORROSIVE, N.O.S.	3305	2	
COMPRESSED GAS, TOXIC, OXIDIZING, N.O.S.	3303	2	
COMPRESSED GAS, TOXIC, OXIDIZING, CORROSIVE, N.O.S.	3306	2	
CONTRIVANCES, WATER-ACTIVATED with burster, expelling charge or propelling charge	0248 0249	1 1	
COPPER ACETOARSENITE	1585	6.1	
COPPER ARSENITE	1586	6.1	
Copper (II) arsenite, see	1586	6.1	
COPPER BASED PESTICIDE, LIQUID, FLAMMABLE, TOXIC, flash-point less than 23 °C	2776	3	
COPPER BASED PESTICIDE, LIQUID, TOXIC	3010	6.1	
COPPER BASED PESTICIDE, LIQUID, TOXIC, FLAMMABLE, flash-point not less than 23 °C	3009	6.1	
COPPER BASED PESTICIDE, SOLID, TOXIC	2775	6.1	
COPPER CHLORATE	2721	5.1	
Copper (II) chlorate, see	2721	5.1	
COPPER CHLORIDE	2802	8	
COPPER CYANIDE	1587	6.1	
Copper selenate, see	2630	6.1	
Copper selenite, see	2630	6.1	
COPRA	1363	4.2	
CORD, DETONATING, flexible	0065 0289	1 1	
CORD, DETONATING, metal clad	0102 0290	1 1	
CORD, DETONATING, MILD EFFECT, metal clad	0104	1	
CORD, IGNITER	0066	1	
Cordite, see	0160 0161	1 1	
CORROSIVE LIQUID, N.O.S.	1760	8	
CORROSIVE LIQUID, ACIDIC, INORGANIC, N.O.S.	3264	8	
CORROSIVE LIQUID, ACIDIC, ORGANIC, N.O.S.	3265	8	

Name and description	UN No.	Class	Remarks
CORROSIVE LIQUID, BASIC, INORGANIC, N.O.S.	3266	8	
CORROSIVE LIQUID, BASIC, ORGANIC, N.O.S.	3267	8	
CORROSIVE LIQUID, FLAMMABLE, N.O.S.	2920	8	
CORROSIVE LIQUID, OXIDIZING, N.O.S.	3093	8	
CORROSIVE LIQUID, SELF-HEATING, N.O.S.	3301	8	
CORROSIVE LIQUID, TOXIC, N.O.S.	2922	8	
CORROSIVE LIQUID, WATER-REACTIVE, N.O.S.	3094	8	
CORROSIVE SOLID, N.O.S.	1759	8	
CORROSIVE SOLID, ACIDIC, INORGANIC, N.O.S.	3260	8	
CORROSIVE SOLID, ACIDIC, ORGANIC, N.O.S.	3261	8	
CORROSIVE SOLID, BASIC, INORGANIC, N.O.S.	3262	8	
CORROSIVE SOLID, BASIC, ORGANIC, N.O.S.	3263	8	
CORROSIVE SOLID, FLAMMABLE, N.O.S.	2921	8	
CORROSIVE SOLID, OXIDIZING, N.O.S.	3084	8	
CORROSIVE SOLID, SELF-HEATING, N.O.S.	3095	8	
CORROSIVE SOLID, TOXIC, N.O.S.	2923	8	
CORROSIVE SOLID, WATER-REACTIVE, N.O.S.	3096	8	
COTTON WASTE, OILY	1364	4.2	
COTTON, WET	1365	4.2	
COUMARIN DERIVATIVE PESTICIDE, LIQUID, FLAMMABLE, TOXIC, flash-point less than 23 °C	3024	3	
COUMARIN DERIVATIVE PESTICIDE, LIQUID, TOXIC	3026	6.1	

Name and description	UN No.	Class	Remarks
COUMARIN DERIVATIVE PESTICIDE, LIQUID, TOXIC, FLAMMABLE, flash-point not less than 23 °C	3025	6.1	
COUMARIN DERIVATIVE PESTICIDE, SOLID, TOXIC	3027	6.1	
Creosote, see	2810	6.1	
Creosote salts, see	1334	4.1	
CRESOLS, LIQUID	2076	6.1	
CRESOLS, SOLID	3455	6.1	
CRESYLIC ACID	2022	6.1	
Crocidolite, see	2212	9	
CROTONALDEHYDE	1143	6.1	
CROTONALDEHYDE, STABILIZED	1143	6.1	
CROTONIC ACID, LIQUID	3472	8	
CROTONIC ACID, SOLID	2823	8	
Crotonic aldehyde / Crotonic aldehyde, stabilized, see	1143	6.1	
CROTONYLENE	1144	3	
Crude naphtha, see	1268	3	
Cumene, see	1918	3	
Cupric chlorate, see	2721	5.1	
CUPRIETHYLENEDIAMINE SOLUTION	1761	8	
Cut backs, with a flash-point not greater than 60 °C, see	1999	3	
Cut backs, with a flash-point above 60 °C, at or above its flash-point, see	3256	3	
Cut backs, at or above 100 °C and below its flash-point, see	3257	9	
CUTTERS, CABLE, EXPLOSIVE	0070	1	
CYANIDE SOLUTION, N.O.S.	1935	6.1	
CYANIDES, INORGANIC, SOLID, N.O.S.	1588	6.1	
Cyanides, organic, flammable, toxic, n.o.s., see	3273	3	
Cyanides, organic, toxic, n.o.s., see	3276 3439	6.1 6.1	

Name and description	UN No.	Class	Remarks
Cyanides, organic, toxic, flammable, n.o.s., see	3275	6.1	
Cyanoacetonitrile, see	2647	6.1	
CYANOGEN	1026	2	
CYANOGEN BROMIDE	1889	6.1	
CYANOGEN CHLORIDE, STABILIZED	1589	2	
CYANURIC CHLORIDE	2670	8	
CYCLOBUTANE	2601	2	
CYCLOBUTYL CHLOROFORMATE	2744	6.1	
1,5,9-CYCLODODECATRIENE	2518	6.1	
CYCLOHEPTANE	2241	3	
CYCLOHEPTATRIENE	2603	3	
1,3,5-Cycloheptatriene, see	2603	3	
CYCLOHEPTENE	2242	3	
1,4-Cyclohexadienedione, see	2587	6.1	
CYCLOHEXANE	1145	3	
Cyclehexanethiol, see	3054	3	
CYCLOHEXANONE	1915	3	
CYCLOHEXENE	2256	3	
CYCLOHEXENYLTRI-CHLOROSILANE	1762	8	
CYCLOHEXYL ACETATE	2243	3	
CYCLOHEXYLAMINE	2357	8	
CYCLOHEXYL ISOCYANATE	2488	6.1	
CYCLOHEXYL MERCAPTAN	3054	3	
CYCLOHEXYLTRI-CHLOROSILANE	1763	8	
CYCLONITE AND CYCLOTETRAMETHYLENE-TETRANITRAMINE MIXTURE, WETTED with not less than 15% water, by mass or DESENSITIZED with not less than 10% phlegmatiser by mass, see	0391	1	
CYCLONITE, DESENSITIZED, see	0483	1	

Name and description	UN No.	Class	Remarks
CYCLONITE, WETTED with not less than 15% water, by mass, see	0072	1	
CYCLOOCTADIENES	2520	3	
CYCLOOCTADIENE PHOSPHINES, see	2940	4.2	
CYCLOOCTATETRAENE	2358	3	
CYCLOPENTANE	1146	3	
CYCLOPENTANOL	2244	3	
CYCLOPENTANONE	2245	3	
CYCLOPENTENE	2246	3	
CYCLOPROPANE	1027	2	
CYCLOTETRAMETHYLENE-TETRANITRAMINE, DESENSITIZED	0484	1	
CYCLOTETRAMETHYLENE-TETRANITRAMINE, WETTED with not less than 15% water, by mass	0226	1	
CYCLOTRIMETHYLENE-TRINITRAMINE AND CYCLOTETRAMETHYLENE-TETRANITRAMINE MIXTURE, DESENSITIZED with not less than 10% phlegmatiser by mass	0391	1	
CYCLOTRIMETHYLENE-TRINITRAMINE AND CYCLOTETRAMETHYLENE-TETRANITRAMINE MIXTURE, WETTED with not less than 15% water, by mass	0391	1	
CYCLOTRIMETHYLENE TRINITRAMINE, DESENSITIZED	0483	1	
CYCLOTRIMETHYLENE-TRINITRAMINE, WETTED with not less than 15% water, by mass	0072	1	
CYMENES	2046	3	
Cymol, see	2046	3	
Deanol, see	2051	8	
Dangerous goods in machinery or dangerous goods in apparatus	3363	9	Not subject to ADN [see also 1.1.3.1 (b)]
DECABORANE	1868	4.1	
DECAHYDRONAPHTHALENE	1147	3	

Name and description	UN No.	Class	Remarks
Decalin, see	1147	3	
n-DECANE	2247	3	
DEFLAGRATING METAL SALTS OF AROMATIC NITRODERIVATIVES, N.O.S.	0132	1	
Depth charge, see	0056	1	
DESENSITIZED EXPLOSIVE, LIQUID, N.O.S.	3379	3	
DESENSITIZED EXPLOSIVE, SOLID, N.O.S.	3380	4.1	
Detonating relays, see	0029 0267 0360 0361 0455 0500	1 1 1 1 1 1	
DETONATOR ASSEMBLIES, NON-ELECTRIC for blasting	0360 0361 0500	1 1 1	
DETONATORS FOR AMMUNITION	0073 0364 0365 0366	1 1 1 1	
DETONATORS, ELECTRIC for blasting	0030 0255 0456	1 1 1	
DETONATORS, NON-ELECTRIC for blasting	0029 0267 0455	1 1 1	
DEUTERIUM, COMPRESSED	1957	2	
DEVICES, SMALL, HYDROCARBON GAS POWERED with release device	3150	2	
DIACETONE ALCOHOL	1148	3	
DIALKYL-(C_{12}-C_{18})-DIMETHYL-AMMONIUM and 2-PROPANOL	3175	4.1	
DIALLYLAMINE	2359	3	
DIALLYL ETHER	2360	3	
4,4'-DIAMINODIPHENYL-METHANE	2651	6.1	
1,2-Diaminoethane, see	1604	8	
Diaminopropylamine, see	2269	8	
DI-n-AMYLAMINE	2841	3	
DIAZODINITROPHENOL, WETTED with not less than 40% water, or mixture of alcohol and water, by mass	0074	1	
Dibenzopyridine, see	2713	6.1	
DIBENZYLDICHLOROSILANE	2434	8	
DIBORANE	1911	2	
1,2-DIBROMOBUTAN-3-ONE	2648	6.1	
DIBROMOCHLOROPROPANES	2872	6.1	
1,2-Dibromo-3-chloropropane, see	2872	6.1	
DIBROMODIFLUORO-METHANE	1941	9	
DIBROMOMETHANE	2664	6.1	
DI-n-BUTYLAMINE	2248	8	
DIBUTYLAMINOETHANOL	2873	6.1	
2-Dibutylaminoethanol, see	2873	6.1	
N,N-Di-n-butylaminoethanol, see	2873	6.1	
DIBUTYL ETHERS	1149	3	
DICHLOROACETIC ACID	1764	8	
1,3-DICHLOROACETONE	2649	6.1	
DICHLOROACETYL CHLORIDE	1765	8	
DICHLOROANILINES, LIQUID	1590	6.1	
DICHLOROANILINES, SOLID	3442	6.1	
o-DICHLOROBENZENE	1591	6.1	
2,2'-DICHLORODIETHYL ETHER	1916	6.1	
DICHLORODIFLUORO-METHANE	1028	2	
DICHLORODIFLUORO-METHANE AND 1,1-DIFLUOROETHANE AZEOTROPIC MIXTURE with approximately 74% dichlorodifluoromethane	2602	2	
Dichlorodifluoromethane and ethylene oxide mixture, see	3070	2	
DICHLORODIMETHYL ETHER, SYMMETRICAL	2249	6.1	Carriage prohibited
1,1-DICHLOROETHANE	2362	3	
1,2-Dichloroethane, see	1184	3	

Name and description	UN No.	Class	Remarks
1,2-DICHLOROETHYLENE	1150	3	
Di(2-chloroethyl) ether, see	1916	6.1	
DICHLOROFLUOROMETHANE	1029	2	
alpha-Dichlorohydrin, see	2750	6.1	
DICHLOROISOCYANURIC ACID, DRY	2465	5.1	
DICHLOROISOCYANURIC ACID SALTS	2465	5.1	
DICHLOROISOPROPYL ETHER	2490	6.1	
DICHLOROMETHANE	1593	6.1	
1,1-DICHLORO-1-NITROETHANE	2650	6.1	
DICHLOROPENTANES	1152	3	
Dichlorophenol, see	2020 2021	6.1 6.1	
DICHLOROPHENYL ISOCYANATES	2250	6.1	
DICHLOROPHENYLTRI-CHLOROSILANE	1766	8	
1,2-DICHLOROPROPANE	1279	3	
1,3-DICHLORO-PROPANOL-2	2750	6.1	
1,3-Dichloro-2-propanone, see	2649	6.1	
DICHLOROPROPENES	2047	3	
DICHLOROSILANE	2189	2	
1,2-DICHLORO-1,1,2,2-TETRAFLUOROETHANE	1958	2	
Dichloro-s-triazine-2,4,6-trione, see	2465	5.1	
1,4-Dicyanobutane, see	2205	6.1	
Dicycloheptadiene, see	2251	3	
DICYCLOHEXYLAMINE	2565	8	
Dicyclohexylamine nitrite, see	2687	4.1	
DICYCLOHEXYLAMMONIUM NITRITE	2687	4.1	
DICYCLOPENTADIENE	2048	3	
1,2-DI-(DIMETHYLAMINO) ETHANE	2372	3	
DIDYMIUM NITRATE	1465	5.1	
DIESEL FUEL	1202	3	
1,1-Diethoxyethane, see	1088	3	
1,2-Diethoxyethane, see	1153	3	
DIETHOXYMETHANE	2373	3	
3,3-DIETHOXYPROPENE	2374	3	
DIETHYLAMINE	1154	3	
2-DIETHYLAMINOETHANOL	2686	8	
3-DIETHYL-AMINOPROPYLAMINE	2684	3	
N,N-DIETHYLANILINE	2432	6.1	
DIETHYLBENZENE	2049	3	
Diethylcarbinol, see	1105	3	
DIETHYL CARBONATE	2366	3	
DIETHYLDICHLOROSILANE	1767	8	
Diethylenediamine, see	2579	8	
DIETHYLENEGLYCOL DINITRATE, DESENSITIZED with not less than 25% non-volatile, water-insoluble phlegmatizer, by mass	0075	1	
DIETHYLENETRIAMINE	2079	8	
N,N-Diethylethanolamine, see	2686	3	
DIETHYL ETHER	1155	3	
N,N-DIETHYLETHYLENE-DIAMINE	2685	8	
Di-(2-ethylhexyl) phosphoric acid, see	1902	8	
DIETHYL KETONE	1156	3	
DIETHYL SULPHATE	1594	6.1	
DIETHYL SULPHIDE	2375	3	
DIETHYLTHIOPHOSPHORYL CHLORIDE	2751	8	
Diethylzinc, see	3394	4.2	
2,4-Difluoroaniline, see	2941	6.1	
Difluorochloroethane, see	2517	2	
1,1-DIFLUOROETHANE	1030	2	
1,1-DIFLUOROETHYLENE	1959	2	
DIFLUOROMETHANE	3252	2	

Name and description	UN No.	Class	Remarks
Difluoromethane, pentafluoroethane, and 1,1,1,2-tetrafluoroethane zeotropic mixture with approximately 10% difluoromethane and 70% pentafluoroethane, see	3339	2	
Difluoromethane, pentafluoroethane, and 1,1,1,2-tetrafluoroethane zeotropic mixture with approximately 20% difluoromethane and 40% pentafluoroethane, see	3338	2	
Difluoromethane, pentafluoroethane, and 1,1,1,2-tetrafluoroethane zeotropic mixture with approximately 23%difluoromethane and 25% pentafluoroethane, see	3340	2	
DIFLUOROPHOSPHORIC ACID, ANHYDROUS	1768	8	
2,3-DIHYDROPYRAN	2376	3	
DIISOBUTYLAMINE	2361	3	
DIISOBUTYLENE, ISOMERIC COMPOUNDS	2050	3	
alpha-Diisobutylene, see	2050	3	
beta-Diisobutylene, see	2050	3	
DIISOBUTYL KETONE	1157	3	
DIISOOCTYL ACID PHOSPHATE	1902	8	
DIISOPROPYLAMINE	1158	3	
DIISOPROPYL ETHER	1159	3	
DIKETENE, STABILIZED	2521	6.1	
1,1-DIMETHOXYETHANE	2377	3	
1,2-DIMETHOXYETHANE	2252	3	
Dimethoxystrychnine, see	1570	6.1	
DIMETHYLAMINE, ANHYDROUS	1032	2	
DIMETHYLAMINE AQUEOUS SOLUTION	1160	3	
2-DIMETHYLAMINO-ACETONITRILE	2378	3	
2-DIMETHYLAMINOETHANOL	2051	8	
2-DIMETHYLAMINOETHYL ACRYLATE	3302	6.1	
2-DIMETHYLAMINOETHYL METHACRYLATE	2522	6.1	
N,N-DIMETHYLANILINE	2253	6.1	
Dimethylarsenic acid, see	1572	6.1	
N,N-Dimethylbenzylamine, see	2619	8	
2,3-DIMETHYLBUTANE	2457	3	
1,3-DIMETHYLBUTYLAMINE	2379	3	
DIMETHYLCARBAMOYL CHLORIDE	2262	8	
DIMETHYL CARBONATE	1161	3	
DIMETHYLCYCLOHEXANES	2263	3	
N,N-DIMETHYLCYCLO-HEXYLAMINE	2264	8	
DIMETHYLDICHLOROSILANE	1162	3	
DIMETHYLDIETHOXYSILANE	2380	3	
DIMETHYLDIOXANES	2707	3	
DIMETHYL DISULPHIDE	2381	3	
Dimethylethanolamine, see	2051	8	
DIMETHYL ETHER	1033	2	
N,N-DIMETHYLFORMAMIDE	2265	3	
DIMETHYLHYDRAZINE, SYMMETRICAL	2382	6.1	
DIMETHYLHYDRAZINE, UNSYMMETRICAL	1163	6.1	
1,1-Dimethylhydrazine, see	1163	6.1	
N,N-Dimethyl-4-nitrosoaniline, see	1369	4.2	
2,2-DIMETHYLPROPANE	2044	2	
DIMETHYL-N-PROPYLAMINE	2266	3	
DIMETHYL SULPHATE	1595	6.1	
DIMETHYL SULPHIDE	1164	3	
DIMETHYL THIOPHOSPHORYL CHLORIDE	2267	6.1	
Dimethylzinc, see	3394	4.2	
DINGU, see	0489	1	
DINITROANILINES	1596	6.1	
DINITROBENZENES, LIQUID	1597	6.1	
DINITROBENZENES, SOLID	3443	6.1	
Dinitrochlorobenzene, see	1577 3441	6.1 6.1	

Name and description	UN No.	Class	Remarks
DINITRO-o-CRESOL	1598	6.1	
DINITROGEN TETROXIDE	1067	2	
DINITROGLYCOLURIL	0489	1	
DINITROPHENOL, dry or wetted with less than 15% water, by mass	0076	1	
DINITROPHENOL SOLUTION	1599	6.1	
DINITROPHENOL, WETTED with not less than 15% water, by mass	1320	4.1	
DINITROPHENOLATES, alkali metals, dry or wetted with less than 15% water, by mass	0077	1	
DINITROPHENOLATES, WETTED with not less than 15% water, by mass	1321	4.1	
DINITRORESORCINOL, dry or wetted with less than 15% water, by mass	0078	1	
DINITRORESORCINOL, WETTED with not less than 15% water, by mass	1322	4.1	
DINITROSOBENZENE	0406	1	
Dinitrotoluene mixed with sodium chlorate, see	0083	1	
DINITROTOLUENES, LIQUID	2038	6.1	
DINITROTOLUENES, MOLTEN	1600	6.1	
DINITROTOLUENES, SOLID	3454	6.1	
DIOXANE	1165	3	
DIOXOLANE	1166	3	
DIPENTENE	2052	3	
DIPHENYLAMINE CHLOROARSINE	1698	6.1	
DIPHENYLCHLOROARSINE, LIQUID	1699	6.1	
DIPHENYLCHLOROARSINE, SOLID	3450	6.1	
DIPHENYLDICHLOROSILANE	1769	8	
DIPHENYLMETHANE-4, 4'-DIISOCYANATE	9004	9	Dangerous in tank vessels only
DIPHENYLMETHYL BROMIDE	1770	8	
DIPICRYLAMINE, see	0079	1	
DIPICRYL SULPHIDE, dry or wetted with less than 10% water, by mass	0401	1	
DIPICRYL SULPHIDE, WETTED with not less than 10% water, by mass	2852	4.1	
DIPROPYLAMINE	2383	3	
Dipropylene triamine, see	2269	8	
DI-n-PROPYL ETHER	2384	3	
DIPROPYL KETONE	2710	3	
DISINFECTANT, LIQUID, CORROSIVE, N.O.S.	1903	8	
DISINFECTANT, LIQUID, TOXIC, N.O.S.	3142	6.1	
DISINFECTANT, SOLID, TOXIC, N.O.S.	1601	6.1	
DISODIUM TRIOXOSILICATE	3253	8	
DIVINYL ETHER, STABILIZED	1167	3	
DODECYLTRICHLOROSILANE	1771	8	
Dry ice, see	1845	9	Not subject to ADN
DYE INTERMEDIATE, LIQUID, CORROSIVE, N.O.S.	2801	8	
DYE INTERMEDIATE, LIQUID, TOXIC, N.O.S.	1602	6.1	
DYE INTERMEDIATE, SOLID, CORROSIVE, N.O.S.	3147	8	
DYE INTERMEDIATE, SOLID, TOXIC, N.O.S.	3143	6.1	
DYE, LIQUID, CORROSIVE, N.O.S.	2801	8	
DYE, LIQUID, TOXIC, N.O.S.	1602	6.1	
DYE, SOLID, CORROSIVE, N.O.S.	3147	8	
DYE, SOLID, TOXIC, N.O.S.	3143	6.1	
Dynamite, see	0081	1	
Electric storage batteries, see	2794 2795 2800 3028	8 8 8 8	

Name and description	UN No.	Class	Remarks
Electrolyte (acid or alkaline) for batteries, see	2796 2797	8 8	
ELEVATED TEMPERATURE LIQUID, N.O.S., at or above 100 °C and below its flash-point (including molten metals, molten salts, etc.)	3257	9	
ELEVATED TEMPERATURE LIQUID, FLAMMABLE, N.O.S. with flash-point above 60 °C, at or above its flash-point	3256	3	
ELEVATED TEMPERATURE SOLID, N.O.S., at or above 240 °C	3258	9	
Empty battery-vehicle, uncleaned			See 4.3.2.4 of ADR, 5.1.3 and 5.4.1.1.6
Empty IBC, uncleaned			See 4.1.1.11 of ADR, 5.1.3 and 5.4.1.1.6
Empty large packaging, uncleaned			See 4.1.1.11 of ADR, 5.1.3 and 5.4.1.1.6
Empty MEGC, uncleaned			See 4.3.2.4 of ADR, 5.1.3 and 5.4.1.1.6
Empty packaging, uncleaned			See 4.1.1.11 of ADR, 5.1.3 and 5.4.1.1.6
Empty receptacle, uncleaned			See 5.1.3 and 5.4.1.1.6
Empty tank, uncleaned			See 4.3.2.4 of ADR, 5.1.3 and 5.4.1.1.6
Empty vehicle, uncleaned			See 5.1.3 and 5.4.1.1.6
Enamel, see	1263 3066 3469 3470	3 8 3 8	
Engines, internal combustion	3166	9	Not subject to ADN
Engines, rocket, see	0250 0322	1 1	
ENVIRONMENTALLY HAZARDOUS SUBSTANCE, LIQUID, N.O.S.	3082	9	
ENVIRONMENTALLY HAZARDOUS SUBSTANCE, SOLID, N.O.S.	3077	9	
EPIBROMOHYDRIN	2558	6.1	
EPICHLOROHYDRIN	2023	6.1	
1,2-Epoxybutane, stabilized, see	3022	3	
Epoxyethane, see	1040	2	
1,2-EPOXY-3-ETHOXYPROPANE	2752	3	
2,3-Epoxy-1-propanal, see	2622	3	
2,3-Epoxypropyl ethyl ether, see	2752	3	
ESTERS, N.O.S.	3272	3	
Ethanal, see	1089	3	
ETHANE	1035	2	
ETHANE, REFRIGERATED LIQUID	1961	2	
Ethanethiol, see	2363	3	
ETHANOL	1170	3	
ETHANOL AND GASOLINE MIXTURE or ETHANOL AND MOTOR SPIRIT MIXTURE or ETHANOL AND PETROL MIXTURE, with more than 10% ethanol	3475	3	
ETHANOL SOLUTION	1170	3	
ETHANOLAMINE	2491	8	
ETHANOLAMINE SOLUTION	2491	8	
Ether, see	1155	3	
ETHERS, N.O.S.	3271	3	
2-Ethoxyethanol, see	1171	3	
2-Ethoxyethyl acetate, see	1172	3	
Ethoxy propane-1, see	2615	3	
ETHYL ACETATE	1173	3	
ETHYLACETYLENE, STABILIZED	2452	2	

Name and description	UN No.	Class	Remarks
ETHYL ACRYLATE, STABILIZED	1917	3	
ETHYL ALCOHOL, see	1170	3	
ETHYL ALCOHOL SOLUTION, see	1170	3	
ETHYLAMINE	1036	2	
ETHYLAMINE, AQUEOUS SOLUTION with not less than 50% but not more than 70% ethylamine	2270	3	
ETHYL AMYL KETONE	2271	3	
N-ETHYLANILINE	2272	6.1	
2-ETHYLANILINE	2273	6.1	
ETHYLBENZENE	1175	3	
N-ETHYL-N-BENZYLANILINE	2274	6.1	
N-ETHYLBENZYLTOLUIDINES, LIQUID	2753	6.1	
N-ETHYLBENZYLTOLUIDINES, SOLID	3460	6.1	
ETHYL BORATE	1176	3	
ETHYL BROMIDE	1891	6.1	
ETHYL BROMOACETATE	1603	6.1	
2-ETHYLBUTANOL	2275	3	
2-ETHYLBUTYL ACETATE	1177	3	
ETHYL BUTYL ETHER	1179	3	
2-ETHYLBUTYRALDEHYDE	1178	3	
ETHYL BUTYRATE	1180	3	
ETHYL CHLORIDE	1037	2	
ETHYL CHLOROACETATE	1181	6.1	
Ethyl chlorocarbonate, see	1182	6.1	
ETHYL CHLOROFORMATE	1182	6.1	
ETHYL 2-CHLOROPROPIONATE	2935	3	
Ethyl-alpha-chloropropionate, see	2935	3	
ETHYL CHLOROTHIOFORMATE	2826	8	
ETHYL CROTONATE	1862	3	
ETHYLDICHLOROARSINE	1892	6.1	
ETHYLDICHLOROSILANE	1183	4.3	
ETHYLENE, ACETYLENE AND PROPYLENE MIXTURE, REFRIGERATED LIQUID containing at least 71.5% ethylene with not more than 22.5% acetylene and not more than 6% propylene	3138	2	
ETHYLENE CHLOROHYDRIN	1135	6.1	
ETHYLENE	1962	2	
ETHYLENEDIAMINE	1604	8	
ETHYLENE DIBROMIDE	1605	6.1	
Ethylene dibromide and methyl bromide, liquid mixture, see	1647	6.1	
ETHYLENE DICHLORIDE	1184	3	
ETHYLENE GLYCOL DIETHYL ETHER	1153	3	
ETHYLENE GLYCOL MONOETHYL ETHER	1171	3	
ETHYLENE GLYCOL MONOETHYL ETHER ACETATE	1172	3	
ETHYLENE GLYCOL MONOMETHYL ETHER	1188	3	
ETHYLENE GLYCOL MONOMETHYL ETHER ACETATE	1189	3	
ETHYLENEIMINE, STABILIZED	1185	6.1	
ETHYLENE OXIDE	1040	2	
ETHYLENE OXIDE AND CARBON DIOXIDE MIXTURE with more than 87% ethylene oxide	3300	2	
ETHYLENE OXIDE AND CARBON DIOXIDE MIXTURE with more than 9% but not more than 87% ethylene oxide	1041	2	
ETHYLENE OXIDE AND CARBON DIOXIDE MIXTURE with not more than 9% ethylene oxide	1952	2	
ETHYLENE OXIDE AND CHLOROTETRAFLUORO-ETHANE MIXTURE with not more than 8.8% ethylene oxide	3297	2	
ETHYLENE OXIDE AND DICHLORODIFLUORO-METHANE MIXTURE with not more than 12.5% ethylene oxide	3070	2	

Name and description	UN No.	Class	Remarks
ETHYLENE OXIDE AND PENTAFLUOROETHANE MIXTURE with not more than 7.9% ethylene oxide	3298	2	
ETHYLENE OXIDE AND PROPYLENE OXIDE MIXTURE, not more than 30% ethylene oxide	2983	3	
ETHYLENE OXIDE AND TETRAFLUOROETHANE MIXTURE with not more than 5.6% ethylene oxide	3299	2	
ETHYLENE OXIDE WITH NITROGEN up to a total pressure of 1 MPa (10 bar) at 50 °C	1040	2	
ETHYLENE, REFRIGERATED LIQUID	1038	2	
ETHYL ETHER, see	1155	3	
ETHYL FLUORIDE	2453	2	
ETHYL FORMATE	1190	3	
2-ETHYLHEXYLAMINE	2276	3	
2-ETHYLHEXYL CHLOROFORMATE	2748	6.1	
Ethylidene chloride, see	2362	3	
ETHYL ISOBUTYRATE	2385	3	
ETHYL ISOCYANATE	2481	3	
ETHYL LACTATE	1192	3	
ETHYL MERCAPTAN	2363	3	
ETHYL METHACRYLATE, STABILIZED	2277	3	
ETHYL METHYL ETHER	1039	2	
ETHYL METHYL KETONE	1193	3	
ETHYL NITRITE SOLUTION	1194	3	
ETHYL ORTHOFORMATE	2524	3	
ETHYL OXALATE	2525	6.1	
ETHYLPHENYL-DICHLOROSILANE	2435	8	
1-ETHYLPIPERIDINE	2386	3	
ETHYL PROPIONATE	1195	3	
ETHYL PROPYL ETHER	2615	3	
Ethyl silicate, see	1292	3	
Ethyl sulphate, see	1594	6.1	
N-ETHYLTOLUIDINES	2754	6.1	
ETHYLTRICHLOROSILANE	1196	3	
EXPLOSIVE, BLASTING, TYPE A	0081	1	
EXPLOSIVE, BLASTING, TYPE B	0082 0331	1 1	
EXPLOSIVE, BLASTING, TYPE C	0083	1	
EXPLOSIVE, BLASTING, TYPE D	0084	1	
EXPLOSIVE, BLASTING, TYPE E	0241 0332	1 1	
Explosives, emulsion, see	0241 0332	1 1	
Explosive, seismic, see	0081 0082 0083 0331	1 1 1 1	
Explosive, slurry, see	0241 0332	1 1	
Explosive, water gel, see	0241 0332	1 1	
EXTRACTS, AROMATIC, LIQUID	1169	3	
EXTRACTS, FLAVOURING, LIQUID	1197	3	
FABRICS, ANIMAL, N.O.S. with oil	1373	4.2	
FABRICS IMPREGNATED WITH WEAKLY NITRATED NITROCELLULOSE, N.O.S.	1353	4.1	
FABRICS, SYNTHETIC, N.O.S. with oil	1373	4.2	
FABRICS, VEGETABLE, N.O.S. with oil	1373	4.2	
FERRIC ARSENATE	1606	6.1	
FERRIC ARSENITE	1607	6.1	
FERRIC CHLORIDE, ANHYDROUS	1773	8	
FERRIC CHLORIDE SOLUTION	2582	8	
FERRIC NITRATE	1466	5.1	
FERROCERIUM	1323	4.1	

Name and description	UN No.	Class	Remarks
FERROSILICON with 30% or more but less than 90% silicon	1408	4.3	
FERROUS ARSENATE	1608	6.1	
FERROUS METAL BORINGS in a form liable to self-heating	2793	4.2	
FERROUS METAL CUTTINGS in a form liable to self-heating	2793	4.2	
FERROUS METAL SHAVINGS in a form liable to self-heating	2793	4.2	
FERROUS METAL TURNINGS in a form liable to self-heating	2793	4.2	
FERTILIZER AMMONIATING SOLUTION with free ammonia	1043	2	
Fertilizer with ammonium nitrate, n.o.s., see	2067	5.1	
Fibres, animal, burnt wet or damp	1372	4.2	Not subject to ADN
FIBRES, ANIMAL, N.O.S. with oil	1373	4.2	
FIBRES IMPREGNATED WITH WEAKLY NITRATED NITROCELLULOSE, N.O.S.	1353	4.1	
FIBRES, SYNTHETIC, N.O.S. with oil	1373	4.2	
Fibres, vegetable, burnt wet or damp	1372	4.2	Not subject to ADN
Fibres, vegetable, dry	3360	4.1	Not subject to ADN
FIBRES, VEGETABLE, N.O.S. with oil	1373	4.2	
Filler, liquid, see	1263 3066 3469 3470	3 8 3 8	
Films, nitrocellulose base, from which gelatin has been removed; film scrap, see	2002	4.2	
FILMS, NITROCELLULOSE BASE, gelatin coated, except scrap	1324	4.1	
FIRE EXTINGUISHER CHARGES, corrosive liquid	1774	8	
Fire extinguisher charges, expelling, explosive, see	0275 0276 0323 0381	1 1 1 1	
FIRE EXTINGUISHERS with compressed or liquefied gas	1044	2	

Name and description	UN No.	Class	Remarks
FIRELIGHTERS, SOLID with flammable liquid	2623	4.1	
FIREWORKS	0333 0334 0335 0336 0337	1 1 1 1 1	See 2.2.1.1.7
FIRST AID KIT	3316	9	
FISH MEAL, STABILIZED	2216	9	
FISH MEAL, UNSTABILIZED	1374	4.2	
FISH SCRAP, STABILIZED, see	2216	9	
FISH SCRAP, UNSTABILIZED, see	1374	4.2	
Flammable gas in lighters, see	1057	2	
FLAMMABLE LIQUID, N.O.S	1993	3	
FLAMMABLE LIQUID, CORROSIVE, N.O.S.	2924	3	
FLAMMABLE LIQUID, TOXIC, N.O.S.	1992	3	
FLAMMABLE LIQUID, TOXIC, CORROSIVE, N.O.S.	3286	3	
FLAMMABLE SOLID, CORROSIVE, INORGANIC, N.O.S.	3180	4.1	
FLAMMABLE SOLID, CORROSIVE, ORGANIC, N.O.S.	2925	4.1	
FLAMMABLE SOLID, INORGANIC, N.O.S.	3178	4.1	
FLAMMABLE SOLID, ORGANIC, N.O.S.	1325	4.1	
FLAMMABLE SOLID, ORGANIC, MOLTEN, N.O.S.	3176	4.1	
FLAMMABLE SOLID, OXIDIZING, N.O.S.	3097	4.1	Carriage prohibited
FLAMMABLE SOLID, TOXIC, INORGANIC, N.O.S.	3179	4.1	
FLAMMABLE SOLID, TOXIC, ORGANIC, N.O.S.	2926	4.1	
FLARES, AERIAL	0093 0403 0404 0420 0421	1 1 1 1 1	

Name and description	UN No.	Class	Remarks
Flares, aeroplane, see	0093	1	
	0403	1	
	0404	1	
	0420	1	
	0421	1	
Flares, highway,	0191	1	
Flares, distress, small,	0373	1	
Flares, railway or highway, see			
FLARES, SURFACE	0092	1	
	0418	1	
	0419	1	
Flares, water-activated, see	0248	1	
	0249	1	
FLASH POWDER	0094	1	
	0305	1	
Flue dusts, toxic, see	1562	6.1	
Fluoric acid, see	1790	8	
FLUORINE, COMPRESSED	1045	2	
FLUOROACETIC ACID	2642	6.1	
FLUOROANILINES	2941	6.1	
2-Fluoroaniline, see	2941	6.1	
4-Fluoroaniline, see	2941	6.1	
o-Fluoroaniline, see	2941	6.1	
p-Fluoroaniline, see	2941	6.1	
FLUOROBENZENE	2387	3	
FLUOROBORIC ACID	1775	8	
Fluoroethane, see	2453	2	
Fluoroform, see	1984	2	
Fluoromethane, see	2454	2	
FLUOROPHOSPHORIC ACID, ANHYDROUS	1776	8	
FLUOROSILICATES, N.O.S.	2856	6.1	
FLUOROSILICIC ACID	1778	8	
FLUOROSULPHONIC ACID	1777	8	
FLUOROTOLUENES	2388	3	
FORMALDEHYDE SOLUTION with not less than 25% formaldehyde	2209	8	
FORMALDEHYDE SOLUTION, FLAMMABLE	1198	3	
Formalin, see	1198	3	
	2209	8	
Formamidine sulphinic acid, see	3341	4.2	
FORMIC ACID with more than 85% acid by mass	1779	8	
FORMIC ACID with not more than 85% acid by mass	3412	8	
Formic aldehyde, see	1198	3	
	2209	8	
2-Formyl-3,4-dihydro-2H-pyran, see	2607	3	
FRACTURING DEVICES, EXPLOSIVE without detonator, for oil wells	0099	1	
FUEL, AVIATION, TURBINE ENGINE	1863	3	
FUEL CELL CARTRIDGES	3478	2	
	3479	2	
	3473	3	
	3476	4.3	
	3477	8	
FUEL CELL CARTRIDGES CONTAINED IN EQUIPMENT	3478	2	
	3479	2	
	3473	3	
	3476	4.3	
	3477	8	
FUEL CELL CARTRIDGES PACKED WITH EQUIPMENT	3478	2	
	3479	2	
	3473	3	
	3476	4.3	
	3477	8	
Fumaroyl dichloride, see	1780	3	
FUMARYL CHLORIDE	1780	8	
FUMIGATED UNIT	3359	9	
FURALDEHYDES	1199	6.1	
FURAN	2389	3	
FURFURYL ALCOHOL	2874	6.1	
FURFURYLAMINE	2526	3	
Furyl carbinol, see	2874	6.1	
FUSE, DETONATING, metal clad	0102	1	
	0290	1	
FUSE, DETONATING, MILD EFFECT, metal clad	0104	1	
FUSE, IGNITER, tubular, metal clad	0103	1	
FUSE, NON-DETONATING	0101	1	
FUSEL OIL	1201	3	

Name and description	UN No.	Class	Remarks
FUSE, SAFETY	0105	1	
Fuze, combination, percussion or time, see	0106	1	
	0107	1	
	0257	1	
	0316	1	
	0317	1	
	0367	1	
	0368	1	
FUZES, DETONATING	0106	1	
	0107	1	
	0257	1	
	0367	1	
FUZES, DETONATING with protective features	0408	1	
	0409	1	
	0410	1	
FUZES, IGNITING	0316	1	
	0317	1	
	0368	1	
GALLIUM	2803	8	
GAS CARTRIDGES without a release device, non-refillable, see	2037	2	
Gas drips, hydrocarbon, see	3295	3	
GAS OIL	1202	3	
GASOLINE	1203	3	
Gasoline and ethanol mixture, with more than 10% ethanol, see	3475	3	
Gasoline, casinghead, see	1203	3	
GAS, REFRIGERATED LIQUID, N.O.S.	3158	2	
GAS, REFRIGERATED LIQUID, FLAMMABLE, N.O.S.	3312	2	
GAS, REFRIGERATED LIQUID, OXIDIZING, N.O.S.	3311	2	
GAS SAMPLE, NON-PRESSURIZED, FLAMMABLE, N.O.S., not refrigerated liquid	3167	2	
GAS SAMPLE, NON-PRESSURIZED, TOXIC, N.O.S., not refrigerated liquid	3169	2	
GAS SAMPLE, NON-PRESSURIZED, TOXIC, FLAMMABLE, N.O.S., not refrigerated liquid	3168	2	
Gelatin, blasting, see	0081	1	
Gelatin, dynamites, see	0081	1	
GENETICALLY MODIFIED MICROORGANISMS	3245	9	
GENETICALLY MODIFIED ORGANISMS	3245	9	
GERMANE	2192	2	
Germanium hydride, see	2192	2	
Glycer-1,3-dichlorohydrin, see	2750	6.1	
GLYCEROL alpha-MONOCHLOROHYDRIN	2689	6.1	
Glyceryl trinitrate, see	0143	1	
	0144	1	
	1204	3	
	3064	3	
GLYCIDALDEHYDE	2622	3	
GRENADES, hand or rifle, with bursting charge	0284	1	
	0285	1	
	0292	1	
	0293	1	
Grenades, illuminating, see	0171	1	
	0254	1	
	0297	1	
GRENADES, PRACTICE, hand or rifle	0110	1	
	0318	1	
	0372	1	
	0452	1	
Grenades, smoke, see	0015	1	
	0016	1	
	0245	1	
	0246	1	
	0303	1	
GUANIDINE NITRATE	1467	5.1	
GUANYLNITROSAMINO-GUANYLIDENE HYDRAZINE, WETTED with not less than 30% water, by mass	0113	1	
GUANYLNITROSAMINO-GUANYLTETRAZENE, WETTED with not less than 30% water, or mixture of alcohol and water, by mass	0114	1	
GUNPOWDER, COMPRESSED, see	0028	1	
GUNPOWDER, granular or as a meal, see	0027	1	
GUNPOWDER, IN PELLETS, see	0028	1	
Gutta percha solution, see	1287	3	

Name and description	UN No.	Class	Remarks
HAFNIUM POWDER, DRY	2545	4.2	
HAFNIUM POWDER, WETTED with not less than 25% water	1326	4.1	
Hay	1327	4.1	Not subject to ADN
HEATING OIL, LIGHT	1202	3	
Heavy hydrogen, see	1957	2	
HELIUM, COMPRESSED	1046	2	
HELIUM, REFRIGERATED LIQUID	1963	2	
HEPTAFLUOROPROPANE	3296	2	
n-HEPTALDEHYDE	3056	3	
n-Heptanal, see	3056	3	
HEPTANES	1206	3	
4-Heptanone, see	2710	3	
n-HEPTENE	2278	3	
HEXACHLOROACETONE	2661	6.1	
HEXACHLOROBENZENE	2729	6.1	
HEXACHLOROBUTADIENE	2279	6.1	
Hexachloro-1,3-butadiene, see	2279	6.1	
HEXACHLOROCYCLO-PENTADIENE	2646	6.1	
HEXACHLOROPHENE	2875	6.1	
Hexachloro-2-propanone, see	2661	6.1	
HEXADECYLTRICHLORO-SILANE	1781	8	
HEXADIENES	2458	3	
HEXAETHYL TETRAPHOSPHATE	1611	6.1	
HEXAETHYL TETRAPHOSPHATE AND COMPRESSED GAS MIXTURE	1612	2	
HEXAFLUOROACETONE	2420	2	
HEXAFLUOROACETONE HYDRATE, LIQUID	2552	6.1	
HEXAFLUOROACETONE HYDRATE, SOLID	3436	6.1	
HEXAFLUOROETHANE	2193	2	
HEXAFLUOROPHOSPHORIC ACID	1782	8	
HEXAFLUOROPROPYLENE	1858	2	
Hexahydrocresol, see	2617	3	
Hexahydromethyl phenol, see	2617	3	
HEXALDEHYDE	1207	3	
HEXAMETHYLENEDIAMINE, SOLID	2280	8	
HEXAMETHYLENEDIAMINE SOLUTION	1783	8	
HEXAMETHYLENE DIISOCYANATE	2281	6.1	
HEXAMETHYLENEIMINE	2493	3	
HEXAMETHYLENETETRAMINE	1328	4.1	
Hexamine, see	1328	4.1	
HEXANES	1208	3	
HEXANITRODIPHENYL-AMINE	0079	1	
HEXANITROSTILBENE	0392	1	
Hexanoic acid, see	2829	8	
HEXANOLS	2282	3	
1-HEXENE	2370	3	
HEXOGEN AND CYCLOTETRA-METHYLENE-TETRANITRAMINE MIXTURE, WETTED with not less than 15% water, by mass or DESENSITIZED with not less than 10% phlegmatiser by mass, see	0391	1	
HEXOGEN, DESENSITIZED, see	0483	1	
HEXOGEN, WETTED with not less than 15% water, by mass, see	0072	1	
HEXOLITE, dry or wetted with less than 15% water, by mass	0118	1	
HEXOTOL, dry or wetted with less than 15% water, by mass, see	0118	1	
HEXOTONAL	0393	1	
HEXOTONAL, cast, see	0393	1	
HEXYL, see	0079	1	

Name and description	UN No.	Class	Remarks
HEXYLTRICHLOROSILANE	1784	8	
HMX, see	0391	1	
HMX, DESENSITIZED, see	0484	1	
HMX, WETTED with not less than 15% water, by mass, see	0226	1	
HYDRAZINE, ANHYDROUS	2029	8	
HYDRAZINE AQUEOUS SOLUTION, with more than 37% hydrazine by mass	2030	8	
HYDRAZINE, AQUEOUS SOLUTION with not more than 37% hydrazine, by mass	3293	6.1	
Hydrides, metal, water-reactive, n.o.s., see	1409	4.3	
Hydriodic acid, anhydrous, see	2197	2	
HYDRIODIC ACID	1787	8	
HYDROBROMIC ACID	1788	8	
HYDROCARBON GAS MIXTURE, COMPRESSED, N.O.S.	1964	2	
HYDROCARBON GAS MIXTURE, LIQUEFIED, N.O.S. such as mixtures A, A01, A02, A0, A1, B1, B2, B or C	1965	2	
HYDROCARBON GAS REFILLS FOR SMALL DEVICES with release device	3150	2	
HYDROCARBONS, LIQUID, N.O.S.	3295	3	
HYDROCHLORIC ACID	1789	8	
HYDROCYANIC ACID, AQUEOUS SOLUTION with not more than 20% hydrogen cyanide	1613	6.1	
HYDROFLUORIC ACID with more than 60% but not more than 85% hydrogen fluoride	1790	8	
HYDROFLUORIC ACID with more than 85% hydrogen fluoride	1790	8	
HYDROFLUORIC ACID with not more than 60% hydrogen fluoride	1790	8	
HYDROFLUORIC ACID AND SULPHURIC ACID MIXTURE	1786	8	
Hydrofluoroboric acid, see	1775	8	
Hydrofluorosilicic acid, see	1778	8	
HYDROGEN AND METHANE MIXTURE, COMPRESSED	2034	2	
Hydrogen arsenide, see	2188	2	
HYDROGEN BROMIDE, ANHYDROUS	1048	2	
Hydrogen bromide solution, see	1788	8	
HYDROGEN CHLORIDE, ANHYDROUS	1050	2	
HYDROGEN CHLORIDE, REFRIGERATED LIQUID	2186	2	Carriage prohibited
HYDROGEN, COMPRESSED	1049	2	
HYDROGEN CYANIDE, AQUEOUS SOLUTION with not more than 20% hydrogen cyanide, see	1613	6.1	
HYDROGEN CYANIDE, SOLUTION IN ALCOHOL with not more than 45% hydrogen cyanide	3294	6.1	
HYDROGEN CYANIDE, STABILIZED containing less than 3% water	1051	6.1	
HYDROGEN CYANIDE, STABILIZED, containing less than 3% water and absorbed in a porous inert material	1614	6.1	
HYDROGENDIFLUORIDES, SOLID, N.O.S.	1740	8	
HYDROGENDIFLUORIDES SOLUTION, N.O.S.	3471	8	
HYDROGEN FLUORIDE, ANHYDROUS	1052	8	
Hydrogen fluoride solution, see	1790	8	
HYDROGEN IN A METAL HYDRIDE STORAGE SYSTEM	3468	2	
HYDROGEN IN A METAL HYDRIDE STORAGE SYSTEM CONTAINED IN EQUIPMENT	3468	2	
HYDROGEN IN A METAL HYDRIDE STORAGE SYSTEM PACKED WITH EQUIPMENT	3468	2	
HYDROGEN IODIDE, ANHYDROUS	2197	2	
Hydrogen iodide solution, see	1787	8	

Name and description	UN No.	Class	Remarks
HYDROGEN PEROXIDE AND PEROXYACETIC ACID MIXTURE with acid(s), water and not more than 5% peroxyacetic acid, STABILIZED	3149	5.1	
HYDROGEN PEROXIDE, AQUEOUS SOLUTION with not less than 8% but less than 20% hydrogen peroxide (stabilized as necessary)	2984	5.1	
HYDROGEN PEROXIDE, AQUEOUS SOLUTION with not less than 20% but not more than 60% hydrogen peroxide (stabilized as necessary)	2014	5.1	
HYDROGEN PEROXIDE, AQUEOUS SOLUTION, STABILIZED with more than 60% hydrogen peroxide and not more than 70% hydrogen peroxide	2015	5.1	
HYDROGEN PEROXIDE, AQUEOUS SOLUTION, STABILIZED with more than 70% hydrogen peroxide	2015	5.1	
HYDROGEN, REFRIGERATED LIQUID	1966	2	
HYDROGEN SELENIDE, ANHYDROUS	2202	2	
Hydrogen silicide, see	2203	2	
HYDROGEN SULPHIDE	1053	2	
Hydroselenic acid, see	2202	2	
Hydrosilicofluoric acid, see	1778	8	
1-HYDROXYBENZOTRIAZOLE, ANHYDROUS, dry or wetted with less than 20% water, by mass	0508	1	
1-HYDROXYBENZOTRIAZOLE, ANHYDROUS, WETTED with not less than 20% water, by mass	3474	4.1	
3-Hydroxybutan-2-one, see	2621	3	
HYDROXYLAMINE SULPHATE	2865	8	
1-Hydroxy-3-methyl-2-penten-4-yne, see	2705	8	
3-Hydroxyphenol, see	2876	6.1	
HYPOCHLORITES, INORGANIC, N.O.S.	3212	5.1	
HYPOCHLORITE SOLUTION	1791	8	
IGNITERS	0121 0314 0315 0325 0454	1 1 1 1 1	
3,3'-IMINODIPROPYLAMINE	2269	8	
India rubber, see	1287	3	
INFECTIOUS SUBSTANCE, AFFECTING ANIMALS only	2900	6.2	
INFECTIOUS SUBSTANCE, AFFECTING HUMANS	2814	6.2	
Ink, printer's, flammable, see	1210	3	
INSECTICIDE GAS, N.O.S.	1968	2	
INSECTICIDE GAS, FLAMMABLE, N.O.S.	3354	2	
INSECTICIDE GAS, TOXIC, N.O.S.	1967	2	
INSECTICIDE GAS, TOXIC, FLAMMABLE, N.O.S.	3355	2	
IODINE MONOCHLORIDE	1792	8	
IODINE PENTAFLUORIDE	2495	5.1	
2-IODOBUTANE	2390	3	
Iodomethane, see	2644	6.1	
IODOMETHYLPROPANES	2391	3	
IODOPROPANES	2392	3	
alpha-Iodotoluene, see	2653	6.1	
I.p.d.i., see	2290	6.1	
Iron chloride, anhydrous, see	1773	8	
Iron (III) chloride, anhydrous, see	1773	8	
Iron chloride solution, see	2582	8	
IRON OXIDE, SPENT obtained from coal gas purification	1376	4.2	
IRON PENTACARBONYL	1994	6.1	
Iron perchloride, anhydrous, see	1773	8	
Iron powder, pyrophoric, see	1383	4.2	
Iron sesquichloride, anhydrous, see	1773	8	

Name and description	UN No.	Class	Remarks
IRON SPONGE, SPENT obtained from coal gas purification	1376	4.2	
Iron swarf, see	2793	4.2	
ISOBUTANE	1969	2	
ISOBUTANOL	1212	3	
Isobutene, see	1055	2	
ISOBUTYL ACETATE	1213	3	
ISOBUTYL ACRYLATE, STABILIZED	2527	3	
ISOBUTYL ALCOHOL, see	1212	3	
ISOBUTYL ALDEHYDE, see	2045	3	
ISOBUTYLAMINE	1214	3	
ISOBUTYLENE	1055	2	
ISOBUTYL FORMATE	2393	3	
ISOBUTYL ISOBUTYRATE	2528	3	
ISOBUTYL ISOCYANATE	2486	3	
ISOBUTYL METHACRYLATE, STABILIZED	2283	3	
ISOBUTYL PROPIONATE	2394	3	
ISOBUTYRALDEHYDE	2045	3	
ISOBUTYRIC ACID	2529	3	
ISOBUTYRONITRILE	2284	3	
ISOBUTYRYL CHLORIDE	2395	3	
ISOCYANATES, FLAMMABLE, TOXIC, N.O.S.	2478	3	
ISOCYANATES, TOXIC, N.O.S.	2206	6.1	
ISOCYANATES, TOXIC, FLAMMABLE, N.O.S.	3080	6.1	
ISOCYANATE SOLUTION, FLAMMABLE, TOXIC, N.O.S.	2478	3	
ISOCYANATE SOLUTION, TOXIC, N.O.S.	2206	6.1	
ISOCYANATE SOLUTION, TOXIC, FLAMMABLE, N.O.S.	3080	6.1	
ISOCYANATO-BENZOTRIFLUORIDES	2285	6.1	
3-Isocyanatomethyl-3,5,5-tri-methylcyclohexyl isocyanate, see	2290	6.1	

Name and description	UN No.	Class	Remarks
Isododecane, see	2286	3	
ISOHEPTENE	2287	3	
ISOHEXENE	2288	3	
Isooctane, see	1262	3	
ISOOCTENE	1216	3	
Isopentane, see	1265	3	
ISOPENTENES	2371	3	
Isopentylamine, see	1106	3	
Isopentyl nitrite, see	1113	3	
ISOPHORONEDIAMINE	2289	8	
ISOPHORONE DIISOCYANATE	2290	6.1	
ISOPRENE, STABILIZED	1218	3	
ISOPROPANOL	1219	3	
ISOPROPENYL ACETATE	2403	3	
ISOPROPENYLBENZENE	2303	3	
ISOPROPYL ACETATE	1220	3	
ISOPROPYL ACID PHOSPHATE	1793	8	
ISOPROPYL ALCOHOL, see	1219	3	
ISOPROPYLAMINE	1221	3	
ISOPROPYLBENZENE	1918	3	
ISOPROPYL BUTYRATE	2405	3	
Isopropyl chloride, see	2356	3	
ISOPROPYL CHLOROACETATE	2947	3	
ISOPROPYL CHLOROFORMATE	2407	6.1	
ISOPROPYL 2-CHLORO-PROPIONATE	2934	3	
Isopropyl-alpha-chloropropionate, see	2934	3	
Isopropyl ether, see	1159	3	
Isopropylethylene, see	2561	3	
Isopropyl formate, see	1281	3	
ISOPROPYL ISOBUTYRATE	2406	3	
ISOPROPYL ISOCYANATE	2483	3	
Isopropyl mercaptan, see	2402	3	

Name and description	UN No.	Class	Remarks
ISOPROPYL NITRATE	1222	3	
ISOPROPYL PROPIONATE	2409	3	
Isolpropyltoluene, see	2046	3	
Isopropyltoluol, see	2046	3	
ISOSORBIDE DINITRATE MIXTURE with not less than 60% lactose, mannose, starch or calcium hydrogen phosphate	2907	4.1	
ISOSORBIDE-5-MONONITRATE	3251	4.1	
Isovaleraldehyde, see	2058	3	
JET PERFORATING GUNS, CHARGED, oil well, without detonator	0124 0494	1 1	
Jet tappers, without detonator, see	0059	1	
KEROSENE	1223	3	
KETONES, LIQUID, N.O.S.	1224	3	
KRYPTON, COMPRESSED	1056	2	
KRYPTON, REFRIGERATED LIQUID	1970	2	
Lacquer, see	1263 3066 3469 3470	3 8 3 8	
Lacquer base, liquid, see	1263 3066 3469 3470	3 8 3 8	
Lacquer base or lacquer chips, nitrocellulose, dry, see	2557	4.1	
Lacquer base or lacquer chips, plastic, wet with alcohol or solvent, see	1263 2059 2555 2556	3 3 4.1 4.1	
LEAD ACETATE	1616	6.1	
Lead (II) acetate, see	1616	6.1	
LEAD ARSENATES	1617	6.1	
LEAD ARSENITES	1618	6.1	
LEAD AZIDE, WETTED with not less than 20% water, or mixture of alcohol and water, by mass	0129	1	
Lead chloride, solid, see	2291	6.1	
LEAD COMPOUND, SOLUBLE, N.O.S.	2291	6.1	

Name and description	UN No.	Class	Remarks
LEAD CYANIDE	1620	6.1	
Lead (II) cyanide	1620	6.1	
LEAD DIOXIDE	1872	5.1	
LEAD NITRATE	1469	5.1	
Lead (II) nitrate	1469	5.1	
LEAD PERCHLORATE, SOLID	1470	5.1	
LEAD PERCHLORATE, SOLUTION	3408	5.1	
Lead (II) perchlorate	1470 3408	5.1 5.1	
Lead peroxide, see	1872	5.1	
LEAD PHOSPHITE, DIBASIC	2989	4.1	
LEAD STYPHNATE, WETTED with not less than 20% water, or mixture of alcohol and water, by mass	0130	1	
LEAD SULPHATE with more than 3% free acid	1794	8	
Lead tetraethyl, see	1649	6.1	
Lead tetramethyl, see	1649	6.1	
LEAD TRINITRORESORCINATE, WETTED with not less than 20% water, or mixture of alcohol and water, by mass, see	0130	1	
LIFE-SAVING APPLIANCES NOT SELF-INFLATING containing dangerous goods as equipment	3072	9	
LIFE-SAVING APPLIANCES, SELF-INFLATING	2990	9	
LIGHTER REFILLS containing flammable gas	1057	2	
LIGHTERS containing flammable gas	1057	2	
LIGHTERS, FUSE	0131	1	
Limonene, inactive, see	2052	3	
LIQUEFIED GAS, N.O.S.	3163	2	
LIQUEFIED GAS, FLAMMABLE, N.O.S.	3161	2	
LIQUEFIED GASES, non-flammable, charged with nitrogen, carbon dioxide or air	1058	2	

Name and description	UN No.	Class	Remarks
LIQUEFIED GAS, OXIDIZING, N.O.S.	3157	2	
LIQUEFIED GAS, TOXIC, N.O.S.	3162	2	
LIQUEFIED GAS, TOXIC, CORROSIVE, N.O.S.	3308	2	
LIQUEFIED GAS, TOXIC, FLAMMABLE, N.O.S.	3160	2	
LIQUEFIED GAS, TOXIC, FLAMMABLE, CORROSIVE, N.O.S.	3309	2	
LIQUEFIED GAS, TOXIC, OXIDIZING, N.O.S.	3307	2	
LIQUEFIED GAS, TOXIC, OXIDIZING, CORROSIVE, N.O.S.	3310	2	
Liquefied petroleum gas, see	1075	2	
Liquid filler, see	1263 3066 3469 3470	3 8 3 8	
Liquid lacquer base, see	1263 3066 3469 3470	3 8 3 8	
LITHIUM	1415	4.3	
Lithium alkyls, liquid, see	3394	4.2	
Lithium alkyls, solid, see	3393	4.2	
LITHIUM ALUMINIUM HYDRIDE	1410	4.3	
LITHIUM ALUMINIUM HYDRIDE, ETHEREAL	1411	4.3	
LITHIUM ION BATTERIES (including lithium ion polymer batteries)	3480	9	
LITHIUM ION BATTERIES CONTAINED IN EQUIPMENT (including lithium ion polymer batteries)	3481	9	
LITHIUM ION BATTERIES PACKED WITH EQUIPMENT (including lithium ion polymer batteries)	3481	9	
LITHIUM METAL BATTERIES (including lithium alloy batteries)	3090	9	
LITHIUM METAL BATTERIES CONTAINED IN EQUIPMENT (including lithium alloy batteries)	3091	9	
LITHIUM METAL BATTERIES PACKED WITH EQUIPMENT (including lithium alloy batteries)	3091	9	
LITHIUM BOROHYDRIDE	1413	4.3	
LITHIUM FERROSILICON	2830	4.3	
LITHIUM HYDRIDE	1414	4.3	
LITHIUM HYDRIDE, FUSED SOLID	2805	4.3	
LITHIUM HYDROXIDE	2680	8	
LITHIUM HYDROXIDE SOLUTION	2679	8	
LITHIUM HYPOCHLORITE, DRY	1471	5.1	
LITHIUM HYPOCHLORITE MIXTURE	1471	5.1	
Lithium in cartouches, see	1415	4.3	
LITHIUM NITRATE	2722	5.1	
LITHIUM NITRIDE	2806	4.3	
LITHIUM PEROXIDE	1472	5.1	
Lithium silicide, see	1417	4.3	
LITHIUM SILICON	1417	4.3	
L.n.g., see	1972	2	
LONDON PURPLE	1621	6.1	
L.p.g., see	1075	2	
Lye, see	1823	8	
Lythene, see	1268	3	
MAGNESIUM in pellets, turnings or ribbons	1869	4.1	
Magnesium alkyls, see	3394	4.2	
MAGNESIUM ALLOYS with more than 50% magnesium in pellets, turnings or ribbons	1869	4.1	
MAGNESIUM ALLOYS POWDER	1418	4.3	
MAGNESIUM ALUMINIUM PHOSPHIDE	1419	4.3	
MAGNESIUM ARSENATE	1622	6.1	
Magnesium bisulphite solution, see	2693	8	
MAGNESIUM BROMATE	1473	5.1	

Name and description	UN No.	Class	Remarks
MAGNESIUM CHLORATE	2723	5.1	
Magnesium chloride and chlorate mixture, see	1459 3407	5.1 5.1	
MAGNESIUM DIAMIDE	2004	4.2	
Magnesium diphenyl, see	3393	4.2	
MAGNESIUM FLUOROSILICATE	2853	6.1	
MAGNESIUM GRANULES, COATED, particle size not less than 149 microns	2950	4.3	
MAGNESIUM HYDRIDE	2010	4.3	
MAGNESIUM NITRATE	1474	5.1	
MAGNESIUM PERCHLORATE	1475	5.1	
MAGNESIUM PEROXIDE	1476	5.1	
MAGNESIUM PHOSPHIDE	2011	4.3	
MAGNESIUM POWDER	1418	4.3	
Magnesium scrap, see	1869	4.1	
MAGNESIUM SILICIDE	2624	4.3	
Magnesium silicofluoride, see	2853	6.1	
Magnetized material	2807	9	Not subject to ADN
MALEIC ANHYDRIDE	2215	8	
MALEIC ANHYDRIDE, MOLTEN	2215	8	
Malonic dinitrile, see	2647	6.1	
Malonodinitrile, see	2647	6.1	
MALONONITRILE	2647	6.1	
MANEB	2210	4.2	
MANEB PREPARATION with not less than 60% maneb	2210	4.2	
MANEB PREPARATION, STABILIZED against self-heating	2968	4.3	
MANEB, STABILIZED against self-heating	2968	4.3	
Manganese ethylene-di-dithiocarbamate, see	2210	4.2	
Manganese ethylene-1,2-dithiocarbamate, see	2210	4.2	
MANGANESE NITRATE	2724	5.1	
Manganese (II) nitrate, see	2724	5.1	
MANGANESE RESINATE	1330	4.1	
Manganous nitrate, see	2724	5.1	
MANNITOL HEXANITRATE, WETTED with not less than 40% water, or mixture of alcohol and water, by mass	0133	1	
MATCHES, FUSEE	2254	4.1	
MATCHES, SAFETY (book, card or strike on box)	1944	4.1	
MATCHES, "STRIKE ANYWHERE"	1331	4.1	
MATCHES, WAX "VESTA"	1945	4.1	
MEDICAL WASTE, N.O.S.	3291	6.2	
MEDICINE, LIQUID, FLAMMABLE, TOXIC, N.O.S.	3248	3	
MEDICINE, LIQUID, TOXIC, N.O.S.	1851	6.1	
MEDICINE, SOLID, TOXIC, N.O.S.	3249	6.1	
p-Mentha-1,8-diene, see	2052	8	
MERCAPTANS, LIQUID, FLAMMABLE, N.O.S.	3336	3	
MERCAPTANS, LIQUID, FLAMMABLE, TOXIC, N.O.S.	1228	3	
MERCAPTANS, LIQUID, TOXIC, FLAMMABLE, N.O.S.	3071	6.1	
MERCAPTAN MIXTURE, LIQUID, FLAMMABLE, N.O.S.	3336	3	
MERCAPTAN MIXTURE, LIQUID, FLAMMABLE, TOXIC, N.O.S.	1228	3	
MERCAPTAN MIXTURE, LIQUID, TOXIC, FLAMMABLE, N.O.S.	3071	6.1	
2-Mercaptoethanol, see	2966	6.1	
2-Mercaptopropionic acid, see	2936	6.1	
5-MERCAPTOTETRAZOL-1-ACETIC ACID	0448	1	
MERCURIC ARSENATE	1623	6.1	
MERCURIC CHLORIDE	1624	6.1	
MERCURIC NITRATE	1625	6.1	

Name and description	UN No.	Class	Remarks
MERCURIC POTASSIUM CYANIDE	1626	6.1	
Mercuric sulphate, see	1645	6.1	
Mercurol, see	1639	6.1	
Mercurous bisulphate, see	1645	6.1	
MERCUROUS NITRATE	1627	6.1	
Mercurous sulphate, see	1645	6.1	
MERCURY	2809	8	
MERCURY ACETATE	1629	6.1	
MERCURY AMMONIUM CHLORIDE	1630	6.1	
MERCURY BASED PESTICIDE, LIQUID, FLAMMABLE, TOXIC, flash-point less than 23 °C	2778	3	
MERCURY BASED PESTICIDE, LIQUID, TOXIC	3012	6.1	
MERCURY BASED PESTICIDE, LIQUID, TOXIC, FLAMMABLE, flash-point not less than 23 °C	3011	6.1	
MERCURY BASED PESTICIDE, SOLID, TOXIC	2777	6.1	
MERCURY BENZOATE	1631	6.1	
Mercury bichloride, see	1624	6.1	
MERCURY BROMIDES	1634	6.1	
MERCURY COMPOUND, LIQUID, N.O.S.	2024	6.1	
MERCURY COMPOUND, SOLID, N.O.S.	2025	6.1	
MERCURY CYANIDE	1636	6.1	
MERCURY FULMINATE, WETTED with not less than 20% water, or mixture of alcohol and water, by mass	0135	1	
MERCURY GLUCONATE	1637	6.1	
MERCURY IODIDE	1638	6.1	
MERCURY NUCLEATE	1639	6.1	
MERCURY OLEATE	1640	6.1	
MERCURY OXIDE	1641	6.1	
MERCURY OXYCYANIDE, DESENSITIZED	1642	6.1	

Name and description	UN No.	Class	Remarks
MERCURY POTASSIUM IODIDE	1643	6.1	
MERCURY SALICYLATE	1644	6.1	
MERCURY SULPHATE	1645	6.1	
MERCURY THIOCYANATE	1646	6.1	
Metal alkyl halides, water-reactive, n.o.s. / Metal aryl halides, water-reactive, n.o.s., see	3394	4.2	
Metal alkyl hydrides, water-reactive, n.o.s. / Metal aryl hydrides, water-reactive, n.o.s., see	3394	4.2	
Metal alkyls, water-reactive, n.o.s. / Metal aryls, water-reactive, n.o.s., see	3393	4.2	
Mesitylene, see	2325	3	
MESITYL OXIDE	1229	3	
METAL CARBONYLS, LIQUID, N.O.S.	3281	6.1	
METAL CARBONYLS, SOLID, N.O.S.	3466	6.1	
METAL CATALYST, DRY	2881	4.2	
METAL CATALYST, WETTED with a visible excess of liquid	1378	4.2	
METALDEHYDE	1332	4.1	
METAL HYDRIDES, FLAMMABLE, N.O.S.	3182	4.1	
METAL HYDRIDES, WATER-REACTIVE, N.O.S.	1409	4.3	
METALLIC SUBSTANCE, WATER-REACTIVE, N.O.S.	3208	4.3	
METALLIC SUBSTANCE, WATER-REACTIVE, SELF-HEATING, N.O.S.	3209	4.3	
METAL POWDER, FLAMMABLE, N.O.S.	3089	4.1	
METAL POWDER, SELF-HEATING, N.O.S.	3189	4.2	
METAL SALTS OF ORGANIC COMPOUNDS, FLAMMABLE, N.O.S.	3181	4.1	
METHACRYLALDEHYDE, STABILIZED	2396	3	
METHACRYLIC ACID, STABILIZED	2531	8	
METHACRYLONITRILE, STABILIZED	3079	3	

Name and description	UN No.	Class	Remarks
METHALLYL ALCOHOL	2614	3	
Methanal, see	1198 2209	3 8	
Methane and hydrogen mixture, see	2034	2	
METHANE, COMPRESSED	1971	2	
METHANE, REFRIGERATED LIQUID	1972	2	
METHANESULPHONYL CHLORIDE	3246	6.1	
METHANOL	1230	3	
2-Methoxyethyl acetate, see	1189	3	
METHOXYMETHYL ISOCYANATE	2605	3	
4-METHOXY-4-METHYLPENTAN-2-ONE	2293	3	
1-Methoxy-2-nitrobenzene, see	2730 3458	6.1 6.1	
1-Methoxy-3-nitrobenzene, see	2730 3458	6.1 6.1	
1-Methoxy-4-nitrobenzene, see	2730 3458	6.1 6.1	
1-METHOXY-2-PROPANOL	3092	3	
METHYL ACETATE	1231	3	
METHYLACETYLENE AND PROPADIENE MIXTURE, STABILIZED such as mixture P1 or mixture P2	1060	2	
beta-Methyl acrolein, see	1143	6.1	
METHYL ACRYLATE, STABILIZED	1919	3	
METHYLAL	1234	3	
Methyl alcohol, see	1230	3	
Methyl allyl alcohol, see	2614	3	
METHYLALLYL CHLORIDE	2554	3	
METHYLAMINE, ANHYDROUS	1061	2	
METHYLAMINE, AQUEOUS SOLUTION	1235	3	
METHYLAMYL ACETATE	1233	3	
Methyl amyl alcohol, see	2053	3	

Name and description	UN No.	Class	Remarks
Methyl amyl ketone, see	1110	3	
N-METHYLANILINE	2294	6.1	
Methylated spirit, see	1986 1987	3 3	
alpha-METHYLBENZYL ALCOHOL, LIQUID	2937	6.1	
alpha-METHYLBENZYL ALCOHOL, SOLID	3438	6.1	
METHYL BROMIDE with not more than 2% chloropicrin	1062	2	
Methyl bromide and chloropicrin mixture, with more than 2% chloropicrin, see	1581	2	
METHYL BROMIDE AND ETHYLENE DIBROMIDE MIXTURE, LIQUID	1647	6.1	
METHYL BROMOACETATE	2643	6.1	
2-METHYLBUTANAL	3371	3	
3-METHYLBUTAN-2-ONE	2397	3	
2-METHYL-1-BUTENE	2459	3	
2-METHYL-2-BUTENE	2460	3	
3-METHYL-1-BUTENE	2561	3	
N-METHYLBUTYLAMINE	2945	3	
METHYL tert-BUTYL ETHER	2398	3	
METHYL BUTYRATE	1237	3	
METHYL CHLORIDE	1063	2	
Methyl chloride and chloropicrin mixture, see	1582	2	
METHYL CHLORIDE AND METHYLENE CHLORIDE MIXTURE	1912	2	
METHYL CHLOROACETATE	2295	6.1	
Methyl chlorocarbonate, see	1238	6.1	
Methyl chloroform, see	2831	6.1	
METHYL CHLOROFORMATE	1238	6.1	
METHYL CHLOROMETHYL ETHER	1239	6.1	
METHYL 2-CHLOROPROPIONATE	2933	3	

Name and description	UN No.	Class	Remarks
Methyl alpha-chloropropionate, see	2933	3	
METHYLCHLOROSILANE	2534	2	
Methyl cyanide, see	1648	3	
METHYLCYCLOHEXANE	2296	3	
METHYLCYCLOHEXANOLS, flammable	2617	3	
METHYLCYCLOHEXANONE	2297	3	
METHYLCYCLOPENTANE	2298	3	
METHYL DICHLOROACETATE	2299	6.1	
METHYLDICHLOROSILANE	1242	4.3	
Methylene bromide, see	2664	6.1	
Methylene chloride, see	1593	6.1	
Methylene chloride and methyl chloride mixture, see	1912	2	
Methylene cyanide, see	2647	6.1	
p,p'-Methylene dianiline, see	2651	6.1	
Methylene dibromide, see	2664	6.1	
2,2'-Methylene-di-(3,4,6-trichlorophenol), see	2875	6.1	
Methyl ethyl ether, see	1039	2	
METHYL ETHYL KETONE, see	1193	3	
2-METHYL-5-ETHYLPYRIDINE	2300	6.1	
METHYL FLUORIDE	2454	2	
METHYL FORMATE	1243	3	
2-METHYLFURAN	2301	3	
Methyl glycol, see	1188	3	
Methyl glycol acetate, see	1189	3	
2-METHYL-2-HEPTANETHIOL	3023	6.1	
5-METHYLHEXAN-2-ONE	2302	3	
METHYLHYDRAZINE	1244	6.1	
METHYL IODIDE	2644	6.1	
METHYL ISOBUTYL CARBINOL	2053	3	
METHYL ISOBUTYL KETONE	1245	3	
METHYL ISOCYANATE	2480	6.1	
METHYL ISOPROPENYL KETONE, STABILIZED	1246	3	
METHYL ISOTHIOCYANATE	2477	6.1	
METHYL ISOVALERATE	2400	3	
METHYL MAGNESIUM BROMIDE IN ETHYL ETHER	1928	4.3	
METHYL MERCAPTAN	1064	2	
Methyl mercaptopropionaldehyde, see	2785	6.1	
METHYL METHACRYLATE MONOMER, STABILIZED	1247	3	
4-METHYLMORPHOLINE	2535	3	
N-METHYLMORPHOLINE, see	2535	3	
METHYL NITRITE	2455	2	Carriage prohibited
METHYL ORTHOSILICATE	2606	6.1	
METHYLPENTADIENE	2461	3	
Methylpentanes, see	1208	3	
2-METHYLPENTAN-2-OL	2560	3	
4-Methylpentan-2-ol, see	2053	3	
3-Methyl-2-penten-4ynol, see	2705	8	
METHYLPHENYL-DICHLOROSILANE	2437	8	
2-Methyl-2-phenylpropane, see	2709	3	
1-METHYLPIPERIDINE	2399	3	
METHYL PROPIONATE	1248	3	
Methylpropylbenzene, see	2046	3	
METHYL PROPYL ETHER	2612	3	
METHYL PROPYL KETONE	1249	3	
Methyl pyridines, see	2313	3	
Methylstyrene, inhibited, see	2618	3	
alpha-Methylstyrene, see	2303	3	
Methyl sulphate, see	1595	6.1	
Methyl sulphide, see	1164	3	
METHYLTETRAHYDROFURAN	2536	3	

Name and description	UN No.	Class	Remarks
METHYL TRICHLOROACETATE	2533	6.1	
METHYLTRICHLOROSILANE	1250	3	
alpha-METHYLVALERAL-DEHYDE	2367	3	
Methyl vinyl benzene, inhibited, see	2618	3	
METHYL VINYL KETONE, STABILIZED	1251	6.1	
M.i.b.c., see	2053	3	
MINES with bursting charge	0136 0137 0138 0294	1 1 1 1	
Mirbane oil, see	1662	6.1	
Missiles, guided, see	0180 0181 0182 0183 0295 0397 0398 0436 0437 0438	1 1 1 1 1 1 1 1 1 1	
Mixtures A, A01, A02, A0, A1, B1, B2, B or C, see	1965	2	
Mixture F1, mixture F2 or mixture F3, see	1078	2	
MIXTURES OF 1,3-BUTADIENE AND HYDROCARBONS, STABILIZED, having a vapour pressure at 70 °C not exceeding 1.1 MPa (11 bar) and a density at 50 °C not lower than 0.525 kg/l	1010	2	
Mixture P1 or mixture P2, see	1060	2	
MOLYBDENUM PENTACHLORIDE	2508	8	
Monochloroacetic acid, see	1750 1751	6.1 6.1	
Monochlorobenzene, see	1134	3	
Monochlorodifluoromethane, see	1018	2	
Monochlorodifluoromethane and monochloro-pentafluoroethane mixture, see	1973	2	
Monochlorodifluoromono-bromomethane, see	1974	2	
Monochloropentafluoroethane and monochloro-difluoromethane mixture, see	1973	2	
Monoethylamine, see	1036	2	
MONONITROTOLUIDINES, see	2660	6.1	
Monopropylamine, see	1277	3	
MORPHOLINE	2054	8	
MOTOR FUEL ANTI-KNOCK MIXTURE	1649	6.1	
MOTOR SPIRIT	1203	3	
Motor spirit and ethanol mixture, with more than 10% ethanol, see	3475	3	
Muriatic acid, see	1789	8	
MUSK XYLENE, see	2956	4.1	
Mysorite, see	2212	9	
Naphta, see	1268	3	
Naphta, petroleum, see	1268	3	
Naphta, solvent, see	1268	3	
NAPHTHALENE, CRUDE	1334	4.1	
NAPHTHALENE, MOLTEN	2304	4.1	
NAPHTHALENE, REFINED	1334	4.1	
alpha-NAPHTHYLAMINE	2077	6.1	
beta-NAPHTHYLAMINE, SOLID	1650	6.1	
beta-NAPHTHYLAMINE, SOLUTION	3411	6.1	
NAPHTHYLTHIOUREA	1651	6.1	
1-Naphthylthiourea, see	1651	6.1	
NAPHTHYLUREA	1652	6.1	
NATURAL GAS, COMPRESSED with high methane content	1971	2	
NATURAL GAS, REFRIGERATED LIQUID with high methane content	1972	2	
Natural gasoline, see	1203	3	
Neohexane, see	1208	3	
NEON, COMPRESSED	1065	2	
NEON, REFRIGERATED LIQUID	1913	2	

Name and description	UN No.	Class	Remarks
Neothyl, see	2612	3	
NICKEL CARBONYL	1259	6.1	
NICKEL CYANIDE	1653	6.1	
Nickel (II) cyanide, see	1653	6.1	
NICKEL NITRATE	2725	5.1	
Nickel (II) nitrate, see	2725	5.1	
NICKEL NITRITE	2726	5.1	
Nickel (II) nitrite, see	2726	5.1	
Nickelous nitrate, see	2725	5.1	
Nickelous nitrite, see	2726	5.1	
Nickel tetracarbonyl, see	1259	6.1	
NICOTINE	1654	6.1	
NICOTINE COMPOUND, LIQUID, N.O.S	3144	6.1	
NICOTINE COMPOUND, SOLID, N.O.S	1655	6.1	
NICOTINE HYDROCHLORIDE, LIQUID	1656	6.1	
NICOTINE HYDROCHLORIDE, SOLID	3444	6.1	
NICOTINE HYDROCHLORIDE SOLUTION	1656	6.1	
NICOTINE PREPARATION, LIQUID, N.O.S.	3144	6.1	
NICOTINE PREPARATION, SOLID, N.O.S.	1655	6.1	
NICOTINE SALICYLATE	1657	6.1	
NICOTINE SULPHATE, SOLID	3445	6.1	
NICOTINE SULPHATE, SOLUTION	1658	6.1	
NICOTINE TARTRATE	1659	6.1	
NITRATES, INORGANIC, N.O.S.	1477	5.1	
NITRATES, INORGANIC, AQUEOUS SOLUTION, N.O.S.	3218	5.1	
NITRATING ACID MIXTURE with more than 50% nitric acid	1796	8	
NITRATING ACID MIXTURE with not more than 50% nitric acid	1796	8	

Name and description	UN No.	Class	Remarks
NITRATING ACID MIXTURE, SPENT, with more than 50% nitric acid	1826	8	
NITRATING ACID MIXTURE, SPENT, with not more than 50% nitric acid	1826	8	
NITRIC ACID, other than red fuming, with at least 65% but not more than 70% nitric acid	2031	8	
NITRIC ACID, other than red fuming, with less than 65% nitric acid	2031	8	
NITRIC ACID, other than red fuming, with more than 70% nitric acid	2031	8	
NITRIC ACID, RED FUMING	2032	8	
NITRIC OXIDE, COMPRESSED	1660	2	
NITRIC OXIDE AND DINITROGEN TETROXIDE MIXTURE	1975	2	
NITRIC OXIDE AND NITROGEN DIOXIDE MIXTURE, see	1975	2	
NITRILES, FLAMMABLE, TOXIC, N.O.S.	3273	3	
NITRILES, TOXIC, LIQUID, N.O.S.	3276	6.1	
NITRILES, TOXIC, SOLID, N.O.S.	3439	6.1	
NITRILES, TOXIC, FLAMMABLE, N.O.S.	3275	6.1	
NITRITES, INORGANIC, N.O.S.	2627	5.1	
NITRITES, INORGANIC, AQUEOUS SOLUTION, N.O.S.	3219	5.1	
NITROANILINES (o-, m-, p-)	1661	6.1	
NITROANISOLES, LIQUID	2730	6.1	
NITROANISOLES, SOLID	3458	6.1	
NITROBENZENE	1662	6.1	
Nitrobenzene bromide, see	2732	6.1	
NITROBENZENESULPHONIC ACID	2305	8	
Nitrobenzol, see	1662	6.1	
5-NITROBENZOTRIAZOL	0385	1	

Name and description	UN No.	Class	Remarks
NITROBENZOTRIFLUORIDES, LIQUID	2306	6.1	
NITROBENZOTRIFLUORIDES, SOLID	3431	6.1	
NITROBROMOBENZENES, LIQUID	2732	6.1	
NITROBROMOBENZENES, SOLID	3459	6.1	
NITROCELLULOSE, dry or wetted with less than 25% water (or alcohol), by mass	0340	1	
NITROCELLULOSE, unmodified or plasticized with less than 18% plasticizing substance, by mass	0341	1	
NITROCELLULOSE MEMBRANE FILTERS, with not more than 12.6% nitrogen, by dry mass	3270	4.1	
NITROCELLULOSE, with not more than 12.6% nitrogen, by dry mass, MIXTURE WITH PLASTICIZER, WITH PIGMENT	2557	4.1	
NITROCELLULOSE, with not more than 12.6% nitrogen, by dry mass, MIXTURE WITH PLASTICIZER, WITHOUT PIGMENT	2557	4.1	
NITROCELLULOSE, with not more than 12.6% nitrogen, by dry mass, MIXTURE WITHOUT PLASTICIZER, WITH PIGMENT	2557	4.1	
NITROCELLULOSE, with not more than 12.6% nitrogen, by dry mass, MIXTURE WITHOUT PLASTICIZER, WITHOUT PIGMENT	2557	4.1	
NITROCELLULOSE, PLASTICIZED with not less than 18% plasticizing substance, by mass	0343	1	
NITROCELLULOSE SOLUTION, FLAMMABLE with not more than 12.6% nitrogen, by dry mass, and not more than 55% nitrocellulose	2059	3	
NITROCELLULOSE, WETTED with not less than 25% alcohol, by mass	0342	1	
NITROCELLULOSE WITH ALCOHOL (not less than 25% alcohol, by mass, and not more than 12.6% nitrogen, by dry mass)	2556	4.1	
NITROCELLULOSE WITH WATER (not less than 25% water, by mass)	2555	4.1	
Nitrochlorobenzenes, see	1578 3409	6.1 6.1	
3-NITRO-4-CHLOROBENZO-TRIFLUORIDE	2307	6.1	
NITROCRESOLS, LIQUID	3434	6.1	
NITROCRESOLS, SOLID	2446	6.1	
NITROETHANE	2842	3	
NITROGEN, COMPRESSED	1066	2	
NITROGEN DIOXIDE, see	1067	2	
NITROGEN, REFRIGERATED LIQUID	1977	2	
NITROGEN TRIFLUORIDE	2451	2	
NITROGEN TRIOXIDE	2421	2	Carriage prohibited
NITROGLYCERIN, DESENSITIZED with not less than 40% non-volatile water-insoluble phlegmatizer, by mass	0143	1	
NITROGLYCERIN MIXTURE, DESENSITIZED, LIQUID, N.O.S. with not more than 30% nitroglycerin, by mass	3357	3	
NITROGLYCERIN MIXTURE, DESENSITIZED, LIQUID, FLAMMABLE, N.O.S. with not more than 30% nitroglycerin, by mass	3343	3	
NITROGLYCERIN MIXTURE, DESENSITIZED, SOLID, N.O.S. with more than 2% but not more than 10% nitroglycerin, by mass	3319	4.1	
NITROGLYCERIN, SOLUTION IN ALCOHOL with more than 1% but not more than 5% nitroglycerin	3064	3	
NITROGLYCERIN SOLUTION IN ALCOHOL with more than 1% but not more than 10% nitroglycerin	0144	1	
NITROGLYCERIN SOLUTION IN ALCOHOL with not more than 1% nitroglycerin	1204	3	
NITROGUANIDINE, dry or wetted with less than 20% water, by mass	0282	1	

Name and description	UN No.	Class	Remarks
NITROGUANIDINE, WETTED with not less than 20% water, by mass	1336	4.1	
NITROHYDROCHLORIC ACID	1798	8	Carriage prohi-bited
NITROMANNITE, WETTED, see	0133	1	
NITROMETHANE	1261	3	
Nitromuriatic acid, see	1798	8	
NITRONAPHTHALENE	2538	4.1	
NITROPHENOLS (o-, m-, p-)	1663	6.1	
4-NITROPHENYLHYDRAZINE, with not less than 30% water, by mass	3376	4.1	
NITROPROPANES	2608	3	
p-NITROSODIMETHYLANILINE	1369	4.2	
NITROSTARCH, dry or wetted with less than 20% water, by mass	0146	1	
NITROSTARCH, WETTED with not less than 20% water, by mass	1337	4.1	
NITROSYL CHLORIDE	1069	2	
NITROSYLSULPHURIC ACID, LIQUID	2308	8	
NITROSYLSULPHURIC ACID, SOLID	3456	8	
NITROTOLUENES, LIQUID	1664	6.1	
NITROTOLUENES, SOLID	3446	6.1	
NITROTOLUIDINES	2660	6.1	
NITROTRIAZOLONE	0490	1	
NITRO UREA	0147	1	
NITROUS OXIDE	1070	2	
NITROUS OXIDE, REFRIGERATED LIQUID	2201	2	
NITROXYLENES, LIQUID	1665	6.1	
NITROXYLENES, SOLID	3447	6.1	
Non-activated carbon, see	1361	4.2	
Non-activated charcoal, see	1361	4.2	
NONANES	1920	3	
NONYLTRICHLOROSILANE	1799	8	
2,5-NORBORNADIENE, STABILIZED, see	2251	3	
Normal propyl alcohol, see	1274	3	
NTO, see	0490	1	
OCTADECYLTRICHLORO-SILANE	1800	8	
OCTADIENE	2309	3	
OCTAFLUOROBUT-2-ENE	2422	2	
OCTAFLUOROCYCLOBUTANE	1976	2	
OCTAFLUOROPROPANE	2424	2	
OCTANES	1262	3	
OCTOGEN, see	0226 0391 0484	1 1 1	
OCTOL, dry or wetted with less than 15% water, by mass, see	0266	1	
OCTOLITE, dry or wetted with less than 15% water, by mass	0266	1	
OCTONAL	0496	1	
OCTYL ALDEHYDES	1191	3	
tert-Octyl mercaptan, see	3023	6.1	
OCTYLTRICHLOROSILANE	1801	8	
Oenanthol, see	3056	3	
OIL GAS, COMPRESSED	1071	2	
Oleum, see	1831	8	
ORGANIC PEROXIDE TYPE B, LIQUID	3101	5.2	
ORGANIC PEROXIDE TYPE B, LIQUID, TEMPERATURE CONTROLLED	3111	5.2	
ORGANIC PEROXIDE TYPE B, SOLID	3102	5.2	
ORGANIC PEROXIDE TYPE B, SOLID, TEMPERATURE CONTROLLED	3112	5.2	
ORGANIC PEROXIDE TYPE C, LIQUID	3103	5.2	

Name and description	UN No.	Class	Remarks
ORGANIC PEROXIDE TYPE C, LIQUID, TEMPERATURE CONTROLLED	3113	5.2	
ORGANIC PEROXIDE TYPE C, SOLID	3104	5.2	
ORGANIC PEROXIDE TYPE C, SOLID, TEMPERATURE CONTROLLED	3114	5.2	
ORGANIC PEROXIDE TYPE D, LIQUID	3105	5.2	
ORGANIC PEROXIDE TYPE D, LIQUID, TEMPERATURE CONTROLLED	3115	5.2	
ORGANIC PEROXIDE TYPE D, SOLID	3106	5.2	
ORGANIC PEROXIDE TYPE D, SOLID, TEMPERATURE CONTROLLED	3116	5.2	
ORGANIC PEROXIDE TYPE E, LIQUID	3107	5.2	
ORGANIC PEROXIDE TYPE E, LIQUID, TEMPERATURE CONTROLLED	3117	5.2	
ORGANIC PEROXIDE TYPE E, SOLID	3108	5.2	
ORGANIC PEROXIDE TYPE E, SOLID, TEMPERATURE CONTROLLED	3118	5.2	
ORGANIC PEROXIDE TYPE F, LIQUID	3109	5.2	
ORGANIC PEROXIDE TYPE F, LIQUID, TEMPERATURE CONTROLLED	3119	5.2	
ORGANIC PEROXIDE TYPE F, SOLID	3110	5.2	
ORGANIC PEROXIDE TYPE F, SOLID, TEMPERATURE CONTROLLED	3120	5.2	
Organic peroxides, see 2.2.52.4 for an alphabetic list of currently assigned organic peroxides and see	3101 to 3120	5.2	
ORGANIC PIGMENTS, SELF-HEATING	3313	4.2	
ORGANOARSENIC COMPOUND, LIQUID, N.O.S.	3280	6.1	
ORGANOARSENIC COMPOUND, SOLID, N.O.S.	3465	6.1	

Name and description	UN No.	Class	Remarks
ORGANOCHLORINE PESTICIDE, LIQUID, FLAMMABLE, TOXIC, flash-point less than 23 °C	2762	3	
ORGANOCHLORINE PESTICIDE, LIQUID, TOXIC	2996	6.1	
ORGANOCHLORINE PESTICIDE, LIQUID, TOXIC, FLAMMABLE, flash-point not less than 23 °C	2995	6.1	
ORGANOCHLORINE PESTICIDE, SOLID, TOXIC	2761	6.1	
ORGANOMETALLIC COMPOUND, TOXIC, LIQUID, N.O.S.	3282	6.1	
ORGANOMETALLIC COMPOUND, TOXIC, SOLID, N.O.S.	3467	6.1	
Organometallic compound, solid, water-reactive, flammable, n.o.s., see	3396	4.3	
Organometallic compound or Organometallic compound solution or Organometallic compound dispersion, water-reactive, flammable, n.o.s., see	3399	4.3	
ORGANOMETALLIC SUBSTANCE, LIQUID, PYROPHORIC	3392	4.2	
ORGANOMETALLIC SUBSTANCE, SOLID, PYROPHORIC	3391	4.2	
ORGANOMETALLIC SUBSTANCE, SOLID, SELF-HEATING	3400	4.2	
ORGANOMETALLIC SUBSTANCE, LIQUID, PYROPHORIC, WATER-REACTIVE	3394	4.2	
ORGANOMETALLIC SUBSTANCE, SOLID, PYROPHORIC, WATER-REACTIVE	3393	4.2	
ORGANOMETALLIC SUBSTANCE, LIQUID, WATER-REACTIVE	3398	4.3	
ORGANOMETALLIC SUBSTANCE, SOLID, WATER-REACTIVE	3395	4.3	
ORGANOMETALLIC SUBSTANCE, LIQUID, WATER-REACTIVE, FLAMMABLE	3399	4.3	

Name and description	UN No.	Class	Remarks
ORGANOMETALLIC SUBSTANCE, SOLID, WATER-REACTIVE, FLAMMABLE	3396	4.3	
ORGANOMETALLIC SUBSTANCE, SOLID, WATER-REACTIVE, SELF-HEATING	3397	4.3	
ORGANOPHOSPHORUS COMPOUND, TOXIC, LIQUID, N.O.S.	3278	6.1	
ORGANOPHOSPHORUS COMPOUND, TOXIC, SOLID, N.O.S.	3464	6.1	
ORGANOPHOSPHORUS COMPOUND, TOXIC, FLAMMABLE, N.O.S.	3279	6.1	
ORGANOPHOSPHORUS PESTICIDE, LIQUID, FLAMMABLE, TOXIC, flash-point less than 23 °C	2784	3	
ORGANOPHOSPHORUS PESTICIDE, LIQUID, TOXIC	3018	6.1	
ORGANOPHOSPHORUS PESTICIDE, LIQUID, TOXIC, FLAMMABLE, flash-point not less than 23 °C	3017	6.1	
ORGANOPHOSPHORUS PESTICIDE, SOLID, TOXIC	2783	6.1	
ORGANOTIN COMPOUND, LIQUID, N.O.S.	2788	6.1	
ORGANOTIN COMPOUND, SOLID, N.O.S.	3146	6.1	
ORGANOTIN PESTICIDE, LIQUID, FLAMMABLE, TOXIC, flash-point less than 23 °C	2787	3	
ORGANOTIN PESTICIDE, LIQUID, TOXIC	3020	6.1	
ORGANOTIN PESTICIDE, LIQUID, TOXIC, FLAMMABLE, flash-point not less than 23 °C	3019	6.1	
ORGANOTIN PESTICIDE, SOLID, TOXIC	2786	6.1	
Orthophospohoric acid, see	1805	8	
OSMIUM TETROXIDE	2471	6.1	
OXIDIZING LIQUID, N.O.S.	3139	5.1	
OXIDIZING LIQUID, CORROSIVE, N.O.S.	3098	5.1	
OXIDIZING LIQUID, TOXIC, N.O.S.	3099	5.1	
OXIDIZING SOLID, N.O.S.	1479	5.1	
OXIDIZING SOLID, CORROSIVE, N.O.S.	3085	5.1	
OXIDIZING SOLID, FLAMMABLE, N.O.S.	3137	5.1	Carriage prohibited
OXIDIZING SOLID, SELF-HEATING, N.O.S.	3100	5.1	Carriage prohibited
OXIDIZING SOLID, TOXIC, N.O.S.	3087	5.1	
OXIDIZING SOLID, WATER-REACTIVE, N.O.S.	3121	5.1	Carriage prohibited
Oxirane, see	1040	2	
OXYGEN, COMPRESSED	1072	2	
OXYGEN DIFLUORIDE, COMPRESSED	2190	2	
OXYGEN GENERATOR, CHEMICAL	3356	5.1	
OXYGEN, REFRIGERATED LIQUID	1073	2	
1-Oxy-4-nitrobenzene, see	1663	6.1	
PAINT (including paint, lacquer, enamel, stain, shellac, varnish, polish, liquid filler and liquid lacquer base)	1263 3066 3469 3470	3 8 3 8	
PAINT RELATED MATERIAL (including paint thinning and reducing compound)	1263 3066 3469 3470	3 8 3 8	
Paint thinning and reducing compound, see	1263 3066 3469 3470	3 8 3 8	
PAPER, UNSATURATED OIL TREATED, incompletely dried (including carbon paper)	1379	4.2	
Paraffin, see	1223	3	
PARAFORMALDEHYDE	2213	4.1	
PARALDEHYDE	1264	3	
PCBs, see	2315 3432	9 9	
PENTABORANE	1380	4.2	
PENTACHLOROETHANE	1669	6.1	

Name and description	UN No.	Class	Remarks
PENTACHLOROPHENOL	3155	6.1	
PENTAERYTHRITE TETRANITRATE with not less than 7% wax, by mass	0411	1	
PENTAERYTHRITE TETRANITRATE, DESENSITIZED with not less than 15% phlegmatizer, by mass	0150	1	
PENTAERYTHRITE TETRANITRATE MIXTURE, DESENSITIZED, SOLID, N.O.S. with more than 10% but not more than 20% PETN, by mass	3344	4.1	
PENTAERYTHRITE TETRANITRATE, WETTED with not less than 25% water, by mass	0150	1	
PENTAERYTHRITOL TETRANITRATE, see	0150 0411 3344	1 1 4.1	
PENTAFLUOROETHANE	3220	2	
Pentafluoroethane, 1,1,1-trifluoroethane, and 1,1,1,2-tetrafluoroethane zeotropic mixture with approximately 44% pentafluoroethane and 52% 1,1,1-trifluoroethane, see	3337	2	
PENTAMETHYLHEPTANE	2286	3	
Pentanal, see	2058	3	
PENTANE-2,4-DIONE	2310	3	
PENTANES, liquid	1265	3	
n-Pentane, see	1265	3	
PENTANOLS	1105	3	
n-Pentanol, see	1105	3	
3-Pentanol, see	1105	3	
1-PENTENE	1108	3	
1-PENTOL	2705	8	
PENTOLITE, dry or wetted with less than 15% water, by mass	0151	1	
Pentyl nitrite, see	1113	3	
PERCHLORATES, INORGANIC, N.O.S.	1481	5.1	
PERCHLORATES, INORGANIC, AQUEOUS SOLUTION, N.O.S.	3211	5.1	
PERCHLORIC ACID with more than 50% but not more than 72% acid, by mass	1873	5.1	
PERCHLORIC ACID with not more than 50% acid, by mass	1802	8	
Perchlorobenzene, see	2729	6.1	
Perchlorocyclopentadiene, see	2646	6.1	
Perchloroethylene, see	1897	6.1	
PERCHLOROMETHYL MERCAPTAN	1670	6.1	
PERCHLORYL FLUORIDE	3083	2	
Perfluoroacetylchloride, see	3057	2	
PERFLUORO(ETHYL VINYL ETHER)	3154	2	
PERFLUORO(METHYL VINYL ETHER)	3153	2	
Perfluoropropane, see	2424	2	
PERFUMERY PRODUCTS with flammable solvents	1266	3	
PERMANGANATES, INORGANIC, N.O.S.	1482	5.1	
PERMANGANATES, INORGANIC, AQUEOUS SOLUTION, N.O.S.	3214	5.1	
PEROXIDES, INORGANIC, N.O.S.	1483	5.1	
PERSULPHATES, INORGANIC, N.O.S.	3215	5.1	
PERSULPHATES, INORGANIC, AQUEOUS SOLUTION, N.O.S.	3216	5.1	
PESTICIDE, LIQUID, FLAMMABLE, TOXIC, N.O.S., flash-point less than 23 °C	3021	3	
PESTICIDE, LIQUID, TOXIC, N.O.S.	2902	6.1	
PESTICIDE, LIQUID, TOXIC, FLAMMABLE, N.O.S., flash-point not less than 23 °C	2903	6.1	
PESTICIDE, SOLID, TOXIC, N.O.S.	2588	6.1	
Pesticide, toxic, under compressed gas, n.o.s, see	1950	2	

Name and description	UN No.	Class	Remarks
PETN, see	0150 0411 3344	1 1 4.1	
PETN/TNT, see	0151	1	
PETROL	1203	3	
Petrol and ethanol mixture, with more than 10% ethanol, see	3475	3	
PETROLEUM CRUDE OIL	1267	3	
PETROLEUM DISTILLATES, N.O.S.	1268	3	
Petroleum ether, see	1268	3	
PETROLEUM GASES, LIQUEFIED	1075	2	
Petroleum naphtha, see	1268	3	
Petroleum oil, see	1268	3	
PETROLEUM PRODUCTS, N.O.S.	1268	3	
Petroleum raffinate, see	1268	3	
Petroleum spirit, see	1268	3	
PHENACYL BROMIDE	2645	6.1	
PHENETIDINES	2311	6.1	
PHENOLATES, LIQUID	2904	8	
PHENOLATES, SOLID	2905	8	
PHENOL, MOLTEN	2312	6.1	
PHENOL, SOLID	1671	6.1	
PHENOL SOLUTION	2821	6.1	
PHENOLSULPHONIC ACID, LIQUID	1803	8	
PHENOXYACETIC ACID DERIVATIVE PESTICIDE, LIQUID, FLAMMABLE, TOXIC, flash-point less than 23 °C	3346	3	
PHENOXYACETIC ACID DERIVATIVE PESTICIDE, LIQUID, TOXIC	3348	6.1	
PHENOXYACETIC ACID DERIVATIVE PESTICIDE, LIQUID, TOXIC, FLAMMABLE, flash-point not less than 23 °C	3347	6.1	
PHENOXYACETIC ACID DERIVATIVE PESTICIDE, SOLID, TOXIC	3345	6.1	

Name and description	UN No.	Class	Remarks
PHENYLACETONITRILE, LIQUID	2470	6.1	
PHENYLACETYL CHLORIDE	2577	8	
Phenylamine, see	1547	6.1	
1-Phenylbutane, see	2709	3	
2-Phenylbutane, see	2709	3	
PHENYLCARBYLAMINE CHLORIDE	1672	6.1	
PHENYL CHLOROFORMATE	2746	6.1	
Phenyl cyanide, see	2224	6.1	
PHENYLENEDIAMINES (o-, m-, p-)	1673	6.1	
Phenylethylene, see	2055	3	
PHENYLHYDRAZINE	2572	6.1	
PHENYL ISOCYANATE	2487	6.1	
Phenylisocyanodichloride, see	1672	6.1	
PHENYL MERCAPTAN	2337	6.1	
PHENYLMERCURIC ACETATE	1674	6.1	
PHENYLMERCURIC COMPOUND, N.O.S.	2026	6.1	
PHENYLMERCURIC HYDROXIDE	1894	6.1	
PHENYLMERCURIC NITRATE	1895	6.1	
PHENYLPHOSPHORUS DICHLORIDE	2798	8	
PHENYLPHOSPHORUS THIODICHLORIDE	2799	8	
2-Phenylpropene, see	2303	3	
PHENYLTRICHLOROSILANE	1804	8	
PHOSGENE	1076	2	
9-PHOSPHABICYCLONONANES	2940	4.2	
PHOSPHINE	2199	2	
Phosphoretted hydrogen, see	2199	2	
PHOSPHORIC ACID, SOLUTION	1805	8	
PHOSPHORIC ACID, SOLID	3453	8	
Phosphoric acid, anhydrous, see	1807	8	
PHOSPHOROUS ACID	2834	8	

Name and description	UN No.	Class	Remarks
PHOSPHORUS, AMORPHOUS	1338	4.1	
Phosphorus bromide, see	1808	8	
Phosphorus chloride, see	1809	6.1	
PHOSPHORUS HEPTASULPHIDE, free from yellow and white phosphorus	1339	4.1	
PHOSPHORUS OXYBROMIDE	1939	8	
PHOSPHORUS OXYBROMIDE, MOLTEN	2576	8	
PHOSPHORUS OXYCHLORIDE	1810	8	
PHOSPHORUS PENTABROMIDE	2691	8	
PHOSPHORUS PENTACHLORIDE	1806	8	
PHOSPHORUS PENTAFLUORIDE	2198	2	
PHOSPHORUS PENTASULPHIDE, free from yellow and white phosphorus	1340	4.3	
PHOSPHORUS PENTOXIDE	1807	8	
PHOSPHORUS SESQUISULPHIDE, free from yellow and white phosphorus	1341	4.1	
Phosphorus (V) sulphide, free from yellow and white phosphorus, see	1340	4.3	
Phosphorus sulphochloride, see	1837	8	
PHOSPHORUS TRIBROMIDE	1808	8	
PHOSPHORUS TRICHLORIDE	1809	6.1	
PHOSPHORUS TRIOXIDE	2578	8	
PHOSPHORUS TRISULPHIDE, free from yellow and white phosphorus	1343	4.1	
PHOSPHORUS, WHITE, DRY	1381	4.2	
PHOSPHORUS, WHITE IN SOLUTION	1381	4.2	
PHOSPHORUS, WHITE, MOLTEN	2447	4.2	
PHOSPHORUS, WHITE, UNDER WATER	1381	4.2	
PHOSPHORUS, YELLOW, DRY	1381	4.2	
PHOSPHORUS, YELLOW, IN SOLUTION	1381	4.2	
PHOSPHORUS, YELLOW, UNDER WATER	1381	4.2	

Name and description	UN No.	Class	Remarks
Phosphoryl chloride, see	1810	8	
PHTHALIC ANHYDRIDE with more than 0.05% of maleic anhydride	2214	8	
PICOLINES	2313	3	
PICRAMIDE, see	0153	1	
PICRIC ACID WETTED, see	1344 3364	4.1 4.1	
PICRITE, see	0282	1	
PICRITE, WETTED, see	1336	4.1	
Picrotoxin, see	3172 3462	6.1 6.1	
PICRYL CHLORIDE, see	0155	1	
PICRYL CHLORIDE, WETTED, see	3365	4.1	
alpha-PINENE	2368	3	
PINE OIL	1272	3	
PIPERAZINE	2579	8	
PIPERIDINE	2401	8	
Pivaloyl chloride, see	2438	6.1	
Plastic explosives , see	0084	1	
PLASTICS MOULDING COMPOUND in dough, sheet or extruded rope form evolving flammable vapour	3314	9	
PLASTICS, NITROCELLULOSE-BASED, SELF-HEATING, N.O.S.	2006	4.2	
Polish, see	1263 3066 3469 3470	3 8 3 8	
POLYAMINES, FLAMMABLE, CORROSIVE, N.O.S.	2733	3	
POLYAMINES, LIQUID, CORROSIVE, N.O.S.	2735	8	
POLYAMINES, LIQUID, CORROSIVE, FLAMMABLE, N.O.S.	2734	8	
POLYAMINES, SOLID, CORROSIVE, N.O.S.	3259	8	
POLYCHLORINATED BIPHENYLS, LIQUID	2315	9	
POLYCHLORINATED BIPHENYLS, SOLID	3432	9	
POLYESTER RESIN KIT	3269	3	

Name and description	UN No.	Class	Remarks
POLYHALOGENATED BIPHENYLS, LIQUID	3151	9	
POLYHALOGENATED BIPHENYLS, SOLID	3152	9	
POLYHALOGENATED TERPHENYLS, LIQUID	3151	9	
POLYHALOGENATED TERPHENYLS, SOLID	3152	9	
POLYMERIC BEADS, EXPANDABLE, evolving flammable vapour	2211	9	
Polystyrene beads, expandable, see	2211	9	
POTASSIUM	2257	4.3	
POTASSIUM ARSENATE	1677	6.1	
POTASSIUM ARSENITE	1678	6.1	
Potassium bifluoride, see	1811	8	
Potassium bisulphate, see	2509	8	
Potassium bisulphite solution, see	2693	8	
POTASSIUM BOROHYDRIDE	1870	4.3	
POTASSIUM BROMATE	1484	5.1	
POTASSIUM CHLORATE	1485	5.1	
POTASSIUM CHLORATE, AQUEOUS SOLUTION	2427	5.1	
Potassium chlorate mixed with mineral oil, see	0083	1	
POTASSIUM CUPROCYANIDE	1679	6.1	
POTASSIUM CYANIDE, SOLID	1680	6.1	
POTASSIUM CYANIDE, SOLUTION	3413	6.1	
Potassium dicyanocuprate (I), see	1679	6.1	
POTASSIUM DITHIONITE	1929	4.2	
POTASSIUM FLUORIDE, SOLID	1812	6.1	
POTASSIUM FLUORIDE, SOLUTION	3422	6.1	
POTASSIUM FLUOROACETATE	2628	6.1	
POTASSIUM FLUOROSILICATE	2655	6.1	
Potassium hexafluorosilicate, see	2655	6.1	
Potassium hydrate, see	1814	8	

Name and description	UN No.	Class	Remarks
POTASSIUM HYDROGENDIFLUORIDE, SOLID	1811	8	
POTASSIUM HYDROGENDIFLUORIDE, SOLUTION	3421	8	
POTASSIUM HYDROGEN SULPHATE	2509	8	
POTASSIUM HYDROSULPHITE, see	1929	4.2	
Potassium hydroxide, liquid, see	1814	8	
POTASSIUM HYDROXIDE, SOLID	1813	8	
POTASSIUM HYDROXIDE SOLUTION	1814	8	
POTASSIUM METAL ALLOYS, LIQUID	1420	4.3	
POTASSIUM METAL ALLOYS, SOLID	3403	4.3	
POTASSIUM METAVANADATE	2864	6.1	
POTASSIUM MONOXIDE	2033	8	
POTASSIUM NITRATE	1486	5.1	
Potassium nitrate and sodium nitrate mixture, see	1499	5.1	
POTASSIUM NITRATE AND SODIUM NITRITE MIXTURE	1487	5.1	
POTASSIUM NITRITE	1488	5.1	
POTASSIUM PERCHLORATE	1489	5.1	
POTASSIUM PERMANGANATE	1490	5.1	
POTASSIUM PEROXIDE	1491	5.1	
POTASSIUM PERSULPHATE	1492	5.1	
POTASSIUM PHOSPHIDE	2012	4.3	
Potassium selenate, see	2630	6.1	
Potassium selenite, see	2630	6.1	
Potassium silicofluoride, see	2655	6.1	
POTASSIUM SODIUM ALLOYS, LIQUID	1422	4.3	
POTASSIUM SODIUM ALLOYS, SOLID	3404	4.3	
POTASSIUM SULPHIDE with less than 30% water of crystallization	1382	4.2	

Name and description	UN No.	Class	Remarks
POTASSIUM SULPHIDE, ANHYDROUS	1382	4.2	
POTASSIUM SULPHIDE, HYDRATED with not less than 30% water of crystallization	1847	8	
POTASSIUM SUPEROXIDE	2466	5.1	
Potassium tetracyano-mercurate (II), see	1626	6.1	
POWDER CAKE, WETTED with not less than 17% alcohol, by mass	0433	1	
POWDER CAKE, WETTED with not less than 25% water, by mass	0159	1	
POWDER PASTE, see	0159	1	
	0433	1	
POWDER, SMOKELESS	0160	1	
	0161	1	
Power devices, explosive, see	0275	1	
	0276	1	
	0323	1	
	0381	1	
PRIMERS, CAP TYPE	0044	1	
	0377	1	
	0378	1	
Primers, small arms, see	0044	1	
PRIMERS, TUBULAR	0319	1	
	0320	1	
	0376	1	
PRINTING INK, flammable or PRINTING INK RELATED MATERIAL (including printing ink thinning or reducing compound), flammable	1210	3	
Projectiles, illuminating, see	0171	1	
	0254	1	
	0297	1	
PROJECTILES, inert with tracer	0345	1	
	0424	1	
	0425	1	
PROJECTILES with burster or expelling charge	0346	1	
	0347	1	
	0426	1	
	0427	1	
	0434	1	
	0435	1	
PROJECTILES with bursting charge	0167	1	
	0168	1	
	0169	1	
	0324	1	
	0344	1	

Name and description	UN No.	Class	Remarks
PROPADIENE, STABILIZED	2200	2	
Propadiene and methyl acetylene mixture, stabilized, see	1060	2	
PROPANE	1978	2	
PROPANETHIOLS	2402	3	
n-PROPANOL	1274	3	
PROPELLANT, LIQUID	0495	1	
	0497	1	
PROPELLANT, SOLID	0498	1	
	0499	1	
	0501	1	
Propellant with a single base, Propellant with a double base, Propellant with a triple base, see	0160	1	
	0161	1	
Propene, see	1077	2	
PROPIONALDEHYDE	1275	3	
PROPIONIC ACID with not less than 10% and less than 90% acid by mass	1848	8	
PROPIONIC ACID with not less than 90% acid by mass	3463	8	
PROPIONIC ANHYDRIDE	2496	8	
PROPIONITRILE	2404	3	
PROPIONYL CHLORIDE	1815	3	
n-PROPYL ACETATE	1276	3	
PROPYL ALCOHOL, NORMAL, see	1274	3	
PROPYLAMINE	1277	3	
n-PROPYLBENZENE	2364	3	
Propyl chloride, see	1278	3	
n-PROPYL CHLOROFORMATE	2740	6.1	
PROPYLENE	1077	2	
PROPYLENE CHLOROHYDRIN	2611	6.1	
1,2-PROPYLENEDIAMINE	2258	8	
Propylene dichloride, see	1279	3	
PROPYLENEIMINE, STABILIZED	1921	3	
PROPYLENE OXIDE	1280	3	

Name and description	UN No.	Class	Remarks
PROPYLENE TETRAMER	2850	3	
Propylene trimer, see	2057	3	
PROPYL FORMATES	1281	3	
n-PROPYL ISOCYANATE	2482	6.1	
Propyl mercaptan, see	2402	3	
n-PROPYL NITRATE	1865	3	
PROPYLTRICHLOROSILANE	1816	8	
Pyrazine hexahydride, see	2579	8	
PYRETHROID PESTICIDE, LIQUID, FLAMMABLE, TOXIC, flash-point less than 23 °C	3350	3	
PYRETHROID PESTICIDE, LIQUID, TOXIC	3352	6.1	
PYRETHROID PESTICIDE, LIQUID, TOXIC, FLAMMABLE, flash-point not less than 23 °C	3351	6.1	
PYRETHROID PESTICIDE, SOLID, TOXIC	3349	6.1	
PYRIDINE	1282	3	
Pyrophoric organometallic compound, water-reactive, n.o.s., liquid, see	3394	4.2	
Pyrophoric organometallic compound, water-reactive, n.o.s., solid, see	3393	4.2	
PYROPHORIC ALLOY, N.O.S.	1383	4.2	
PYROPHORIC LIQUID, INORGANIC, N.O.S.	3194	4.2	
PYROPHORIC LIQUID, ORGANIC, N.O.S.	2845	4.2	
PYROPHORIC METAL, N.O.S.	1383	4.2	
PYROPHORIC SOLID, INORGANIC, N.O.S.	3200	4.2	
PYROPHORIC SOLID, ORGANIC, N.O.S.	2846	4.2	
PYROSULPHURYL CHLORIDE	1817	8	
Pyroxylin solution, see	2059	3	
PYRROLIDINE	1922	3	
QUINOLINE	2656	6.1	
Quinone, see	2587	6.1	
RADIOACTIVE MATERIAL, EXCEPTED PACKAGE - ARTICLES MANUFACTURED FROM NATURAL URANIUM or DEPLETED URANIUM or NATURAL THORIUM	2909	7	
RADIOACTIVE MATERIAL, EXCEPTED PACKAGE - EMPTY PACKAGING	2908	7	
RADIOACTIVE MATERIAL, EXCEPTED PACKAGE - INSTRUMENTS or ARTICLES	2911	7	
RADIOACTIVE MATERIAL, EXCEPTED PACKAGE - LIMITED QUANTITY OF MATERIAL	2910	7	
RADIOACTIVE MATERIAL, LOW SPECIFIC ACTIVITY (LSA-I), non fissile or fissile-excepted	2912	7	
RADIOACTIVE MATERIAL, LOW SPECIFIC ACTIVITY (LSA-II), FISSILE	3324	7	
RADIOACTIVE MATERIAL, LOW SPECIFIC ACTIVITY (LSA-II), non fissile or fissile-excepted	3321	7	
RADIOACTIVE MATERIAL, LOW SPECIFIC ACTIVITY, (LSA-III), FISSILE	3325	7	
RADIOACTIVE MATERIAL, LOW SPECIFIC ACTIVITY (LSA-III), non fissile or fissile-excepted	3322	7	
RADIOACTIVE MATERIAL, SURFACE CONTAMINATED OBJECTS (SCO-I or SCO-II), FISSILE	3326	7	
RADIOACTIVE MATERIAL, SURFACE CONTAMINATED OBJECTS (SCO-I or SCO-II), non fissile or fissile-excepted	2913	7	
RADIOACTIVE MATERIAL, TRANSPORTED UNDER SPECIAL ARRANGEMENT, FISSILE	3331	7	
RADIOACTIVE MATERIAL, TRANSPORTED UNDER SPECIAL ARRANGEMENT, non fissile or fissile-excepted	2919	7	

Name and description	UN No.	Class	Remarks
RADIOACTIVE MATERIAL, TYPE A PACKAGE, FISSILE, non-special form	3327	7	
RADIOACTIVE MATERIAL, TYPE A PACKAGE, non-special form, non fissile or fissile-excepted	2915	7	
RADIOACTIVE MATERIAL, TYPE A PACKAGE, SPECIAL FORM, FISSILE	3333	7	
RADIOACTIVE MATERIAL, TYPE A PACKAGE, SPECIAL FORM, non fissile or fissile-excepted	3332	7	
RADIOACTIVE MATERIAL, TYPE B(M) PACKAGE, FISSILE	3329	7	
RADIOACTIVE MATERIAL, TYPE B(M) PACKAGE, non fissile or fissile-excepted	2917	7	
RADIOACTIVE MATERIAL, TYPE B(U) PACKAGE, FISSILE	3328	7	
RADIOACTIVE MATERIAL, TYPE B(U) PACKAGE, non fissile or fissile-excepted	2916	7	
RADIOACTIVE MATERIAL, TYPE C PACKAGE, FISSILE	3330	7	
RADIOACTIVE MATERIAL, TYPE C PACKAGE, non fissile or fissile-excepted	3323	7	
RADIOACTIVE MATERIAL, URANIUM HEXAFLUORIDE, FISSILE	2977	7	
RADIOACTIVE MATERIAL, URANIUM HEXAFLUORIDE, non fissile or fissile-excepted	2978	7	
Rags, oily	1856	4.2	Not subject to ADN
RDX, see	0072 0391 0483	1 1 1	
RECEPTACLES, SMALL, CONTAINING GAS without a release device, non-refillable	2037	2	
Red phosphorus, see	1338	4.1	
REFRIGERANT GAS, N.O.S., such as mixture F1, mixture F2 or mixture P2	1078	2	
REFRIGERANT GAS R 12, see	1028	2	
REFRIGERANT GAS R 12B1, see	1974	2	

Name and description	UN No.	Class	Remarks
REFRIGERANT GAS R 13, see	1022	2	
REFRIGERANT GAS R 13B1, see	1009	2	
REFRIGERANT GAS R 14, see	1982	2	
REFRIGERANT GAS R 21, see	1029	2	
REFRIGERANT GAS R 22, see	1018	2	
REFRIGERANT GAS R 23, see	1984	2	
REFRIGERANT GAS R 32, see	3252	2	
REFRIGERANT GAS R 40, see	1063	2	
REFRIGERANT GAS R 41, see	2454	2	
REFRIGERANT GAS R 114, see	1958	2	
REFRIGERANT GAS R 115, see	1020	2	
REFRIGERANT GAS R 116, see	2193	2	
REFRIGERANT GAS R 124, see	1021	2	
REFRIGERANT GAS R 125, see	3220	2	
REFRIGERANT GAS R 133a, see	1983	2	
REFRIGERANT GAS R 134a, see	3159	2	
REFRIGERANT GAS R 142b, see	2517	2	
REFRIGERANT GAS R 143a, see	2035	2	
REFRIGERANT GASR 152a, see	1030	2	
REFRIGERANT GAS R 161, see	2453	2	
REFRIGERANT GAS R 218, see	2424	2	
REFRIGERANT GAS R 227, see	3296	2	
REFRIGERANT GAS R 404A	3337	2	
REFRIGERANT GAS R 407A	3338	2	
REFRIGERANT GAS R 407B	3339	2	
REFRIGERANT GAS R 407C	3340	2	
REFRIGERANT GAS R 500, see	2602	2	
REFRIGERANT GAS R 502, see	1973	2	
REFRIGERANT GAS R 503, see	2599	2	
REFRIGERANT GAS R 1132a, see	1959	2	
REFRIGERANT GAS R 1216, see	1858	2	
REFRIGERANT GAS R 1318, see	2422	2	
REFRIGERANT GAS RC 318, see	1976	2	

Name and description	UN No.	Class	Remarks
REFRIGERATING MACHINES containing flammable, non-toxic, liquefied gas	3358	2	
REFRIGERATING MACHINES containing non-flammable, non-toxic, gases or ammonia solutions (UN 2672)	2857	2	
REGULATED MEDICAL WASTE, N.O.S.	3291	6.2	
RELEASE DEVICES, EXPLOSIVE	0173	1	
RESIN SOLUTION, flammable	1866	3	
Resorcin, see	2876	6.1	
RESORCINOL	2876	6.1	
RIVETS, EXPLOSIVE	0174	1	
ROCKET MOTORS	0186 0280 0281	1 1 1	
ROCKET MOTORS, LIQUID FUELLED	0395 0396	1 1	
ROCKET MOTORS WITH HYPERGOLIC LIQUIDS with or without expelling charge	0250 0322	1 1	
ROCKETS with bursting charge	0180 0181 0182 0295	1 1 1 1	
ROCKETS with expelling charge	0436 0437 0438	1 1 1	
ROCKETS with inert head	0183 0502	1 1	
ROCKETS, LINE-THROWING	0238 0240 0453	1 1 1	
ROCKETS, LIQUID FUELLED with bursting charge	0397 0398	1 1	
ROSIN OIL	1286	3	
RUBBER SCRAP, powdered or granulated	1345	4.1	
RUBBER SHODDY, powdered or granulated	1345	4.1	
RUBBER SOLUTION	1287	3	
RUBIDIUM	1423	4.3	
RUBIDIUM HYDROXIDE	2678	8	
RUBIDIUM HYDROXIDE SOLUTION	2677	8	
Saltpetre, see	1486	5.1	
SAMPLES, EXPLOSIVE, other than initiating explosive	0190	1	
Sand acid, see	1778	8	
SEAT-BELT PRETENSIONERS	0503 3268	1 9	
SEED CAKE with more than 1.5% oil and not more than 11% moisture	1386	4.2	
SEED CAKE with not more than 1.5% oil and not more than 11% moisture	2217	4.2	
Seed expellers, see	1386 2217	4.2 4.2	
SELENATES	2630	6.1	
SELENIC ACID	1905	8	
SELENITES	2630	6.1	
SELENIUM COMPOUND, LIQUID, N.O.S.	3440	6.1	
SELENIUM COMPOUND, SOLID, N.O.S.	3283	6.1	
SELENIUM DISULPHIDE	2657	6.1	
SELENIUM HEXAFLUORIDE	2194	2	
SELENIUM OXYCHLORIDE	2879	8	
SELF-HEATING LIQUID, CORROSIVE, INORGANIC, N.O.S.	3188	4.2	
SELF-HEATING LIQUID, CORROSIVE, ORGANIC, N.O.S.	3185	4.2	
SELF-HEATING LIQUID, INORGANIC, N.O.S.	3186	4.2	
SELF-HEATING LIQUID, ORGANIC, N.O.S.	3183	4.2	
SELF-HEATING LIQUID, TOXIC, INORGANIC, N.O.S.	3187	4.2	
SELF-HEATING LIQUID, TOXIC, ORGANIC, N.O.S.	3184	4.2	
SELF-HEATING SOLID, CORROSIVE, INORGANIC, N.O.S.	3192	4.2	
SELF-HEATING SOLID, CORROSIVE, ORGANIC, N.O.S.	3126	4.2	

Name and description	UN No.	Class	Remarks
SELF-HEATING SOLID, INORGANIC, N.O.S.	3190	4.2	
SELF-HEATING SOLID, ORGANIC, N.O.S.	3088	4.2	
SELF-HEATING SOLID, OXIDIZING, N.O.S	3127	4.2	Carriage prohibited
SELF-HEATING SOLID, TOXIC, INORGANIC, N.O.S.	3191	4.2	
SELF-HEATING SOLID, TOXIC, ORGANIC, N.O.S.	3128	4.2	
SELF-REACTIVE LIQUID TYPE B	3221	4.1	
SELF-REACTIVE LIQUID TYPE B, TEMPERATURE CONTROLLED	3231	4.1	
SELF-REACTIVE LIQUID TYPE C	3223	4.1	
SELF-REACTIVE LIQUID TYPE C, TEMPERATURE CONTROLLED	3233	4.1	
SELF-REACTIVE LIQUID TYPE D	3225	4.1	
SELF-REACTIVE LIQUID TYPE D, TEMPERATURE CONTROLLED	3235	4.1	
SELF-REACTIVE LIQUID TYPE E	3227	4.1	
SELF-REACTIVE LIQUID TYPE E, TEMPERATURE CONTROLLED	3237	4.1	
SELF-REACTIVE LIQUID TYPE F	3229	4.1	
SELF-REACTIVE LIQUID TYPE F, TEMPERATURE CONTROLLED	3239	4.1	
SELF-REACTIVE SOLID TYPE B	3222	4.1	
SELF-REACTIVE SOLID TYPE B, TEMPERATURE CONTROLLED	3232	4.1	
SELF-REACTIVE SOLID TYPE C	3224	4.1	
SELF-REACTIVE SOLID TYPE C, TEMPERATURE CONTROLLED	3234	4.1	
SELF-REACTIVE SOLID TYPE D	3226	4.1	
SELF-REACTIVE SOLID TYPE D, TEMPERATURE CONTROLLED	3236	4.1	
SELF-REACTIVE SOLID TYPE E	3228	4.1	
SELF-REACTIVE SOLID TYPE E, TEMPERATURE CONTROLLED	3238	4.1	
SELF-REACTIVE SOLID TYPE F	3230	4.1	
SELF-REACTIVE SOLID TYPE F, TEMPERATURE CONTROLLED	3240	4.1	
SHALE OIL	1288	3	
Shaped charges, see	0059	1	
	0439	1	
	0440	1	
	0441	1	
Shellac, see	1263	3	
	3066	8	
	3469	3	
	3470	8	
SIGNAL DEVICES, HAND	0191	1	
	0373	1	
SIGNALS, DISTRESS, ship	0194	1	
	0195	1	
	0505	1	
	0506	1	
Signals, distress, ship, water-activated, see	0249	1	
SIGNALS, RAILWAY TRACK, EXPLOSIVE	0192	1	
	0193	1	
	0492	1	
	0493	1	
SIGNALS, SMOKE	0196	1	
	0197	1	
	0313	1	
	0487	1	
	0507	1	
SILANE	2203	2	
Silicofluoric acid, see	1778	8	
Silicofluorides, n.o.s., see	2856	6.1	
Silicon chloride, see	1818	8	
SILICON POWDER, AMORPHOUS	1346	4.1	
SILICON TETRACHLORIDE	1818	8	
SILICON TETRAFLUORIDE	1859	2	
SILVER ARSENITE	1683	6.1	

Name and description	UN No.	Class	Remarks
SILVER CYANIDE	1684	6.1	
SILVER NITRATE	1493	5.1	
SILVER PICRATE, WETTED with not less than 30% water, by mass	1347	4.1	
SLUDGE ACID	1906	8	
SODA LIME with more than 4% sodium hydroxide	1907	8	
SODIUM	1428	4.3	
Sodium aluminate, solid	2812	8	Not subject to ADN
SODIUM ALUMINATE SOLUTION	1819	8	
SODIUM ALUMINIUM HYDRIDE	2835	4.3	
SODIUM AMMONIUM VANADATE	2863	6.1	
SODIUM ARSANILATE	2473	6.1	
SODIUM ARSENATE	1685	6.1	
SODIUM ARSENITE, AQUEOUS SOLUTION	1686	6.1	
SODIUM ARSENITE, SOLID	2027	6.1	
SODIUM AZIDE	1687	6.1	
Sodium bifluoride, see	2439	8	
Sodium binoxide, see	1504	5.1	
Sodium bisulphite solution, see	2693	8	
SODIUM BOROHYDRIDE	1426	4.3	
SODIUM BOROHYDRIDE AND SODIUM HYDROXIDE SOLUTION, with not more than 12% sodium borohydride and not more than 40% sodium hydroxide by mass	3320	8	
SODIUM BROMATE	1494	5.1	
SODIUM CACODYLATE	1688	6.1	
SODIUM CARBONATE PEROXYHYDRATE	3378	5.1	
SODIUM CHLORATE	1495	5.1	
SODIUM CHLORATE, AQUEOUS SOLUTION	2428	5.1	
Sodium chlorate mixed with dinitrotoluene, see	0083	1	
SODIUM CHLORITE	1496	5.1	
SODIUM CHLOROACETATE	2659	6.1	
SODIUM CUPROCYANIDE, SOLID	2316	6.1	
SODIUM CUPROCYANIDE SOLUTION	2317	6.1	
SODIUM CYANIDE, SOLID	1689	6.1	
SODIUM CYANIDE, SOLUTION	3414	6.1	
Sodium dicyanocuprate (I), solid, see	2316	6.1	
Sodium dicyanocuprate (I) solution, see	2317	6.1	
Sodium dimethylarsenate, see	1688	6.1	
SODIUM DINITRO-o-CRESOLATE, dry or wetted with less than 15% water, by mass	0234	1	
SODIUM DINITRO-o-CRESOLATE, WETTED with not less than 10% water, by mass	3369	4.1	
SODIUM DINITRO-o-CRESOLATE, WETTED with not less than 15% water, by mass	1348	4.1	
Sodium dioxide, see	1504	5.1	
SODIUM DITHIONITE	1384	4.2	
SODIUM FLUORIDE, SOLID	1690	6.1	
SODIUM FLUORIDE, SOLUTION	3415	6.1	
SODIUM FLUOROACETATE	2629	6.1	
SODIUM FLUOROSILICATE	2674	6.1	
Sodium hexafluorosilicate, see	2674	6.1	
Sodium hydrate, see	1824	8	
SODIUM HYDRIDE	1427	4.3	
Sodium hydrogen 4-amino-phenylarsenate, see	2473	6.1	
SODIUM HYDROGENDIFLUORIDE	2439	8	
SODIUM HYDROSULPHIDE with less than 25% water of crystallization	2318	4.2	

Name and description	UN No.	Class	Remarks
SODIUM HYDROSULPHIDE, HYDRATED with not less than 25% water of crystallization	2949	8	
SODIUM HYDROSULPHITE, see	1384	4.2	
SODIUM HYDROXIDE, SOLID	1823	8	
SODIUM HYDROXIDE SOLUTION	1824	8	
Sodium metasilicate pentahydrate, see	3253	8	
SODIUM METHYLATE	1431	4.2	
SODIUM METHYLATE SOLUTION in alcohol	1289	3	
SODIUM MONOXIDE	1825	8	
SODIUM NITRATE	1498	5.1	
SODIUM NITRATE AND POTASSIUM NITRATE MIXTURE	1499	5.1	
SODIUM NITRITE	1500	5.1	
Sodium nitrite and potassium nitrate mixture, see	1487	5.1	
SODIUM PENTACHLOROPHENATE	2567	6.1	
SODIUM PERBORATE MONOHYDRATE	3377	5.1	
SODIUM PERCHLORATE	1502	5.1	
SODIUM PERMANGANATE	1503	5.1	
SODIUM PEROXIDE	1504	5.1	
SODIUM PEROXOBORATE, ANHYDROUS	3247	5.1	
SODIUM PERSULPHATE	1505	5.1	
SODIUM PHOSPHIDE	1432	4.3	
SODIUM PICRAMATE, dry or wetted with less than 20% water, by mass	0235	1	
SODIUM PICRAMATE, WETTED with not less than 20% water, by mass	1349	4.1	
Sodium potassium alloys, liquid, see	1422	4.3	
Sodium selenate, see	2630	6.1	
Sodium selenite, see	2630	6.1	
Sodium silicofluoride, see	2674	6.1	
SODIUM SULPHIDE, ANHYDROUS	1385	4.2	
SODIUM SULPHIDE with less than 30% water of crystallization	1385	4.2	
SODIUM SULPHIDE, HYDRATED with not less than 30% water	1849	8	
SODIUM SUPEROXIDE	2547	5.1	
SOLIDS CONTAINING CORROSIVE LIQUID, N.O.S.	3244	8	
SOLIDS or mixtures of solids (such as preparations and wastes) CONTAINING FLAMMABLE LIQUID, N.O.S. having a flash-point up to 60°C	3175	4.1	
SOLIDS CONTAINING TOXIC LIQUID, N.O.S.	3243	6.1	
Solvents, flammable, n.o.s., see	1993	3	
Solvents, flammable, toxic, n.o.s., see	1992	3	
SOUNDING DEVICES, EXPLOSIVE	0204 0296 0374 0375	1 1 1 1	
Squibs, see	0325 0454	1 1	
Stain, see	1263 3066 3469 3470	3 8 3 8	
STANNIC CHLORIDE, ANHYDROUS	1827	8	
STANNIC CHLORIDE PENTAHYDRATE	2440	8	
STANNIC PHOSPHIDES	1433	4.3	
Steel swarf, see	2793	4.2	
STIBINE	2676	2	
Straw	1327	4.1	Not subject to ADN
Strontium alloys, pyrophoric, see	1383	4.2	
STRONTIUM ARSENITE	1691	6.1	
STRONTIUM CHLORATE	1506	5.1	
Strontium dioxide, see	1509	5.1	
STRONTIUM NITRATE	1507	5.1	

Name and description	UN No.	Class	Remarks
STRONTIUM PERCHLORATE	1508	5.1	
STRONTIUM PEROXIDE	1509	5.1	
STRONTIUM PHOSPHIDE	2013	4.3	
STRYCHNINE	1692	6.1	
STRYCHNINE SALTS	1692	6.1	
STYPHNIC ACID, see	0219 0394	1 1	
STYRENE MONOMER, STABILIZED	2055	3	
SUBSTANCES, EVI, N.O.S., see	0482	1	
SUBSTANCES, EXPLOSIVE, N.O.S.	0357 0358 0359 0473 0474 0475 0476 0477 0478 0479 0480 0481 0485	1 1 1 1 1 1 1 1 1 1 1 1 1	
SUBSTANCES, EXPLOSIVE, VERY INSENSITIVE, N.O.S.	0482	1	
Substances liable to spontaneous combustion, n.o.s., see	2845 2846 3194 3200	4.2 4.2 4.2 4.2	
SUBSTANCES WITH A FLASH-POINT ABOVE 60 °C which are carried heated within a limiting range of 15K below their flash-point	9001	3	Danger-ous in tank vessels only
SUBSTANCES WITH A FLASH-POINT ABOVE 60 °C AND NOT MORE THAN 100 °C, which do not belong to another Class	9003	9	Danger-ous in tank vessels only
SUBSTANCES WITH AN AUTO-IGNITION TEMPERATURE OF 200 °C AND BELOW, n.o.s.	9002	3	Danger-ous in tank vessels only
SUBSTITUTED NITROPHENOL PESTICIDE, LIQUID, FLAMMABLE, TOXIC, flash-point less than 23 °C	2780	3	
SUBSTITUTED NITROPHENOL PESTICIDE, LIQUID, TOXIC	3014	6.1	
SUBSTITUTED NITROPHENOL PESTICIDE, LIQUID, TOXIC, FLAMMABLE, flash-point not less than 23 °C	3013	6.1	

Name and description	UN No.	Class	Remarks
SUBSTITUTED NITROPHENOL PESTICIDE, SOLID, TOXIC	2779	6.1	
SULPHAMIC ACID	2967	8	
SULPHUR	1350	4.1	
SULPHUR CHLORIDES	1828	8	
Sulphur dichloride, see	1828	8	
SULPHUR DIOXIDE	1079	2	
Sulphuretted hydrogen, see	1053	2	
SULPHUR HEXAFLUORIDE	1080	2	
SULPHURIC ACID with more than 51% acid	1830	8	
SULPHURIC ACID with not more than 51% acid	2796	8	
SULPHURIC ACID, FUMING	1831	8	
SULPHURIC ACID, SPENT	1832	8	
Sulphuric and hydrofluoric acid mixture, see	1786	8	
SULPHUR, MOLTEN	2448	4.1	
Sulphur monochloride, see	1828	8	
SULPHUROUS ACID	1833	8	
SULPHUR TETRAFLUORIDE	2418	2	
SULPHUR TRIOXIDE, STABILIZED	1829	8	
SULPHURYL CHLORIDE	1834	8	
SULPHURYL FLUORIDE	2191	2	
Talcum with tremolite and/or actinolite, see	2590	9	
TARS, LIQUID, including road asphalt and oils, bitumen and cut backs, with a flash-point not greater than 60 °C	1999	3	
Tars, liquid, with a flash-point above 60 °C, at or above its flash-point, see	3256	3	
Tars, liquid, at or above 100 °C and below its flash-point, see	3257	9	
Tartar emetic, see	1551	6.1	
TEAR GAS CANDLES	1700	6.1	
TEAR GAS SUBSTANCE, LIQUID, N.O.S.	1693	6.1	

Name and description	UN No.	Class	Remarks
TEAR GAS SUBSTANCE, SOLID, N.O.S.	3448	6.1	
TELLURIUM COMPOUND, N.O.S.	3284	6.1	
TELLURIUM HEXAFLUORIDE	2195	2	
TERPENE HYDROCARBONS, N.O.S.	2319	3	
TERPINOLENE	2541	3	
TETRABROMOETHANE	2504	6.1	
1,1,2,2-TETRACHLOROETHANE	1702	6.1	
TETRACHLOROETHYLENE	1897	6.1	
TETRAETHYL DITHIO-PYROPHOSPHATE	1704	6.1	
TETRAETHYLENEPENTAMINE	2320	8	
Tetraethyl lead, see	1649	6.1	
TETRAETHYL SILICATE	1292	3	
Tetraethyoxysilane, see	1292	3	
Tetrafluorodichloroethane, see	1958	2	
1,1,1,2-TETRAFLUOROETHANE	3159	2	
TETRAFLUOROETHYLENE, STABILIZED	1081	2	
TETRAFLUOROMETHANE	1982	2	
1,2,3,6-TETRAHYDRO-BENZALDEHYDE	2498	3	
TETRAHYDROFURAN	2056	3	
TETRAHYDRO-FURFURYLAMINE	2943	3	
Tetrahydro-1,4-oxazine, see	2054	3	
TETRAHYDROPHTHALIC ANHYDRIDES with more than 0.05% of maleic anhydride	2698	8	
1,2,3,6-TETRAHYDROPYRIDINE	2410	3	
TETRAHYDROTHIOPHENE	2412	3	
Tetramethoxysilane, see	2606	6.1	
TETRAMETHYLAMMONIUM HYDROXIDE, SOLID	3423	8	
TETRAMETHYLAMMONIUM HYDROXIDE, SOLUTION	1835	8	
Tetramethylene, see	2601	2	

Name and description	UN No.	Class	Remarks
Tetramethylene cyanide, see	2205	6.1	
Tetramethyl lead, see	1649	6.1	
TETRAMETHYLSILANE	2749	3	
TETRANITROANILINE	0207	1	
TETRANITROMETHANE	1510	5.1	
TETRAPROPYL ORTHOTITANATE	2413	3	
TETRAZENE, WETTED with not less than 30% water, or mixture of alcohol and water, by mass, see	0114	1	
TETRAZOL-1-ACETIC ACID	0407	1	
1H-TETRAZOLE	0504	1	
TETRYL, see	0208	1	
Textile waste, wet	1857	4.2	Not subject to ADN
THALLIUM CHLORATE	2573	5.1	
Thallium (I) chlorate, see	2573	5.1	
THALLIUM COMPOUND, N.O.S.	1707	6.1	
THALLIUM NITRATE	2727	6.1	
Thallium (I) nitrate, see	2727	6.1	
Thallous chlorate, see	2573	5.1	
4-THIAPENTANAL	2785	6.1	
Thia-4-pentanal, see	2785	6.1	
THIOACETIC ACID	2436	3	
THIOCARBAMATE PESTICIDE, LIQUID, FLAMMABLE, TOXIC, flash-point less than 23 °C	2772	3	
THIOCARBAMATE PESTICIDE, LIQUID, TOXIC	3006	6.1	
THIOCARBAMATE PESTICIDE, LIQUID, TOXIC, FLAMMABLE, flash-point not less than 23 °C	3005	6.1	
THIOCARBAMATE PESTICIDE, SOLID, TOXIC	2771	6.1	
THIOGLYCOL	2966	6.1	
THIOGLYCOLIC ACID	1940	8	
THIOLACTIC ACID	2936	6.1	
THIONYL CHLORIDE	1836	8	

Name and description	UN No.	Class	Remarks
THIOPHENE	2414	3	
Thiophenol, see	2337	6.1	
THIOPHOSGENE	2474	6.1	
THIOPHOSPHORYL CHLORIDE	1837	8	
THIOUREA DIOXIDE	3341	4.2	
Tin (IV) chloride, anhydrous, see	1827	8	
Tin (IV) chloride pentahydrate, see	2440	8	
TINCTURES, MEDICINAL	1293	3	
Tin tetrachloride, see	1827	8	
TITANIUM DISULPHIDE	3174	4.2	
TITANIUM HYDRIDE	1871	4.1	
TITANIUM POWDER, DRY	2546	4.2	
TITANIUM POWDER, WETTED with not less than 25% water	1352	4.1	
TITANIUM SPONGE GRANULES	2878	4.1	
TITANIUM SPONGE POWDERS	2878	4.1	
TITANIUM TETRACHLORIDE	1838	8	
TITANIUM TRICHLORIDE MIXTURE	2869	8	
TITANIUM TRICHLORIDE MIXTURE, PYROPHORIC	2441	4.2	
TITANIUM TRICHLORIDE, PYROPHORIC	2441	4.2	
TNT, see	0209 0388 0389	1 1 1	
TNT mixed with aluminium, see	0390	1	
TNT, WETTED with not less than 30% water, by mass, see	1356	4.1	
TNT, WETTED with not less than 10% water, by mass, see	3366	4.1	
Toe puffs, nitrocellulose base, see	1353	4.1	
TOLUENE	1294	3	
TOLUENE DIISOCYANATE	2078	6.1	
TOLUIDINES, LIQUID	1708	6.1	
TOLUIDINES, SOLID	3451	6.1	
Toluol, see	1294	3	

Name and description	UN No.	Class	Remarks
2,4-TOLUYLENEDIAMINE, SOLID	1709	6.1	
2,4-TOLUYLENEDIAMINE, SOLUTION	3418	6.1	
Toluylene diisocyanate, see	2078	6.1	
Tolylene diisocyanate, see	2078	6.1	
Tolylethylene, inhibited, see	2618	3	
TORPEDOES with bursting charge	0329 0330 0451	1 1 1	
TORPEDOES, LIQUID FUELLED with inert head	0450	1	
TORPEDOES, LIQUID FUELLED with or without bursting charge	0449	1	
TOXIC BY INHALATION LIQUID, N.O.S. with an inhalation toxicity lower than or equal to 200 ml/m^3 and saturated vapour concentration greater than or equal to 500 LC_{50}	3381	6.1	
TOXIC BY INHALATION LIQUID, N.O.S. with an inhalation toxicity lower than or equal to 1000 ml/m^3 and saturated vapour concentration greater than or equal to 10 LC_{50}	3382	6.1	
TOXIC BY INHALATION LIQUID, CORROSIVE, N.O.S. with an inhalation toxicity lower than or equal to 200 ml/m^3 and saturated vapour concentration greater than or equal to 500 LC_{50}	3389	6.1	
TOXIC BY INHALATION LIQUID, CORROSIVE, N.O.S. with an inhalation toxicity lower than or equal to 1000 ml/m^3 and saturated vapour concentration greater than or equal to 10 LC_{50}	3390	6.1	
TOXIC BY INHALATION LIQUID, FLAMMABLE, N.O.S. with an inhalation toxicity lower than or equal to 200 ml/m^3 and saturated vapour concentration greater than or equal to 500 LC_{50}	3383	6.1	
TOXIC BY INHALATION LIQUID, FLAMMABLE, N.O.S. with an inhalation toxicity lower than or equal to 1000 ml/m^3 and saturated vapour concentration greater than or equal to 10 LC_{50}	3384	6.1	

Name and description	UN No.	Class	Remarks
TOXIC BY INHALATION LIQUID, OXIDIZING, N.O.S. with an inhalation toxicity lower than or equal to 200 ml/m^3 and saturated vapour concentration greater than or equal to 500 LC_{50}	3387	6.1	
TOXIC BY INHALATION LIQUID, OXIDIZING, N.O.S. with an inhalation toxicity lower than or equal to 1000 ml/m^3 and saturated vapour concentration greater than or equal to 10 LC_{50}	3387	6.1	
TOXIC BY INHALATION LIQUID, OXIDIZING, N.O.S. with an inhalation toxicity lower than or equal to 1000 ml/m^3 and saturated vapour concentration greater than or equal to 10 LC_{50}	3388	6.1	
TOXIC BY INHALATION LIQUID, WATER-REACTIVE, N.O.S. with an inhalation toxicity lower than or equal to 200 ml/m^3 and saturated vapour concentration greater than or equal to 500 LC_{50}	3385	6.1	
TOXIC BY INHALATION LIQUID, WATER-REACTIVE, N.O.S. with an inhalation toxicity lower than or equal to 1000 ml/m^3 and saturated vapour concentration greater than or equal to 10 LC_{50}	3386	6.1	
TOXIC LIQUID, CORROSIVE, INORGANIC, N.O.S.	3289	6.1	
TOXIC LIQUID, CORROSIVE, ORGANIC, N.O.S.	2927	6.1	
TOXIC LIQUID, FLAMMABLE, ORGANIC, N.O.S.	2929	6.1	
TOXIC LIQUID, INORGANIC, N.O.S.	3287	6.1	
TOXIC LIQUID, ORGANIC, N.O.S.	2810	6.1	
TOXIC LIQUID, OXIDIZING, N.O.S.	3122	6.1	
TOXIC LIQUID, WATER-REACTIVE, N.O.S.	3123	6.1	
TOXIC SOLID, CORROSIVE, INORGANIC, N.O.S.	3290	6.1	
TOXIC SOLID, CORROSIVE, ORGANIC, N.O.S.	2928	6.1	
TOXIC SOLID, FLAMMABLE, ORGANIC, N.O.S.	2930	6.1	
TOXIC SOLID, INORGANIC, N.O.S.	3288	6.1	
TOXIC SOLID, ORGANIC, N.O.S.	2811	6.1	
TOXIC SOLID, OXIDIZING, N.O.S.	3086	6.1	
TOXIC SOLID, SELF-HEATING, N.O.S.	3124	6.1	
TOXIC SOLID, WATER-REACTIVE, N.O.S.	3125	6.1	
TOXINS, EXTRACTED FROM LIVING SOURCES, LIQUID, N.O.S.	3172	6.1	
TOXINS, EXTRACTED FROM LIVING SOURCES, SOLID, N.O.S.	3462	6.1	
TRACERS FOR AMMUNITION	0212 0306	1 1	
Tremolite, see	2590	9	
TRIALLYLAMINE	2610	3	
TRIALLYL BORATE	2609	6.1	
TRIAZINE PESTICIDE, LIQUID, FLAMMABLE, TOXIC, flash-point less than 23 °C	2764	3	
TRIAZINE PESTICIDE, LIQUID, TOXIC	2998	6.1	
TRIAZINE PESTICIDE, LIQUID, TOXIC, FLAMMABLE, flash-point not less than 23 °C	2997	6.1	
TRIAZINE PESTICIDE, SOLID, TOXIC	2763	6.1	
Tribromoborane, see	2692	8	
TRIBUTYLAMINE	2542	6.1	
TRIBUTYLPHOSPHANE	3254	4.2	
Trichloroacetaldehyde, see	2075	6.1	
TRICHLOROACETIC ACID	1839	8	
TRICHLOROACETIC ACID SOLUTION	2564	8	
Trichlororaceticaldehyde, see	2075	6.1	
TRICHLOROACETYL CHLORIDE	2442	8	
TRICHLOROBENZENES, LIQUID	2321	6.1	
TRICHLOROBUTENE	2322	6.1	

Name and description	UN No.	Class	Remarks
1,1,1-TRICHLOROETHANE	2831	6.1	
TRICHLOROETHYLENE	1710	6.1	
TRICHLOROISOCYANURIC ACID, DRY	2468	5.1	
Trichloronitromethane, see	1580	6.1	
TRICHLOROSILANE	1295	4.3	
1,3,5-Trichloro-s-triazine-2,4,6-trione, see	2468	5.1	
2,4,6-Trichloro-1,3,5- triazine, see	2670	8	
TRICRESYL PHOSPHATE with more than 3% ortho isomer	2574	6.1	
TRIETHYLAMINE	1296	3	
Triethyl borate, see	1176	3	
TRIETHYLENETETRAMINE	2259	8	
Triethyl orthoformate, see	2524	3	
TRIETHYL PHOSPHITE	2323	3	
TRIFLUOROACETIC ACID	2699	8	
TRIFLUOROACETYL CHLORIDE	3057	2	
Trifluorobromomethane, see	1009	2	
Trifluorochloroethane, see	1983	2	
TRIFLUOROCHLORO-ETHYLENE, STABILIZED	1082	2	
Trifluorochloromethane, see	1022	2	
1,1,1-TRIFLUOROETHANE	2035	2	
TRIFLUOROMETHANE	1984	2	
TRIFLUOROMETHANE, REFRIGERATED LIQUID	3136	2	
2-TRIFLUOROMETHYLANILINE	2942	6.1	
3-TRIFLUOROMETHYLANILINE	2948	6.1	
TRIISOBUTYLENE	2324	3	
TRIISOPROPYL BORATE	2616	3	
TRIMETHYLACETYL CHLORIDE	2438	6.1	
TRIMETHYLAMINE, ANHYDROUS	1083	2	
TRIMETHYLAMINE, AQUEOUS SOLUTION, not more than 50% trimethylamine, by mass	1297	3	

Name and description	UN No.	Class	Remarks
1,3,5-TRIMETHYL-BENZENE	2325	3	
TRIMETHYL BORATE	2416	3	
TRIMETHYLCHLORO-SILANE	1298	3	
TRIMETHYLCYCLO-HEXYLAMINE	2326	8	
Trimethylene chlorobromide, see	2688	6.1	
TRIMETHYLHEXA-METHYLENEDIAMINES	2327	8	
TRIMETHYLHEXAMETHYLENE DIISOCYANATE	2328	6.1	
2,4,4-Trimethylpentene-1, see	2050	3	
2,4,4-Trimethylpentene-2, see	2050	3	
TRIMETHYL PHOSPHITE	2329	3	
TRINITROANILINE	0153	1	
TRINITROANISOLE	0213	1	
TRINITROBENZENE, dry or wetted with less than 30% water, by mass	0214	1	
TRINITROBENZENE, WETTED with not less than 10% water, by mass	3367	4.1	
TRINITROBENZENE, WETTED with not less than 30% water, by mass	1354	4.1	
TRINITROBENZENE-SULPHONIC ACID	0386	1	
TRINITROBENZOIC ACID, dry or wetted with less than 30% water, by mass	0215	1	
TRINITROBENZOIC ACID, WETTED with not less than 10% water, by mass	3368	4.1	
TRINITROBENZOIC ACID, WETTED with not less than 30% water, by mass	1355	4.1	
TRINITROCHLOROBENZENE	0155	1	
TRINITROCHLOROBENZENE WETTED with not less than 10% water, by mass	3365	4.1	
TRINITRO-m-CRESOL	0216	1	
TRINITROFLUORENONE	0387	1	
TRINITRONAPHTHALENE	0217	1	

Name and description	UN No.	Class	Remarks
TRINITROPHENETOLE	0218	1	
TRINITROPHENOL, dry or wetted with less than 30% water, by mass	0154	1	
TRINITROPHENOL (PICRIC ACID), WETTED with not less than 30% water, by mass	1344	4.1	
TRINITROPHENOL WETTED with not less than 10% water, by mass	3364	4.1	
TRINITROPHENYL-METHYLNITRAMINE	0208	1	
TRINITRORESORCINOL, dry or wetted with less than 20% water, or mixture of alcohol and water, by mass	0219	1	
TRINITRORESORCINOL, WETTED with not less than 20% water, or mixture of alcohol and water, by mass	0394	1	
TRINITROTOLUENE (TNT), dry or wetted with less than 30% water, by mass	0209	1	
TRINITROTOLUENE AND HEXANITROSTILBENE MIXTURE	0388	1	
TRINITROTOLUENE MIXTURE CONTAINING TRINITROBENZENE AND HEXANITROSTILBENE	0389	1	
TRINITROTOLUENE AND TRINITROBENZENE MIXTURE	0388	1	
TRINITROTOLUENE, WETTED with not less than 10% water, by mass	3366	4.1	
TRINITROTOLUENE, WETTED with not less than 30% water, by mass	1356	4.1	
TRIPROPYLAMINE	2260	3	
TRIPROPYLENE	2057	3	
TRIS-(1-AZIRIDINYL) PHOSPHINE OXIDE SOLUTION	2501	6.1	
TRITONAL	0390	1	
Tropilidene, see	2603	3	
TUNGSTEN HEXAFLUORIDE	2196	2	
TURPENTINE	1299	3	
TURPENTINE SUBSTITUTE	1300	3	

Name and description	UN No.	Class	Remarks
UNDECANE	2330	3	
UREA HYDROGEN PEROXIDE	1511	5.1	
UREA NITRATE, dry or wetted with less than 20% water, by mass	0220	1	
UREA NITRATE, WETTED with not less than 10% water, by mass	3370	4.1	
UREA NITRATE, WETTED with not less than 20% water, by mass	1357	4.1	
Valeral, see	2058	3	
VALERALDEHYDE	2058	3	
n-Valeraldehyde, see	2058	3	
Valeric aldehyde, see	2058	3	
VALERYL CHLORIDE	2502	8	
VANADIUM COMPOUND, N.O.S.	3285	6.1	
Vanadium (IV) oxide sulphate, see	2931	6.1	
Vanadium oxysulphate, see	2931	6.1	
VANADIUM OXYTRICHLORIDE	2443	8	
VANADIUM PENTOXIDE, non-fused form	2862	6.1	
VANADIUM TETRACHLORIDE	2444	8	
VANADIUM TRICHLORIDE	2475	8	
VANADYL SULPHATE	2931	6.1	
Varnish, see	1263 3066 3469 3470	3 8 3 8	
Vehicle, flammable gas powered or vehicle, flammable liquid powered	3166	9	Not subject to ADN
Villiaumite, see	1690	6.1	
VINYL ACETATE, STABILIZED	1301	3	
Vinylbenzene, see	2055	3	
VINYL BROMIDE, STABILIZED	1085	2	
VINYL BUTYRATE, STABILIZED	2838	3	
VINYL CHLORIDE, STABILIZED	1086	2	
VINYL CHLOROACETATE	2589	6.1	
VINYL ETHYL ETHER, STABILIZED	1302	3	
VINYL FLUORIDE, STABILIZED	1860	2	

Name and description	UN No.	Class	Remarks
VINYLIDENE CHLORIDE, STABILIZED	1303	3	
VINYL ISOBUTYL ETHER, STABILIZED	1304	3	
VINYL METHYL ETHER, STABILIZED	1087	2	
VINYLPYRIDINES, STABILIZED	3073	6.1	
VINYLTOLUENES, STABILIZED	2618	3	
VINYLTRICHLOROSILANE	1305	3	
Warheads for guided missiles, see	0286 0287 0369 0370 0371	1 1 1 1 1	
WARHEADS, ROCKET with burster or expelling charge	0370 0371	1 1	
WARHEADS, ROCKET with bursting charge	0286 0287 0369	1 1 1	
WARHEADS, TORPEDO with bursting charge	0221	1	
WATER-REACTIVE LIQUID, N.O.S.	3148	4.3	
WATER-REACTIVE LIQUID, CORROSIVE, N.O.S.	3129	4.3	
WATER-REACTIVE LIQUID, TOXIC, N.O.S.	3130	4.3	
WATER-REACTIVE SOLID, N.O.S.	2813	4.3	
WATER-REACTIVE SOLID, CORROSIVE, N.O.S.	3131	4.3	
WATER-REACTIVE SOLID, FLAMMABLE, N.O.S.	3132	4.3	
WATER-REACTIVE SOLID, OXIDIZING, N.O.S.	3133	4.3	Carriage prohibited
WATER-REACTIVE SOLID, SELF-HEATING, N.O.S.	3135	4.3	
WATER-REACTIVE SOLID, TOXIC, N.O.S.	3134	4.3	
White arsenic, see	1561	6.1	
WHITE ASBESTOS (chrysotile, actinolite, anthophyllite, tremolite)	2590	9	
White spirit, see	1300	3	
WOOD PRESERVATIVES, LIQUID	1306	3	
Wool waste, wet	1387	4.2	Not subject to ADN
XANTHATES	3342	4.2	
XENON	2036	2	
XENON, REFRIGERATED LIQUID	2591	2	
XYLENES	1307	3	
XYLENOLS, LIQUID	3430	6.1	
XYLENOLS, SOLID	2261	6.1	
XYLIDINES, LIQUID	1711	6.1	
XYLIDINES, SOLID	3452	6.1	
Xylols, see	1307	3	
XYLYL BROMIDE, LIQUID	1701	6.1	
XYLYL BROMIDE, SOLID	3417	6.1	
ZINC AMMONIUM NITRITE	1512	5.1	
ZINC ARSENATE	1712	6.1	
ZINC ARSENATE AND ZINC ARSENITE MIXTURE	1712	6.1	
ZINC ARSENITE	1712	6.1	
ZINC ASHES	1435	4.3	
Zinc bisulphite solution, see	2693	8	
ZINC BROMATE	2469	5.1	
ZINC CHLORATE	1513	5.1	
ZINC CHLORIDE, ANHYDROUS	2331	8	
ZINC CHLORIDE SOLUTION	1840	8	
ZINC CYANIDE	1713	6.1	
ZINC DITHIONITE	1931	9	
ZINC DUST	1436	4.3	
ZINC FLUOROSILICATE	2855	6.1	
Zinc hexafluorosilicate, see	2855	6.1	
ZINC HYDROSULPHITE, see	1931	9	
ZINC NITRATE	1514	5.1	
ZINC PERMANGANATE	1515	5.1	

Name and description	UN No.	Class	Remarks
ZINC PEROXIDE	1516	5.1	
ZINC PHOSPHIDE	1714	4.3	
ZINC POWDER	1436	4.3	
ZINC RESINATE	2714	4.1	
Zinc selenate, see	2630	4.1	
Zinc selenite, see	2630	4.1	
Zinc silicofluoride, see	2855	6.1	
ZIRCONIUM, DRY, coiled wire, finished metal sheets, strip (thinner than 254 microns but not thinner than 18 microns)	2858	4.1	
ZIRCONIUM, DRY, finished sheets, strip or coiled wire	2009	4.2	
ZIRCONIUM HYDRIDE	1437	4.1	
ZIRCONIUM NITRATE	2728	5.1	
ZIRCONIUM PICRAMATE, dry or wetted with less than 20% water, by mass	0236	1	
ZIRCONIUM PICRAMATE, WETTED with not less than 20% water, by mass	1517	4.1	
ZIRCONIUM POWDER, DRY	2008	4.2	
ZIRCONIUM POWDER, WETTED with not less than 25% water	1358	4.1	
ZIRCONIUM SCRAP	1932	4.2	
ZIRCONIUM SUSPENDED IN A FLAMMABLE LIQUID	1308	3	
ZIRCONIUM TETRACHLORIDE	2503	8	

CHAPTER 3.3

SPECIAL PROVISIONS APPLICABLE TO CERTAIN ARTICLES OR SUBSTANCES

3.3.1 When Column (6) of Table A of Chapter 3.2 indicates that a special provision is relevant to a substance or article, the meaning and requirements of that special provision are as set forth below.

16 Samples of new or existing explosive substances or articles may be carried as directed by the competent authorities (see 2.2.1.1.3) for purposes including: testing, classification, research and development, quality control, or as a commercial sample. Explosive samples which are not wetted or desensitised shall be limited to 10 kg in small packages as specified by the competent authorities. Explosive samples which are wetted or desensitised shall be limited to 25 kg.

23 Even though this substance has a flammability hazard, it only exhibits such hazard under extreme fire conditions in confined areas.

32 This substance is not subject to the requirements of ADN when in any other form.

37 This substance is not subject to the requirements of ADN when coated.

38 This substance is not subject to the requirements of ADN when it contains not more than 0.1% calcium carbide.

39 This substance is not subject to the requirements of ADN when it contains less than 30% or not less than 90% silicon.

43 When offered for carriage as pesticides, these substances shall be carried under the relevant pesticide entry and in accordance with the relevant pesticide provisions (see 2.2.61.1.10 to 2.2.61.1.11.2).

45 Antimony sulphides and oxides which contain not more than 0.5% of arsenic calculated on the total mass are not subject to the requirements of ADN.

47 Ferricyanides and ferrocyanides are not subject to the requirements of ADN.

48 The carriage of this substance, when it contains more than 20% hydrocyanic acid, is prohibited.

59 These substances are not subject to the requirements of ADN when they contain not more than 50% magnesium.

60 If the concentration is more than 72%, the carriage of this substance is prohibited.

61 The technical name which shall supplement the proper shipping name shall be the ISO common name (see also ISO 1750:1981 "*Pesticides and other agrochemicals - common names*", as amended), other name listed in the WHO "*Recommended Classification of Pesticides by Hazard and Guidelines to Classification*" or the name of the active substance (see also 3.1.2.8.1 and 3.1.2.8.1.1).

62 This substance is not subject to the requirements of ADN when it contains not more than 4% sodium hydroxide.

65 Hydrogen peroxide aqueous solutions with less than 8% hydrogen peroxide are not subject to the requirements of ADN.

103 The carriage of ammonium nitrites and mixtures of an inorganic nitrite with an ammonium salt is prohibited.

105 Nitrocellulose meeting the descriptions of UN No. 2556 or UN No. 2557 may be classified in Class 4.1.

113 The carriage of chemically unstable mixtures is prohibited.

119 Refrigerating machines include machines or other appliances which have been designed for the specific purpose of keeping food or other items at a low temperature in an internal compartment, and air conditioning units. Refrigerating machines and refrigerating machine components are not subject to the provisions of ADN if they contain less than 12 kg of gas in Class 2, group A or O according to 2.2.2.1.3, or if they contain less than 12 litres ammonia solution (UN No. 2672).

122 The subsidiary risks, control and emergency temperatures if any, and the UN number (generic entry) for each of the currently assigned organic peroxide formulations are given in 2.2.52.4.

127 Other inert material or inert material mixture may be used, provided this inert material has identical phlegmatizing properties.

131 The phlegmatized substance shall be significantly less sensitive than dry PETN.

135 The dihydrated sodium salt of dichloroisocyanuric acid is not subject to the requirements of ADN.

138 p-Bromobenzyl cyanide is not subject to the requirements of ADN.

141 Products which have undergone sufficient heat treatment so that they present no hazard during carriage are not subject to the requirements of ADN.

142 Solvent extracted soya bean meal containing not more than 1.5% oil and 11% moisture, which is substantially free of flammable solvent, is not subject to the requirements of ADN.

144 An aqueous solution containing not more than 24% alcohol by volume is not subject to the requirements of ADN.

145 Alcoholic beverages of packing group III, when carried in receptacles of 250 litres or less, are not subject to the requirements of ADN.

152 The classification of this substance will vary with particle size and packaging, but borderlines have not been experimentally determined. Appropriate classifications shall be made in accordance with 2.2.1.

153 This entry applies only if it is demonstrated, on the basis of tests, that the substances when in contact with water are not combustible nor show a tendency to auto-ignition and that the mixture of gases evolved is not flammable.

163 A substance mentioned by name in Table A of Chapter 3.2 shall not be carried under this entry. Substances carried under this entry may contain 20% or less nitrocellulose provided the nitrocellulose contains not more than 12.6% nitrogen (by dry mass).

168 Asbestos which is immersed or fixed in a natural or artificial binder (such as cement, plastics, asphalt, resins or mineral ore) in such a way that no escape of hazardous quantities of respirable asbestos fibres can occur during carriage is not subject to the requirements of ADN. Manufactured articles containing asbestos and not meeting this provision are nevertheless not subject to the requirements of ADN when packed so that no escape of hazardous quantities of respirable asbestos fibres can occur during carriage.

169 Phthalic anhydride in the solid state and tetrahydrophthalic anhydrides, with not more than 0.05% maleic anhydride, are not subject to the requirements of ADN. Phthalic anhydride molten at a temperature above its flash-point, with not more than 0.05% maleic anhydride, shall be classified under UN No. 3256.

172 For radioactive material with a subsidiary risk:

(a) The packages shall be labelled with a label corresponding to each subsidiary risk exhibited by the material; corresponding placards shall be affixed to vehicles, wagons or containers in accordance with the relevant provisions of 5.3.1;

(b) The radioactive material shall be allocated to packing groups I, II or III, as and if appropriate, by application of the grouping criteria provided in Part 2 corresponding to the nature of the predominant subsidiary risk.

The description required in 5.4.1.2.5.1 (b) shall include a description of these subsidiary risks (e.g. "Subsidiary risk: 3, 6.1"), the name of the constituents which most predominantly contribute to this (these) subsidiary risk(s), and where applicable, the packing group.

177 Barium sulphate is not subject to the requirements of ADN.

178 This designation shall be used only when no other appropriate designation exists in Table A of Chapter 3.2, and only with the approval of the competent authority of the country of origin (see 2.2.1.1.3).

181 Packages containing this type of substance shall bear a label conforming to model No. 1 (see 5.2.2.2.2) unless the competent authority of the country of origin has permitted this label to be dispensed with for the specific packaging employed because test data have proved that the substance in this packaging does not exhibit explosive behaviour (see 5.2.2.1.9).

182 The group of alkali metals includes lithium, sodium, potassium, rubidium and caesium.

183 The group of alkaline earth metals includes magnesium, calcium, strontium and barium.

186 In determining the ammonium nitrate content, all nitrate ions for which a molecular equivalent of ammonium ions is present in the mixture shall be calculated as ammonium nitrate.

188 Cells and batteries offered for carriage are not subject to other provisions of ADN if they meet the following:

(a) For a lithium metal or lithium alloy cell, the lithium content is not more than 1 g, and for a lithium-ion cell, the Watt-hour rating is not more than 20 Wh;

(b) For a lithium metal or lithium alloy battery the aggregate lithium content is not more than 2 g, and for a lithium-ion battery, the Watt-hour rating is not more than 100 Wh. Lithium ion batteries subject to this provision shall be marked with the Watt-hour rating on the outside case, except those manufactured before 1 January 2009 which may be carried in accordance with this special provision and without this marking until 31 December 2010;

(c) Each cell or battery is of the type proved to meet the requirements of each test in the *Manual of Tests and Criteria*, Part III, sub-section 38.3;

(d) Cells and batteries, except when installed in equipment, shall be packed in inner packagings that completely enclose the cell or battery. Cells and batteries shall be protected so as to prevent short circuits. This includes protection against contact with conductive materials within the same packaging that could lead to a short circuit. The inner packagings shall be packed in strong outer packagings which conform to the provisions of 4.1.1.1, 4.1.1.2 and 4.1.1.5 of ADR;

(e) Cells and batteries when installed in equipment shall be protected from damage and short circuit, and the equipment shall be equipped with an effective means of preventing accidental activation. When batteries are installed in equipment, the equipment shall be packed in strong outer packagings constructed of suitable material of adequate strength and design in relation to the packaging's capacity and its intended use unless the battery is afforded equivalent protection by the equipment in which it is contained;

(f) Except for packages containing no more than four cells installed in equipment or no more than two batteries installed in equipment, each package shall be marked with the following:

(i) an indication that the package contains "lithium metal" or "lithium ion" cells or batteries, as appropriate;

(ii) an indication that the package shall be handled with care and that a flammability hazard exists if the package is damaged;

(iii) an indication that special procedures shall be followed in the event the package is damaged, to include inspection and repacking if necessary; and

(iv) a telephone number for additional information;

(g) Each consignment of one or more packages marked in accordance with paragraph (f) shall be accompanied with a document including the following:

(i) an indication that the package contains "lithium metal" or "lithium ion" cells or batteries, as appropriate;

(ii) an indication that the package shall be handled with care and that a flammability hazard exists if the package is damaged;

(iii) an indication that special procedures shall be followed in the event the package is damaged, to include inspection and repacking if necessary; and

(iv) a telephone number for additional information;

(h) Except when batteries are installed in equipment, each package shall be capable of withstanding a 1.2 m drop test in any orientation without damage to cells or batteries contained therein, without shifting of the contents so as to allow battery to battery (or cell to cell) contact and without release of contents; and

(i) Except when batteries are installed in or packed with equipment, packages shall not exceed 30 kg gross mass.

As used above and elsewhere in ADN, "lithium content" means the mass of lithium in the anode of a lithium metal or lithium alloy cell.

Separate entries exist for lithium metal batteries and lithium ion batteries to facilitate the carriage of these batteries for specific modes of carriage and to enable the application of different emergency response actions.

190 Aerosol dispensers shall be provided with protection against inadvertent discharge. Aerosols with a capacity not exceeding 50 ml containing only non-toxic constituents are not subject to the requirements of ADN.

191 Receptacles, small, with a capacity not exceeding 50 ml, containing only non-toxic constituents are not subject to the requirements of ADN.

193 This entry may only be used for uniform ammonium nitrate based fertilizer mixtures of the nitrogen, phosphate or potash type, containing not more than 70% ammonium nitrate and not more than 0.4% total combustible/organic material calculated as carbon or with not more than 45% ammonium nitrate and unrestricted combustible material. Fertilizers within these composition limits are not subject to the requirements of ADN if shown by a Trough Test (see *Manual of Tests and Criteria*, Part III, sub-section 38.2) not to be liable to self-sustaining decomposition.

194 The control and emergency temperatures, if any, and the UN number (generic entry) for each of the currently assigned self-reactive substances are given in 2.2.41.4.

196 Formulations which in laboratory testing neither detonate in the cavitated state nor deflagrate, which show no effect when heated under confinement and which exhibit no explosive power may be carried under this entry. The formulation must also be thermally stable (i.e. the SADT is 60 °C or higher for a 50 kg package). Formulations not meeting these criteria shall be carried under the provisions of Class 5.2, (see 2.2.52.4).

198 Nitrocellulose solutions containing not more than 20 % nitrocellulose may be carried as paint or printing ink, as applicable (see UN Nos. 1210, 1263, 3066, 3469 and 3470).

199 Lead compounds which, when mixed in a ratio of 1:1000 with 0.07M hydrochloric acid and stirred for one hour at a temperature of 23 °C ± 2 °C, exhibit a solubility of 5 % or less (see ISO 3711:1990 *"Lead chromate pigments and lead chromate-molybdate pigments – Specifications and methods of test"*) are considered insoluble and are not subject to the requirements of ADN unless they meet the criteria for inclusion in another class.

201 Lighters and lighter refills shall comply with the provisions of the country in which they were filled. They shall be provided with protection against inadvertent discharge. The liquid portion of the gas shall not exceed 85% of the capacity of the receptacle at 15 °C. The receptacles, including the closures, shall be capable of withstanding an internal pressure of twice the pressure of the liquefied petroleum gas at 55 °C. The valve mechanisms and ignition devices shall be securely sealed, taped or otherwise fastened or designed to prevent operation or leakage of the contents during carriage. Lighters shall not contain more than 10 g of liquefied petroleum gas. Lighter refills shall not contain more than 65 g of liquefied petroleum gas.

NOTE: For waste lighters collected separately see Chapter 3.3, special provision 654.

203 This entry shall not be used for polychlorinated biphenyls, liquid, UN No. 2315 and polychlorinated biphenyls, solid, UN No. 3432.

205 This entry shall not be used for UN No. 3155 PENTACHLOROPHENOL.

207 Polymeric beads and moulding compounds may be made from polystyrene, poly(methyl methacrylate) or other polymeric material.

208 The commercial grade of calcium nitrate fertilizer, when consisting mainly of a double salt (calcium nitrate and ammonium nitrate) containing not more than 10% ammonium nitrate and at least 12% water of crystallization, is not subject to the requirements of ADN.

210 Toxins from plant, animal or bacterial sources which contain infectious substances, or toxins that are contained in infectious substances, shall be classified in Class 6.2.

215 This entry only applies to the technically pure substance or to formulations derived from it having an SADT higher than 75 °C and therefore does not apply to formulations which are self-reactive substances (for self-reactive substances, see 2.2.41.4). Homogeneous mixtures containing not more than 35 % by mass of azodicarbonamide and at least 65 % of inert substance are not subject to the requirements of ADN unless criteria of other classes are met.

216 Mixtures of solids which are not subject to the requirements of ADN and flammable liquids may be carried under this entry without first applying the classification criteria of Class 4.1, provided there is no free liquid visible at the time the substance is loaded or at the time the packaging, vehicle, wagon or container is closed. Sealed packets and articles containing less than 10 ml of a packing group II or III flammable liquid absorbed into a solid material are not subject to ADN provided there is no free liquid in the packet or article.

217 Mixtures of solids which are not subject to the requirements of ADN and toxic liquids may be carried under this entry without first applying the classification criteria of Class 6.1, provided there is no free liquid visible at the time the substance is loaded or at the time the packaging, vehicle, wagon or container is closed. This entry shall not be used for solids containing a packing group I liquid.

218 Mixtures of solids which are not subject to the requirements of ADN and corrosive liquids may be carried under this entry without first applying the classification criteria of Class 8, provided there is no free liquid visible at the time the substance is loaded or at the time the packaging, vehicle, wagon or container is closed.

219 Genetically modified micro-organisms and genetically modified organisms which meet the definition of an infectious substance and the criteria for inclusion in Class 6.2 in accordance with section 2.2.62 shall be carried as UN No. 2814, UN No. 2900 or UN No. 3373, as appropriate.

220 Only the technical name of the flammable liquid component of this solution or mixture shall be shown in parentheses immediately following the proper shipping name.

221 Substances included under this entry shall not be of packing group I.

224 Unless it can be demonstrated by testing that the sensitivity of the substance in its frozen state is no greater than in its liquid state, the substance shall remain liquid during normal transport conditions. It shall not freeze at temperatures above -15 °C.

225 Fire extinguishers under this entry may include installed actuating cartridges (cartridges, power device of classification code 1.4C or 1.4S), without changing the classification of Class 2, group A or O according to 2.2.2.1.3 provided the total quantity of deflagrating (propellant) explosives does not exceed 3.2 g per extinguishing unit.

226 Formulations of this substance containing not less than 30% non-volatile, non-flammable phlegmatizer are not subject to the requirements of ADN.

227 When phlegmatized with water and inorganic inert material the content of urea nitrate may not exceed 75% by mass and the mixture shall not be capable of being detonated by the Series 1, type (a), test in the *Manual of Tests and Criteria*, Part 1.

228 Mixtures not meeting the criteria for flammable gases (see 2.2.2.1.5) shall be carried under UN No. 3163.

230 This entry applies to cells and batteries containing lithium in any form, including lithium polymer and lithium ion cells and batteries.

Lithium cells and batteries may be carried under this entry if they meet the following provisions:

(a) Each cell or battery is of the type proved to meet the requirements of each test of the *Manual of Tests and Criteria*, Part III, sub-section 38.3;

(b) Each cell and battery incorporates a safety venting device or is designed to preclude a violent rupture under normal conditions of carriage;

(c) Each cell and battery is equipped with an effective means of preventing external short circuits;

(d) Each battery containing cells or series of cells connected in parallel is equipped with effective means as necessary to prevent dangerous reverse current flow (e.g. diodes, fuses, etc.).

235 This entry applies to articles which contain Class 1 explosive substances and which may also contain dangerous goods of other classes. These articles are used as life-saving vehicle air bag inflators or air bag modules or seat-belt pretensioners.

236 Polyester resin kits consist of two components: a base material (Class 3, packing group II or III) and an activator (organic peroxide). The organic peroxide shall be type D, E or F, not requiring temperature control. Packing group shall be II or III, according to the criteria for Class 3, applied to the base material. The quantity limit referred to in Column (7a) of Table A of Chapter 3.2 applies to the base material.

237 The membrane filters, including paper separators, coating or backing materials, etc., that are present in carriage, shall not be liable to propagate a detonation as tested by one of the tests described in the *Manual of Tests and Criteria*, Part I, Test series 1 (a).

In addition the competent authority may determine, on the basis of the results of suitable burning rate tests taking account of the standard tests in the *Manual of Tests and Criteria*, Part III, sub-section 33.2.1, that nitrocellulose membrane filters in the form in which they are to be carried are not subject to the requirements applicable to flammable solids in Class 4.1.

238 (a) Batteries can be considered as non-spillable provided that they are capable of withstanding the vibration and pressure differential tests given below, without leakage of battery fluid.

Vibration test: The battery is rigidly clamped to the platform of a vibration machine and a simple harmonic motion having an amplitude of 0.8 mm (1.6 mm maximum total excursion) is applied. The frequency is varied at the rate of 1 Hz/min between the limits of 10 Hz and 55 Hz. The entire range of frequencies and return is traversed in 95 ± 5 minutes for each mounting position (direction of vibration) of the battery. The battery is tested in three mutually perpendicular positions (to include testing with fill openings and vents, if any, in an inverted position) for equal time periods.

Pressure differential test: Following the vibration test, the battery is stored for six hours at 24 °C ± 4 °C while subjected to a pressure differential of at least 88 kPa. The battery is tested in three mutually perpendicular positions (to include testing with fill openings and vents, if any, in an inverted position) for at least six hours in each position.

(b) Non-spillable batteries are not subject to the requirements of ADN if, at a temperature of 55 °C, the electrolyte will not flow from a ruptured or cracked case and there is no free liquid to flow and if, as packaged for carriage, the terminals are protected from short circuit.

239 Batteries or cells shall not contain dangerous substances other than sodium, sulphur and/or polysulphides. Batteries or cells shall not be offered for carriage at a temperature such that liquid elemental sodium is present in the battery or cell unless approved and under the conditions established by the competent authority of the country of origin. If the country of origin is not a Contracting Party to ADN, the approval and conditions of carriage shall be recognized by the competent authority of the first country Contracting Party to ADN reached by the consignment.

Cells shall consist of hermetically sealed metal casings which fully enclose the dangerous substances and which are so constructed and closed as to prevent the release of the dangerous substances under normal conditions of carriage.

Batteries shall consist of cells secured within and fully enclosed by a metal casing so constructed and closed as to prevent the release of the dangerous substances under normal conditions of carriage.

241 The formulation shall be prepared so that it remains homogeneous and does not separate during carriage. Formulations with low nitrocellulose contents and not showing dangerous properties when tested for their liability to detonate, deflagrate or explode when heated under defined confinement by tests of Test series 1 (a), 2 (b) and 2 (c) respectively in the *Manual of Tests and Criteria*, Part I and not being a flammable solid when tested in accordance with test No. 1 in the *Manual of Tests and Criteria*, Part III, sub-section 33.2.1.4 (chips, if necessary, crushed and sieved to a particle size of less than 1.25 mm) are not subject to the requirements of ADN.

242 Sulphur is not subject to the requirements of ADN when it has been formed to a specific shape (e.g. prills, granules, pellets, pastilles or flakes).

243 Gasoline, motor spirit and petrol for use in spark-ignition engines (e.g. in automobiles, stationary engines and other engines) shall be assigned to this entry regardless of variations in volatility.

244 This entry includes e.g. aluminium dross, aluminium skimmings, spent cathodes, spent potliner, and aluminium salt slags.

247 Alcoholic beverages containing more than 24% alcohol but not more than 70% by volume, when carried as part of the manufacturing process, may be carried in wooden barrels with a capacity of more than 250 litres and not more than 500 litres meeting the general requirements of 4.1.1 of ADR, as appropriate, on the following conditions:

(a) The wooden barrels shall be checked and tightened before filling;

(b) Sufficient ullage (not less than 3%) shall be left to allow for the expansion of the liquid;

(c) The wooden barrels shall be carried with the bungholes pointing upwards;

(d) The wooden barrels shall be carried in containers meeting the requirements of the CSC. Each wooden barrel shall be secured in custom-made cradles and be wedged by appropriate means to prevent it from being displaced in any way during carriage.

249 Ferrocerium, stabilized against corrosion, with a minimum iron content of 10% is not subject to the requirements of ADN.

250 This entry may only be used for samples of chemicals taken for analysis in connection with the implementation of the Convention on the Prohibition of the Development, Production, Stockpiling and Use of Chemical Weapons and on their Destruction. The carriage of substances under this entry shall be in accordance with the chain of custody and security procedures specified by the Organisation for the Prohibition of Chemical Weapons.

The chemical sample may only be carried providing prior approval has been granted by the competent authority or the Director General of the Organisation for the Prohibition of Chemical Weapons and providing the sample complies with the following provisions:

(a) It shall be packed according to packing instruction 623 in the ICAO Technical Instructions (see S-3-8 of the Supplement); and

(b) During carriage, a copy of the document of approval for transport, showing the quantity limitations and the packing provisions shall be attached to the transport document.

251 The entry CHEMICAL KIT or FIRST AID KIT is intended to apply to boxes, cases etc. containing small quantities of various dangerous goods which are used for example for medical, analytical or testing or repair purposes. Such kits may not contain dangerous goods for which the code "LQ0" has been indicated in Column (7a) of Table A of Chapter 3.2.

Components shall not react dangerously (see "dangerous reaction" in 1.2.1). The total quantity of dangerous goods in any one kit shall not exceed either 1 *l* or 1 kg. The packing group assigned to the kit as a whole shall be the most stringent packing group assigned to any individual substance in the kit.

Kits which are carried on board vessels for first-aid or operating purposes are not subject to the requirements of ADN.

Chemical kits and first aid kits containing dangerous goods in inner packagings which do not exceed the quantity limits for limited quantities applicable to individual substances as specified in Column (7a) of Table A of Chapter 3.2 in accordance with the LQ code defined in 3.4.6 may be carried in accordance with Chapter 3.4.

252 Provided the ammonium nitrate remains in solution under all conditions of carriage, aqueous solutions of ammonium nitrate, with not more than 0.2% combustible material, in a concentration not exceeding 80%, are not subject to the requirements of ADN.

266 This substance, when containing less alcohol, water or phlegmatizer than specified, shall not be carried unless specifically authorized by the competent authority (see 2.2.1.1).

267 Any explosives, blasting, type C containing chlorates shall be segregated from explosives containing ammonium nitrate or other ammonium salts.

270 Aqueous solutions of Class 5.1 inorganic solid nitrate substances are considered as not meeting the criteria of Class 5.1 if the concentration of the substances in solution at the minimum temperature encountered during carriage is not greater than 80% of the saturation limit.

271 Lactose or glucose or similar materials, may be used as a phlegmatizer provided that the substance contains not less than 90%, by mass, of phlegmatizer. The competent authority may authorize these mixtures to be classified in Class 4.1 on the basis of a test Series 6 (c) of Section 16 of Part I of the *Manual of Tests and Criteria* on at least three packages as prepared for carriage. Mixtures containing at least 98%, by mass, of phlegmatizer are not subject to the requirements of ADN. Packages containing mixtures with not less than 90%, by mass, of phlegmatizer need not bear a label conforming to model No. 6.1.

272 This substance shall not be carried under the provisions of Class 4.1 unless specifically authorized by the competent authority (see UN No. 0143).

273 Maneb and maneb preparations stabilized against self-heating need not be classified in Class 4.2 when it can be demonstrated by testing that a cubic volume of 1 m^3 of substance does not self-ignite and that the temperature at the centre of the sample does

not exceed 200 °C, when the sample is maintained at a temperature of not less than 75 °C ± 2 °C for a period of 24 hours.

274 The provisions of 3.1.2.8 apply.

278 These substances shall not be classified and carried unless authorized by the competent authority on the basis of results from Series 2 tests and a Series 6(c) test of Part I of the *Manual of Tests and Criteria* on packages as prepared for carriage (see 2.2.1.1). The competent authority shall assign the packing group on the basis of 2.2.3 criteria and the package type used for the Series 6(c) test.

279 The substance is assigned to this classification or packing group based on human experience rather than the strict application of classification criteria set out in ADN.

280 This entry applies to articles which are used as life-saving vehicle air bag inflators, or air bag modules or seat-belt pretensioners and which contain dangerous goods of Class 1 or dangerous goods of other classes and when carried as component parts and when these articles as presented for carriage have been tested in accordance with Test series 6 (c) of Part I of the *Manual of Tests and Criteria*, with no explosion of the device, no fragmentation of device casing or pressure vessel, and no projection hazard nor thermal effect which would significantly hinder fire-fighting or other emergency response efforts in the immediate vicinity.

283 Articles, containing gas, intended to function as shock absorbers, including impact energy-absorbing devices, or pneumatic springs are not subject to the requirements of ADN provided:

(a) Each article has a gas space capacity not exceeding 1.6 litres and a charge pressure not exceeding 280 bar where the product of the capacity (litres) and charge pressure (bars) does not exceed 80 (i.e. 0.5 litres gas space and 160 bar charge pressure, 1 litre gas space and 80 bar charge pressure, 1.6 litres gas space and 50 bar charge pressure, 0.28 litres gas space and 280 bar charge pressure);

(b) Each article has a minimum burst pressure of 4 times the charge pressure at 20 °C for products not exceeding 0.5 litres gas space capacity and 5 times charge pressure for products greater than 0.5 litres gas space capacity;

(c) Each article is manufactured from material which will not fragment upon rupture;

(d) Each article is manufactured in accordance with a quality assurance standard acceptable to the competent authority; and

(e) The design type has been subjected to a fire test demonstrating that the article relieves its pressure by means of a fire degradable seal or other pressure relief device, such that the article will not fragment and that the article does not rocket.

See also 1.1.3.2 (d) of ADR for equipment used for the operation of the vehicle.

284 An oxygen generator, chemical, containing oxidizing substances shall meet the following conditions:

(a) The generator when containing an explosive actuating device shall only be carried under this entry when excluded from Class 1 in accordance with the NOTE under paragraph 2.2.1.1.1 (b);

(b) The generator, without its packaging, shall be capable of withstanding a 1.8 m drop test onto a rigid, non-resilient, flat and horizontal surface, in the position most likely to cause damage, without loss of its contents and without actuation;

(c) When a generator is equipped with an actuating device, it shall have at least two positive means of preventing unintentional actuation.

286 Nitrocellulose membrane filters covered by this entry, each with a mass not exceeding 0.5 g, are not subject to the requirements of ADN when contained individually in an article or a sealed packet.

288 These substances shall not be classified and carried unless authorized by the competent authority on the basis of results from Series 2 tests and a Series 6 (c) test of Part I of the *Manual of tests and Criteria* on packages as prepared for carriage (see 2.2.1.1).

289 Air bag inflators, air bag modules or seat-belt pretensioners installed in conveyances or in completed conveyance components such as steering columns, door panels, seats, etc. are not subject to the requirements of ADN.

290 When this material meets the definitions and criteria of other classes as defined in Part 2, it shall be classified in accordance with the predominant subsidiary risk. Such material shall be declared under the proper shipping name and UN number appropriate for the material in that predominant Class, with the addition of the name applicable to this material according to Column (2) of Table A of Chapter 3.2, and shall be carried in accordance with the provisions applicable to that UN number. In addition, all other requirements specified in 1.7.1.5 shall apply, except 5.2.1.7.2.

291 Flammable liquefied gases shall be contained within refrigerating machine components. These components shall be designed and tested to at least three times the working pressure of the machinery. The refrigerating machines shall be designed and constructed to contain the liquefied gas and preclude the risk of bursting or cracking of the pressure retaining components during normal conditions of carriage. Refrigerating machines and refrigerating-machine components are not subject to the requirements of ADN if they contain less than 12 kg of gas.

292 Mixtures containing not more than 23.5% oxygen by volume may be carried under this entry when no other oxidizing gases are present. A label conforming to model 5.1 is not required for any concentrations within this limit.

293 The following definitions apply to matches:

(a) Fusee matches are matches the heads of which are prepared with a friction-sensitive igniter composition and a pyrotechnic composition which burns with little or no flame, but with intense heat;

(b) Safety matches are matches which are combined with or attached to the box, book or card that can be ignited by friction only on a prepared surface;

(c) Strike anywhere matches are matches that can be ignited by friction on a solid surface;

(d) Wax Vesta matches are matches that can be ignited by friction either on a prepared surface or on a solid surface.

295 Batteries need not be individually marked and labelled if the pallet bears the appropriate mark and label.

296 These entries apply for life-saving appliances such as life rafts, personal flotation devices and self-inflating slides. UN No. 2990 applies to self-inflating appliances and UN No. 3072 applies to life-saving appliances that are not self-inflating. Life-saving appliances may contain:

(a) Signal devices (Class 1) which may include smoke and illumination signal flares packed in packagings that prevent them from being inadvertently activated;

(b) For UN No. 2990 only, cartridges, power device of Division 1.4, compatibility group S, may be contained for purposes of the self-inflating mechanism and provided that the quantity of explosives per appliance does not exceed 3.2 g;

(c) Class 2 compressed gases, group A or O, according to 2.2.2.1.3;

(d) Electric storage batteries (Class 8) and lithium batteries (Class 9);

(e) First aid kits or repair kits containing small quantities of dangerous goods (e.g.: substances of Class 3, 4.1, 5.2, 8 or 9); or

(f) "Strike anywhere" matches packed in packagings that prevent them from being inadvertently activated.

300 Fish meal or fish scrap shall not be loaded if the temperature at the time of loading exceeds 35 °C or 5 °C above the ambient temperature whichever is higher.

302 In the proper shipping name, the word "UNIT" means: a vehicle, a wagon, a container or a tank.

Fumigated vehicles, wagons, containers and tanks are only subject to the provisions of 5.5.2.

303 Receptacles shall be assigned to the classification code of the gas or mixture of gases contained therein determined in accordance with the provisions of section 2.2.2.

304 Batteries, dry, containing corrosive electrolyte which will not flow out of the battery if the battery case is cracked are not subject to the requirements of ADN provided the batteries are securely packed and protected against short-circuits. Examples of such batteries are: alkali-manganese, zinc-carbon, nickel-metal hydride and nickel-cadmium batteries.

305 These substances are not subject to the requirements of ADN when in concentrations of not more than 50 mg/kg.

306 This entry may only be used for substances that do not exhibit explosive properties of Class 1 when tested in accordance to Test Series 1 and 2 of Class 1 (see *Manual of Tests and Criteria*, Part I).

307 This entry may only be used for uniform mixtures containing ammonium nitrate as the main ingredient within the following composition limits:

(a) Not less than 90% ammonium nitrate with not more than 0.2% total combustible/organic material calculated as carbon and with added matter, if any, which is inorganic and inert towards ammonium nitrate; or

(b) Less than 90% but more than 70% ammonium nitrate with other inorganic materials or more than 80% but less than 90% ammonium nitrate mixed with calcium carbonate and/or dolomite and/or mineral calcium sulphate and not more than 0.4% total combustible/organic material calculated as carbon; or

(c) Nitrogen type ammonium nitrate based fertilizers containing mixtures of ammonium nitrate and ammonium sulphate with more than 45% but less than 70% ammonium nitrate and not more than 0.4% total combustible/organic material calculated as carbon such that the sum of the percentage compositions of ammonium nitrate and ammonium sulphate exceeds 70%.

309 This entry applies to non sensitized emulsions, suspensions and gels consisting primarily of a mixture of ammonium nitrate and fuel, intended to produce a Type E blasting explosive only after further processing prior to use.

The mixture for emulsions typically has the following composition: 60-85% ammonium nitrate, 5-30% water, 2-8% fuel, 0.5-4% emulsifier agent, 0-10% soluble flame suppressants, and trace additives. Other inorganic nitrate salts may replace part of the ammonium nitrate.

The mixture for suspensions and gels typically has the following composition: 60-85% ammonium nitrate, 0-5% sodium or potassium perchlorate, 0-17% hexamine nitrate or monomethylamine nitrate, 5-30% water, 2-15% fuel, 0.5-4% thickening agent, 0-10% soluble flame suppressants, and trace additives. Other inorganic nitrate salts may replace part of the ammonium nitrate.

Substances shall satisfactorily pass Test Series 8 of the Manual of Tests and Criteria, Part I, Section 18 and be approved by the competent authority.

310 The testing requirements in sub-section 38.3 of the *Manual of Tests and Criteria* do not apply to production runs consisting of not more than 100 cells and batteries, or to pre-production prototypes of cells and batteries when these prototypes are carried for testing, if:

(a) the cells and batteries are carried in an outer packaging that is a metal, plastics or plywood drum or a metal, plastics or wooden box and that meets the criteria for packing group I; and

(b) each cell and battery is individually packed in an inner packaging inside an outer packaging and is surrounded by cushioning material that is non-combustible, and non-conductive.

311 Substances shall not be carried under this entry unless approved by the competent authority on the basis of the results of appropriate tests according to Part I of the *Manual of Tests and Criteria*. Packaging shall ensure that the percentage of diluent does not fall below that stated in the competent authority approval, at any time during carriage.

312 *(Reserved)*

313 Substances and mixtures meeting the criteria for Class 8 shall bear a subsidiary risk label conforming to model No. 8 (see 5.2.2.2.2).

314 (a) These substances are liable to exothermic decomposition at elevated temperatures. Decomposition can be initiated by heat or by impurities (e.g. powdered metals (iron, manganese, cobalt, magnesium) and their compounds);

(b) During the course of carriage, these substances shall be shaded from direct sunlight and all sources of heat and be placed in adequately ventilated areas.

315 This entry shall not be used for Class 6.1 substances which meet the inhalation toxicity criteria for packing group I described in 2.2.61.1.8.

316 This entry applies only to calcium hypochlorite, dry, when carried in non friable tablet form.

317 "Fissile-excepted" applies only to those packages complying with 6.4.11.2 of ADR.

318 For the purposes of documentation, the proper shipping name shall be supplemented with the technical name (see 3.1.2.8). When the infectious substances to be carried are unknown, but suspected of meeting the criteria for inclusion in category A and assignment to UN No. 2814 or 2900, the words "suspected category A infectious substance" shall be shown, in parentheses, following the proper shipping name on the transport document.

319 Substances packed and packages marked in accordance with packing instruction P650 of ADR are not subject to any other requirements of ADN.

321 These storage systems shall always be considered as containing hydrogen.

322 When carried in non-friable tablet form, these goods are assigned to packing group III.

323 *(Reserved)*

324 This substance needs to be stabilized when in concentrations of not more than 99%.

325 In the case of non-fissile or fissile excepted uranium hexafluoride, the material shall be classified under UN No 2978.

326 In the case of fissile uranium hexafluoride, the material shall be classified under UN No 2977.

327 Waste aerosols consigned in accordance with 5.4.1.1.3 may be carried under this entry for the purposes of reprocessing or disposal. They need not be protected against inadvertent discharge provided that measures to prevent dangerous build up of pressure and dangerous atmospheres are addressed. Waste aerosols, other than those leaking or severely deformed, shall be packed in accordance with packing instruction P003 of ADR and special provision PP87 of ADR, or packing instruction LP02 of ADR and special packing provision L2 of ADR. Leaking or severely deformed aerosols shall be carried in salvage packagings provided appropriate measures are taken to ensure there is no dangerous build up of pressure.

NOTE*: For maritime carriage, waste aerosols shall not be carried in closed containers.*

328 This entry applies to fuel cell cartridges including when contained in equipment or packed with equipment. Fuel cell cartridges installed in or integral to a fuel cell

system are regarded as contained in equipment. Fuel cell cartridge means an article that stores fuel for discharge into the fuel cell through (a) valve(s) that control(s) the discharge of fuel into the fuel cell. Fuel cell cartridges, including when contained in equipment, shall be designed and constructed to prevent fuel leakage under normal conditions of carriage.

Fuel cell cartridge design types using liquids as fuels shall pass an internal pressure test at a pressure of 100 kPa (gauge) without leakage.

Except for fuel cell cartridges containing hydrogen in metal hydride which shall be in compliance with special provision 339, each fuel cell cartridge design type shall be shown to pass a 1.2 meter drop test onto an unyielding surface in the orientation most likely to result in failure of the containment system with no loss of contents.

329 *(Reserved)*

331 *(Reserved)*

332 Magnesium nitrate hexahydrate is not subject to the requirements of ADN.

333 Ethanol and gasoline, motor spirit or petrol mixtures for use in spark-ignition engines (e.g. in automobiles, stationary engines and other engines) shall be assigned to this entry regardless of variations in volatility.

334 A fuel cell cartridge may contain an activator provided it is fitted with two independent means of preventing unintended mixing with the fuel during carriage.

335 Mixtures of solids which are not subject to the requirements of ADN and environmentally hazardous liquids or solids shall be classified as UN 3077 and may be carried under this entry provided there is no free liquid visible at the time the substance is loaded or at the time the packaging or vehicle, wagon or container is closed. Each vehicle or container shall be leakproof when used for carriage in bulk. If free liquid is visible at the time the mixture is loaded or at the time the packaging or vehicle, wagon or container is closed, the mixture shall be classified as UN 3082. Sealed packets and articles containing less than 10 ml of an environmentally hazardous liquid, absorbed into a solid material but with no free liquid in the packet or article, or containing less than 10 g of an environmentally hazardous solid, are not subject to the requirements of ADN.

336 A single package of non-combustible solid LSA-II or LSA-III material, if carried by air, shall not contain an activity greater than 3 000 A_2.

337 Type B(U) and Type B(M) packages, if carried by air, shall not contain activities greater than the following:

(a) For low dispersible radioactive material: as authorized for the package design as specified in the certificate of approval;

(b) For special form radioactive material: 3 000 A_1 or 100 000 A_2, whichever is the lower; or

(c) For all other radioactive material: 3 000 A_2.

338 Each fuel cell cartridge carried under this entry and designed to contain a liquefied flammable gas shall:

(a) Be capable of withstanding, without leakage or bursting, a pressure of at least two times the equilibrium pressure of the contents at 55 °C;

(b) Not contain more than 200 ml of liquefied flammable gas with a vapour pressure not exceeding 1 000 kPa at 55 °C; and

(c) Pass the hot water bath test prescribed in 6.2.6.3.1 of ADR.

339 Fuel cell cartridges containing hydrogen in a metal hydride carried under this entry shall have a water capacity less than or equal to 120 ml.

The pressure in the fuel cell cartridge shall not exceed 5 MPa at 55 °C. The design type shall withstand, without leaking or bursting, a pressure of twice the design pressure of the cartridge at 55 °C or 200 kPa more than the design pressure of the cartridge at 55 °C, whichever is greater. The pressure at which this test is conducted is referred to in the drop test and the hydrogen cycling test as the "minimum shell burst pressure".

Fuel cell cartridges shall be filled in accordance with procedures provided by the manufacturer. The manufacturer shall provide the following information with each fuel cell cartridge:

(a) Inspection procedures to be carried out before initial filling and before refilling of the fuel cell cartridge;

(b) Safety precautions and potential hazards to be aware of;

(c) Method for determining when the rated capacity has been achieved;

(d) Minimum and maximum pressure range;

(e) Minimum and maximum temperature range; and

(f) Any other requirements to be met for initial filling and refilling including the type of equipment to be used for initial filling and refilling.

The fuel cell cartridges shall be designed and constructed to prevent fuel leakage under normal conditions of carriage. Each cartridge design type, including cartridges integral to a fuel cell, shall be subjected to and shall pass the following tests:

Drop test

A 1.8 metre drop test onto an unyielding surface in four different orientations:

(a) Vertically, on the end containing the shut-off valve assembly;

(b) Vertically, on the end opposite to the shut-off valve assembly;

(c) Horizontally, onto a steel apex with a diameter of 38 mm, with the steel apex in the upward position; and

(d) At a 45° angle on the end containing the shut-off valve assembly.

There shall be no leakage, determined by using a soap bubble solution or other equivalent means on all possible leak locations, when the cartridge is charged to its rated charging pressure. The fuel cell cartridge shall then be hydrostatically pressurized to destruction. The recorded burst pressure shall exceed 85% of the minimum shell burst pressure.

Fire test

A fuel cell cartridge filled to rated capacity with hydrogen shall be subjected to a fire engulfment test. The cartridge design, which may include a vent feature integral to it, is deemed to have passed the fire test if :

(a) The internal pressure vents to zero gauge pressure without rupture of the cartridge; or

(b) The cartridge withstands the fire for a minimum of 20 minutes without rupture.

Hydrogen cycling test

This test is intended to ensure that a fuel cell cartridge design stress limits are not exceeded during use.

The fuel cell cartridge shall be cycled from not more than 5% rated hydrogen capacity to not less than 95% rated hydrogen capacity and back to not more than 5% rated hydrogen capacity. The rated charging pressure shall be used for charging and temperatures shall be held within the operating temperature range. The cycling shall be continued for at least 100 cycles.

Following the cycling test, the fuel cell cartridge shall be charged and the water volume displaced by the cartridge shall be measured. The cartridge design is deemed to have passed the hydrogen cycling test if the water volume displaced by the cycled cartridge does not exceed the water volume displaced by an uncycled cartridge charged to 95% rated capacity and pressurized to 75% of its minimum shell burst pressure.

Production leak test

Each fuel cell cartridge shall be tested for leaks at 15 °C ± 5 °C, while pressurized to its rated charging pressure. There shall be no leakage, determined by using a soap bubble solution or other equivalent means on all possible leak locations.

Each fuel cell cartridge shall be permanently marked with the following information:

(a) The rated charging pressure in MPa;

(b) The manufacturer's serial number of the fuel cell cartridges or unique identification number; and

(c) The date of expiry based on the maximum service life (year in four digits; month in two digits).

340 Chemical kits, first aid kits and polyester resin kits containing dangerous substances in inner packagings which do not exceed the quantity limits for excepted quantities applicable to individual substances as specified in column 7b of Table A of Chapter 3.2, may be carried in accordance with Chapter 3.5. Class 5.2 substances, although

not individually authorized as excepted quantities in column 7b of Table A of Chapter 3.2, are authorized in such kits and are assigned Code E2 (see 3.5.1.2).

341-499 *(Reserved)*

500 UN No. 3064 nitroglycerin, solution in alcohol with more than 1% but not more than 5% nitroglycerin, packed in accordance with packing instruction P300 of 4.1.4.1 of ADR, is a substance of Class 3.

501 For naphthalene, molten, see UN No. 2304.

502 UN No. 2006 plastics, nitrocellulose-based, self-heating, n.o.s., and UN No. 2002 celluloid scrap are substances of Class 4.2.

503 For phosphorus, white or yellow, molten, see UN No. 2447.

504 UN No. 1847 potassium sulphide, hydrated with not less than 30% water of crystallization, UN No. 1849 sodium sulphide, hydrated with not less than 30% water of crystallization and UN No. 2949 sodium hydrosulphide with not less than 25% water of crystallization are substances of Class 8.

505 UN No. 2004 magnesium diamide is a substance of Class 4.2.

506 Alkaline earth metals and alkaline earth metal alloys in pyrophoric form are substances of Class 4.2.

UN No. 1869 magnesium or magnesium alloys containing more than 50% magnesium as pellets, turnings or ribbons, are substances of Class 4.1.

507 UN No. 3048 aluminium phosphide pesticides, with additives inhibiting the emission of toxic flammable gases are substances of Class 6.1.

508 UN No. 1871 titanium hydride and UN No. 1437 zirconium hydride are substances of Class 4.1. UN No. 2870 aluminium borohydride is a substance of Class 4.2.

509 UN No. 1908 chlorite solution is a substance of Class 8.

510 UN No. 1755 chromic acid solution is a substance of Class 8.

511 UN No. 1625 mercuric nitrate, UN No. 1627 mercurous nitrate and UN No. 2727 thallium nitrate are substances of Class 6.1. Thorium nitrate, solid, uranyl nitrate hexahydrate solution and uranyl nitrate, solid are substances of Class 7.

512 UN No. 1730 antimony pentachloride, liquid, UN No. 1731 antimony pentachloride solution, UN No. 1732 antimony pentafluoride and UN No. 1733 antimony trichloride are substances of Class 8.

513 UN No. 0224 barium azide, dry or wetted with less than 50% water, by mass, is a substance of Class 1. UN No. 1571 barium azide, wetted with not less than 50% water, by mass, is a substance of Class 4.1. UN No. 1854 barium alloys, pyrophoric, are substances of Class 4.2. UN No. 1445 barium chlorate, solid, UN No. 1446 barium nitrate, UN No. 1447 barium perchlorate, solid, UN No. 1448 barium permanganate, UN No. 1449 barium peroxide, UN No. 2719 barium bromate, UN No. 2741 barium hypochlorite with more than 22% available chlorine, UN No. 3405 barium chlorate, solution and UN No. 3406 barium perchlorate, solution, are substances of Class 5.1.

UN No. 1565 barium cyanide and UN No. 1884 barium oxide are substances of Class 6.1.

514 UN No. 2464 beryllium nitrate is a substance of Class 5.1.

515 UN No. 1581 chloropicrin and methyl bromide mixture and UN No. 1582 chloropicrin and methyl chloride mixture are substances of Class 2.

516 UN No. 1912 methyl chloride and methylene chloride mixture is a substance of Class 2.

517 UN No. 1690 sodium fluoride, solid, UN No. 1812 potassium fluoride, solid, UN No. 2505 ammonium fluoride, UN No. 2674 sodium fluorosilicate, UN No. 2856 fluorosilicates, n.o.s., UN No. 3415 sodium fluoride, solution and UN No. 3422 potassium fluoride, solution, are substances of Class 6.1.

518 UN No. 1463 chromium trioxide, anhydrous (chromic acid, solid) is a substance of Class 5.1.

519 UN No. 1048 hydrogen bromide, anhydrous, is a substance of Class 2.

520 UN No. 1050 hydrogen chloride, anhydrous, is a substance of Class 2.

521 Solid chlorites and hypochlorites are substances of Class 5.1.

522 UN No. 1873 perchloric acid aqueous solution with more than 50% but not more than 72% pure acid, by mass are substances of Class 5.1. Perchloric acid solutions containing more than 72% pure acid, by mass, or mixtures of perchloric acid with any liquid other than water, are not to be accepted for carriage.

523 UN No. 1382 anhydrous potassium sulphide and UN No. 1385 anhydrous sodium sulphide and their hydrates with less than 30% water of crystallization, and UN No. 2318 sodium hydrosulphide with less than 25% water of crystallization are substances of Class 4.2.

524 UN No. 2858 finished zirconium products of a thickness of 18 μm or more are substances of Class 4.1.

525 Solutions of inorganic cyanides with a total cyanide ion content of more than 30% shall be classified in packing group I, solutions with a total cyanide ion content of more than 3% and not more than 30% in packing group II and solutions with a cyanide ion content of more than 0.3% and not more than 3% in packing group III.

526 UN No. 2000 celluloid is assigned to Class 4.1.

527 (*Reserved*)

528 UN No. 1353 fibres or fabrics impregnated with weakly nitrated cellulose, non-self heating are articles of Class 4.1.

529 UN No. 0135 mercury fulminate, wetted with not less than 20% water, or mixture of alcohol and water, by mass, is a substance of Class 1. Mercurous chloride (calomel) is a substance of Class 9 (UN No. 3077).

530 UN No. 3293 hydrazine, aqueous solution with not more than 37% hydrazine, by mass, is a substance of Class 6.1.

531 Mixtures having a flash-point below 23 °C and containing more than 55% nitrocellulose, whatever its nitrogen content or containing not more than 55% nitrocellulose with a nitrogen content above 12.6% (by dry mass), are substances of Class 1 (see UN Nos. 0340 or 0342) or of Class 4.1.

532 UN No. 2672 ammonia solution containing not less than 10% but not more than 35% ammonia is a substance of Class 8.

533 UN No. 1198 formaldehyde solutions, flammable are substances of Class 3. Formaldehyde solutions, non-flammable, with less than 25% formaldehyde are not subject to the requirements of ADN.

534 While in some climatic conditions, petrol (gasoline) may have a vapour pressure at 50 °C of more than 110 kPa (1.10 bar) but not more than 150 kPa (1.50 bar) it is to continue to be considered as a substance having a vapour pressure at 50 °C of not more than 110 kPa (1.10 bar).

535 UN No. 1469 lead nitrate, UN No. 1470 lead perchlorate, solid and UN No. 3408 lead perchlorate, solution are substances of Class 5.1.

536 For naphthalene, solid, see UN No. 1334.

537 UN No. 2869 titanium trichloride mixture, not pyrophoric, is a substance of Class 8.

538 For sulphur (in the solid state), see UN No. 1350.

539 Solutions of isocyanates having a flash-point of not less than 23 °C are substances of Class 6.1.

540 UN No. 1326 hafnium powder, wetted, UN No. 1352 titanium powder, wetted or UN No. 1358 zirconium powder, wetted, with not less than 25% water, are substances of Class 4.1.

541 Nitrocellulose mixtures with a water content, alcohol content or plasticizer content lower than the stated limits are substances of Class 1.

542 Talc containing tremolite and/or actinolite is covered by this entry.

543 UN No. 1005 ammonia, anhydrous, UN No. 3318 ammonia solution with more than 50% ammonia and UN No. 2073 ammonia solution, with more than 35% but not more than 50% ammonia, are substances of Class 2. Ammonia solutions with not more than 10% ammonia are not subject to the requirements of ADN.

544 UN No. 1032 dimethylamine, anhydrous, UN No. 1036 ethylamine, UN No. 1061 methylamine, anhydrous and UN No. 1083 trimethylamine, anhydrous, are substances of Class 2.

545 UN No. 0401 dipicryl sulphide, wetted with less than 10% water by mass is a substance of Class 1.

546 UN No. 2009 zirconium, dry, finished sheets, strip or coiled wire, in thicknesses of less than 18 µm, is a substance of Class 4.2. Zirconium, dry, finished sheets, strip or coiled wire, in thicknesses of 254 µm or more, is not subject to the requirements of ADN.

547 UN No. 2210 maneb or UN No. 2210 maneb preparations in self-heating form are substances of Class 4.2.

548 Chlorosilanes which, in contact with water, emit flammable gases, are substances of Class 4.3.

549 Chlorosilanes having a flash-point of less than 23 °C and which, in contact with water, do not emit flammable gases are substances of Class 3. Chlorosilanes having a flash-point equal to or greater than 23 °C and which, in contact with water, do not emit flammable gases are substances of Class 8.

550 UN No. 1333 cerium in slabs, rods or ingots is a substance of Class 4.1.

551 Solutions of these isocyanates having a flash-point below 23 °C are substances of Class 3.

552 Metals and metal alloys in powdered or other flammable form, liable to spontaneous combustion, are substances of Class 4.2. Metals and metal alloys in powdered or other flammable form which, in contact with water, emit flammable gases are substances of Class 4.3.

553 This mixture of hydrogen peroxide and peroxyacetic acid shall, in laboratory testing (see *Manual of Tests and Criteria*, Part II, section 20), neither detonate in the cavitated state nor deflagrate at all and shall show no effect when heated under confinement nor any explosive power. The formulation shall be thermally stable (self-accelerating decomposition temperature 60 °C or higher for a 50 kg package), and a liquid compatible with peroxyacetic acid shall be used for desensitization. Formulations not meeting these criteria are to be regarded as substances of Class 5.2 (see *Manual of Tests and Criteria*, Part II, paragraph 20.4.3 (g)).

554 Metal hydrides which, in contact with water, emit flammable gases are substances of Class 4.3. UN No. 2870 aluminium borohydride or UN No. 2870 aluminium borohydride in devices is a substance of Class 4.2.

555 Dust and powder of metals in non-spontaneously combustible form, non-toxic which nevertheless, in contact with water, emit flammable gases, are substances of Class 4.3.

556 Organometallic compounds and their solutions which ignite spontaneously are substances of Class 4.2. Flammable solutions with organometallic compounds in concentrations which, in contact with water, neither emit flammable gases in dangerous quantities nor ignite spontaneously are substances of Class 3.

557 Dust and powder of metals in pyrophoric form are substances of Class 4.2.

558 Metals and metal alloys in pyrophoric form are substances of Class 4.2. Metals and metal alloys which, in contact with water, do not emit flammable gases and are not pyrophoric or self-heating, but which are easily ignited, are substances of Class 4.1.

559 Mixtures of a hypochlorite with an ammonium salt are not to be accepted for carriage. UN No. 1791 hypochlorite solution is a substance of Class 8.

560 UN No. 3257 elevated temperature liquid, n.o.s., at or above 100 °C and, for a substance with a flash-point, below its flash-point (including molten metals and molten salts) is a substance of Class 9.

561 Chloroformates having predominantly corrosive properties are substances of Class 8.

562 Spontaneously combustible organometallic compounds are substances of Class 4.2. Water-reactive organometallic compounds, flammable, are substances of Class 4.3.

563 UN No. 1905 selenic acid is a substance of Class 8.

564 UN No. 2443 vanadium oxytrichloride, UN No. 2444 vanadium tetrachloride and UN No. 2475 vanadium trichloride are substances of Class 8.

565 Unspecified wastes resulting from medical/veterinary treatment of humans/animals or from biological research, and which are unlikely to contain substances of Class 6.2 shall be assigned to this entry. Decontaminated clinical wastes or wastes resulting from biological research which previously contained infectious substances are not subject to the requirements of Class 6.2.

566 UN No. 2030 hydrazine aqueous solution, with more than 37% hydrazine, by mass, is a substance of Class 8.

567 Mixtures containing more than 21% oxygen by volume shall be classified as oxidizing.

568 Barium azide with a water content lower than the stated limit is a substance of Class 1, UN No. 0224.

569-579 *(Reserved)*

580 Tank-vehicles, tank-wagons, specialized vehicles, specialized wagons and specially equipped vehicles and wagons for carriage in bulk shall bear on both sides and at the rear the mark referred to in 5.3.3. Tank-containers, portable tanks, special containers and specially equipped containers for carriage in bulk shall bear this mark on both sides and at each end.

581 This entry covers mixtures of methylacetylene and propadiene with hydrocarbons, which as:

Mixture P1, contain not more than 63% methylacetylene and propadiene by volume and not more than 24% propane and propylene by volume, the percentage of C_4-saturated hydrocarbons being not less than 14% by volume, and as

Mixture P2, contain not more than 48% methylacetylene and propadiene by volume and not more than 50% propane and propylene by volume, the percentage of C_4-saturated hydrocarbons being not less than 5% by volume,

as well as mixtures of propadiene with 1 to 4% methylacetylene.

When relevant, in order to meet the requirements for the transport document (5.4.1.1), the term "Mixture P1" or "Mixture P2" may be used as technical name.

582 This entry covers, <u>inter alia</u>, mixtures of gases indicated by the letter R ..., which as

Mixture F1, have a vapour pressure at 70° C not exceeding 1.3 MPa (13 bar) and a density at 50 °C not lower than that of dichlorofluoromethane (1.30 kg/l);

Mixture F2, have a vapour pressure at 70 °C not exceeding 1.9 MPa (19 bar) and a density at 50 °C not lower than that of dichloridifluoromethane (1.21 kg/l);

Mixture F3, have a vapour pressure at 70 °C not exceeding 3 MPa (30 bar) and a density at 50 °C not lower than that of chlorodifluoromethane (1.09 kg/l).

NOTE: *Trichlorofluoromethane (refrigerant R 11), 1,1,2-trichloro-1,2,2-trifluoroethane (refrigerant R 113), 1,1,1-trichloro-2,2,2-trifluoroethane (refrigerant R 113a), 1-chloro-1,2,2-trifluoroethane (refrigerant R 133) and 1-chloro-1,1,2-trifluoroethane (refrigerant R 133 b) are not substances of Class 2. They may, however, enter into the composition of mixtures F1 to F3.*

When relevant, in order to meet the requirements for the transport document (5.4.1.1), the term "Mixture F1", "Mixture F2" or "Mixture F3" may be used as technical name.

583 This entry covers, inter alia, mixtures which as:

Mixture A, have a vapour pressure at 70 °C not exceeding 1.1 MPa (11 bar) and a density at 50 °C not lower than 0.525 kg/l;

Mixture A01, have a vapour pressure at 70 °C not exceeding 1.6 MPa (16 bar) and a density at 50 °C not lower than 0.516 kg/l;

Mixture A02, have a vapour pressure at 70 °C not exceeding 1.6 MPa (16 bar) and a density at 50 °C not lower than 0.505 kg/l;

Mixture A0, have a vapour pressure at 70 °C not exceeding 1.6 MPa (16 bar) and a density at 50 °C not lower than 0.495 kg/l;

Mixture A1, have a vapour pressure at 70 °C not exceeding 2.1 MPa (21 bar) and a density at 50 °C not lower than 0.485 kg/l;

Mixture B1, have a vapour pressure at 70 °C not exceeding 2.6 MPa (26 bar) and a density at 50 °C not lower than 0.474 kg/l;

Mixture B2, have a vapour pressure at 70 °C not exceeding 2.6 MPa (26 bar) and a density at 50 °C not lower than 0.463 kg/l;

Mixture B, have a vapour pressure at 70 °C not exceeding 2.6 MPa (26 bar) and a density at 50 °C not lower than 0.450 kg/l;

Mixture C, have a vapour pressure at 70 °C not exceeding 3.1 MPa (31 bar) and a density at 50 °C not lower than 0.440 kg/l;

When relevant, in order to meet the requirements for the transport document (5.4.1.1), the following terms may be used as technical name:

- "Mixture A" or "Butane";
- "Mixture A01" or "Butane";
- "Mixture A02" or "Butane";
- "Mixture A0" or "Butane";
- "Mixture A1";
- "Mixture B1";

- "Mixture B2";
- "Mixture B";
- "Mixture C" or "Propane".

For carriage in tanks, the trade names "butane" or "propane" may be used only as a complement.

584 This gas is not subject to the requirements of ADN when:

- it is in the gaseous state;
- it contains not more than 0.5% air;
- it is contained in metal capsules (sodors, sparklets) free from defects which may impair their strength;
- the leakproofness of the closure of the capsule is ensured;
- a capsule contains not more than 25 g of this gas;
- a capsule contains not more than 0.75 g of this gas per cm^3 of capacity.

585 Cinnabar is not subject to the requirements of ADN.

586 Hafnium, titanium and zirconium powders shall contain a visible excess of water. Hafnium, titanium and zirconium powders, wetted, mechanically produced, of a particle size of 53 µm and over, or chemically produced, of a particle size of 840 µm and over, are not subject to the requirements of ADN.

587 Barium stearate and barium titanate are not subject to the requirements of ADN.

588 Solid hydrated forms of aluminium bromide and aluminium chloride are not subject to the requirements of ADN.

589 Calcium hypochlorite mixtures, dry, containing not more than 10% available chlorine are not subject to the requirements of ADN.

590 Ferric chloride hexahydrate is not subject to the requirements of ADN.

591 Lead sulphate with not more than 3% free acid is not subject to the requirements of ADN.

592 Uncleaned empty packagings (including empty IBCs and large packagings), empty tank-vehicles, empty demountable tanks, empty portable tanks, empty tank-containers and empty small containers which have contained this substance are not subject to the requirements of ADN.

593 This gas, intended for the cooling of e.g. medical or biological specimens, if contained in double wall receptacles which comply with the provisions of packing instruction P203 (12) of 4.1.4.1 of ADR is not subject to the requirements of ADN.

594 The following articles, manufactured and filled according to the regulations of the manufacturing State and packaged in strong outer packagings, are not subject to the requirements of ADN:

– UN No. 1044 fire extinguishers provided with protection against inadvertent discharge;

– UN No. 3164 articles, pressurized pneumatic or hydraulic, designed to withstand stresses greater than the internal gas pressure by virtue of transmission of force, intrinsic strength or construction.

596 Cadmium pigments, such as cadmium sulphides, cadmium sulphoselenides and cadmium salts of higher fatty acids (e.g. cadmium stearate), are not subject to the requirements of ADN.

597 Acetic acid solutions with not more than 10% pure acid by mass, are not subject to the requirements of ADN.

598 The following are not subject to the requirements of ADN:

(a) New storage batteries when:

– they are secured in such a way that they cannot slip, fall or be damaged;

– they are provided with carrying devices, unless they are suitably stacked, e.g. on pallets;

– there are no dangerous traces of alkalis or acids on the outside;

– they are protected against short circuits;

(b) Used storage batteries when:

– their cases are undamaged;

– they are secured in such a way that they cannot leak, slip, fall or be damaged, e.g. by stacking on pallets;

– there are no dangerous traces of alkalis or acids on the outside of the articles;

– they are protected against short circuits.

"Used storage batteries" means storage batteries carried for recycling at the end of their normal service life.

599 Manufactured articles or instruments containing not more than 1 kg of mercury are not subject to the requirements of ADN.

600 Vanadium pentoxide, fused and solidified, is not subject to the requirements of ADN.

601 Pharmaceutical products (medicines) ready for use, which are substances manufactured and packaged for retail sale or distribution for personal or household consumption are not subject to the requirements of ADN.

602 Phosphorus sulphides which are not free from yellow and white phosphorus are not to be accepted for carriage.

603 Anhydrous hydrogen cyanide not meeting the description for UN No. 1051 or UN No. 1614 is not to be accepted for carriage. Hydrogen cyanide (hydrocyanic acid) containing less than 3% water is stable, if the pH-value is 2.5 ± 0.5 and the liquid is clear and colourless.

604 Ammonium bromate and its aqueous solutions and mixtures of a bromate with an ammonium salt are not to be accepted for carriage.

605 Ammonium chlorate and its aqueous solutions and mixtures of a chlorate with an ammonium salt are not to be accepted for carriage.

606 Ammonium chlorite and its aqueous solutions and mixtures of a chlorite with an ammonium salt are not to be accepted for carriage.

607 Mixtures of potassium nitrate and sodium nitrite with an ammonium salt are not to be accepted for carriage.

608 Ammonium permanganate and its aqueous solutions and mixtures of a permanganate with an ammonium salt are not to be accepted for carriage.

609 Tetranitromethane not free from combustible impurities is not to be accepted for carriage.

610 The carriage of this substance, when it contains more than 45% hydrogen cyanide is prohibited.

611 Ammonium nitrate containing more than 0.2% combustible substances (including any organic substance calculated as carbon) is not to be accepted for carriage unless it is a constituent of a substance or article of Class 1.

612 *(Reserved)*

613 Chloric acid solution containing more than 10% chloric acid and mixtures of chloric acid with any liquid other than water is not to be accepted for carriage.

614 2,3,7,8-tetrachlorodibenzo-p-dioxin (TCDD) in concentrations considered highly toxic according to the criteria in 2.2.61.1 is not to be accepted for carriage.

615 *(Reserved)*

616 Substances containing more than 40% liquid nitric esters shall satisfy the exudation test specified in 2.3.1.

617 In addition to the type of explosive, the commercial name of the particular explosive shall be marked on the package.

618 In receptacles containing 1,2-butadiene, the oxygen concentration in the gaseous phase shall not exceed 50 ml/m^3.

619-622 *(Reserved)*

623 UN No. 1829 sulphur trioxide shall be inhibited. Sulphur trioxide, 99.95% pure or above, may be carried without inhibitor in tanks provided that its temperature is maintained at or above 32.5 °C. For the carriage of this substance without inhibitor in tanks at a minimum temperature of 32.5 °C, the specification "**Transport under**

minimum temperature of the product of 32.5 °C" shall appear in the transport document.

625 Packages containing these articles shall be clearly marked as follows:
"UN 1950 AEROSOLS"

626-631 *(Reserved)*

632 Considered to be spontaneously flammable (pyrophoric).

633 Packages and small containers containing this substance shall bear the following marking: "**Keep away from any source of ignition**". This marking shall be in an official language of the forwarding country, and also, if that language is not English, French or German, in English, French or German, unless any agreements concluded between the countries concerned in the transport operation provide otherwise.

635 Packages containing these articles need not bear a label conforming to model No. 9 unless the article is fully enclosed by packaging, crates or other means that prevent the ready identification of the article.

636 (a) Cells contained in equipment shall not be capable of being discharged during carriage to the extent that the open circuit voltage falls below 2 volts or two thirds of the voltage of the undischarged cell, whichever is the lower.

(b) Used lithium cells and batteries with a gross mass of not more than 500 g each collected and presented for carriage for disposal between the consumer collecting point and the intermediate processing facility, together with other non-lithium cells or batteries, are not subject to the other provisions of ADN if they meet the following conditions:

(i) The provisions of packing instruction P903b of ADR are complied with;

(ii) A quality assurance system is in place to ensure that the total amount of lithium cells or batteries in each wagon or large container/transport unit does not exceed 333 kg;

(iii) Packages shall bear the inscription: "USED LITHIUM CELLS".

637 Genetically modified microorganisms and genetically modified organisms are those which are not dangerous for humans and animals, but which could alter animals, plants, microbiological substances and ecosystems in such a way as cannot occur naturally. Genetically modified microorganisms and genetically modified organisms are not subject to the requirements of ADN when authorized for use by the competent authorities of the countries of origin, transit and destination.[1] Live vertebrate or invertebrate animals shall not be used to carry these substances classified under this UN number unless the substance can be carried in no other way. For the carriage of easily perishable substances under this UN number appropriate information shall be given, e.g.: "**Cool at +2 °/+4 °C**" or "**Carry in frozen state**" or "**Do not freeze**".

638 Substances related to self-reactive substances (see 2.2.41.1.19).

---

[1] *See in particular Part C of Directive 2001/18/EC of the European Parliament and of the Council on the deliberate release into the environment of genetically modified organisms and repealing Council Directive 90/220/EEC (Official Journal of the European Communities, No. L 106, of 17 April 2001, pp. 8-14), which sets out the authorization procedures for the European Community.*

639 See 2.2.2.3, classification code 2F, UN No. 1965, Note 2.

640 The physical and technical characteristics mentioned in column (2) of Table A of Chapter 3.2 determine different tank codes for the carriage of substances of the same packing group in tanks conforming to Chapter 6.8 of RID or ADR.

In order to identify these physical and technical characteristics of the product carried in the tank, the following shall be added, to the particulars required in the transport document, only in case of carriage in tanks conforming to Chapter 6.8 of ADR or RID:

"Special provision 640X" where "X" is the applicable capital letter appearing after the reference to special provision 640 in column (6) of Table A of Chapter 3.2.

These particulars may, however, be dispensed with in the case of carriage in the type of tank which, for substances of a specific packing group of a specific UN number, meets at least the most stringent requirements.

643 Stone or aggregate asphalt mixture is not subject to the requirements for Class 9.

644 This substance is admitted for carriage provided that:

- The pH is between 5 and 7 measured in an aqueous solution of 10% of the substance carried;

- The solution does not contain more than 0.2% combustible material or chlorine compounds in quantities such that the chlorine level exceeds 0.02%.

645 The classification code as mentioned in Column (3b) of Table A of Chapter 3.2 shall be used only with the approval of the competent authority of a Contracting Party to ADN prior to carriage. When assignment to a division is made in accordance with the procedure in 2.2.1.1.7.2, the competent authority may require the default classification to be verified on the basis of test data derived from Test Series 6 of the Manual of Tests and Criteria, Part I, Section 16.

646 Carbon made by steam activation process is not subject to the requirements of ADN.

647 Except for carriage in tank vessels, the carriage of vinegar and acetic acid with not more than 25 % pure acid by mass is subject only to the following requirements:

(a) Packagings, including IBCs and large packagings, and tanks shall be manufactured from stainless steel or plastic material which is permanently resistant to corrosion of vinegar/acetic acid food grade;

(b) Packagings, including IBCs and large packagings, and tanks shall be subjected to a visual inspection by the owner at least once a year. The results of the inspections shall be recorded and the records kept for at least one year. Damaged packagings, including IBCs and large packagings, and tanks shall not be filled;

(c) Packagings, including IBCs and large packagings, and tanks shall be filled in a way that no product is spilled or adheres to the outer surface;

(d) Seals and closures shall be resistant to vinegar/acetic acid food grade. Packagings, including IBCs and large packagings, and tanks shall be hermetically sealed by the person in charge of packaging and/or filling so that under normal conditions of carriage there will be no leakage;

(e) Combination packagings with inner packaging made of glass or plastic (see packing instruction P001 in 4.1.4.1 of ADR) which fulfil the general packing requirements of 4.1.1.1, 4.1.1.2, 4.1.1.4, 4.1.1.5, 4.1.1.6, 4.1.1.7 and 4.1.1.8 of ADR may be used;

The other provisions of ADN do not apply except those relating to carriage in tank vessels.

648 Articles impregnated with this pesticide, such as fibreboard plates, paper strips, cotton-wool balls, sheets of plastics material, in hermetically closed wrappings, are not subject to the provisions of ADN.

649 To determine the initial boiling point, as mentioned under 2.2.3.1.3, packing group I, the test method according to standard ASTM D86-01[2] is suitable.

Substances which have an initial boiling point above 35 °C determined with this method are substances of packing group II and shall be classified in accordance with the applicable entry of this packing group.

650 Waste consisting of packaging residues, solidified residues and liquid residues of paint may be carried under the conditions of packing group II. In addition to the provisions of UN No. 1263, packing group II, the waste may also be packed and carried as follows:

(a) The waste may be packed in accordance with packing instruction P002 of 4.1.4.1 of ADR or to packing instruction IBC006 of 4.1.4.2 of ADR;

(b) The waste may be packed in flexible IBCs of types 13H3, 13H4 and 13H5 in overpacks with complete walls;

(c) Testing of packagings and IBCs indicated under (a) or (b) may be carried out in accordance with the requirements of Chapters 6.1 or 6.5 of ADR, as appropriate, in relation to solids, at the packing group II performance level.

The tests shall be carried out on packagings and IBCs, filled with a representative sample of the waste, as prepared for carriage;

(d) Carriage in bulk in sheeted wagons, movable roof wagons/sheeted vehicles, closed containers or sheeted large containers, all with complete walls is allowed. The wagons, containers or body of vehicles shall be leakproof or rendered leakproof, for example by means of a suitable and sufficiently stout inner lining;

(e) If the waste is carried under the conditions of this special provision, the goods shall be declared in accordance with 5.4.1.1.3 in the transport document, as follows: "WASTE, UN 1263 PAINT, 3, II.

651 Special provision V2 (1) of ADR is only applicable for a net explosive content of more than 3,000 kg (4,000 kg with trailer).

652 (*Reserved*)

---

[2] *Standard Test Method for Distillation of Petroleum Products at Atmospheric Pressure, published September 2001 by ASTM International.*

653 The carriage of this gas in cylinders with a maximum capacity of 0.5 litres, is not subject to the other provisions of ADN if the following conditions are met:

- The provisions for construction and testing of cylinders are observed;

- The cylinders are contained in outer packagings which at least meet the requirements of Part 4 for combination packagings. The general provisions of packing of 4.1.1.1, 4.1.1.2 and 4.1.1.5 to 4.1.1.7 of ADR shall be observed;

- The cylinders are not packed together with other dangerous goods;

- The total gross mass of a package does not exceed 30 kg; and

- Each package is clearly and durably marked with "UN 1013". This marking is displayed within a diamond-shaped area surrounded by a line that measures at least 100 mm by 100 mm.

654 Waste lighters collected separately and consigned in accordance with 5.4.1.1.3 may be carried under this entry for the purposes of disposal. They need not be protected against inadvertent discharge provided that measures are taken to prevent the dangerous build up of pressure and dangerous atmospheres.

Waste lighters, other than those leaking or severely deformed, shall be packed in accordance with packing instruction P003 of ADR. In addition the following provisions shall apply:

– only rigid packagings of a maximum capacity of 60 litres shall be used;

– the packagings shall be filled with water or any other appropriate protection material to avoid any ignition;

– under normal conditions of carriage all ignition devices of the lighters shall fully be covered by the protection material;

– the packagings shall be adequately vented to prevent the creation of flammable atmosphere and the build up of pressure;

– the packages shall only be carried in ventilated or open wagons/vehicles or containers.

Leaking or severely deformed lighters shall be carried in salvage packagings, provided appropriate measures are taken to ensure there is no dangerous build up of pressure.

NOTE: *Special provision 201 and special packing provisions PP84 and RR5 of packing instruction P002 in 4.1.4.1 of ADR do not apply to waste lighters.*

800 Oil seeds, crushed seeds and seedcake containing vegetable oil, treated with solvents, not subject to spontaneous combustion, are allocated to UN No. 3175. These substances are not subject to ADN when they have been prepared or treated to ensure that they cannot give off dangerous gases in dangerous quantities (no risk of explosion) during carriage and when this is mentioned in the transport document.

801 Ferrosilicon with between 25 and 30% or more than 90% silicon content by mass is a dangerous substance of Class 4.3 for carriage in bulk or without packaging by inland navigation vessel.

802 See 7.1.4.10.

CHAPTER 3.4

DANGEROUS GOODS PACKED IN LIMITED QUANTITIES

3.4.1 General requirements

3.4.1.1 Packagings used in accordance with 3.4.3 to 3.4.6 below, need only to conform to the general provisions of 4.1.1.1, 4.1.1.2 and 4.1.1.4 to 4.1.1.8 of ADR.

3.4.1.2 The maximum gross mass of a combination packaging shall not exceed 30 kg and for shrink and stretched wrapped trays shall not exceed 20 kg.

NOTE: *The limit for combination packagings does not apply when LQ5 is issued.*

3.4.1.3 Subject to the maximum limits in 3.4.1.2 and individual limits in table 3.4.6, dangerous goods may be packed together with other articles or substances, provided they will not react dangerously in the event of leakage.

3.4.2 When the code "LQ0" is shown in Column (7a) of Table A in Chapter 3.2 for a given substance or article, that substance or article is not exempted from any of the applicable provisions of ADN when it is packed in limited quantities, unless otherwise specified in these annexed Regulations.

3.4.3 Unless otherwise provided in this Chapter, when one of the codes "LQ1" or "LQ2" is shown in Column (7a) of Table A in Chapter 3.2 for a given substance or article, the provisions of other Chapters of ADN do not apply to the carriage of that substance or article, provided:

(a) the provisions of 3.4.5 (a) to (c) are observed; with respect to these provisions, articles are considered to be inner packagings;

(b) inner packagings meet the conditions of 6.2.5.1 and 6.2.6.1 to 6.2.6.3 of ADR.

3.4.4 Unless otherwise provided in this Chapter, when the code "LQ3" is shown in Column (7a) of Table A in Chapter 3.2 for a given substance, the provisions of other Chapters of ADN do not apply to the carriage of that substance, provided:

(a) The substance is carried in combination packagings, the following outer packagings being allowed:

- steel or aluminium drums with removable head;
- steel or aluminium jerricans with removable head;
- plywood or fibre drums;
- plastics drums or jerricans with removable head;
- boxes of natural wood, plywood, reconstituted wood, fibreboard, plastics, steel or aluminium;

and be so designed that they meet the relevant construction requirements of 6.1.4 of ADR;

(b) The maximum net quantities per inner packaging shown in columns (2) or (4) and per package in columns (3) or (5), where indicated, of table 3.4.6 are not exceeded;

(c) Each package is clearly and durably marked with :

(i) the UN number of the goods contained therein, as given in Column (1) of Table A in Chapter 3.2, preceded by the letters "UN";

(ii) in the case of different goods with different UN numbers within a single package:

– the UN numbers of the goods contained therein, preceded by the letters "UN", or

– the letters "LQ"[1].

These markings shall be displayed within a diamond-shaped area surrounded by a line that measures at least 100 mm × 100 mm. The width of line forming the diamond shall be at least 2 mm; the number shall be at least 6 mm high. Where more than one substance assigned to different UN numbers are included in the package, the diamond shall be large enough to include each relevant UN number. If the size of the package so requires, the dimension may be reduced, provided the markings remain clearly visible.

3.4.5 Unless otherwise provided in this Chapter, when one of the codes "LQ4" to "LQ19" and "LQ22" to "LQ28" is shown in Column (7a) of Table A in Chapter 3.2 for a given substance, the provisions of other Chapters of ADN do not apply to the carriage of that substance, provided:

(a) The substance is carried:

– in combination packagings, corresponding to the prescriptions of 3.4.4 (a), or

– in metal or plastics inner packagings which are not liable to break or be easily punctured, placed in shrink-wrapped or stretch-wrapped trays;

(b) The maximum net quantities per inner packaging shown in columns (2) or (4) and per package in columns (3) or (5), where indicated, of table 3.4.6 are not exceeded;

(c) Each package is clearly and durably marked as indicated in 3.4.4 (c).

[1] *The letters "LQ" are an abbreviation of the English words "Limited Quantities". The letters "LQ" are not permitted by the IMDG Code or the ICAO Technical Instructions.*

3.4.6 **Table**

Code	Combination packagings[a] Maximum net quantity		Inner packagings placed in shrink-wrapped or stretch-wrapped trays[a] Maximum net quantity	
	per inner packaging	per package[b]	per inner packaging	per package[b]
(1)	(2)	(3)	(4)	(5)
LQ0	No exemption under the conditions of 3.4.2.			
LQ1	120 ml		120 ml	
LQ2	1 *l*		1 *l*	
LQ3 [c]	500 ml	1 *l*	Not allowed	Not allowed
LQ4 [c]	3 *l*		1 *l*	
LQ5 [c]	5 *l*	Unlimited	1 *l*	
LQ6 [c]	5 *l*		1 *l*	
LQ7 [c]	5 *l*		5 *l*	
LQ8	3 kg		500 g	
LQ9	6 kg		3 kg	
LQ10	500 ml		500 ml	
LQ11	500 g		500 g	
LQ12	1 kg		1 kg	
LQ13	1 *l*		1 *l*	
LQ14	25 ml		25 ml	
LQ15	100 g		100 g	
LQ16	125 ml		125 ml	
LQ17	500 ml	2 *l*	100 ml	2 *l*
LQ18	1 kg	4kg	500 g	4 kg
LQ19	5 kg		5 kg	
LQ20	Reserved	Reserved	Reserved	Reserved
LQ21	Reserved	Reserved	Reserved	Reserved
LQ22	1 *l*		500 ml	
LQ23	3 kg		1 kg	
LQ24	6 kg		2 kg	
LQ25[d]	1 kg		1 kg	
LQ26[d]	500 ml	2 *l*	500 ml	2 *l*
LQ27	6 kg		6 kg	
LQ28	3 *l*		3 *l*	

[a] *See 3.4.1.2.*

[b] *See 3.4.1.3.*

[c] *In the case of homogenous mixtures of Class 3 containing water, the quantities specified relate only to the substance of Class 3 contained in those mixtures.*

[d] *For UN Nos. 2315, 3151, 3152 and 3432 when carried in apparatus, the inner packaging quantities shall not be exceeded per piece of apparatus. The apparatus shall be carried in a leakproof packaging and the complete package shall conform to 3.4.4 (c). Shrink-wrapped and stretch-wrapped trays shall not be used for apparatus.*

3.4.7 Overpacks containing packages conforming to 3.4.3, 3.4.4 or 3.4.5 shall be marked, as required by 3.4.4 (c) for each item of dangerous goods contained in the overpack, unless markings representative of all dangerous goods contained in the overpack are visible.

3.4.8 The requirements

(a) of sub-section 5.2.1.9 on the placement of orientation arrows on packages;

(b) of sub-section 5.1.2.1 (b) on the placement of orientation arrows on overpacks; and

(c) of sub-section 7.5.1.5 of ADR on the orientation of packages

shall be applicable also to packages and overpacks carried in accordance with this chapter.

3.4.9 Consignors of dangerous goods packed in limited quantities shall inform the carrier of the total gross mass of such goods to be consigned, in advance of carriage not involving maritime transport.

3.4.10 (a) Transport units with a maximum mass exceeding 12 tonnes carrying packages with dangerous goods in limited quantities shall be marked in accordance with 3.4.12 at the front and at the rear except when orange-coloured plate marking is displayed in accordance with 5.3.2.

(b) Wagons carrying packages with dangerous goods in limited quantities shall be marked in accordance with 3.4.12 on both sides except when placards in accordance with section 5.3.1 are already affixed.

(c) Containers carrying packages with dangerous goods in limited quantities shall be marked in accordance with 3.4.12 on all four sides except.

– when placards in accordance with section 5.3.1 are already affixed;

– for small containers loaded on a wagon;

– for containers loaded on a transport unit with a maximum mass less than or equal to 12 tonnes.

If the containers are loaded on a transport unit or wagon, the carrying transport unit or wagon need not be marked, except when the marking affixed to the containers is not visible from the outside of this carrying transport unit or wagon. In this latter case, the same marking shall also be affixed at the front and the rear of the carrying transport unit, or on both sides of the carrying wagon.

3.4.11 Markings specified in 3.4.10 may be dispensed with, if the total gross mass of the packages containing dangerous goods packed in limited quantities carried does not exceed 8 tonnes per transport unit, wagon or large container.

3.4.12 The marking shall consist of "LTD QTY"[2] in black letters not less than 65 mm high on a white background.

3.4.13 Markings according to chapter 3.4 of the IMDG Code are also acceptable for carriage in a transport chain including maritime carriage.

[2] *The letters "LTD QTY" are an abbreviation of the English words "Limited Quantity".*

CHAPTER 3.5

DANGEROUS GOODS PACKED IN EXCEPTED QUANTITIES

3.5.1 Excepted quantities

3.5.1.1 Excepted quantities of dangerous goods of certain classes, other than articles, meeting the provisions of this Chapter are not subject to any other provisions of ADN except for:

(a) The training requirements in Chapter 1.3;

(b) The classification procedures and packing group criteria in Part 2;

(c) The packaging requirements of 4.1.1.1, 4.1.1.2, 4.1.1.4 and 4.1.1.6 of ADR.

NOTE: *In the case of radioactive material, the requirements for radioactive material in excepted packages in 1.7.1.5 apply.*

3.5.1.2 Dangerous goods which may be carried as excepted quantities in accordance with the provisions of this Chapter are shown in column 7b of Table A of Chapter 3.2 list by means of an alphanumeric code as follows:

Code	**Maximum net quantity per inner packaging** (in grams for solids and ml for liquids and gases)	**Maximum net quantity per outer packaging** (in grams for solids and ml for liquids and gases, or sum of grams and ml in the case of mixed packing)
E0	Not permitted as Excepted Quantity	
E1	30	1000
E2	30	500
E3	30	300
E4	1	500
E5	1	300

For gases, the volume indicated for inner packagings refers to the water capacity of the inner receptacle and the volume indicated for outer packagings refers to the combined water capacity of all inner packagings within a single outer packaging.

3.5.1.3 Where dangerous goods in excepted quantities for which different codes are assigned are packaged together the total quantity per outer packaging shall be limited to that corresponding to the most restrictive code.

3.5.2 Packagings

Packagings used for the carriage of dangerous goods in excepted quantities shall be in compliance with the following:

(a) There shall be an inner packaging and each inner packaging shall be constructed of plastic (with a minimum thickness of 0.2 mm when used for liquids), or of glass, porcelain, stoneware, earthenware or metal (see also 4.1.1.2 of ADR) and the closure of each inner packaging shall be held securely in place with wire, tape or other positive means; any receptacle having a neck with moulded screw threads shall have a leak proof threaded type cap. The closure shall be resistant to the contents;

(b) Each inner packaging shall be securely packed in an intermediate packaging with cushioning material in such a way that, under normal conditions of carriage, they cannot break, be punctured or leak their contents. The intermediate packaging shall completely contain the contents in case of breakage or leakage, regardless of package orientation. For liquids, the intermediate packaging shall contain sufficient absorbent material to absorb the entire contents of the inner packaging. In such cases, the absorbent material may be the cushioning material. Dangerous goods shall not react dangerously with cushioning, absorbent material and packaging material or reduce the integrity or function of the materials;

(c) The intermediate packaging shall be securely packed in a strong, rigid outer packaging (wooden, fibreboard or other equally strong material);

(d) Each package type shall be in compliance with the provisions in 3.5.3;

(e) Each package shall be of such a size that there is adequate space to apply all necessary markings; and

(f) Overpacks may be used and may also contain packages of dangerous goods or goods not subject to the requirements of ADN.

3.5.3 Tests for packages

3.5.3.1 The complete package as prepared for carriage, with inner packagings filled to not less than 95% of their capacity for solids or 98% for liquids, shall be capable of withstanding, as demonstrated by testing which is appropriately documented, without breakage or leakage of any inner packaging and without significant reduction in effectiveness:

(a) Drops onto a rigid, non-resilient flat and horizontal surface from a height of 1.8 m:

(i) Where the sample is in the shape of a box, it shall be dropped in each of the following orientations:

– flat on the base;
– flat on the top;
– flat on the longest side;
– flat on the shortest side;
– on a corner;

(ii) Where the sample is in the shape of a drum, it shall be dropped in each of the following orientations:

– diagonally on the top chime, with the centre of gravity directly above the point of impact;
– diagonally on the base chime;
– flat on the side;

NOTE: *Each of the above drops may be performed on different but identical packages.*

(b) A force applied to the top surface for a duration of 24 hours, equivalent to the total weight of identical packages if stacked to a height of 3 m (including the sample).

3.5.3.2 For the purposes of testing, the substances to be carried in the packaging may be replaced by other substances except where this would invalidate the results of the tests. For solids, when another substance is used, it must have the same physical characteristics (mass, grain size, etc.) as the substance to be carried. In the drop tests for liquids, when another substance is used, its relative density (specific gravity) and viscosity should be similar to those of the substance to be carried.

3.5.4 Marking of packages

3.5.4.1 Packages containing excepted quantities of dangerous goods prepared in accordance with this Chapter shall be durably and legibly marked with the mark shown in 3.5.4.2. The first or only label number indicated in column (5) of Table A of Chapter 3.2 for each of the dangerous goods contained in the package shall be shown in the mark. Where the name of the consignor or consignee is not shown elsewhere on the package this information shall be included within the mark.

3.5.4.2 The dimensions of the mark shall be a minimum of 100 mm × 100 mm.

Excepted quantities mark

Hatching and symbol of the same colour, black or red,
on white or suitable contrasting background

* The first or only label number indicated in column (5) of Table A of Chapter 3.2 shall be shown in this location.

** The name of the consignor or of the consignee shall be shown in this location if not shown elsewhere on the package.

3.5.4.3 An overpack containing dangerous goods in excepted quantities shall display the markings required by 3.5.4.1, unless such markings on packages within the overpack are clearly visible.

3.5.5 Maximum number of packages in any vehicle, wagon or container

The number of packages in any vehicle, wagon or container shall not exceed 1 000.

3.5.6 Documentation

If a document or documents (such as a bill of lading, air waybill or CMR/CIM consignment note) accompanies(y) dangerous goods in excepted quantities, at least one of these documents shall include the statement "Dangerous Goods in Excepted Quantities" and indicate the number of packages.